THE BLOOMSBURY HANDBOOK TO THE DIGITAL HUMANITIES

THE BLOOMSBURY HANDBOOK TO THE DIGITAL HUMANITIES

Edited by James O'Sullivan

BLOOMSBURY ACADEMIC
LONDON • NEW YORK • OXFORD • NEW DELHI • SYDNEY

BLOOMSBURY ACADEMIC
Bloomsbury Publishing Plc
50 Bedford Square, London, WC1B 3DP, UK
1385 Broadway, New York, NY 10018, USA
29 Earlsfort Terrace, Dublin 2, Ireland

BLOOMSBURY, BLOOMSBURY ACADEMIC and the Diana logo
are trademarks of Bloomsbury Publishing Plc

First published in Great Britain 2022
Paperback edition published 2024

Cover design: Rebecca Heselton
Cover image © Vasilyev Alexandr/ Shutterstock

A catalogue record for this book is available from the British Library.

A catalog record for this book is available from the Library of Congress.

ISBN: HB: 978-1-3502-3211-2
PB: 978-1-3504-5257-2
PDF: 978-1-3502-3212-9
eBook: 978-1-3502-3213-6

Series: Bloomsbury Handbooks

Typeset by Integra Software Services Private Limited.
Printed and bound in Great Britain

To find out more about our authors and books visit www.bloomsbury.com
and sign up for our newsletters.

*The editor would like to dedicate his work on
this volume to Dr. Órla Murphy, a valued mentor and friend*

CONTENTS

Part 2 Methods, Tools, & Techniques

Part 3 Public Digital Humanities

Part 4 Institutional Contexts

Part 5 DH Futures

FIGURES

ABOUT THE EDITOR

JAMES O'SULLIVAN lectures in the Department of Digital Humanities at University College Cork, Ireland. He has previously held faculty positions at Pennsylvania State University and the University of Sheffield, as well as visiting fellowships at NOVA University, Lisbon, and Trinity College Dublin.

He is the author of *Towards a Digital Poetics* (Palgrave Macmillan, 2019) and has also co-edited several collections, including *Electronic Literature as Digital Humanities* (Bloomsbury, with Grigar, 2021) and *Reading Modernism with Machines* (Palgrave Macmillan, with Ross, 2016). His scholarship has appeared in a variety of international publications, most notably *Digital Scholarship in the Humanities*, *Digital Studies/Le Champ Numérique*, *Poetics*, the *International Journal of Humanities and Arts Computing*, and *Digital Humanities Quarterly*. James has also had writing in *The Guardian*, *The Irish Times*, and *LA Review of Books*.

He is a Principal Investigator for the *C21 Editions* project, an international collaboration between institutions in Ireland and the UK, jointly funded by the Irish Research Council (IRC) and Arts & Humanities Research Council (AHRC).

See jamesosullivan.org for more on his work, or follow him on Twitter @jamescosullivan.

CONTRIBUTORS

TITILOLA BABALOLA AIYEGBUSI is a PhD student in the English department at the University of Toronto in Canada. She has researched and published in the area of digital humanities. Her latest publication is *Decolonizing Digital Humanities* (2019).

GRAHAM ALLEN is a Professor in the School of English, University College Cork, Ireland. He has published numerous books and articles on literary and cultural theory, Romantic Studies, and adaptation and Film Studies, including the defining book on intertextual theory, *Intertextuality* (3rd ed. 2022). He is also an award-winning poet. His poetry collections *The One That Got Away* and *The Madhouse System* are published with New Binary Press, as is his ongoing e-poem *Holes* (holesbygrahamallen.org).

DANIEL ALLINGTON is Reader in Social Analytics in the Department of Digital Humanities at King's College London. His research has primarily dealt with the history and sociology of publishing and readership, with his works including *Communicating in English: Talk, Text, Technology* (Routledge, 2012; co-edited with Barbara Mayor) and *The Book in Britain: A Historical Introduction* (Wiley-Blackwell, 2018; co-authored with David Brewer, Stephen Colclough, and Sian Echard under the editorial direction of Zachary Lesser). However, in recent years, he has pivoted to a focus on misinformation, extremism, and hate on digital media, which he studies with a wide variety of tools. His current teaching centers on research methods.

RAFAEL ALVARADO is Associate Professor, general faculty, in the School of Data Science at the University of Virginia, USA. He teaches courses and advises on capstone projects in the field of text and cultural analytics. He is currently co-director of the Multepal Project, an international NSF-funded initiative to create a digital collection of indigenous Mayan texts, including the Popol Wuj. A career digital humanist, he has taught and published in the field, and has developed the software for several digital collections, including the Charrette, Shahnameh, and Geniza Projects at Princeton University and the House Divided Project at Dickinson College, both USA.

ALYSSA ARBUCKLE (alyssaarbuckle.com) is the Associate Director of the Electronic Textual Cultures Lab at the University of Victoria, British Columbia, Canada, where she is Operational Lead for the Implementing New Knowledge Environments Partnership and a co-facilitator of its *Connection* cluster. With her colleagues Randa El Khatib and Ray Siemens she is a Co-Director of the Digital Humanities Summer Institute. She holds an interdisciplinary PhD from the University of Victoria; her dissertation focused on open social scholarship and its implementation.

TAYLOR ARNOLD is Associate Professor of Statistics at the University of Richmond, Virginia, USA. He is the author of *Humanities Data in R* (Springer, 2015) and *A Computational Approach to Statistical Learning* (CRC Press, 2019). His work has appeared in journals such as *Digital Scholarship in the Humanities*, *Journal of Computational and Graphical Statistics*, and *Proceedings of the National Academy of Sciences*. He is chief architect of the Distant Viewing Toolkit. He is Director of the Distant Viewing Lab.

TULLY BARNETT is a Senior Lecturer in Creative Industries at Flinders University in South Australia. She works across digital humanities, looking at the policies, practices, and implications of digitization and digital reading, and cultural policy studies looking at how value is understood in arts and culture. She is the co-author of *What Matters? Talking Value in Australian Culture* (Monash University Publishing, 2018) and author of the articles "Read in Browser: Reading Platforms, Frames, Interfaces, and Infrastructure" in *Participations* and "Distributed Reading: Literary Reading in Diverse Environments" in *DHQ*. She serves on the board of the Australasian Association for Digital Humanities.

DAVID M. BERRY is Professor of Digital Humanities, University of Sussex, and a Visiting Professor of Critical Theory and Digital Humanities at King's College London. He writes widely on the theory and philosophy of computation and algorithms. His most recent book is *Digital Humanities: Knowledge and Critique in a Digital Age* (Polity, 2017; with Anders Fagerjord). His forthcoming work is concerned with the idea of a university in a digital age, particularly in relation to the notion of a data-intensive university.

JASON BOYD is Associate Professor in the Department of English and Director of the Centre for Digital Humanities at Toronto Metropolitan University, Canada. He is the co-author (with Bo Ruberg and James Howe) of "Towards a Queer Digital Humanities" (2018) and has co-taught the "Queer Digital Humanities" course at the Digital Humanities Summer Institute (DHSI) in Victoria, Canada several times, and is interested in intersections of queerness and literature, text encoding, biography, playable stories and procedural creativity.

JOANNA BYSZUK is a full-time researcher and PhD candidate at the Institute of Polish Language of the Polish Academy of Sciences, with a background in Translation Studies and Computer Science. A member of the Computational Stylistics Group and Leader of Working Group 2, "Methods and Tools" at the COST Action "Distant Reading," her work concerns computational methods of text and media analysis, primarily cross-lingual computational stylistics and advancing stylometric methodology and its understanding, especially locating method limitations and developing evaluation procedures. She is also interested in discourse analysis and audiovisual media, and will soon finish her dissertation proposing solutions for multimodal stylometry.

BRIAN CROXALL is an Assistant Research Professor of Digital Humanities at Brigham Young University, Utah, USA. He is the co-editor (with Diane K. Jakacki) of the forthcoming *What We Teach When We Teach DH: Digital Humanities in the Classroom*, from the University of Minnesota Press, and co-founder of the Pedagogy and Training special interest group in the international Alliance of Digital Humanities Organizations (ADHO). He is also the co-editor (with Rachel A. Bowser) of *Like Clockwork: Steampunk Pasts, Presents, and Futures* (2016), again from the University of Minnesota Press.

JAMES CUMMINGS is the Senior Lecturer of Late-Medieval Literature and Digital Humanities for the School of English at Newcastle University, UK. He studies digital editing and medieval drama. From 2005 to 2019 he was elected to the TEI Technical Council, and since then the TEI Board of Directors. He has a BA in Medieval Studies (University of Toronto), an MA in Medieval Studies (University of Leeds), and a PhD in Medieval English Drama (University of Leeds). He was the founder of the Digital Humanities at Oxford Summer School, and annual DH Awards, and in 2021 became a Fellow of the Alan Turing Institute, the UK's national institute for data science and artificial intelligence.

SHAWN DAY (shawnday.com) is a Lecturer in Digital Humanities at University College Cork. He was formerly Project Manager of the Digital Humanities Observatory (Royal Irish Academy). His research in digital humanities ranges from information visualization to project management and audience engagement. His current research interest in the construction of identity in Irish craft beer extends earlier work in the socioeconomic dimensions of the hospitality trade in Victorian Canada. He has published in the spatial and medical humanities blending a background in management economics and digital history with an entrepreneurial ethos from a number of successful software development ventures prior to academic pursuits.

JENNIFER DEBIE (jenniferdebie.com) is a recent graduate of, and postdoctoral researcher at, University College Cork, Ireland, where she researches Romantic literature and leads seminars on plague and apocalypse fiction. She is also a poet and novelist, with work appearing in anthologies by PactPress, Kallisto Gaia Press, and Raven Chronicles Press, among many others. Her second novel, *Heretic*, was published by Wild Wolf Publishing in 2022.

QUINN DOMBROWSKI is the Academic Technology Specialist in the Division of Literatures, Cultures, and Languages, and in the Library, at Stanford University, USA. Prior to coming to Stanford in 2018, their many digital humanities (DH) adventures included supporting the high-performance computing cluster at UC Berkeley, running the DiRT tool directory with support from the Mellon Foundation, writing books on Drupal for Humanists and University of Chicago library graffiti, and working on the program staff of Project Bamboo, a failed digital humanities cyberinfrastructure initiative. They have a BA/MA in Slavic Linguistics from the University of Chicago, and an MLIS from the University of Illinois at Urbana-Champaign. Since coming to Stanford, they have supported numerous non-English DH projects, taught courses on non-English DH, started a Textile Makerspace, developed a tabletop role-playing game to teach DH project management, explored trends in multilingual Harry Potter fanfic, and started the Data-Sitters Club, a feminist DH pedagogy and research group focused on Ann M. Martin's 1990s girls' series "The Baby-Sitters Club." They are currently co-VP of the Association for Computers and the Humanities along with Roopika Risam, and advocates for better support for DH in languages other than English.

JOHANNA DRUCKER is the Breslauer Professor of Bibliographical Studies and Distinguished Professor of Information Studies at UCLA, USA. She is internationally known for her work in the history of graphic design, typography, artists' books, experimental poetry, aesthetics, and digital humanities. Recent publications include *Visualizing Interpretation* (MIT, 2020), *Iliazd: Metabiography of a Modernist* (Johns Hopkins University Press, 2021), *The Digital Humanities Coursebook* (Routledge, 2021), and *Inventing the Alphabet* (Chicago University Press, 2022). Her work has been translated into Korean, Italian, Catalan, Chinese, Spanish, French, Hungarian, Danish, and Portuguese. She was the recipient of the AIGA's 2021 Steven Heller Award for Cultural Criticism.

STUART DUNN is Professor of Spatial Humanities at King's College London, where he is also Head of the Department of Digital Humanities. He teaches, and has written extensively on, the study of space and place in the human record, especially the application of GIS in literary and archaeological research. He is author of *A History of Place in the Digital Age* (2019) and co-editor of the *Routledge International Handbook of Digital Humanities Research Methods.* He holds honorific positions at Riga Technical University, the Australian National University, and Beijing Normal University's campus at Zhuhai.

AMY E. EARHART is Associate Professor of English at Texas A&M University, USA. She is the recipient of a NEH-Mellon Fellowship for Digital Publication for her book-length digital project "Digital Humanities and the Infrastructures of Race in African-American Literature." She has published scholarship on a variety of digital humanities topics, with work that includes a monograph

Traces of Old, Uses of the New: The Emergence of Digital Literary Studies (University of Michigan Press, 2015), a co-edited collection *The American Literature Scholar in the Digital Age* (University of Michigan Press, 2010), and a number of articles and book chapters.

PATRICK EGAN (Pádraig Mac Aodhgáin), PhD, is an international scholar specializing in digital humanities, ethnomusicology, and metadata. His interests have generally focused on how digital practice intersects with research in ethnomusicology. He examined the career of 1960s Irish artist Seán Ó Riada using ethnography and digital visualization, and continues to research more broadly on digital investigations with cultural heritage in archival institutions. As an interdisciplinary scholar, he is also a performer of Irish traditional music on concertina. He is an Irish language speaker, and presents papers at conferences in Ireland, England, Europe, and North America.

LORI EMERSON is Associate Professor in the English Department and the Intermedia Arts, Writing, and Performance Program (IAWP) at the University of Colorado Boulder, USA. She is also Director of IAWP and the Media Archaeology Lab. Emerson is co-author of *THE LAB BOOK: Situated Practices in Media Studies* (University of Minnesota Press, 2022) with Jussi Parikka and Darren Wershler, author of *Reading Writing Interfaces: From the Digital to the Bookbound* (University of Minnesota Press, 2014), and editor of numerous collections.

MARTIN PAUL EVE is the Professor of Literature, Technology and Publishing at Birkbeck, University of London and the Visiting Professor of Digital Humanities at Sheffield Hallam University, UK. He is the winner of the Philip Leverhulme Prize, the KU Leuven Medal of Honour in the Humanities and Social Sciences, and was included in the Shaw Trust's 2021 list of the most influential people with disabilities in the UK.

DOMENICO FIORMONTE (PhD University of Edinburgh) is a Lecturer in the Sociology of Communication and Culture in the Department of Political Sciences at University Roma Tre, Italy. He has edited and co-edited a number of collections of digital humanities texts, and has published books and articles on digital philology, digital writing, text encoding, and cultural criticism of digital humanities. With Sukanta Chaudhuri and Paola Ricaurte, he has recently edited the *Global Debates in the Digital Humanities* (Minnesota University Press, 2022). His latest monograph is *Per una critica del testo digitale. Filologia, letteratura e rete* (Rome, 2018).

JULIE M. FUNK is a PhD candidate in English and Cultural, Social, and Political Thought (CSPT) at the University of Victoria, Canada. They are a member of the Praxis Studio for Comparative Media Studies and have work published in CHI '21 conference proceedings.

RAHUL K. GAIROLA (University of Washington, Seattle) is The Krishna Somers Senior Lecturer of English and Postcolonial Literature and Fellow of the Asia Research Centre at Murdoch University, Australia. He has published five books, and is working on a new monograph tentatively titled *Digital Homes: Technology and Sexuality in the Indiaspora* to be published by Routledge. He is widely published and has delivered many international research talks, and earned several internal and external grants. He was appointed Digital Champion for the State of Western Australia in 2019 by The Digital Studio, University of Melbourne, Australia, and is Editor of the Routledge/Asian Studies Association of Australia (ASAA) South Asia Book Series.

MARY GALVIN is Assistant Professor and Head of Department of Design Innovation at Maynooth University, Ireland. She has held previous positions at Trinity College Dublin and University College Dublin. Her research has appeared in a number of international peer-reviewed publications, including *Digital Scholarship in the Humanities*, *Digital Humanities Quarterly*, *Leonardo*, and *Research Involvement and Engagement*. She holds a PhD in Psychology (Digital Arts and Humanities) from University College Cork and undertakes research across behavioral design, inclusion health design and public patient involvement.

KENT GERBER, the Digital Initiatives Manager at Bethel University, Minnesota, USA, is responsible for the library's digital collections, institutional repository, Makerspace, and collaborative digital scholarship projects. He holds an MLIS and Certificate of Advanced Studies in Digital Libraries from Syracuse University and focuses on how libraries engage with technology, teaching, research, cultural heritage, and digital humanities. His interests within digital humanities include digital archives, digital libraries, open and public scholarship, cyberinfrastructure, and pedagogy. He also co-designed Bethel's Digital Humanities major and teaches portions of the courses in Introduction to Digital Humanities and Advanced Digital Humanities.

SARAH RUTH JACOBS teaches online courses in digital media production and the humanities as an Adjunct Assistant Professor in the Communication and Media and Liberal Studies departments at the School of Professional Studies at the City University of New York, USA. Her work examines how older and emerging forms of media cooperate and compete, as well as long-term demographic shifts in who consumes various types of media. She received her doctorate in American Literature and New Media from the Graduate Center of the City University of New York in February 2020. Her writing has appeared in *The New York Times* and *The Chronicle of Higher Education*.

DIANE K. JAKACKI is Digital Scholarship Coordinator and Affiliated Faculty in Comparative and Digital Humanities at Bucknell University, Pennsylvania, USA. She is the lead investigator of the REED London Online Project and principal investigator of the NHPRC-Andrew W. Mellon Foundation-funded Liberal Arts Based Digital Publishing Cooperative project. She is the co-editor (with Brian Croxall) of the forthcoming *What We Teach When We Teach DH: Digital Humanities in the Classroom*, from the University of Minnesota Press and (with Laura Estill and Michael Ullyot) of *Early Modern Studies after the Digital Turn* (2016) from Iter Press.

ANNE KARHIO is Associate Professor in English Literature at the Inland Norway University of Applied Sciences, Hamar, and holds a PhD from the National University of Ireland, Galway. She has worked as a lecturer and researcher in Finland, Ireland, Norway, and France, and has published widely on topics related to her research interests, including digital literature and art, contemporary Irish poetry, and the esthetics of space and landscape. Her publications also include the monograph *"Slight Return": Paul Muldoon's Poetics of Place* (Peter Lang, 2017) and a co-edited collection titled *Crisis and Contemporary Poetry* (Palgrave Macmillan, 2011).

MAX KEMMAN is a Senior Researcher with Dialogic, conducting policy research on science policies, open science, digitalization of higher education and digital heritage, among other topics. He received his PhD from the University of Luxembourg in 2019 with his study on digital history as trading zones. During his PhD he developed and taught a course Introduction to Digital History, mandatory to all students of the history master program, which inspired his contribution to this volume.

LANGA KHUMALO is Professor and Executive Director of the South African Center for Digital Language Resources (SADiLaR) at the North West University (NWU) in South Africa. He holds a PhD in Linguistics from the University of Oslo, Norway, and an MPhil in Linguistics from Cambridge University, UK. He is an expert in corpus linguistics and computational solutions for the African Academy of Languages (ACALAN), which is a Secretariat of the African Union (AU). He is an award-winning author in Intelligent Text Processing and Computational Linguistics. His current work involves developing digital resources and tools for the under-resourced (African) languages.

ABIGAIL MORESHEAD is a PhD candidate in the Texts and Technology program at the University of Central Florida, USA. Her research looks at the intersection of book studies and feminist media studies, focusing on gendered labor in textual production and knowledge creation. She is currently the assistant editor of the *James Joyce Literary Supplement* and the social media manager for *Johnson's Dictionary Online*.

ÓRLA MURPHY is Department Head for Digital Humanities, School of English and Digital Humanities, University College Cork, Ireland. Her EU international leadership as service roles include National Coordinator of the Digital Research Infrastructure for the Arts and Humanities, National Representative and vice chair on the Scientific Committee of CoST-EU; Co-operation in Science and Technology, and National Representative on the Social Science and Humanities Strategy Working Group of the European Strategy Forum on Research Infrastructures. Nationally, she is a board member of the Digital Repository of Ireland and co-chair of The Arts and Culture in Education Research Repository.

PEDRO NILSSON-FERNÀNDEZ teaches in the Department of Spanish, Portuguese and Latin American Studies, University College Cork, Ireland, where he completed a doctoral thesis titled *Mapping the Works of Manuel de Pedrolo in Relation to the Post-Civil War Catalan landscape* (2019). His research looks at literary spaces in the context of Iberian minoritized languages and cultures, particularly through the use of text-mining and GIS to map post-war Catalan literary and cultural heritage. He is also interested in computer-assisted text analysis and the use of digital technologies to enhance learning and teaching.

MICHAEL PIDD is Director of the Digital Humanities Institute at the University of Sheffield, UK. He has over twenty-five years' experience in developing, managing, and delivering large collaborative research projects and technology R&D in the humanities and heritage subject domains. His responsibilities include overseeing the development and delivery of its research projects and the implementation of its postgraduate degree programs.

ELENA PIERAZZO is Professor in Digital Humanities at the University of Tours, France, specializing in Renaissance Italian texts, digital editions, and text encoding. She obtained her doctorate in Italian philology at the Scuola Normale in Pisa, then held positions including those at the University of Grenoble and the Department of Digital Humanities at King's College London. Her present and former professional roles include Chair of the Text Encoding Initiative (TEI), Co-Chair of the Programme Committee for the DH2019 conference (Utrecht), and Co-Chair of the Working group on Digital Editions for the Network for Digital Methods in the Arts and Humanities (NeDiMAH).

ANDREW PILSCH is an Associate Professor of English at Texas A&M University, USA, where he teaches and researches rhetoric and the digital humanities. His first book, *Transhumanism: Evolutionary Futurism and the Human Technologies of Utopia*, was released by University of Minnesota Press in 2017, when it won the Science Fiction and Technoculture Studies Book Prize for that year. His research has been published in *Amodern*, *Philosophy & Rhetoric*, and *Science Fiction Studies*.

ANDREW PRESCOTT is Professor of Digital Humanities in the School of Critical Studies, University of Glasgow, UK. He trained as a medieval historian and from 1979 to 2000 was a Curator in the Department of Manuscripts of the British Library, where he was the principal curatorial contact for Kevin Kiernan's *Electronic Beowulf*. He was from 2012 to 2019 Theme

Leader Fellow for the AHRC strategic theme of *Digital Transformations*. He has also worked in libraries, archives, and digital humanities units at the University of Sheffield, King's College London, and the University of Wales Lampeter.

GIMENA DEL RIO RIANDE is Associate Researcher at the Instituto de Investigaciones Bibliográficas y Crítica Textual (IIBICRIT-CONICET, Argentina). She holds an MA and Summa Cum Laude PhD in Romance Philology (Universidad Complutense de Madrid). Her main academic interests deal with digital humanities, digital scholarly editions, and open research practices in the Humanities. She is the Director of the Laboratorio de Humanidades Digitales HD CAICYT LAB (CONICET), and the Revista de Humanidades Digitales (RHD). She also serves as president of the Asociación Argentina de Humanidades Digitales (AAHD), and as member of the board of directors at Poblaciones, FORCE11, and the Text Encoding Initiative Consortium. Since 2020, she is one of the DOAJ (Directory of Open Access Journals) Ambassadors for Latin America. She is the Director of the first postgraduate training in digital humanities in Argentina (UCES) and co-director of the Digital Humanities Master at UNED (Spain) and she also coordinates a Global Classrooms Program for Digital Minimal Editions with Raffaele Viglianti (2020–2023, University of Maryland-Universidad del Salvador).

ROOPIKA RISAM is Associate Professor of Film and Media Studies and Comparative Literature and Faculty of Digital Humanities and Social Engagement at Dartmouth College, USA. Her first monograph, *New Digital Worlds: Postcolonial Digital Humanities in Theory, Praxis, and Pedagogy*, was published by Northwestern University Press in 2018. She is the co-editor of *Intersectionality in Digital Humanities* (Arc Humanities/Amsterdam University Press, 2019) and *South Asian Digital Humanities: Postcolonial Mediations Across Technology's Cultural Canon* (Routledge, 2020). Her co-edited collection *The Digital Black Atlantic* in the Debates in the Digital Humanities series (University of Minnesota Press) was published in 2021.

SHAWNA ROSS is an Associate Professor of English at Texas A&M University, USA, where she specializes in Victorian literature, transatlantic modernism, and the digital humanities. Her book *Charlotte Brontë at the Anthropocene* was published by SUNY Press in 2021, and she is currently at work on a monograph about the history of modernism online. Her digital humanities work centers on pedagogy, digital editions, and social media use. She co-wrote *Using Digital Humanities in the Classroom* (Bloomsbury, 2017) and is co-editor of the essay collections *Humans at Work in the Digital Age* (Routledge, 2019) and *Reading Modernism with Machines* (Palgrave Macmillan, 2016).

BO RUBERG, PhD, is an Associate Professor in the Department of Film and Media Studies and an affiliate faculty member in the Department of Informatics at the University of California, Irvine, USA. Their research explores gender and sexuality in digital media and video games. They are the author of *Video Games Have Always Been Queer* (New York University Press, 2019), *The Queer Games Avant-Garde: How LGBTQ Game Makers Are Reimagining the Medium of Video Games* (Duke University Press, 2020), and *Sex Dolls at Sea: Imagined Histories of Sexual Technologies* (MIT Press, 2022).

ANASTASIA SALTER is the Director of Graduate Programs for the College of Arts and Humanities and Associate Professor of English, overseeing the innovative interdisciplinary doctoral program in Texts & Technology at the University of Central Florida, and author of seven books that draw on humanities methods alongside computational discourse and subjects. These include *Twining: Critical and Creative Approaches to Hypertext Narratives* (Amherst College Press 2021; with Stuart Moulthrop), and *Adventure Games: Playing the Outsider* (Bloomsbury 2020; with Aaron Reed and John Murray).

JENTERY SAYERS is an Associate Professor of English and Director of the Praxis Studio for Comparative Media Studies at the University of Victoria, Canada. He is the editor of three books: *Making Things and Drawing Boundaries* (University of Minnesota Press, 2017), *The Routledge Companion to Media Studies and Digital Humanities* (Routledge, 2018), and *Digital Pedagogy in the Humanities* (Modern Language Association, with Davis, Gold, and Harris, 2020).

ALEXANDRA SCHOFIELD is an Assistant Professor at Harvey Mudd College in Claremont, California, USA, where she teaches courses in natural language processing and algorithms. Her research focuses on empirical approaches to improve the application of topic models to text analysis in the Humanities and Social Sciences. She earned her PhD in Computer Science at Cornell University. She also serves as a co-organizer of Widening NLP, an organization dedicated to improving access to the natural language processing community for researchers from underrepresented groups.

ROXANNE SHIRAZI is Assistant Professor and dissertation research librarian in the Mina Rees Library at the Graduate Center, City University of New York, USA. She was a founding co-editor of *dh+lib: where the digital humanities and librarianship meet* and has taught digital humanities at Pratt Institute's School of Information.

ANNA-MARIA SICHANI is a cultural and media historian and a digital humanist. She is currently a Postdoctoral Research Associate in Digital Humanities at the School of Advanced Study, University of London, working with The Congruence Engine project, funded by the Arts and Humanities Research Council under the Towards a National Collection funding stream. She previously held a Research Fellowship in Media History and Historical Data Modelling at University of Sussex, working on the AHRC-funded Connected Histories of the BBC project. Her research interests include literary and cultural history, cultural and social aspects of media changes, born-digital archives, digital scholarly editing and publishing of archival assets, research infrastructure, and digital pedagogy. She has collaborated with a number of Digital Humanities projects, including COST Action "Distant Reading for European Literary History," Digital Scholarly Editing Initial Training Network (DiXiT), Transcribe Bentham, and DARIAH.

RAY SIEMENS (web.uvic.ca/~siemens/) directs the Electronic Textual Cultures Lab, the Implementing New Knowledge Environments Partnership, and the Digital Humanities Summer Institute. He is Distinguished Professor in the Faculty of Humanities at the University of Victoria, Canada, in English and Computer Science, and past Canada Research Chair in Humanities Computing (2004–2015). In 2019–2020, he was also Leverhulme Visiting Professor at Loughborough University and he is the current Global Innovation Chair in Digital Humanities at the University of Newcastle (2019–2022).

JAMES SMITHIES is Professor of Digital Humanities in the Department of Digital Humanities, King's College London. He was previously founding director of King's Digital Lab and Deputy Director of King's eResearch. Before working at King's, he worked at the University of Canterbury in New Zealand as a Senior Lecturer in Digital Humanities and Associate Director of the UC CEISMIC earthquake archive. He has also worked in the government and commercial IT sector as a technical writer and editor, business analyst, and project manager. His approach to digital humanities is presented in *The Digital Humanities and the Digital Modern* (Palgrave Macmillan, 2017).

PETER STOKES is Professor at the École Pratique des Hautes Études – Université PSL (Paris), specializing in Digital Humanities applied to historical writing. After degrees in Classics and Medieval English and in Computer Engineering, he obtained his doctorate in Palaeography at the University of Cambridge and held posts there and the Department of Digital Humanities at King's College London before moving to France. He has led or co-led major projects including an ERC Starting Grant for

DigiPal; other roles include the Bureau of the Comité International de Paléographie Latine and the Medieval Academy of America's Committee on Digital Humanities and Multimedia Studies.

MELISSA TERRAS is Professor of Digital Cultural Heritage at the University of Edinburgh, UK, leading digital aspects of research as Director of the Edinburgh Centre for Data, Culture and Society. Her research interest is the digitization of cultural heritage, including advanced digitization techniques, usage of large-scale digitization, and the mining and analysis of digitized content. She previously directed UCL Centre for Digital Humanities in UCL Department of Information Studies, where she was employed from 2003 to 2017. She is a Fellow of the Alan Turing Institute, and Expert Advisor to the UK Government's Department of Digital, Culture, Media and Sport.

LAUREN TILTON is Assistant Professor of Digital Humanities at the University of Richmond, Virginia, USA. She is the author of *Humanities Data in R* (Springer, 2015) and co-editing the forthcoming *Debates in the Digital Humanities*: *Computational Humanities*. Her work has appeared in journals such as *American Quarterly*, *Journal of Cultural Analytics*, and *Digital Humanities Quarterly*. She directs *Photogrammar* (photogrammar.org) and the Distant Viewing Lab.

NAOMI WELLS is a Lecturer in Italian and Spanish with Digital Humanities at the Institute of Modern Languages Research and the Digital Humanities Research Hub (School of Advanced Study, University of London). Her current research focuses on the Internet and social media, particularly in relation to online multilingualism, and contemporary and archived web content produced by and for migrant communities. She has a broader interest in promoting and facilitating cross-languages digital research and is joint editor of the Digital Modern Languages section of the Modern Languages Open journal with Liverpool University Press.

JANE WINTERS is Professor of Digital Humanities and Director of the Digital Humanities Research Hub at the School of Advanced Study, University of London. She is editor of a series on academic publishing, which is part of Cambridge University Press Elements: Publishing and Book Culture. Her research interests include digital history, born-digital archives (particularly the archived web), the use of social media by cultural heritage institutions, and open access publishing. She has published most recently on non-print legal deposit and web archives, born-digital archives and the problem of search, and the archiving and analysis of national web domains.

ACKNOWLEDGMENTS

This book would not have been possible without the work of a great many people.

But as editor, I would like to draw particular attention to the efforts of Ben Doyle, Amy Brownbridge, Laura Cope, and all their colleagues at Bloomsbury, as well as Deborah Maloney, Joanne Rippin, Dawn Cunneen, and Lisa Scholey for their role in production. Ben has often been kind enough to provide a home for my research, so I am very pleased that we are able to collaborate once more on this major collection of essays. Looking at his back catalog as a publisher and editor, it is quite evident that Ben has long championed interdisciplinary research in the arts and humanities—I hope this book can act as some small tribute to such advocacy.

I also want to thank this collection's contributors for their resilience and professionalism—I am deeply grateful to each of them for being so generous with their expertise and producing such remarkable work in times such as these. In addition to the bleak societal situation in which we now find ourselves, I am conscious that many of this book's authors faced significant personal trials in recent months. Collections of this sort do not happen without contributors—both seen and unseen—so, once again, sincerest thanks to everyone who has an essay in this finished volume. I am equally indebted to those whose writing didn't quite make it into these pages for one reason or another.

This project has benefited from the support of the Irish Research Council, as well as the College of Arts, Celtic Studies and Social Sciences at University College Cork. The latter provided me with a research sabbatical during which parts of this volume were compiled. Thanks to Daniel Alves, Teresa Araújo, Joana Paulino, and others at NOVA University, Lisbon, who facilitated a visiting fellowship during the aforementioned research leave. Warmest thanks to Leonor de Oliveira for making my trip possible through her kind hospitality.

Thanks to Patricia Hswe for allowing me to pick her brain on suitable contributors, and to my colleagues at UCC—particularly Órla Murphy, Shawn Day, Máirín MacCarron, Mike Cosgrave, Patrick Egan, Ann Riordan, Max Darby, Claire Connolly, Lee Jenkins, and Graham Allen—for their collegiality.

James O'Sullivan

Introduction: Reconsidering the Present and Future of the Digital Humanities

JAMES O'SULLIVAN (UNIVERSITY COLLEGE CORK)

I do not think it unreasonable to suggest that "digital humanities" (DH) is something of a charged term. Many of its opponents wish it would just go away and stop neoliberalizing and cannibalizing the humanities, while even some of its advocates have grown weary of the "what is/are DH?" debate, on reifying its value to the humanities, and figuring out where their "new" sorts of practices fit in long-established disciplinary and institutional structures. We might see a day when arguments about DH, its parameters and intellectual and pedagogical purposes are no longer in vogue, but we are not there yet. In fact, I would hold that the "we no longer need to talk about what the digital humanities are" perspective is quite premature. It has been over a decade now since Stephen Ramsay got himself into all sorts of trouble for asking—at a 2011 MLA panel on the future of DH—"who's in and who's out?" (2013, 240). But is that question really so untimely?

Nobody wants to talk about who's in and who's out, because to do so will inevitably involve exclusion, and if there is anything that research culture and education could do without, its more exclusion. Consequently, the conversation that Ramsay attempted to start never really progressed, the consensus now being that one does not waste their time navel-gazing about what the digital humanities look like, or even better, what they should look like, but rather, get on with the actual work of doing research in the humanities that relies heavily on some new-fangled digital tool or technique. Eventually, we will all just come to know good DH work when we see it, and the field will naturally become a "semi-normal thing" (Underwood 2019). Everyone will start to see that DH is not at all about "building project websites" because sophisticated applications of cultural analytics will become more commonplace, and the "charlatans" (Unsworth 2002) who think they are doing DH because they are blogging or tweeting, will simply fade into obscurity.

It seems reasonable to expect the "real" digital humanities to just become known in time, but the last two decades or so—what might be thought of as the "DH moment" (Gold 2012)—have seen a lot of resources, both fiscal and human, committed to this thing called DH. It is not, to further quote Ramsay, "some airy Lyceum," but rather, it is "money, students, funding agencies, big schools, little schools, programs, curricula, old guards, new guards, gatekeepers, and prestige" (2013, 240). And yet, despite all this investment, all these "concrete instantiations," there are still people who think DH is putting pictures of books on WordPress sites. Perhaps it is time we stopped asking what DH is, and returned to the contentious task of saying what it is not.

We have given so much to the digital humanities that it seems almost reckless to argue that we should halt all typological discussions of the field, that policymakers, institutions, and educators should simply continue to invest in DH for the sake of keeping up with the Joneses. Whatever DH has been—and there are some compelling treatments of its histories (Jones 2013; Terras, Nyhan, and Vanhoutte 2013; Nyhan and Flinn 2016; Kim and Koh 2021)—it is imperative that the discipline's present moments are continuously reassessed, not so that we can carry on with the navel-gazing, but rather, so that we might learn from the mistakes of present DH in an attempt to better shape our DH futures.

Re-engaging with the question, "What are the digital humanities?," has never been timelier. DH is everywhere, across all continents and cultures, all intellectual communities and research practices. In any institution that does research or teaching in the arts and humanities, if a concerted effort to develop some form of work in the digital humanities is already underway, the "what is/are DH?" conversation has probably been had. This is a natural consequence of the (perhaps unfortunate) fact that most human experiences are now mediated through software and screens—it was inevitable that the digital would eventually start to radically change the arts and humanities. This change is not always positive, but it is happening, and closing our eyes to it and hoping that it works out for the best—that our disciplinary practices, methodologies, knowledge, and learners will just settle into some new shape that suits everyone—is, to reiterate, quite reckless. Whether one loves or loathes the digital humanities (whatever one thinks that term to mean), it is intellectually dishonest for any researcher, administrator, cultural practitioner, or teacher to simply turn their gaze from the substantive social and cultural shifts that we are seeing in this age of machines. If we are going to invest public funds in DH, if we are going to research and teach it; if we are going to re-design cultural spaces so that they better incorporate digital ways of representation, and if we are going to at least attempt to reclaim some "cultural authority" (Drucker 2012) from those who build the tools and technologies on which we have become utterly reliant, then we need to at least talk about DH. And then we need to talk about DH again, and again.

This book attempts to reconsider the digital humanities. As a "handbook," it acts as an essential guide to reconsidering the digital humanities, and that is what DH needs right now; not more tools and methods, but renewed engagement with a very basic question—"to what end?" Conscious of DH's past and present, this is a future-focused handbook, one that attempts to envisage where the dominant sentiments and practices of the day might take us in the coming times, what should be protected, and what might be better jettisoned. This handbook is a reference to the present, but it also provides instructions on how we might get to some other future.

The text is divided into five thematic clusters: "Perspectives & Polemics," "Methods, Tools & Techniques," "Public Digital Humanities," "Institutional Contexts," and "DH Futures." These clusters represent those areas thought most essential to the act of disciplinary reconsideration: epistemology, methodology, knowledge and representation, pedagogy, labor, future possibilities and their preconditions. Like any such volume it is incomplete and imperfect, but it nonetheless provides a set of comprehensive provocations on some of the most important topics in the digital humanities, and indeed, wider arts and humanities as a whole.

In her essay, "Digital Humanities Futures," Amy Earhart cites Rosanne G. Potter, who, all the way back in 1989, called for "a new kind of literary study absolutely comfortable with scientific methods yet completely suffused with the values of the humanities" (Earhart 2016). The DH moment has produced some remarkable scholarship and public resources, but Potter's vision seems as distant now as it did in the late 1980s. Across all disciplines in the humanities, digital or otherwise, we find scholars who are uncomfortable with methods beyond those with which they are familiar, and the values of the humanities have never seemed more absent from those institutions from which they first emerged. Self-preservation and progression have become central to the academic psyche: overheads from funding are valued over the making of meaning, students are treated like customers, and scholars compete among themselves for professional capital, either to escape precarity, or to further their own individual careers.

The computer, an immensely powerful instrument that has radically reshaped culture and society, has not produced the type of interdisciplinarity Potter imagined. Instead, the computer has just become a thing that must be learned; rather than broadly facilitating new approaches to the great questions of culture, it has made such pursuits seem redundant in a world where critical thinking is seen as less relevant to the workplace than programming. Everyone is frustrated, but everyone must survive so nothing changes—no one wishes to reconsider DH, because DH secures enrollment, it makes students feel less panicked about the perils of studying philosophy and literature. It is a grim situation and it is one that nobody wants: those in the digital humanities simply want to do humanities research, but using new computer-assisted methods, and those beyond DH want to do what they have always done, and be left alone to get on with that (hugely important) work. Interdisciplinarity will happen where it needs to happen, and some of it will involve DH, some of it will not. But the aforementioned frustration has further fueled the antagonism between differing disciplinary and cultural perspectives, and instead of a humanities comfortable with scientific methods but still suffused with its own values, we have division and confusion. Dismissive op-eds and aggressive Twitter spats are commonplace in assessments of the digital humanities, and all the while it is becoming increasingly difficult for scholars of DH to separate themselves from the market mentality that compels administrators to shut down archaeology and classics departments while opening new humanities "laboratories" and creative "exploratoriums."

Whatever one might say about the future of digital humanities, it seems reasonable to assert that we should strive for something different to its present. Earhart believes that such a future needs to be built on "shared spaces," that scholars need to collaborate so as to "develop working models that best articulate our hopes" (2016). Articulating our hopes seems like such a simple starting point, but we even seem incapable of that. Ramsay argued that future digital humanists should all be coders—a position not too dissimilar from Potter's—and his brief speech has been roundly pilloried ever since. It is a brave person who now dares to envision a future for the digital humanities, or even just praise or critique what we have at present. Anyone who disagrees will do so publicly and vehemently, because that is what intellectual exchange has become in our age—a reflection of everyone's anxieties and fears, a part of that desperate attempt to survive in socio-economic contexts where most perish. If only everyone could get along, bring shared values (Spiro 2012) to the shared spaces, then everything would be a great deal better.

This book will attract several kinds of reader: dyed-in-the-wool DHers, the DH-curious, and naysayers. I suspect each will find something to satisfy whatever urge brought them here to begin with, and each will take umbrage with some part or another. A good handbook should provide ready-made reference materials, a set of intuitive treatments of a consistent topic. Regardless of how readers of this book view the digital humanities, my hope is that the provocations contained within facilitate some further progress, however small, towards Potter's "new kind" of humanities—a humanities in which there is no tension between misplaced senses of old and new, but simply a desire to produce and sustain cultural knowledge using all the capacities of both the human and the machine.

REFERENCES

Drucker, Johanna. 2012. "Humanistic Theory and Digital Scholarship." In *Debates in the Digital Humanities*, edited by Matthew K. Gold, 35–45. Minneapolis, MN: University of Minnesota Press. http://dhdebates.gc.cuny.edu/debates/text/34.

Earhart, Amy. 2016. "Digital Humanities Futures: Conflict, Power, and Public Knowledge." *Digital Studies/Le Champ Numérique* 6 (1). https://doi.org/10.16995/dscn.1.

Gold, Matthew K. 2012. "The Digital Humanities Moment." In *Debates in the Digital Humanities*, edited by Matthew K. Gold, ix–xvi. Minneapolis, MN: University of Minnesota Press. https://dhdebates.gc.cuny.edu/projects/debates-in-the-digital-humanities.

Jones, Steven E. 2013. *The Emergence of the Digital Humanities*. New York, NY: Routledge.

Kim, Dorothy and Adeline Koh, eds. 2021. *Alternative Historiographies of the Digital Humanities*. Santa Barbara, CA: Punctum Books. http://www.jstor.org/stable/j.ctv1r7878x.

Nyhan, Julianne and Andrew Flinn. 2016. *Computation and the Humanities: Towards an Oral History of Digital Humanities*. Cham: Springer.

Ramsay, Stephen. 2013. "Who's In and Who's Out." In *Defining Digital Humanities*, edited by Melissa Terras, Julianne Nyhan, and Edward Vanhoutte, 239–41. Farnham, UK: Burlington, VT: Ashgate.

Spiro, Lisa. 2012. "'This Is Why We Fight': Defining the Values of the Digital Humanities." In *Debates in the Digital Humanities*, edited by Matthew K. Gold, 16–34. Minneapolis: University of Minnesota Press. http://dhdebates.gc.cuny.edu/debates/text/13.

Terras, Melissa, Julianne Nyhan, and Edward Vanhoutte, eds. 2013. *Defining Digital Humanities: A Reader*, new edn. Farnham, UK: Burlington, VT: Ashgate.

Underwood, Ted. 2019. "Digital Humanities as a Semi-Normal Thing." In *Debates in the Digital Humanities 2019*, edited by Matthew K. Gold and Lauren F. Klein, 96–8. Minneapolis, MN: University of Minnesota Press. https://doi.org/10.5749/j.ctvg251hk.13.

Underwood, Ted. 2020. "Machine Learning and Human Perspective." *PMLA* 135 (1): 92–109. https://doi.org/10.1632/pmla.2020.135.1.92.

Unsworth, John. 2002. "What Is Humanities Computing and What Is Not?" Forum Computerphilologie. 2002.

PART ONE

Perspectives & Polemics

CHAPTER ONE

Normative Digital Humanities

JOHANNA DRUCKER
(UNIVERSITY OF CALIFORNIA, LOS ANGELES)

To characterize digital humanities as normative suggests that it is experiencing a flattened intellectual curve, or reaching a state of equilibrium in ongoing research and pedagogy. The implications are more dulling than damning, and the specifics need to be detailed before a full assessment of liabilities and benefits is possible.

The innovative fervor that launched a boom in the digital humanities more than twenty years ago has abated. The intellectual challenges that galvanized the exchange between computational methods and humanities projects have been tamed to some degree through clarification of their scope and ambitions. In addition, professional competence has increased. The numbers of individuals who identify as "digital humanists" and the institutional sites of activity associated with the field have multiplied. This saturation has resulted in a critical mass of programs and projects, many associated with libraries or separate academic centers with dedicated staff, faculty, or fellowship positions. In short, the field has become institutionalized and is no longer a novelty.

As this growth has occurred, the once vague implications of the modifiers "digital" and "computational" attached to the humanities have assumed fairly clear outlines. Adjectives that seemed to gesture towards a broad, generalized realm are now linked to distinct activities. These go beyond repository building and online collections to include computationally assisted methods performed on files that are stored as binary code. Humanities documents and artifacts are now regularly abstracted into "data" and subjected to quantitative and statistical analyses that were originally developed in the natural and social sciences. Distant reading techniques designed for summarizing corpora of professional literature in law, medicine, and other fields have been reimagined for literary study. Other activities include qualitative approaches to markup and metadata that keep hermeneutics in the mix. In addition, ever more sophisticated understandings of data models and statistical methods continue to add depth and dimensions to automated processes within humanistic research and inquiry. Virtual, augmented, and mixed-reality applications are increasingly integrated into museums, exhibitions, and research environments. And, most recently, work in AI, machine learning, and emergent systems has introduced new acronyms and processes—GANs, GPT-3, and so on.

In short, digital humanities have become a flourishing academic industry that draws on a rich and varied range of approaches to produce work that is equally varied in its quality. This makes the digital humanities just like any other field of intellectual inquiry. Some work is good (insightful, well-documented, measured in its claims), some work is not. The question of what criteria distinguish these deserves deliberate attention, but wholesale dismissal of computational methods in the humanities is clearly foolish. Computational methods are integrated into research, but even more subtly and significantly into infrastructures and systems for access, use, publication, and work with primary and scholarly materials. The time is long past when any scholar working within a networked environment of any kind could declare themselves "non-digital." Nonetheless, as computational methods have become increasingly integral to humanities work, the criticism lodged against them has also become well established—and, likewise, normatively formulaic.[1]

The larger question posed here is how, in becoming so integral to humanities work, computational and digital methods enact epistemological (and sociological) normalization. The terms *normativity* and *normalization* refer to processes by which activities become habitual and thus lose their conspicuous identity. They are conditions and processes by which a cultural field ceases to provoke innovation—and loses awareness of its conceptual limitations. Normativity is the process by which what *can* be thought comes to pass for what *is*—and the possibility space for imagining otherwise collapses, flat-lined, but operational. The digital humanities are not alone in assuming normative identity within disciplinary fields. But because of the short cycle of the uptake and integration of relatively new techniques as a defining feature of their activity, the digital humanities provide a study in the emergence of a normative scholarship and pedagogy. The benefits and liabilities of this phenomenon are pertinent to the digital humanities, but also to academic work more generally.

DIGITAL HUMANITIES RESEARCH

To begin this assessment, consider what has changed in computationally based research practices over the course of these recent decades (automated text analysis has been around for more than half a century, machine-readable data for almost twice that long) so that their use no longer requires esoteric or specialized skills.

Even twenty years ago, writing a script to produce a visualization from the structured data of an archive was a non-trivial task. Now this can be accomplished in a few clicks without any programming knowledge. More significantly, the conceptual frameworks for understanding the translation of such analysis into a graph or visualization were not sufficiently familiar at that time to make this work legible within the humanities. The intellectual challenge was not merely to find a way to *make* something in the digital environment by acquiring (or hiring) the technical skill, but to learn how to *think* it, enter into its epistemological insights at a formative stage of research. The conception, not just the implementation, had to be informed by computational processes in what were unfamiliar approaches. The formats of structured data were not forms of thought within which humanists had been trained, and neither were the graphic visualizations. What did it even mean, one wondered, to identify a *topic* in computational terms, or to understand *collocation* and *proximity* as features of a text or corpus of materials processed automatically?

These and other concepts had to be integrated into the intellectual framework on which research is initiated for them to be meaningful. Even a very basic concept, like *sampling*, has to be internalized and digested before it can be used in research design. Acquiring more than superficial understanding of these concepts takes time, experimentation, and direct involvement in design and implementation—but the first challenge was simply to "get our heads around" these ideas. Computational research and pedagogy both depend on being initiated into these modes of thought, but the pace and mode of that initiation have changed dramatically as digital humanities have become codified.

My own introduction to this work began in the early 2000s. The first digital project in which I was involved, Temporal Modeling, required conceptualizing its structure in a formalized and analytic way. We began by determining its fundamental components and then designing their various characteristics. The three categories of components we defined were objects, behaviors, and actions. The first was fairly self-evident, the objects had to constitute the temporal primitives (points, intervals, events, and their attributes). The second, the behaviors of objects, was more challenging. Should the objects bounce, snap to a grid-line, be able to stretch, fall, move, or have other properties? What semantic value did these behaviors have as part of a temporal system? The third, the actions of a user, also required innovation since at that moment, the early 2000s, interface design was still evolving. We had become trained by various text and graphics programs or applications to think about "grabbing" and "dragging" or otherwise manipulating objects. But applying that experience to the design of a platform for modeling temporal relations meant linking a set of concepts to a range of actions. This was quite a stretch at the time—even making the distinction among objects, behaviors, and actions was not immediately intuitive. But the activity was thrilling. A kind of vertigo was produced by learning to think in this meta-design mode. This work continues in our Heterochronologies and TimeCapsule projects, where it still feels exploratory and innovative.[2] But now this work proceeds with the assumption that we know how to approach the technical and conceptual design of the projects.

In the last few years, data-related scholarship has come to be supported in software that exists as off-the-shelf applications and packaged tutorials. Professional-level tools in visualization, mapping, or presentation in augmented, virtual, or other modes, though not immediately accessible to the novice, do not have to be *designed*, only *learned*. That is a crucial distinction. *Design* required fundamental conception of the components and processes. *Use* merely requires *training*. The distance from the initial steep learning curve required for creating Temporal Modeling and that of learning a new program or design environment marks the shift from innovation to normalization.

The vast digital *terra incognita* of two decades ago has been mapped well beyond its coastlines and anyone with a search engine can find their way through the terrain. The change from the phase in which technical frameworks had to be conceived and designed to one in which one is tasked instead to learn existing protocols is more than a sign of maturity. It is a shift in the fundamental epistemological foundation on which the engagement with technological methods proceeds. The work of practitioners with higher levels of technological savvy may engage with probabilistic natural language processing, neural networks, and deep-learning algorithms. Innovative work still proceeds. But for the most part, the aspiring digital humanist is being trained *by* the programs and applications to conform research concepts and execution to their functional operations. This continues the long traditions of numeracy and literacy, where the sums to be added and the vocabulary to be learned might change under different circumstances, but the procedures and syntaxes remain the same conceptually and operationally. This activity of being-trained-into digital humanities is one of the instruments for normalization.

These normalizing processes are also at work in pedagogy, scholarly discourse, and criticism of the field from within and outside. Before addressing how normalization works, what its impacts and implications might be, a few more general statements about the state of digital humanities are in order.

The early excitement of the intersection of computation and humanistic work was generated in part by the need to make explicit assumptions that had long been implicit. The tasks of metadata description and markup required making deliberate choices. Creating a list of terms, tags, pick-lists or fields into which to put parsed character data felt like disciplining the open-ended practices of close reading—which had been amplified in the encounters with post-structural and deconstructionist infusions of "the play of signification." However quaint such formulations feel in the present, their place in early digital humanities was crucial. Rather than a turn away from the possibilities offered by such theoretical interventions, digital humanities was often seen as a way to extend their potential. Thus, Edward Ayers proposed his *Valley of the Shadow* project as a way to position a scholarly argument as *an* interpretation of primary materials, not *the* reading of them.[3] Jerome McGann conceived of *The Rossetti Archive* as an opportunity to expose the multiplicity of the identities of this artist's works in their dual image-text formats, with digital infrastructure opening the editorial process in previously unforeseen ways.[4] Willard McCarty sought to encode the variable dimensions of interpreting mutable language in Ovid's poetics (1993). In project after project from the 1990s and early 2000s onward, the lessons of deconstruction were frequently the start point for digital project design. True, a positivist streak ran through some projects in a desire for "certainty" in regard to digitization practices and/or the representation of complex objects in digital surrogates. Other scholars were motivated by straightforward desires to use the rapidly expanding capacities of the web to make research and collections accessible.

Most of the early scholar-practitioners in the sphere around the Institute for Advanced Technology in the Humanities at the University of Virginia, my point of entry into the field, were endeavoring to implement theoretical complexities, not escape them. The intellectual excitement was high. Whole new epistemological vistas seemed to open. In retrospect, it has become clear that a far more developed scaffolding already existed in the information professions than was evident to the humanists cheerfully creating their own metadata schemes and infrastructure in the 1990s and early 2000s. But considerable intellectual imagination was put to conceiving what computational work meant, how it could be implemented, and what specific tasks were involved. To reiterate the argument sketched above, this included figuring out such things as what "parsing" meant and why "tokenization" was an essential concept in remediating analog materials into digital formats. Initiation into these practices was not merely a matter of being taught to tread unfamiliar pathways, it also created new patterns of thought, ways of addressing what the path was for, where it led, what was required for negotiating the unknown domain. This meant learning the intellectual foundations on which systems, infrastructure, or applications operated in order to begin to understand what it meant to build them.

This background is essential for making clear the distance between those moments and the present. Consider a few common platforms—Voyant, Cytoscape, Omeka, Leaflet, and Tableau. Each of these, designed within the digital humanities community, is now an off-the-shelf, readily accessible, platform to which a novice can be introduced in a few lessons. Becoming adept takes time, practice, and the commitment to explore all features of the platforms to understand customization and effective use. But the intellectual engagement with these platforms now requires submission, not invention. The platforms have their protocols, and the user must follow them, disciplining mind and imagination in a practice of alignment with what is possible in the programs. Few humanities scholars create new protocols. Data modeling remains an act of interpretation, but running data through any of the above-mentioned platforms to create analytics, network graphs, digital exhibits and asset management, maps, or other visualizations occurs within prescribed protocols.

The prescriptive and the normative are not identical, but they align in the resulting conformance to procedures that constrain imagination through habit—and through the delusion of creative control.

DIGITAL HUMANITIES PEDAGOGY

Consider, as well, the changes in pedagogy in digital humanities. Also at the University of Virginia in the early 2000s, in the context of a grant from the National Endowment for the Humanities' newly established programs in support of such activity, I convened a seminar to produce a blueprint for an introductory course in digital humanities. The group of participants was felicitous—Jerome McGann, Geoffrey Rockwell, Daniel Pitti, Bethany Nowviskie, Steve Ramsay, Worthy Martin, John Unsworth, Andrea Laue, and several others all participated. Each week, one or more persons would present on a topic and the challenge was to combine hands-on exercises and critical issues. Rockwell's exemplary digitization discussion made a deep impression. He explained that the rationale and goals for the lesson were to make students understand that even the most basic task, like creating a digital surrogate through scanning, had critical theoretical implications for its identity, use, accessibility, and longevity. We had no roadmap. Even creating the list of topics that comprised the course (which later served as the basis for *Digital Humanities 101* when I came to UCLA and was finally able to implement it) was an exercise in conjuring a world into being.[5] After the fact, the inclusion of topics like digitization, markup, programming languages, data visualization, topic modeling, interface design, intellectual property, and other fundamentals makes perfect sense. But this material had never been taught before, and no syllabi existed on which to build. The exercise was exhilarating. We struggled to keep our core principles foremost in designing the class—that technical skills and theoretical discussion should always be taught in tandem. That course became codified, first in an online format, then in my *Digital Humanities Coursebook* (Routledge, 2021).

The ease of learning digital tools has expanded. Students bring a familiarity with platforms and applications and how to use them. A few lessons in Voyant, for instance, makes students exuberant as they compare the sentence length of Terms of Service Agreements with those of a beloved nineteenth-century novel, track the keywords in context, and struggle to understand what vocabulary density might indicate. The black box effect disappears in the accessible dashboard display of results. So much information is offered, how could anything not be apparent? Such ease of use blinds us to the processes at work and their transformations of inputs into outputs. The results in Cytoscape are even more prone to this easy obfuscation, the display masked by the speed and efficiency through which the network graphs are generated. The processes by which basic chart visualizations are constructed are easier to recover, because plotting points can be readily imitated in analog form. But once the scale or complexity of a visualization or graph outstrips the ability of an individual to plot data against axes, the dazzle effect takes over.

This is one of the liabilities of normativity—acclimatization to the acceptance of output without any knowledge of the way it was generated. This disconnect from the analog processes permeates most digital humanities methods. Reverse engineering the algorithms of production, forcing students to create step-by-step instructions for their work, has some ameliorating effect, but only for very low-level processes. The operations of the algorithms remain unseen, taken for granted, unexamined, and out of range of intervention.

The relation between normativity and black-box effects is one downside to the established platforms and protocols in current digital humanities. This is true at the research level as well as in classroom pedagogy. Higher-level projects involving sophisticated statistical manipulation, specialized analytics, and other work where the methods are part of the design, being tested and refined, evaluated in terms of how they produce results, exist at the frontier of new inquiry. In those cases, the innovative methods *are* the research as much as is their application to specific problems. But the level at which these operate is often so complex that, like the Google engineers who claim their machine-learning algorithms outstrip their authors' control, they cannot be understood in all of their particulars.

CRITICISM OF DIGITAL HUMANITIES

If research methods and pedagogical practices are often, now, following protocols baked into the platforms and programs in regular use, then to what extent are the criticisms raised against digital humanities also falling into normative patterns? This criticism takes two forms: methodological and cultural. The methodological criticisms reiterate the same tired formulations—computational processing is not reading, digital analysis is reductive, and humanists are bad statisticians. In other words, they assert that computation is not useful in literary studies and if it is to be used it needs to be done better.

This trashing of the digital humanities has become an armchair sport with one dismissal after another of distant reading, text processing, cultural analytics, data mining, and other techno-assisted interventions. The arguments do not change much, even though they are applied with regularity to ever more powerful procedures. This attack began almost as soon as digital work and its computational counterparts garnered public attention.[6] The critical skill required for these criticisms is at the level of shooting simulacral fish in a bowl-sized barrel or bobbing for virtual apples in a teacup. A survey of these dismissals reveals a combination of crankiness, bitterness, often thinly concealed personal vendettas—or opportunistic efforts to gain visibility with an across-the-board dismissal—as if all of "digital art history" or "computational humanities" were without value. The wholesale terms of dismissal do not align with the claims made by the practitioners, most of whom are realistically aware of the ways computational work augments, rather than replaces, conventional methods of research.

Meanwhile, the infrastructural underpinnings of research expand exponentially, providing ever greater access to resources. The very tools so glibly dismissed out of hand as trivial and useless are baked into the networked environments. These offer up detailed and precise search results within seconds so that the proceedings of an obscure nineteenth-century scholarly society or an artist's sketchbook can be located in instants for online use or download. The intellectual work that undergirds this system of access disappears entirely from view as the consumer behavior modeled on online shopping sets the bar for search results and interactions—even in scholarly environments. The normative condition of humanities scholarship is now permeated with digital tools and services that embody intellectual and cultural values crucial to humanistic study—but which have to meet expectations set by commercial marketplace standards of delivery.

Repeating yet again the same old litany of plaints hardly warrants the price of ink on the page or pixels on the screen, though the discourse of dissing digital humanities still creates cultural career capital for those who launch the attacks. One recent high-profile example created a puny furor in the columns of *Critical Inquiry*, where some new as well as familiar issues were raised on both sides (Da 2019). The stated terms of the argument were computational, but the intention was the same—to discredit digital humanities wholesale, this time by suggesting that its techniques could not be replicated (thus were not truly empirical, as if that were the claim). The usual respondents pointed out, yet again, that digital humanities never threatened traditional practices, but offered an augmentation of them. The goal of this criticism is unclear, given that these methods and techniques are not only here to stay, but, as noted, undergird the basic infrastructure of access and use of materials in networked environments while providing useful insights at scale. Claims for any work in any field always need qualification. That, too, is normative.

The other aspect of criticism is from a cultural perspective. Having asserted that the field of digital humanities is intellectually trivial, it can then be fully dismissed as an expression of greedy, unethical complicity with the supposedly newly emerging neoliberal university. This criticism is also based on ignorance that has also become normative through repetition. Some very bright people think that humanists become digital to get "lots" of money.

This argument suggests that the digital humanities absorbed massive resources (money, faculty lines, space) that supposedly would have gone to "traditional" humanities in their absence. Several fallacies are inherent in these assertions. The first is that somehow the humanities until the advent of the digital were part of some pure and altruistic educational enterprise. But the research university as it is currently configured came into being in the post-Second World War era in a funding nexus of military–government, private–corporate, and university interests. The humanities lived from the revenues of these contracts in the university system, and earned their keep by preserving a then-honored tradition of knowledge of history, languages, and liberal arts. This was the bit of education that provided cultural cachet, a kind of intellectual finish for the gentlemen (mainly) going into business, government, or science. The humanities were always complicit with these arrangements, but could compartmentalize their activity and pretend they served as the conscience of culture and its institutions. But the idea that the humanities were somehow pure before they became digital is a myth. They lived on revenues they did not generate whose politics they conveniently ignored. They were also the direct legacy of patriarchal, white, Anglo-European culture with mythic connections to a classical past.

At the University of Virginia, nearly a decade into the existence of digital projects, what was clear was that the revenue generated from grants from public and private foundations outstripped that brought into the university by all other humanities activities by a significant factor. But that money would never have come to the university had it not been for the digital dimensions of its scholarship. Cluster hires to bring those with digital competence into departments became a trend. But so had the impulses to hire deconstructionists, or feminists, or queer theorists had their vogue—as now the rush to "decolonize" is pushing other agendas deemed urgent and timely. This is the history of institutions, where shifts in intellectual fashions form a long-standing pattern.

The final anti-digital humanities argument is that teaching students to use tools and platforms that make them competitive in the workplace does not have the value of teaching history, literature, or philosophy, or other subject-specific knowledge. This argument is a bit thin if these are supplementary components of a curriculum. But it does gain credibility when digital skills become a standalone subject. Current moves to make digital humanities a field, rather than an auxiliary suite of competencies, is cheating students—and future generations—out of knowledge of culture, language, ideas, imaginative expressions, and experience. One normative criticism is that workforce preparedness is a malevolent force driving skill-building pedagogy over critical approaches. This is met by an equally normalizing assertion that the digital humanities has sufficient intellectual substance to stand alone. Neither of these is an acceptable position. The fundamental need for digital fluency prevails in a culture permeated with computational technology. But the need to expand knowledge of the cultural record only expands with new discoveries and diversification. Anyone who thinks digital humanities constitutes a full field of study is participating in intellectual deception. Its tools are discipline agnostic and only make sense when used in collaboration with expertise in a humanistic field. We still need the substantive humanities.

New flourishes get added to these criticisms which follow fashionable intellectual trends. Thus the digital humanities get characterized as masculinist, an exclusive boys' club, which is white, patriarchal, colonialist, and racist. Since these latter criticisms are being leveled against every academic practice and institution, the digital humanities cannot feel particularly singled out. Important female pioneers, Marilyn Deegan and Susan Hockey, for instance, are left out in this account as is the reality that anyone was welcome to join in these digital activities. We were all ignorant and had to learn new methods—and still do.

The criticism from inside digital humanities is more interesting, since it derives from individuals well-versed in knowledge of the field and its limitations. Questions from this critical perspective often focus on how a corpus was selected and defined, and how its limitations determine or influence results. The selection of terms, texts, and their situatedness within institutional conditions that need careful qualification if they are to be read as indicators of cultural phenomena are subject to careful, sometimes contentious, examination. This kind of criticism, even if routine, has the virtue of pushing arguments toward greater specificity and claims toward more careful modification. Data models, definitions, readings of value-laden terminology in areas of gender, race, and sexual identity are all brought into sharp relief through critical discussion. This is work that aligns with the humanities, where debates about *how* to read are usually coupled with *what* to read in cycles of generative and productive interpretation. Some of this work, closer to the realms of information studies professions, addresses the biases of legacy terms in description and access, the politics of classification, and/or the biases of algorithms and the hegemony of computational norms as they are integrated into infrastructure. Some of this work merges with critical digital studies, and usefully so.

NORMATIVITY

So, what is left to say about digital humanities that is not an out-of-hand dismissal or a qualified defense given the extent to which its practices have become normalized? Normative activity has an agency of its own in which habits become divorced from original impulses or activities. The normative state of digital work makes it generally, though not exclusively, more mechanistic now than during its earlier innovative phase. Tools and processes often get applied without intimate knowledge of their workings as results are produced through black-box technologies. This does not negate the value of the tools for doing certain kinds of analysis, but it does mean that the outcomes are based on procedures that are not sufficiently understood for them to be explained or manipulated effectively. The need to integrate humanistic and hermeneutic approaches into the mechanistic practices of computation persists. The challenges are many, and the concessions made letting computation set the terms of digital humanities have largely been forgotten as, again, the normative processes have assumed a habitual familiarity.

Here the discussion circles back to where it began. What is the force and effect of normativity? It tends to produce blindness, dullness, and a diminishment of innovation and intellectual risk-taking. We become inured to the procedures and processes, distracted by outcomes, dazzled or disappointed by results. If the practice of the humanities has at its core an engagement with reading and re-reading as a conversation with a text (image, artifact, other human expression)—and a community of readers across large historical and geographical spans—then have digital methods threatened that activity? Digital methods have not replaced the specificity of individual readings with the certainty of machine outputs. They have merely added another kind of probabilistic method to the mix. Insisting on the conviction that interpretation is a co-constitutive process, not a product, is not the same as conflating hermeneutics with mechanistic processing.

A larger question could be raised about the role of the humanities within our current culture and the mass-entertainment complex with its easy modes of access, short cycles of consumption and fashion, quick fix uptake of conspiracies and falsified histories. Once, the standard response was that humanistic expertise was needed to establish the authenticity and diversity of cultural record. Humanists promoted modes of reading, practices of ethical engagement, or at least, foundations for assessing these dimensions of human behavior. These may not hold against the prevalent assaults on the capacity to pay attention without distraction or care about the authority of knowledge—or assaults on legacy cultural traditions in which diversification is equated with destruction. We may need a better paradigm for asserting the value of the humanities, but digital methods are neither their salvation nor damnation. The benefits of those methods are circumscribed, but not null.

The positive alternative to normative digital humanities is not a matter of increased capacity. It is not as if getting the tools right will make the work align with humanistic values. Instead, the effort requires a rework of foundational assumptions about the role, value, and use of epistemological inquiry. Computational methods are neither the cause nor result of normativity, they are merely instruments that like others become part of a normative condition through habitual use. The normative is at once useful and destructive. It makes certain practices possible without overwhelming effort—such as finding a phrase in an ancient or out-of-print text through a simple string-search. But it also inures us to the looming issues of sustainability and costs. These are issues of sustainability that cannot be ignored.

Finally, if the humanities are about attending to those aspects of culture that seem to make us human, then digital humanities have their own potential to contribute to the non-normative and compelling dimensions of experience. Kim Gallon's *Black Covid*, a work of memorials, testimony, and mourning, and Manuel Portela's Archive of Fernando Pessoa's *Book of Disquiet* each provide evidence of this, in their different ways. Gallon's work is premised in part on the need to humanize data. This pushes back on more than two centuries of the legacy of political arithmetic, that formulation of the eighteenth-century statisticians, John Graunt and William Petty. Gallon's digital interventions try to cut through the fog of bureaucracy so the obfuscating effects of computational methods get re-humanized. A face is attached to each data point, a story told and an individual remembered even as the pandemic figures and dashboards of a bureaucratic new modern state interested in the management of its citizenry readily conceal these issues in their efficient aggregation of information. The normative aspect of reducing human beings to statistics gets contravened in Gallon's work—called to attention and also reworked into an alternative. The death of humanity, Gallon's project suggests, is not only in conflict and struggle, but in the reduction of human beings to statistics that negate their individual identity. Digital humanities have a role in offering alternatives by showing the vulnerability and precarity of black communities.

By contrast, Portela's project is a custom-built platform to serve the work of the great modern Portuguese writer, Fernando Pessoa. The Archive for the *Book of Disquiet* provides a platform for reading, editing, writing across the fragments of this unique work. The project brings attention to and engagement with a set of practices appropriate to this complex modern work comprising many fragments left un-organized at the author's death. The archive also documents the various editions of Pessoa's work, calling attention to scholarly practices relevant to constructing an authoritative record when no fixed or stable version can ever be determined. Enabling the open-ended and variable reading of a complex text enacts values that are difficult, if not impossible, to support in a fixed print format. The platform provokes reflection on the identity of the documents in relation to their place within a variable frame—and all that this implies for meaning production as reading.

In both cases, the structure of the platforms is meant to awaken awareness, insist on encounters in which reading or viewing is thought production, made anew in each instance. Both projects identify their speakers and situate their viewers within contemporary communities and values. Each, in their own way, addresses the dubiousness of data, the need for the life cycle of data collection and digital surrogacy to be exposed. Portela salvages Pessoa from obscurity, from the realms of fixed editions that belie the combinatoric flexibility of his imagination, offering the reader a chance to thread through the fragments that comprise the massive archive. Gallon brings attention to the inequitable vulnerability of the black community with recognition of the complexity of this condition as well as its fundamental, incontrovertible, reality. The design of these platforms embodies and enacts alternatives to habits of thought and work. They are not normative in spite of building on established precedents and some familiar methods.

As computational practices have become standardized in platforms, applications, programs, or frequently used methods, the need for innovation has become a matter of refinement rather than invention. Extant tools and platforms are a useful start point for pedagogy and research, easily defensible on these grounds. The benefits of processing at scale offered insights into patterns across corpora too large for familiar reading practices. These and other early-stage engagements matured into standard tools and platforms that are now the normative activities of digital humanities: topic modeling, visualization, cultural analytics, text analysis, mapping, data mining, and web presentation are supported with off-the-shelf software that is discipline agnostic. Little specialized technical knowledge is required to use these, and a general acceptance of the black-box processors or invisible code has prevailed. Is the answer to this condition to insist on higher technical skill levels—or to increase critical reflection and skepticism? What dimensions of awareness do we believe are essential within digital practices? Ethical ones? Intellectual? Critical reflections on knowledge and its assumptions? Or should we simply go on using the tools and platforms as if no consequences are attached to this but a certain dullness of mind produced by using processes inaccessible to our intervention or understanding? To take something as a given is to be robbed of comprehension.

Normativity blinds us to the impacts and effects of actions. Habits of consumption are woven into the cultural patterns. The biggest challenge is to keep normativity from becoming orthodoxy, a set of unexamined claims that assert authority on the basis of habit, institutionalized priorities, and work repressively to maintain a status quo that resists skeptical investigation or has to conform to constraints of particular formulaic agendas. That is the moment when oxygen leaves the room.

NOTES

1. The term "digital" is used here to indicate media based on binary code; computational refers to processes that can be carried out through algorithmic automation.
2. The Heterochronologies and TimeCapsule projects extend the work originally done in Temporal Modeling, but with attention to graphic expressions of chronologies.
3. See *The Valley of the Shadow: Two Communities in the American Civil War*, valley.lib.virginia.edu/.
4. See *The Rossetti Archive*, http://www.rossettiarchive.org/.
5. The original version of that project remains online on the site of the research assistant and designer who helped craft the online PDF; see issuu.com/imansalehian/docs/introductiontodigitalhumanities_tex It can also be found here in its entirety: stacks.stanford.edu/file/druid:vw253bt6706/IntroductionToDigitalHumanities_Textbook.pdf.
6. One landmark is Franco Moretti's "Conjectures on World Literature" (2000) and the response it generated; see also "A Genealogy of Distant Reading" (Underwood 2017).

REFERENCES

Da, Nan Z. 2019. "The Computational Case against Computational Literary Studies." *Critical Inquiry* 45 (3): 601–39. https://doi.org/10.1086/702594.

McCarty, Willard. 1993. "A Potency of Life: Scholarship in an Electronic Age." In *If We Build It*, ed. North American Serials Interest Group, 79–98. New York, NY: Routledge.

Moretti, Franco. 2000. "Conjectures on World Literature." *New Left Review*, no. 1. https://newleftreview.org/issues/II1/articles/franco-moretti-conjectures-on-world-literature.

Underwood, Ted. 2017. "A Genealogy of Distant Reading." *Digital Humanities Quarterly* 11 (2). http://www.digitalhumanities.org/dhq/vol/11/2/000317/000317.html.

CHAPTER TWO

The Peripheries and Epistemic Margins of Digital Humanities

DOMENICO FIORMONTE (UNIVERSITÀ DEGLI STUDI ROMA TRE) AND GIMENA DEL RIO RIANDE (IIBICRIT-CONICET, ARGENTINA)

EPISTEMIC MARGINS AND CULTURAL DIVERSITY

From the second half of the first decade of the twenty-first century, due to global changes in the historical, cultural, technological, and geopolitical situation, Digital Humanities (DH) began to spread and establish itself beyond its *original* center in North America and Europe. This expansion drew the attention of Northern digital humanists—including those belonging to historically hegemonic communities—to the problem posed by the variety of languages, expressive forms, methods, and tools that these *other* DH used to represent, build, and disseminate culture and knowledge. Consequently, the practices of interpersonal diversity, inclusion, and cultural pluralism emerged among the preoccupations of many DH communities (Priego 2012; ADHO 2015; O'Donnell et al. 2016; Ortega 2016; Fiormonte and del Rio Riande 2017), though mainly as a need for representation of non-Global North scholars in international consortia, projects, or conferences.

In this chapter we aim to reflect on what we understand by epistemic diversity (Solomon 2006; Gobbo and Russo 2020), in order to analyze its impact on knowledge production in the field of DH. We also explore the concept of "South" as an alternative way of examining local and global questions about DH and a framework for critical episteme and reflection.

We define epistemic diversity as the possibility of developing diverse and rich epistemic apparatuses that could help in the building of knowledge as a distributed, embodied, and situated phenomenon.

Information and communication technologies have contributed to the production of collective knowledge in the last twenty years. Many societies have started experiencing an epistemic shift towards open and participatory ways of collaborating and learning. Citizen science projects, makerspaces, and open research practices are becoming part of our life inside and outside academia. In this sense, even though epistemic diversity may have many definitions, we understand it in line with the 2001 UNESCO Universal Declaration on Cultural Diversity as "the dialogue between different knowledge holders, that recognizes the richness of diverse knowledge systems and epistemologies and diversity of knowledge producers" (UNESCO 2021, 9).

Although this essay is the result of a constant exchange and dialogue between the authors, the content of sections 1 and 2 (second half) are by G. del Rio Riande and the first half of section 2, section 3, and 4 are by D. Fiormonte. The English translation is by Desmond Schmidt.

We believe epistemic diversity can contribute to a better understanding of DH from a geopolitical perspective, as it moves from the individual representation of scholars and/or their culture and focuses on the inequities in social distribution of knowledge.

However, any such reflection on cultural, interpersonal, or epistemic diversity must recognize all the forces in play, and attempt to confront the evolutionary history, forms of knowledge production, institutional structures, and geopolitical interests that have gradually formed in the course of the already considerable history of DH.[1] All of these are part of what we could call the problem of representation. Representation (Latin *representatĭo*) is the action and effect of representing (making something present with figures or words, referring, replacing someone, executing a work in public). A representation, therefore, can be the idea or image that replaces reality. Representation is also the way in which the world is socially constructed and represented at a textual and contextual level. The idea of representation is linked to the way in which the idea of subject is produced, that is, the descriptions and sets of values within which the different groups and/or individuals are identified. Moreover, following Davis, Shrobe, and Szolovits (1993, 19) we can also claim that:

> [S]electing a representation means making a set of ontological commitments. The commitments are in effect a strong pair of glasses that determine what we can see, bringing some part of the world into sharp focus, at the expense of blurring other parts. These commitments and their focusing/blurring effect are not an incidental side effect of a representation choice; they are of the essence ...

Mapping projects produced by the DH community, for example, can help us understand this problem of representation and epistemic diversity. Most of them were produced by the mainstream DH community, so the "South"—or rather "the regions of the South" (Sousa Santos and Mendes 2017, 68)—are underrepresented or absent.[3] Although this is not a comprehensive list, let's revise some of these projects of the last ten years to understand how DH has been and, in some cases, is still represented. For example, one of the most comprehensive maps of DH courses, maintained by the European consortium DARIAH, although currently in progress and open to the community, still excludes many initiatives.[2] The reason is simple, many communities outside Europe don't know what DARIAH is or have never heard about this map, so they haven't added their courses to it. Two other much-cited maps are those by Melissa Terras (2012) and Alan Liu (2015). However, the first one is mainly devoted to North American and European DH and the second one just focuses on disciplines and fields. Other initiatives from outside the Northern academies or its margins include Shanmugapriya and Menon's (2020) mapping of Sneha's report (2016), MapaHD (Ortega and Gutiérrez De la Torre 2014), and AroundDH 2020 (Global Outlook Digital Humanities 2020[3]). This last map follows Alex Gil's first attempt to give a global, diverse, and inclusive perspective of DH (Gil 2014[4]) and aims to highlight "Digital humanities projects in languages other than English or from minoritized groups worldwide." We can also mention initiatives halfway between maps and surveys, like Colabora_HD[5] and Atlas de Ciencias Sociales y Humanidades Digitales (Romero Frías 2013[6]), which have aimed at a more extended view of DH.

We think that these examples show that the geography of DH is complex and has never been understood globally. If we want to understand the importance of epistemic diversity within DH, we must contextualize it as part of the issues that are defined beyond geography and under the label geopolitics of knowledge.[7]

GEOPOLITICS OF KNOWLEDGE AND EPISTEMIC DIVERSITY IN THE MARGINS OF DH

As Vinay Lal (2005) argues, battles in the twenty-first century will be for the domination of knowledge and information, which cases like Wikileaks, the Snowden datagate, and the Cambridge Analytica-Facebook scandal demonstrate. These and other recent events, like the artificial intelligence race between China and the US, confirm that the Internet has become the terrain on which the geopolitical balances of the planet will be played out, and reaffirm the radical epistemic violence of digital devices and the cultures that have developed them. And, as noted above, the center of this "new world" will be data. The question is: How was data obtained and managed? We believe the role of DH can no longer be limited to the application of information technology to humanistic problems and objects, DH must also ask itself about the nature and the profound purposes of the technologies they use, and the new objects, traces, and memories they are assembling.

Cultural hegemony and domination over forms of knowledge can be expressed on many levels, but here we will consider those related to the problem of representation and epistemic diversity. The first arises from the linguistic and rhetorical-discursive advantages of the global Anglophone North in the creation of academic knowledge, and the second from the inequities inherent in the infrastructures of knowledge production and communication. Obviously these two levels are closely related, but it is still important to distinguish them.

Inequities of the first kind have been studied in the pioneering work of Suresh Canagarajah (2002) and the questions he raised help to summarize the main points of the problem: What role does writing play in peripheral academic communities? What kind of representation challenges do they face when adopting the epistemological standards and conventions of the "center"? And above all, how are the experiences and knowledge of these communities molded and reformulated by this process? (Canagarajah 2002). Although Canagarajah's approach is sometimes based on a vision of a dichotomous center/periphery model, which today, following China's intrusion into the "market" of scientific production (Veugelers 2017), among other global issues, is weakening, we know that in the academic communities of ex-colonial countries there is a kind of intellectual dependency that has its roots in the education system:

> Periphery students are taught to be consumers of center knowledge rather than producers of knowledge. Often this attitude of dependency develops very early in a periphery subject's educational life ... Furthermore, Western-based (nonindigenous) literacy practices exacerbate this intellectual dependency ... From the above perspective it is easy to understand the feeling of many that the democratization of academic literacy should start in schools.
>
> (Canagarajah 2002, 283–4)

Obtaining funds, directing research, developing a project, writing an essay or an article, etc., are intellectual and discursive practices that depend on precise forms of representation and undisputed standards set once and for all by the great "knowledge centers" of the Global North (Bhattacharyya 2017, 32). These self-proclaimed "centers" (universities, research centers, etc.) of excellence[8] base their persuasive and imposing powers on structures and infrastructures of dissemination such as the large private oligopolies of scientific publishing (Larivière et al. 2015) dominated by English-speaking countries.

This is closely related to what Gobbo and Russo have stated about the role of English and the impossibility of epistemic diversity in monolingual contexts. Even though their article is mainly related to philosophy, Gobbo and Russo (2020) demonstrate English is a communicative strategy and the concept of lingua franca is no longer useful. The authors believe that this communication strategy hinders linguistic justice, a concept developed in the 1990s to refer to the uneven conditions that different speakers encounter while speaking the same language. That is to say that in academic communication, clearly English native speakers remain advantaged over non-native speakers and that English, contrary to what is usually thought, is no lingua franca. Epistemic diversity refers, in this context, to the consequences of a monolingual context at the level of the epistemic apparatuses that speakers have or produce, in order to make sense of their world. Language also impacts on knowledge creation. As they clearly state: "English is neither ethnically nor epistemological neutral."

But the problem is also how hegemonic forces define (and manage) knowledge for their own benefit. Already in 1983, Eugene Garfield, the inventor of the impact factor, argued that Western journals controlled the flow of scientific communication almost as much as Western news agencies monopolized the agenda of international news (Guédon 2008, 9). Fernanda Beigel (2014) makes it clear when she discusses the arbitrariness of the current structure of global academic research and shows how scholars, forced into an international pseudo-competition, are evaluated on the basis of their contribution to mainstream journals:

> Over the second half of the 20th century, American academia played a main role in "universalizing" a set of criteria to define scientific "quality" as a research agenda. The Institute of Scientific Information (ISI, now the Web of Science-Thomson Reuters) created citation indexes and journal rankings supposedly based on objective procedures. It was mainly due to the Science Citation Index (SCI) that *international publishing* became the most valorized academic capital and the most relevant indicator for institutional evaluations worldwide. This perpetuated the notion of "core journals" and the impact factor became a yardstick for "excellence" in a publishing system in which the English language became progressively dominant.

The strategy of knowledge monopolies (Fiormonte 2021) is not to close or limit these flows, but to perpetuate their own hegemony, institutionalizing dependence on the ways in which knowledge is represented in those journals, and maintaining the subordination and invisibility of locally produced knowledge.

In the field of DH, the results of this epistemic disparity have been highlighted by some quantitative studies that reveal the linguistic, cultural, institutional, and gender imbalances and inequalities in ADHO conferences (Weingart and Eichmann-Kalwara 2017). Another more restricted study (Fiormonte 2021) deals with an often overlooked aspect: the languages of the sources of articles published in the main DH journals. These sources, cited in the form of bibliographic references and notes, are a key indicator of how humanists and social scientists work. Even more than the language in which an article is written, the sources reveal valuable information about the content of the research and the author's skills (known languages, theoretical trends, methodological choices, etc.). The use of sources, which shows itself in the references cited, can reveal the skills of a medievalist, the cultural background of a new media expert, the geopolitical orientation of a historian, and so on. The aim of the experiment was to collect information about the language(s) of the sources used by the authors who published in six journals, which in 2014, at the time of data collection, represented a mixed sample reflecting both linguistic and scientific interests: *Caracteres* (*CA*); *Digital Humanities Quarterly* (*DHQ*); *Digital Medievalist* (*DM*); *Digital Studies/Le champ numérique* (*DSCN*); *Jahrbuch für Computerphilologie* (*JCP*); *Informatica Umanistica* (*IU*); and *Literary and Linguistic Computing* (*LLC*) (since 2014 *Digital Scholarship in the Humanities, DSH*). All journals, except *LLC/DSH*, are freely available online. Only *CA*, *IU*, and *JCP* have a defined geolinguistic location, but all frequently publish articles in English. Unfortunately, the majority of DH journals are published in English-speaking contexts and this limits the possibilities for comparison. To try to correct this, *Caracteres* (*Estudios culturales y críticos de la esfera digital*—Cultural studies and reviews in the digital sphere) was added to the group, which at the time was one of the few online Spanish-speaking journals that dedicated some space to DH.[9]

The conclusions are clear enough: while English-speaking journals are essentially monolingual (94 percent of *LLC/DSH* sources are in English; 97 percent in *DHQ* and *DSCN*; 83 percent in *DM*), the authors who published in Italian, German, and Spanish journals cite sources in multiple languages. What does it mean to be "international" for an author who publishes in a DH journal? Contrary to popular perception, the data show that the more "local" a journal is, the more attention it pays to the outside world, demolishing the myth of "cultural provincialism." We believe all this strengthens the idea that, as in biocultural diversity (Maffi and Woodley 2010, 8), there is greater diversity in the margins than in the center.[10]

THE SOUTH AS A METAPHOR OF A MORE DIVERSE DH

In *Theory from the South*, South African scholars Jean and John L. Comaroff (2012) argue that contemporary historical, social, and economic processes are altering received geographies, placing in the South (and East) of the world "some of the most innovative and energetic modes of producing value" (Comaroff and Comaroff 2012, 7). In other words, it is no longer possible to consider the South as the periphery of knowledge, because it is precisely the South that is developing as the place where new assemblages of capital and labor are taking shape to prefigure the future of the Global North (Comaroff and Comaroff 2012, 12). In this line of thought, we can recall the epistemologies of the South, by Portuguese sociologist Boaventura de Sousa Santos. The epistemologies of the South are a set of analyses, methods, and practices developed as a resistance against the systematic injustices and oppressions caused by capitalism, colonialism, and patriarchy. This approach implies opening the way creatively to new forms of organization and social knowledge (Sousa Santos 2018).

The South thus emerges for us as a critical category able to adopt horizontal approaches and draw on the imagination of a transnational political actor and producer of *resistant* culture rather on the margins than the periphery. This reflection leads us to consider the category "South" from a geopolitical point of view, and as a discourse of resistance and innovation from the margins.

We are aware that the relations between margins and center, between center and peripheries, between North and South, can legitimize hegemonic discourses and systems, but also reinforce dualistic visions. Reality, however, almost always turns out to be profoundly ambiguous. Even technology, as Walter Benjamin had predicted in the 1930s (Benjamin 2006), plays a double and ambiguous role: on the one hand, the processes of concentration and standardization that arise from it (Bowker and Leigh Star 1999) are accompanied by the exploitation of resources and by the absorption of marginal or subordinate cultures. On the other hand, it can also provide an opportunity to preserve, spread, and make Southern cultures more aware of their own strengths.

Tensions always exist wherever different cultures and knowledge come into contact, whether from a geographical, cultural, or epistemological perspective. The field of DH is no exception to this rule. What can be seen today is a potential alliance between the ambitions of the South in their search for legitimacy (the oppressed who seeks to resemble the oppressor, as Fanon [2004] and Freire [2017] [11,12] observed) and those who want to continue to be a "center," by extending and consolidating their own hegemony. It is a well-known pattern in postcolonial studies where one speaks of reabsorption of subordinate subjectivity or appropriation of the emergent by the dominant (Kwet 2019).

It may be instructive at this point to recall a little-known episode in the technological relations between North and South. In 1985, at a historic moment when Brazil supported a policy of protection and incentives for its national technology industry, a local company, Unitron, designed and produced the *Mac de la periferia* (Suburban Mac), the first clone in the world of the Macintosh 512 KB (also known as *Fat Mac*). Unitron obtained about ten million dollars in funding from the Brazilian government and managed to produce five hundred machines at a low cost for the national market and the rest of Latin America. However, when the computers were ready to be sent, the project was blocked by the local arm of Apple:

> Though Apple had no intellectual property protection for the Macintosh in Brazil, the American corporation was able to pressure government and other economic actors within Brazil to reframe Unitron's activities, once seen as nationalist and anti-colonial, as immoral piracy. In exerting political pressure through its economic strength, Apple was able to reshape notions of authorship to exclude reverse engineering and modification, realigning Brazilian notions of intellectual authorship with American notions that privilege designated originators over maintainers and modifiers of code and hardware architecture.
>
> (Philip-Irani-Dourish 2010, 9–10)

This story suggests some considerations (and questions) that would add up to a "classic" postcolonial critique, and connects also to the "informatics of the oppressed," an expression coined by Rodrigo Ochigame on Freire's pedagogy of the oppressed to describe the counterhegemonic beginnings of computer science in Latin America (Ochigame 2020). The early death of the Suburban Mac not only cemented the victory of the (future) giant from Cupertino and the defeat of a pioneering project in the South, but marked the beginning of the leaking of talent, creativity, and resources from the South to the North (the intellectual emigration of the 1980s and 1990s). But above all, the case is emblematic of a loss of cultural, technological and epistemological sovereignty which implies the refusal (and in some cases denial) to invest in technology and innovation. What would Brazil be like today (but in reality the same argument could apply to Southern European countries, including Italy [Secrest 2019]) if thirty years ago it had the opportunity to develop its own IT industry? What would Brazilian DH be like today if a whole industry could support research? Reflecting on this point means starting to think about a historical revision of scientific-technological innovation and, above all, laying the foundations for a geopolitics of the digitization of knowledge (Fiormonte 2017).

Recovering cultural and epistemological sovereignty is directly connected with the role of DH in the South. Southern digital humanists cannot excuse themselves from their responsibility to address the geopolitical implications of digital knowledge and its decolonization. What kind of knowledge are we building with the technological tools we use? What are the social, political, cultural, and other implications of the digital tools we use every day? Is it possible to create independent models that are sustainable from a socioeconomic perspective, or are we bound to always incorporate the paradigms and standards of the Global North—to be part of mainstream journals and media?

REFLECTIONS FROM THE EPISTEMIC MARGINS OF DH

There are many examples of Southern innovation that do not easily find a parallel in a world crushed by the GAFAM oligopoly.[13] From the recovery of indigenous communities in Peru (Chan 2014) to the decolonization of university curricula in South Africa (Adriansen et al. 2017), from movements for open knowledge and technologies (Barandiaran and Vila-Viñas 2015) to community networking experiments in Latin America (Aguaded and Contreras-Pulido 2020), from the Africa of digital rights[14] to the India of knowledge commons and Net Neutrality, a grassroots movement that in 2016 led to the "defeat" of Mark Zuckerberg (Mukerjee 2016). One could easily add many more examples. The *peripheries* are transforming themselves into models of proactive resistance for the rest of the world. This is also witnessed by various educational projects that aim at a genuine revision of policies, technologies, and new models of community education, as evidenced by experiments and real projects in Latin America (Guilherme and Dietz 2017) and Asia (Alvares and Shad 2012).

Digital humanists of the South today have the opportunity not so much to replace or overlay the technological realities that still dominate, but to become the reference point for multiple and sustainable models of conservation, access, and communication of knowledge in digital format. In this sense, the epistemic margins of DH can become its center.

NOTES

1. Nyhan and Flinn (2016) offer a historical perspective based on "oral histories," interviewing European and North American pioneers. Although it is an extremely interesting work, not only is there no reference to the non-Western world, but pioneers from southern Europe are absent, such as the Spanish Francisco Marcos Marín (the term "Spain" occurs twice in the entirety of the 285-page volume and it's related to the chapter on the Hispanic Seminary of Medieval Studies at the University of Wisconsin).
2. Boaventura de Sousa Santos is right when he states that "There is therefore a dual modern cartography: a legal cartography and an epistemological cartography" (Sousa Santos 2018, 195).
3. DARIAH Course Registry, https://dhcr.clarin-dariah.eu/courses.
4. http://arounddh.org/.
5. http://arounddh.elotroalex.com/.
6. https://emapic.es/survey/bCGfKO4.
7. *Atlas de Ciencias Sociales y Humanidades Digitales*, https://grinugr.org/proyectos_internos/atlas-de-ciencias-sociales-y-humanidades-digitales/.
8. We hope that the forthcoming collective volume *Global Debates in the Digital Humanities* will help change the current imbalances by publishing global DH research results in English with little or no visibility in the Global North (https://dhdebates.gc.cuny.edu/page/cfps-gddh-eng).
9. There has been a heated debate on how to build, measure, and certify this excellence for some time; we limit ourselves to pointing out some contributions, such as Hazelkorn (2017).
10. The maximum time frame taken into consideration was five years, preferably 2009–2014 if all were available, but *CA* started publishing in 2012 and *IU* and *JCP* stopped in 2011, so in the first and second cases we extracted references from all available numbers. In the case of the third journal (*JCP*) five years were chosen but in the range 2004–2010, since some years were not present. While it is true that

the total number of references extracted from each journal varied greatly (more than six thousand from *LLC/DSH*, but less than three hundred for *IU*), percentages based on the extracted totals still give a fairly representative picture of the linguistic tendencies of each journal.

11. Trend confirmed by the creation in 2017 of the *Revista de Humanidades Digitales* (*RHD*; http://revistas.uned.es/index.php/RHD/index). The *RHD*, founded by a group of researchers and professors from UNED (Spain), UNAM (Mexico), and CONICET (Argentina), is the first open access academic journal entirely dedicated to DH that has adopted Spanish as its main scientific language. The *RHD* also accepts contributions in English and other Romance languages, such as Portuguese, French, and Italian.
12. At some point in its marginal existence, the oppressed feels an irresistible attraction to the oppressor: "Taking part in these standards is an unconstrained aspiration. In their alienation, they seek, at all costs, to resemble their oppressor" (Freire 2017, 62).
13. https://gafam.info/.
14. https://africaninternetrights.org/en.

REFERENCES

ADHO. 2015. "Revised Protocol for the Standing Committee on Multi-lingualism & Multi-culturalism." May. http://adho.org/administration/multi-lingualism-multi-culturalism/revised-protocol-standing-committee-multi.

Adriansen, Hanne Kristine, Lene Møller Madsen, and Rajani Naidoo. 2017. "Khanya College: A South African Story of Decolonisation." *The Conversation*, 10 October. https://theconversation.com/khanya-college-a-south-african-story-of-decolonisation-85005.

Aguaded, Ignacio and Paloma Contreras-Pulido, eds. 2020. "Acceso universal y empoderamiento digital de los pueblos frente a la brecha desigual. Nuevas formas de diálogo y participación." *Trípodos* (número monográfico) 46. http://www.tripodos.com/index.php/Facultat_Comunicacio_Blanquerna/issue/view/22/showToc.

Alvares, Claude A. and Shad Saleem Faruqui, eds. 2012. *Decolonising the University: The Emerging Quest for Non-Eurocentric Paradigms*. Pulau Pinang: Penerbit Universiti Sains Malaysia.

Barandiaran, Xabier E. and David Vila-Viñas. 2015. "The Flok Doctrine." *Journal of Peer Production* 7. http://peerproduction.net/issues/issue-7-policies-for-the-commons/the-flok-doctrine/.

Beigel, F. 2014. "Publishing from the Periphery: Structural Heterogeneity and Segmented Circuits. The Evaluation of Scientific Publications for Tenure in Argentina's CONICET." *Current Sociology*, 62 (5): 743–65. https://doi.org/10.1177/0011392114533977.

Benjamin, Walter. 2006. *Selected Writings*, vol. 4, *1938–1940*, edited by Howard Eiland and Michael W. Jennings. Cambridge, MA and London: Belknap Press of Harvard University Press.

Bhattacharyya, Sayan. 2017. "Words in a World of Scaling-up: Epistemic Normativity and Text as Data." *Sanglap: Journal of Literary and Cultural Inquiry* 4 (1): 31–41. http://sanglap-journal.in/index.php/sanglap/article/view/86/144.

Bowker, Geoffrey C. and Susan Leigh Star. 1999. *Sorting Things Out: Classification and its Consequences*. Cambridge, MA: MIT Press.

Canagarajah, A. Suresh. 2002. *A Geopolitics of Academic Writing*. Pittsburgh, PA: University of Pittsburgh Press.

Chan, Anita Say. 2014. *Networking Peripheries: Technological Futures and the Myth of Digital Universalism*. Cambridge, MA: MIT Press.

Comaroff, Jean and John L. Comaroff. 2012. *Theory from the South: Or, How Euro-America Is Evolving Toward Africa*. Boulder, CO: Paradigm Publishers.

Davis, Randall, Howard Shrobe, and Peter Szolovits.1993. "What is a Knowledge Representation? *AI Magazine* 14 (1): 17–33. https://groups.csail.mit.edu/medg/ftp/psz/k-rep.html.

Fanon, Franz. 2004. *The Wretched of the Earth*. New York, NY: Grove Press.

Fiormonte, Domenico. 2017. "Digital Humanities and the Geopolitics of Knowledge." *Digital Studies/Le Champ Numérique* 7 (1): 5. http://doi.org/10.16995/dscn.274.

Fiormonte, Domenico. 2021. "Taxation against Overrepresentation? The Consequences of Monolingualism for Digital Humanities." In Dorothy Kim and Adeline Koh, *Alternative Historiographies of Digital Humanities*, 325–66. New York, NY: Punctum Books.

Fiormonte, D. and Gimena del Rio Riande. 2017. "Por unas Humanidades Digitales Globales." *Infolet. Cultura e Critica dei Media Digitali*, October 10. https://infolet.it/2017/10/09/humanidades-digitales-globales/.

Freire, Paulo. 2017. *Pedagogy of the Oppressed*. London: Penguin.

Gil, Alex. 2014. *Around DH in 80 Days*. http://arounddh.elotroalex.com/.

Global Outlook Digital Humanities. 2020. AroundDH 2020. Digital humanities projects in languages other than English or from minoritized groups worldwide. http://arounddh.org/.

Gobbo, Federico and Federica Russo. 2020. "Epistemic Diversity and the Question of Lingua Franca in Science and Philosophy." *Foundations of Science* 25, 185–207. https://doi.org/10.1007/s10699-019-09631-6.

Guédon, Jean-Claude. 2008. "Open Access and the Divide between 'Mainstream' and 'Peripheral' Science." In *Como gerir e qualificar revistas científicas* [pre-print version]. http://eprints.rclis.org/10778/.

Guilherme, Manuela and Gunther Dietz. 2017. "Introduction. Winds of the South: Intercultural University Models for the 21st Century." *Winds of the South: Intercultural University Models for the 21st Century*. Special Issue of *Arts & Humanities in Higher Education* 16 (1): 7–16.

Hazelkorn, Ellen, ed. 2017. *Global Rankings and the Geopolitics of Higher Education: Understanding the Influence and Impact of Rankings on Higher Education, Policy and Society*. London and New York, NY: Routledge.

Kwet, Michael. 2019. "Digital Colonialism: US Empire and the New Imperialism in the Global South." *Race & Class* 60 (4): 3–26. https://doi.org/10.1177/0306396818823172.

Lal, Vinay. 2005. *Empire of Knowledge Culture and Plurality in the Global Economy*. New Delhi: Vistaar Publications.

Larivière, Vincent, Stefanie Haustein, and Philippe Mongeon. 2015. "The Oligopoly of Academic Publishers in the Digital Era." *PLOS ONE* 10 (6). http://dx.doi.org/10.1371/journal.pone.0127502.

Liu, Alan. 2015. "Map of Digital Humanities." Prezi. https://prezi.com/hjkj8ztj-clv/map-of-digital-humanities/.

Maffi, Luisa and Ellen Woodley, eds. 2010. *Biocultural Diversity Conservation: A Global Sourcebook*. Washington, DC and London: Earthscan.

Mukerjee, Subhayan. 2016. "Net Neutrality, Facebook, and India's Battle to #SaveTheInternet." *Communication and the Public* 1 (3): 356–61. doi: 10.1177/2057047316665850.

Nyhan, Julianne and Andrew Flinn. 2016. *Computation and the Humanities: Towards an Oral History of Digital Humanities*. Cham: Springer International Publishing. http://dx.doi.org/10.1007/978-3-319-20170-2.

Ochigame, Rodrigo. 2020. "Informatics of the Oppressed." *Logic* 11, August 31. https://logicmag.io/care/informatics-of-the-oppressed/.

O'Donnell, Daniel, Barbara Bordalejo, Padmini Murray Ray, Gimena del Rio, and Elena González-Blanco. 2016. "Boundary Land: Diversity as a Defining Feature of the Digital Humanities." In *Digital Humanities 2016: Conference Abstracts*, 76–82. Jagiellonian University & Pedagogical University, Kraków. http://dh2016.adho.org/abstracts/406.

Ortega, Élika. 2016. "Crisscrossing Borders: GO::DH Regional Networks in Dialogue," January 13. https://elikaortega.net/blog/2016/dh-at-the-borders/.

Ortega, Élika and Silvia E. Gutiérrez De la Torre. 2014. "MapaHD. Una exploración de las Humanidades Digitales en español y portugués." *En Ciencias Sociales y Humanidades Digitales. Técnicas, herramientas y experiencias de e-Research e investigación en colaboración*, 101–28. CAC, Cuadernos Artesanos de Comunicación. https://grinugr.org/wp-content/uploads/libro-ciencias-sociales-y-humanidades-digitales-completo.pdf.

Philip, Kavita, Lilly Irani, and Paul Dourish. 2010. "Postcolonial Computing: A Tactical Survey." *Science, Technology, & Human Values* 37 (1): 3–29. doi: 10.1177/0162243910389594.

Priego, Ernesto. 2012. "Globalisation of Digital Humanities: An Uneven Promise." *Inside Higher Education*. January 26. https://www.insidehighered.com/blogs/university-venus/globalisation-digital-humanities-uneven-promise.

Romero Frías, Esteban. 2013. "Atlas de las Ciencias Sociales y Humanidades Digitales." grinugr. https://grinugr.org/proyectos_internos/atlas-de-ciencias-sociales-y-humanidades-digitales/.

Secrest, Meryle. 2019. *The Mysterious Affair at Olivetti. IBM, the CIA, and the Cold War Conspiracy to Shut Down Production of the World's First Desktop Computer*. New York, NY: Knopf Doubleday.

Shanmugapriya, T. and Nirmala Menon. 2020. "Infrastructure and Social Interaction: Situated Research Practices in Digital Humanities in India." *Digital Humanities Quarterly* 14(3). http://www.digitalhumanities.org/dhq/vol/14/3/000471/000471.html.

Sneha, Puthiya Purayil. 2016. *Mapping Digital Humanities in India*. The Centre for Internet & Society, December 30. https://cis-india.org/papers/mapping-digital-humanities-in-india.

Solomon, Miriam. 2006. "Norms of Epistemic Diversity." *Episteme* 3 (1–2): 23–36. doi: 10.3366/epi.2006.3.1-2.23.

Sousa Santos, Boaventura de. 2018. "Introducción a las Epistemologías del sur." En *Epistemologías del sur*, 25–62. Buenos Aires: CLACSO/Coímbra: CES.

Sousa Santos, Boaventura de and José Manuel Mendes, eds. 2017. *Demodiversidad. Imaginar nuevas posibilidades democráticas*. Madrid: Akal.

Terras, Melissa. 2012. "Infographic: Quantifying Digital Humanities," January 20. http://melissaterras.blogspot.it/2012/01/infographic-quanitifying-digital.html.

UNESCO. 2021. "UNESCO Recommendation on Open Science." https://en.unesco.org/science-sustainable-future/open-science/recommendation.

Veugelers, Reinhilde. 2017. "The Challenge of China's Rise as a Science and Technology Powerhouse." *Policy Contribution* 19, July. http://bruegel.org/wp-content/uploads/2017/07/PC-19-2017.pdf.

Weingart, Scott B. and Nickoal Eichmann-Kalwara. 2017. "What's under the Big Tent?: A Study of ADHO Conference Abstracts." *Digital Studies/Le Champ Numérique* 7 (1): 6. http://doi.org/10.16995/dscn.284.

CHAPTER THREE

Digital Humanities Outlooks beyond the West

TITILOLA BABALOLA AIYEGBUSI
(UNIVERSITY OF TORONTO) AND LANGA KHUMALO (SADILAR)

As Digital Humanities (DH) continues to grow, its landscape has become more inclusive. This is evidenced through its fast and vast reach across disciplinary and geographical borders. At the heart of DH discourse are themes of connectivity, access, infrastructure, cultural sensitivity, and inclusion, principles that are necessary to achieve the collaboration so essential to DH. The contribution of DH programs to the global knowledge economy has been noted, both in the humanities and beyond (Berry 2012; Warwick, Terras, and Nyhan 2012). Yet there remains some exigency to reflect on DH work from a global, rather than Western, perspective. Does the purpose of DH programs as observed in the West align with what the field achieves and seeks to achieve at the peripheries? What is the implication of studying or practicing DH for communities outside the West? Beyond the epistemological and the pedagogical, what is common or different in the conceptualization and operationalization of DH across spaces? And furthermore, why do these differences exist?

Our purpose here is to consider how DH engages with cultures and societies beyond the West and proffer some explanations for some identified deviations. As such, a fundamental question from a macro (global) perspective is "Is DH intrinsically inclusive?" There are several factors that have made DH what it is today. One is the impact of globalization—a flow of people, culture, ideas, values, knowledge, technology, and economy across borders resulting in a more interconnected and interdependent world (Teferra and Knight 2008). Globalization, in this sense, has enabled the spread of DH beyond Europe and North America, and it is through this lens that DH is viewed as borderless both in terms of disciplinary boundaries and geographical (*also continental*) borders. Another is the deliberate effort by the DH community to build an inclusive field, one that is particularly focused on bridging the gap between the center and the peripheries (Fiormonte 2014), and that is representative of diverse cultures and languages (Risam 2018). It is unclear, however, how these originating forces have impacted DH's inclusiveness at the meso (country) and micro (individual communities) levels. Our focus is to situate DH in community contexts where it is still gaining roots with the intention to examine how as a field, it maps out, or remaps, its new domain. We will emphasize common traits and differences regarding how it is perceived in these communities with the aim of teasing out ways of thinking that could hinder the field's actualization of a truly inclusive space.

Digital Humanities provides a bridge between the traditional practices of research and technology-driven research to scholars straddling quintessential humanist approaches and modern digital methodologies, tools, and frameworks to support them in novel avenues of enquiry. It is a growing scholarship in the humanities and social sciences that is spurred by advances in computing and digital spheres and providing new collaboration with engineering and computer (and techno) sciences.

The impact and adoption of DH beyond the West is complex because the field strives for global presence at a time when there is great scrutiny for appearances of neocolonialism in former Western-dominated territories. Many of the communities occupying the peripheries are still emerging from a history of colonial subjugation and apartheid. It is a history that saw the subjugation and supplanting of existing knowledge systems and ways of knowing, itself a form of epistemic violence that still exists in most education systems across Africa and Asia. While our focus in this chapter is to provide a critical examination of the interaction of DH with cultures beyond the West, we might, intermittently, discuss the implication of such interactions for pedagogy. Education is in any case an engine of development, and how it is pursued is critical to the success of DH in developing states.

INCLUSION AND DIVERSITY WITHIN DH

Recently, there has been a drive towards inclusion and diversity within digital humanities. Conference themes, presentations, and scholarly publications have included titles like "Towards an Open Digital Humanities" (THATCamp 2011), "Digital Humanities from a Global Perspective" (Fiormonte 2014), "Decolonizing Digital Humanities" (Aiyegbusi 2018), "Gender and Digital Humanities" (Smyth et al. 2020), "Minority Report: The Myth of Equality in the Digital Humanities" (Bordalejo 2018), "Towards a Queer Digital Humanities" (Ruberg et al. 2018), etc. These topics point to a shift in subject matter from one that centers primarily around pedagogical approaches, theory, practice, and tools, to a consciousness that reflects social trends, advances, and changes within the field. Thus, inclusion and diversity have been two of the major drivers of discourse at DH conferences and events. But for DH to really function as a "big tent," we need to appreciate what inclusion means and requires at the peripheries.

For the DH scholar living in the Southern Hemisphere, inclusion suggests the opportunity to present periphery narratives without fear of exclusion based on writing, accent, thematic relevance, gender, or geographical proximity; it is knowing that every voice counts and matters to the achievement of an inclusive digital humanities field. Ironically, the discourse about inclusion and diversity is happening mostly at the center, and while efforts to correct the imbalance seem to be working in this space, the picture in the Global South continues to reflect a divide that is not just based on digital technology but also on cultural differences.

This difference is captured by Nickoal Eichmann-Kalwara, Jeana Jorgensen, and Scott Weingart in their paper on representation at DH conferences between 2000–2015 in which they note that gender inequality and the underrepresentation of non-white people in DH discourse are more visible at the periphery (2018, 72–91). They suggest that perhaps names and topics may play a role in the marginalization of this group at scholarly events. However, due to limited information, they could not ascertain if the bias against non-Western authors results from peer reviewers' disinterest in topics related to the peripheries or is just a function of unfamiliar names.

We propose that apart from non-Western names and research interests, there could be a few other factors that may contribute to bias in digital humanities: language, location, and misunderstanding of cultural milieu.

DIGITAL HUMANITIES AND THE LANGUAGE CONUNDRUM

In attempting to define language we refrain from using a litany of definitions found in the literature on Language and Linguistics that are drawn from Eurocentric theories. In appreciating what language means, we defer to the explication of mother tongue in isiZulu[1] as simply *ulimi lwebele* (literally, the language that one sucks from their mother's breast). Mother tongue in this sense means one's first language. In IsiNdebele,[2] when a child utters their first word, the elders exclaim: *Sekungumuntu!* (meaning, It is now a human being). Thus, language is an essential part of that which makes us *human*; it is at the heart of our human existence. The mother tongue is inextricably linked to our cultural identity. It formulates and records our personal and social experiences. As argued by Khumalo, the use of the mother tongue in education stimulates and enables the child's literacy skills, enables the child's cognitive development, stimulates the child's critical thinking, and stimulates the child's imagination and innovation (Khumalo 2020). Thus, the organization of knowledge and its transmission is facilitated through language. Education, which involves the training of citizens, and intellectual growth is conducted through language using various models and discourses (see Mchombo 2017). As argued by some scholars, with the end of the Second World War, language imperialism became a new tactical method for global domination (Khumalo 2016; Philipson 2017). Tragically because of Africa's colonial past, which saw the subjugation of her peoples and the proscription of African languages and African knowledge systems and ways of knowing from the (Western) school system, the curriculum in most African countries is packaged and parceled out in English. This effectively limits African learners both in terms of the language of learning and teaching, and the concepts that are explained in the foreign language. Learning in such schools, if any, is thereby reduced to rote memorization.

Since, as discussed above, the spread of English is a form of postcolonial or neocolonial movement (Majhanovich 2013; Pennycook 2002), the continued swift global expansion of DH beyond the West, particularly into parts of Africa and Asia, has therefore initiated discussions about language and ways of knowing, which hitherto have been suppressed. English hegemony, the colorful history of the British colonial Empire, and modern day British-American global domination invoke critical perspectives in the context of DH expansion into (and its impact in) the Southern Hemisphere. For instance, African languages present with remarkably interesting linguistic structures, as a result posing very interesting challenges to computational solutions (Keet and Khumalo 2017, 2018). African epistemologies and worldviews, that were previously exempt from the global knowledge economy, provide "new knowledge" that challenges and enriches our collective understanding. Given the condescending perception of Africa's indigenous knowledge systems (IKS) by the West, the question arises as to whether these African IKS are viewed as new centers of knowledge from the peripheries or merely sites with potential to enrich the center. DH approaches have the capacity to challenge the norms of knowledge generation, sharing, and access. While in the West African traditional IKS the griot[3] is the repository of knowledge and customs, DH provides new affordances. The use of productivity tools, often subject to continuous updating, puts student and teacher on equal footing, and often allows the student to find solutions to problems as yet unimagined by the teacher. This brings into question the entire university system as conceived in the nineteenth-century West. Accustomed to authority accumulated on layers of analyses of cultural objects and their meanings as produced by history, humanists can only wonder about their future. Equally, the availability of traditional IKS in digital platforms through various licenses challenges in a novel way the authority of the griot.

The impact of English on Africa's epistemology and output of knowledge is quite similar to Asia's. In her paper "English as a Tool of Neo-Colonialism and Globalization in Asian Contexts," Majhanovich details the responses of several Asian countries to the adoption of English as the de facto lingua franca of the modern world, and the efforts, though sometimes unsuccessful, by some Asian countries like China, Japan, and Malaysia to overturn English pedagogical hegemony, and emphasize indigenous national languages. Writing about the effects of English Language Teaching (ELT) in Asia and how the adopted Western pedagogy perpetuates colonialism and weakens indigenous ways of learning, Alastair Pennycook states that "theories and practices that emanate from the former colonial powers still carry the traces of those colonial histories both because of the long history of direct connections between ELT and colonialism and because such theories and practices derive from broader European cultures and ideologies that themselves are products of colonialization" (Pennycook 1998, 19). It is important for DH scholars therefore to acknowledge this power relationship and the fact that although English has attained a global status, it does not possess the ability to provide a complete representation of native cultures in Southern Hemispheric communities despite being the national and sometimes *official* language. Taking such inadequacies of English into consideration would necessitate an embrace of diverse languages and the acceptance of accents at the peripheries, thereby enriching the base from which knowledge is created (see Risam 2018, 79–83 for more on the impact of international accent on DH research).

GEOGRAPHICAL LOCATION

Cultural and linguistic diversity have been topics at the fore of global DH discourse (Fiormonte 2014; Mahony 2018; Risam 2018). Until the last couple of years, the English language has occupied the lead position in DH conversations. The impact of such heavy reliance on English is more visible in locations far from the center. For example, Africa has two lingua francas: English and French. A Google search of DH presence in Africa yielded results for only Anglophone African countries: Ghana, Kenya, Nigeria, and South Africa. There is an absence of DH in Francophone Africa. The question that comes to mind then is: Are there no DH activities in these regions? Given that most DH projects in Africa exist outside the umbrella of the field, as witnessed during the African Digital Storytelling Symposium organized by James Yeku and Brian Rosenblum at Institute for Digital Research University of Kansas, where several participants admitted not knowing about digital humanities as a field even though their works fall within its scope, it can be assumed that the answer to this question is yes. Therefore, the issue of inclusivity goes beyond simple fixes like providing interpretation at conferences, issuing calls in multiple languages, or translating publications to those at the peripheries; it is also a problem of proximity.

For clarity, in this chapter, the center consists of developed countries in Europe and North America, and the peripheries are developing countries with specific focus on Africa and Asia. The center has championed the growth of DH to this point and has consequently defined what DH is and should be across the globe despite its distance and disconnection from the peripheries.

The marginal representation of fringe DH perspectives is also mostly only accessible from the center: digital humanists moving to the center to tell their stories. As Isabel Galina Russell suggests, it is important that scholars "don't just move from the periphery to the center (for meetings, conferences, research visits, etc.) but that scholars move from the center to the periphery as well" (2011). If the purpose of digital humanities is to challenge the methodology of research by proposing new and different ways of accessing, presenting, and interpreting knowledge and culture through digital instruments (Fiormonte 2014, 4), then, observing how DH is taught or practiced in spaces beyond the West establishes that its objective is universal and that its purpose aligns with the realities of the communities across the globe. But, as Fiormonte asks, "are centers ready to learn from peripheries?" (4).

Learning from the peripheries comes with an understanding and acknowledgment of cultural differences, and a willingness to engage with communities outside the West with an unbiased perspective. There are quite a few scholars already bridging the geographical divide and engaging with the perspective of DH at the periphery in a bid to promote inclusivity and gain insights about practices and nuances prevalent at the fringes. They actively participate in DH conferences in Nigeria and across Africa, and they are useful resources that provide context to critical and relevant discourse. However, such engagements should be approached with caution and careful consideration in order to prevent a reinforcement of enduring global power dynamics whereby the center superimposes a Eurocentric DH discourse on Global South communities. Thus, as DH continues to struggle to gain legitimacy in places like Africa due to several reasons (Aiyegbusi 2018), it is important that the center engages with scholars across the globe in ways that impart knowledge to the peripheries while striving to gain perspective on realities from these local communities. An efficient way to achieve such a level of engagement might be to rotate the Big DH tent around the world. Hosting the Alliance for Digital Humanities Organization (ADHO) conference in China, India, Nigeria, or South Africa, for instance, allows DH scholars the opportunity to experience cultures at the periphery, and hopefully, such an understanding may impact how periphery narratives are viewed or reviewed. Therefore, it is important that rotated hosting should reflect topics and themes that reflect the key DH issues for that location. Such undertaking is feasible given recent changes in how we host conferences due to the impact of Covid-19. It is now possible to organize a virtual conference, which can be attended by delegates from any location in the world thus eliminating issues of logistics, visa, and security. Similarly, providing funding and opportunities for scholars at the periphery to speak at conferences in the center would increase output of knowledge from these spaces.

MISUNDERSTANDING CULTURAL MILIEU

At the core of appreciating the center-periphery dynamic is grasping how DH interacts with its host community's culture. DH practice can be adversely impacted by subjective behavioral patterns. For example, in an African university, the average student has no strong connection with the library, and this indifference affects how such students approach DH practice. This kind of disinterest reflects how mindsets, and not just infrastructure, influence scholars' stance on digital scholarship. But why would such a relationship matter for inclusion and diversity in the global discourse of digital humanities? To answer this question, we consider the relationship between Library Information Science (LIS) and DH in the center and compare our findings with the interaction between these fields at the periphery. A common feature of DH at its North American/British center is the presence of LIS research and practitioners. Globally, Library Information Science's focus, though distinct, interweaves with that of digital humanities. As many scholars have observed, there is significant overlap between both fields and research interests and questions tend to drive scholars in both fields towards parallel ends (Russell 2011; Warwick, Terras, and Nyhan 2012; Sula 2013; Robinson, Priego, and Bawden 2015; Koltay 2016). In his paper about a conceptual model for both DH and LIS, Chris Sula notes that "Within library and information science (LIS), there is a corresponding (if more dispersed) discussion of DH" (2013, 11). No doubt, there is a symbiotic relationship between LIS and DH. It is a relationship that is carved out of reliance on the expertise both fields exhibit. From the management and retrieval of records, to processing information and guidelines, or navigating the complex web of material ownership, management, and credits, digital humanists have benefited, and continue to benefit, from this affiliation. But this relationship is not exempt from the center-periphery effect. The strong link between LIS and DH is mostly seen in the Global North; at the periphery, the picture tells a different tale.

While LIS in the West is entwined with DH, the relationship between these disciplines appears to be hardly existent in the Global South. This absence can be accounted for along two lines: pedagogy and praxis. Along the lines of pedagogy, there are many papers written in the last decade that discuss the challenges LIS faces in the Global South (Kumar and Sharma 2010; Mittal 2011; Okeji and Mayowa-Adebara 2020). Many of these essays focus on the state of LIS in parts of Asia and Africa, and the problems they identify are common to both continents: lack of qualified staff with digital competencies in LIS, outdated curriculum that does not account for the changing technological trends nor make provision for the management of digital records, inadequate infrastructure for practice, record keeping, and retrieval of records, and lastly, standardized format for teaching LIS in schools. These problems reflect challenges with the outdated and unregulated pedagogical and epistemological approaches used in LIS programs. If, as expressed by Lyn Robinson et al., DH's fate appears to be entwined with LIS (2015: 19), then the practice of DH considerably, though not entirely, relies on the existing structures in LIS to function. But what happens at the peripheries where existing LIS structures are inadequate to support digital research, where libraries are struggling to gain relevance? As Russell notes in her paper on the role of LIS in DH:

> The arrival of the Web, the extensive use of search engines to find information, the integration of library-type services into academic journal publishing platforms, among others, compete with libraries to become the starting point for users in their search for information. These new search patterns are found both among researchers and students, and thus, traditional academic libraries are struggling to remain relevant.
>
> (2011, 1–2)

It might be worthwhile in the future to examine how this new search pattern that Russell discusses impacts pedagogical approaches in the peripheries, and perhaps also investigate the extent to which it renders LIS redundant in these spaces.

Record-keeping institutions like museums, archives, and libraries still retain a traditional outlook to the process of record management, making them unequipped to support the kind of work necessary for the practice of digital humanities. A part of the problem with libraries in these countries can be directly linked to pedagogy. For example, in an evaluation of LIS curriculum and digital literacy in Nigerian universities, Chukwuma Clement Okeji and Okeoghene Mayowa-Adebara observed that out of the thirty-one universities that offer LIS, only twelve schools include digital libraries as a standalone course in their curricula, and these courses are taught without a standardized model, leaving schools to offer it at different levels (2020, 9). Similarly, Jagtar Singh and Pradeepa Wijetunge extensively examine the state of LIS across seven South Asian countries. They explain that while courses in this field are offered across all levels at tertiary institutions, access to advanced learning is dependent on country of residence. India, Bangladesh, and Pakistan for instance have provided learning from diploma (pre-entry) to PhD levels, Sri Lanka on the other hand only offers courses at the Bachelor's and Master's levels (Singh and Wijetunge 2006). Notably common to all is a lack of uniform approach both to learning and accreditation.

The second problem, which lies along the lines of praxis, is crucial to understanding how the mindset of students about LIS directly impacts digital humanities in the Global South. In Africa for instance, there is a remarkable lack of interest in libraries among students (Baro and Keboh 2012). In a study on the use of libraries among undergraduate students in three Nigerian universities, Iyabo Mabawonku notes that only about half of the respondents use libraries on a regular basis (2004). Some of these admit accessing libraries for personal work, the majority of the students had no training in the use of libraries and concluded that "a student could graduate without having used libraries" (2004, 151). The implication of such disinterest is rooted in the general belief among students that the library has no relevance to the quality of their education. Occasional use of libraries among students could impact the level of funding available to the development of LIS, which could affect the kind of upgrades made to libraries, which in turn would affect the DH practice.

More importantly, the ability of libraries to support DH research influences the narratives at the peripheries. The inability to format records required by DH scholars limits the quality and quantity of research produced and ultimately portrays a skewed picture of DH practice at the peripheries. Also, many DH projects are being carried out outside the walls of academia. While such works could be impactful to the DH narrative on the peripheries, they may be lost to the field unless efforts are made to bring them under the DH canopy. Therefore, it is important that DH scholars take into consideration the cultural milieu of the community within which DH exists for a clear depiction and evaluation of the work the field produces. This understanding is necessary because digital tools, platforms, and other representations of the digital are not neutral—they embody biases and idiosyncrasies reflective in the space it occupies, be it gender, race, religion, linguistic, or, as in the case of the relationship between students and libraries, just plain disinterest. Consequently, the process through which DH, whether as an academic field or practice, engages with established tradition and customs of a given community is crucial to understanding the purpose of DH, its impact, and how it evolves and develops within such new spaces.

Looking beyond the centered West, what is the purpose of the field? The overarching objective of DH, which is an integration of quantitative and qualitative approaches into humanistic research, and the enhancement of teaching and other practices in ways that improve learning, experience, and collaboration among scholars and people outside the walls of academia, appears to align across the globe. The difference, however, lies in its interaction with the cultural worldviews and priorities of each geographical region. Whether in the development of its pedagogy, praxis, or theory within a new domain, DH engages with pre-existing conditions, and must, therefore, morph around and within such conditions in order to thrive.

Given our analysis of the relationship between LIS and DH in the peripheries, and the current deficiencies in both disciplines, one promising option for DH may be to adopt a collaborative approach at the peripheries; one that integrates LIS and DH work to extract obvious synergies. Adopting such an approach creates avenues for both fields to support each other. Integrating DH into the field of LIS would also open the boundaries of DH which is currently mostly limited to the humanities in the Global South, while infusing LIS with new life.

DISRUPTING THE CENTER, EMBRACING THE PERIPHERIES

When it comes to amplifying otherwise-silenced voices in global DH, some fundamental questions to ask are: What counts as knowledge? Is there a single ontology? Is there a single way of knowing? Whose knowledge dominates the current global knowledge economies? For example, the decolonization of curricula in South African Higher Education is viewed in the context of the current dominant Eurocentric theories, thinkers, and ideas, at the expense of African worldview, centeredness and relevance (Fataar 2018). It is argued that societal relevance is a crucial factor for education to be meaningful. Gayatri Spivak (2016) has argued about the notion of epistemic violence, a form of silencing that has left those existing on the fringe cognitively damaged. This notion of epistemic violence expresses the Eurocentric and Western domination and the subjugation of former colonial subjects through knowledge systems. The curricula, packaged in European languages and expressing Eurocentric worldviews, were designed to degrade, exploit, and subjugate Western colonies. The learning is reduced to rote memorization devoid of critical thinking and cognitive development, which has resulted in what Spivak calls cognitive damage. The tragedy is that these Eurocentric worldviews persist today in (South) African Higher Education (Prah 2017), and DH may have been unwittingly ensnared by the legacy of colonial infrastructure.

In order to exploit the full content and context of decolonization of DH in academia within the peripheries, there are a few things that can be done to amplify silenced voices. These include deepened understanding, encouragement of diversity, and provision of training and technical support. The center may explore how to understand and contextualize global differences; digital projects that exhibit Afrocentric, Oriental, or South Asian reflections should be encouraged; acknowledgment of linguistic diversity, especially of nuance at the periphery, and its influence on the presentation of ideas. Perhaps, then we can begin to experience what Anne Burdick et al. describe when they write: "The decolonization of knowledge in the profound sense will arrive only when we enable people to express their otherness, their difference, and their selves, through truly social and participatory forms of cultural creation" (2012, 91). DH platforms must not be exclusionary; they must not be prescriptive but must be accommodative in embracing the peripheries and their complexities. In this way, the center is disrupted in a positive way in order to embrace the new insights and complexities of the peripheries.

The infrastructure in the Global South adds to the rich diversity of the DH field. South Africa, for instance, has recently established the South African Center for Digital Language Resources (SADiLaR[4]). SADiLaR is one of the thirteen National Research Infrastructures established in the South African Research Infrastructure Roadmap (SARIR), that are funded by the Department of Science and Innovation (DSI). SARIR is intended to provide a strategic, rational, medium- to long-term framework for planning, implementing, monitoring, and evaluating the provision of research infrastructures (RIs) necessary for a competitive and sustainable national system of innovation. SADiLaR has a strategic function to stimulate and support research and development in the domains of human language technologies (HLT), natural language processing (NLP), language-related studies, and Digital Humanities (DH) in the humanities and social sciences. It is thus far the first and only such infrastructure on the African continent.

CONCLUSION

Delving into the conversation about DH and what it looks like beyond the West gives context to some of the ways we engage with people and cultures within the field across the globe, but more importantly, such discourse pushes us to inquire about how the peripheries speak, and in turn investigate how the center listens. There are several other factors that impact how the peripheries speak that we are unable to uncover due to space constraints, but we hope to continue with this thread in future dialogues as we believe it is germane to the actualization of an inclusive digital humanities field. But to conclude this chapter, it is important that we emphasize, especially to DH scholars belonging to spaces far from the center, that the freedom to think and create knowledge is the hallmark of epistemic freedom. To attain epistemic freedom, the peripheries need to break free from mindsets that hinder the march towards intellectual sovereignty. Such sovereignty is achieved through constant dialogue with stakeholders with other worldviews in ways that demand mutual respect for perspectives and content, and access to support by all parties involved. The decolonization and transformation of the curricula in Africa and Asia therefore would require the incorporation of epistemic perspectives, knowledge, and thinking from the Global South into the growing global knowledge economy from which everyone draws. Decolonization is, thus, a call for plurality in sources of knowledge, and not the current negation of "other" worldviews. By embracing rather than rejecting, a unified DH may realize a noble purpose of enriching and harmonizing knowledge and learning and supporting equity in development.

NOTES

1. IsiZulu is one of the eleven official languages in South Africa.
2. IsiNdebele is one of the sixteen official languages in Zimbabwe.
3. A griot is the traditional storyteller in Africa. She preserves cultural and traditional histories through oratory narratives that are often presented in the context of recent events. In essence, the griot maintains connection between the past and present and serves as the primary source of knowledge.
4. SADiLaR has a Center status within the Common Language Resources and Technology Infrastructure (CLARIN), which is a European Research Infrastructure for Languages Resources and Technology, www.sadilar.org. As a digital resource center it is set to expand and add to the complexity, diversity, and richness of DH programs globally. Current resources are available from https://www.sadilar.org/index.php/en/resources.

REFERENCES

Aiyegbusi, Titilola B. 2018. "Decolonizing Digital Humanities: Africa in Perspective." In *Bodies of Information: Intersectional Feminism and Digital Humanities*, edited by Elizabeth Losh and Jacqueline Wernimont, 434–46. Minneapolis, MN: University of Minnesota Press.

Baro, Emmanuel E. and Tarela Keboh. 2012. "Teaching and Fostering Information Literacy Programmes: a Survey of Five University Libraries in Africa." *The Journal of Academic Librarianship* 38 (5): 311–15. doi: 10.1016/j.acalib.2012.07.001.

Berry, David M. 2012. "Introduction: Understanding the Digital Humanities." In *Understanding Digital Humanities*, edited by David M. Berry, 1–20. London: Palgrave Macmillan.

Burdick, Anne, Johanna Drucker, Peter Lunenfeld, Todd Presner, and Jeffrey Schnapp. 2012. *Digital Humanities*. Cambridge, MA: MIT Press.

Bordalejo, Barbara. 2018. "Minority Report: The Myth of Equality in the Digital Humanities." In *Bodies of Information: Intersectional Feminism and Digital Humanities*, edited by Elizabeth Losh and Jacqueline Wernimont, 320–43. Minneapolis, MN: University of Minnesota Press.

Eichmann-Kalwara, Nickoal, Jenna Jorgensen, and Scott B. Weingart. 2018. "Representation at Digital Humanities Conferences (2000–2015)." In *Bodies of Information: Intersectional Feminism and Digital Humanities*, edited by Elizabeth E. Losh and Jacqueline Wernimont, 72–92. Minneapolis, MN: University of Minnesota Press.

Fataar, Aslam. 2018. "Placing Students at the Centre of the Decolonizing Education Imperative: Engaging the (Mis) Recognition Struggles of Students at the Postapartheid University." *Educational Studies* 54 (6): 595–608. https://doi.org/10.1080/00131946.2018.1518231.

Fiormonte, Domenico. 2014. "Digital Humanities from a Global Perspective." *Laboratorio dell'ISPF* 11 (10.12862): 121–36. https://www.academia.edu/9476859/Digital_Humanities_from_a_global_perspective.

Keet, C. Maria and Langa Khumalo. 2017. "Grammar Rules for the IsiZulu Complex Verb." *Southern African Linguistics and Applied Language Studies* 35 (2): 183–200.

Keet, C. Maria and Langa Khumalo. 2018. "On the Ontology of Part-Whole Relations in Zulu Language and Culture." In *Formal Ontology in Information Systems: Proceedings of the 10th International Conference*, edited by S. Stefano Borgo, Pascal Hitzler, and Oliver Kutzpp, 225–38. Amsterdam: IOS Press.

Khumalo, Langa. 2016. "Disrupting Language Hegemony." In *Disrupting Higher Education Curriculum*. Leiden: Brill. https://brill.com/view/book/edcoll/9789463008969/BP000016.xml.

Khumalo, Langa. 2020. "African Languages and Indigenous Knowledge Systems." Keynote Address, October 27. *First Annual African Languages and Indigenous Knowledge Systems Seminar*. Pretoria: University of South Africa.

Koltay, Tibor. 2016, "Library and Information Science and the Digital Humanities: Perceived and Real Strengths and Weaknesses." *Journal of Documentation* 72 (4): 781–92. https://doi.org/10.1108/JDOC-01-2016-0008.

Kumar, Khrishan and Jaideep Sharma. 2010. "Library and Information Science Education in India: A Historical Perspective." *DESIDOC Journal of Library & Information Technology* 30 (5): 3–8. https://citeseerx.ist.psu.edu/viewdoc/download?doi=10.1.1.1028.9865&rep=rep1&type=pdf.

Mabawonku, Iyabo. 2004. "Library Use in Distance Learning: A Survey of Undergraduates in Three Nigerian Universities." *African Journal of Library, Archives and Information Science* 14 (2): 151–65. https://www.ajol.info/index.php/ajlais/article/view/26151.

Mahony, Simon. 2018. "Cultural Diversity and the Digital Humanities." *Fudan Journal of the Humanities and Social Sciences* 11 (3): 371–88. https://link.springer.com/article/10.1007/s40647-018-0216-0.

Majhanovich, Suzanne. 2013. "English as a Tool of Neo-colonialism and Globalization in Asian Contexts." In *Critical Perspectives on International Education*, edited by Yvonne Hébert and Ali A. Abdi, 249–61. Leiden: Brill Sense.

Mchombo, Sam. 2017. "Politics of Language Choice in African Education: The Case of Kenya and Malawi." *International Relations and Diplomacy Journal* 5 (4): 181–204. doi: 10.17265/2328-2134/2017.04.001.

Mittal, Rekha. 2011. "Library and Information Science Research Trends in India." *NISCARI Online Periodicals Repository* 58 (4): 319–25. http://nopr.niscair.res.in/handle/123456789/13481.

Okeji, Chukwuma Clement and Okeoghene Mayowa-Adebara. 2020. "An Evaluation of Digital Library Education in Library and Information Science Curriculum in Nigerian Universities." *Digital Library Perspectives* (ahead of print). https://www.emerald.com/insight/content/doi/10.1108/DLP-04-2020-0017/full/html.

Pennycook, Alastair. 1998. *English and the Discourses of Colonialism*. London: Routledge.
Pennycook, Alastair. 2002. "Mother Tongues, Governmentality, and Protectionism." *International Journal of the Sociology of Language* 154: 11–28. https://opus.lib.uts.edu.au/handle/10453/1097.
Phillipson, Robert. 2017. "Myths and Realities of European Union Language Policy." *World Englishes* 36 (3): 347–49. doi: 10.1111/weng.12270.
Prah, Kwesi. 2017. "The Intellectualisation of African Languages for Higher Education." *Alternation Journal* 24 (2): 215–25. https://journals.ukzn.ac.za/index.php/soa/article/view/1327.
Risam, Roopika. 2018. *New Digital Worlds: Postcolonial Digital Humanities in Theory, Praxis, and Pedagogy*. Evanston, IL: Northwestern University Press.
Robinson, Lyn, Ernesto Priego, and David Bawden. 2015. "Library and Information Science and Digital Humanities: Two Disciplines, Joint Future?" Paper presented at the 14th International Symposium on Information Science, Zadar, Croatia, May 19–21. https://openaccess.city.ac.uk/id/eprint/11889/.
Ruberg, Bonnie, Jason Boyd, and James Howe. 2018. "Toward a Queer Digital Humanities." In *Bodies of Information: Intersectional Feminism and Digital Humanities*, edited by Elizabeth Losh and Jacqueline Wernimont, 108–28. Minneapolis, MN: University of Minnesota Press.
Russell, Galina Isabel. 2011. "The Role of Libraries in Digital Humanities." Paper presented at the 77th IFLA World Library and Information Congress, San Juan, Puerto Rico, August 13–18. https://www.ifla.org/past-wlic/2011/104-russell-en.pdf.
Singh, Jagtar and Pradeepa Wijetunge. 2006."Library and Information Science Education in South Asia: Challenges and Opportunities." Presented at the *Asia-Pacific Conference on Library & Information Education & Practice* (A-LIEP), Singapore, April 3–6. http://hdl.handle.net/10150/106432.
Smyth, Hannah, Julianne Nyhan, and Andrew Flinn. 2020. "Opening the 'black box' of Digital Cultural Heritage Processes: Feminist Digital Humanities and Critical Heritage Studies." In *Routledge International Handbook of Research Methods in Digital Humanities*, edited by Kirsten Schuster and Stuart Dunn, 295–308. New York: Routledge.
Spivak, Gayatri Chakravorty. 2016. "Histories, Democracy and the Politics of Knowledge in Higher Education." In *Disrupting Higher Education Curriculum: Undoing Cognitive Damage*, edited by Michael Anthony Samuel, Rubby Dhunpath, and Nyna Amin, 17–30. Rotterdam: Sense Publishers.
Sula, Chris Alen. 2013. "Digital Humanities and Libraries: A Conceptual Model." *Journal of Library Administration* 53 (1): 10–26. https://www.tandfonline.com/doi/abs/10.1080/01930826.2013.756680.
THATCamp SoCal. 2011. "Towards an Open Digital Humanities." January 11–12. http://socal2011.thatcamp.org/01/11/opendh.
Teferra, Damtew and Jane Knight, eds. 2008. *Higher Education in Africa: The International Dimension*. Accra: Association of African Universities; Chestnut Hill, MA: Center for International Higher Education, Lynch School of Education, Boston College.
Warwick, Claire, Melissa Terras, and Julianne Nyhan, eds. 2012. *Digital Humanities in Practice*. London: Facet Publishing.

CHAPTER FOUR

Postcolonial Digital Humanities Reconsidered

ROOPIKA RISAM (DARTMOUTH COLLEGE)

In November 2018, my book, *New Digital Worlds: Postcolonial Digital Humanities in Theory, Praxis, and Pedagogy*, was published by Northwestern University Press. A culmination of years of thinking alongside friends and interlocutors about the ways digital humanities reproduces the hallmarks of colonialism in the digital cultural record, the book offered an opening salvo for a conversation about how our digital humanities practices need to change to more fully realize the promises of and attend to the challenges of open access to knowledge. *New Digital Worlds* offered a broad look at the full supply chain of digital knowledge production, from methods to tools, from archives to professional organizations, from research to teaching. My primary goal was to call for an urgent rethinking of digital humanities methodologies to ensure that our work is not unthinkingly reproducing and amplifying the gaps and omissions of the cultural record as we construct a digital one. Never intended to be prescriptive, the book is a space-clearing gesture, a provocation to build upon and to quibble with, and, as is the ambition for my scholarship, to eventually be rendered obsolete and unnecessary as practitioners adopt the values for which it advocates.

New Digital Worlds has enjoyed a warm welcome from many corners of digital humanities. My intention, when writing the book, was to address digital humanities practitioners of the Global North. I wanted to urge those of us who work in these spaces to examine how the methodological, technical, and design choices we make reinstantiate the canons of literature, history, and culture (Risam 2018). While many of us have been engaging in conversations and offering advocacy on these points for many years through hashtags like #DHPoco and #TransformDH, as well as organizations like Global Outlook::Digital Humanities, even *digital* humanists are more likely to believe something in monograph form. And, indeed, *New Digital Worlds* found this audience in the Global North.

As a scholar trained in postcolonial studies in the US, I was writing in that tradition, concerned that those of us who practice digital humanities in the academy of the Global North are doing harm, however unintentionally. I hoped that we might collectively consider alternate ways of thinking and making with digital humanities and recognize our role in decentering digital humanities practices of the Global North. I was urging practitioners of the Global North to rethink forms of building, collaboration, organization, and teaching that could resist the reification of colonial histories, practices, and politics in the digital scholarship we produce through attention to not only the content represented in our projects but also *how* we develop our research projects (Risam 2018).

The adoption of *New Digital Worlds* in courses in the Global North, as well as its wide citation in the framing of both digital and analog scholarship, suggests that it has reached my intended audience.

But *New Digital Worlds* and the concepts behind postcolonial digital humanities found another audience among scholars from the Global South. The feedback I've received from them is entirely unwarranted. I've been told that they can do their work because of my book—when they were already doing important work well before my book. They've thanked me for inspiring them—when their work has long inspired me. And they've said that I've given them a voice—when they already had one. These exceedingly kind responses bring me both surprise and discomfort. These practitioners were not my intended audience because *they* are the experts on digital humanities in the Global South—not me. I wasn't actually speaking to them—and I certainly wasn't speaking *for* them.

I recognize that the symbolic value of my identity as a Kashmiri woman and an immigrant to the US clearly plays a role in this enthusiastic reception of *New Digital Worlds* from my unintended audience. However, the responses from my colleagues in the Global South rehearse a troubling trend in digital humanities scholarship that, ironically, I identify in the book: the way that well-circulated writing from the Global North becomes the frequently taught, cited, and discussed scholarship. The danger of this phenomenon is that it overshadows local expertise and threatens to replace or forestall the growth of local digital humanities practices with a fictive "global" that masquerades as "universal." As I point out in *New Digital Worlds*, such practices are not universal at all—they are particular practices of the Global North masquerading as "universal" (Risam 2018). An unintended consequence, *New Digital Worlds* has been functioning in a role I was explicitly warning *against* in the book.

Thus, this unanticipated reception for *New Digital Worlds* has made me wonder and even reconsider: what kind of harm has postcolonial digital humanities been doing? This essay reconsiders several key ideas in postcolonial digital humanities and seeks to clarify how we might better understand them to mitigate potential harm: 1) overdetermination of postcolonial thought by the US and European academies, 2) discursive postcoloniality, and 3) the centering of the postcolonial elite in digital humanities discourses. These critiques are ones that could as easily be made—and have been made—of postcolonial studies more generally, and I explore them here in the context of South Asian studies. Yet, they are issues of critical importance to consider in the broader contexts of postcolonial digital humanities to ensure that it does not become yet another hegemonic discourse that loses sight of the needs of the most vulnerable communities. And, perhaps, elaborating on these concerns may more fully realize the promise of *New Digital Worlds.*

DISPLACING THE US AND EUROPE

Postcolonial studies has been an inextricable part of the emergence of the academic star system in the academy of the Global North in the 1980s and 1990s and has overly emphasized voices of those working in the Global North, often as émigrés (Slemon 1994; Williams 2001). As a result, the geospatial terrain of postcolonial studies scholarship is heavily skewed towards scholars in the US and Europe, who are among the most widely read and cited (Huggan 2002; Giles 2011). This is a phenomenon born from the uneven distribution of institutional resources in the context of global capitalism that has shaped the landscape of higher education and knowledge production more generally. Despite the financial challenges faced by many higher education institutions in the Global North, the relative wealth of institutions in the US and Europe has allowed them to attract and retain postcolonial scholars, positioning them as the putative centers of postcolonial knowledge production as they contribute to "brain drain" in postcolonial states (Sunder Rajan 1997; Ray 2005).

In the context of digital humanities, my thinking on these issues has been deeply influenced by conversations with James Smithies and Paul Barrett, who have raised crucial questions about how postcolonial digital humanities risks reproducing these dynamics. In 2013, when first engaging with the concept, Smithies wrote in horror:

> My first reaction to seeing what #dhpoco [postcolonial digital humanities] stood for was – understandably according to the colleagues I've spoken to – to recoil in confusion. It didn't represent anything I understood postcolonialism to be, or speak to the postcolonial world(s) I know. Critical Theory, yes. A powerful melange of identity theory and feminisms, queer theories and activist politics, but not "my" postcolonialism. Frankly, my first thought was "The US is colonising postcolonialism!" In the context of recent US invasions of the (actual geographic) postcolonial world this apparent intellectual appropriation was jarring, to say the least. Coupled with what less charitable people view as a global program of US cultural colonisation it felt at best wrong-headed and at worst an example of (unconscious) neo-colonialism. The baser side of me initially concluded it was simply the intellectual follow-on to [George W.] Bush's adventures [in Iraq]; an inevitable smothering of postcolonialism by the only remaining superpower. Domination by intellectual appropriation. Then I wondered if maybe I needed to catch up on my reading, that in my years away from academe postcolonial discourse had radically altered. Wikipedia and fleeting conversations with some random colleagues suggested this isn't the case, leaving me with the view that all I'm seeing is a provincial (I don't mean that in a pejorative way) US instantiation of the term that requires localization for New Zealand (and probably Australia and the rest of the postcolonial world).
>
> (2013)

Ouch. And yet, Smithies identified the crucial pitfalls of undertaking postcolonial studies—its overdetermination by scholars located within the US and its need for provincialization. Postcolonial digital humanities as articulated through a US frame cannot be neatly overlaid onto the digital humanities practices in any cultural context. Rather, it requires localization to the histories—and presents—of colonialism in the contexts in which it is invoked. In the US, we contend with the tensions between how postcolonialism—as an instance of critical theory—offers crucial insight on questions of representation, materialism, and globalization without recognizing that the US itself is a settler colony. Even critiques of US imperialism posed by US-based postcolonialists tend to look abroad—indeed, to Bush's adventures—while failing to look inward at how the universities that produce the conditions in which we theorize have benefited from Indigenous dispossession and genocide. Smithies' observations were thus crucial to my thinking and to what would become a central tenet of *New Digital Worlds*: the need for attention to local, not global, digital humanities practices.

Several years later, in 2015, Paul Barrett asked an important question: "Where is the nation in postcolonial digital humanities?" He noted that "One element of the postcolonial that seems absent from a postcolonial digital humanities approach, however, is the continued salience of the nation as an organizing structure and category of analysis." Barrett offered a series of probing questions for digital humanities practitioners to consider:

> How is the nation present, in however ghostly or marginal a form, within our digital humanities work? Does digital humanities work operate in a post-national space or is that ... just the latest form of nationalism? ... Do digital humanities and nationalism inherit shared notions of humanism and if so how does that humanism structure our work? Does digital humanities occur in some transnational space, speaking across borders through the power of the internet? Or is our work invisibly yet meaningfully indebted and structured by the very state institutions that fund it? Is digital humanities a form of American hegemony masquerading as transnationalism? (2014)

These questions were equally salient to my thinking as Smithies' critiques. Therefore, in the book, in addition to the importance of local digital humanities, I explicitly address the phenomenon where the "prominent" voices of digital humanities in the US and Europe, frequently cited scholars like Matthew Kirschenbaum, Susan Schreibman, Steven Ramsay, and Franco Moretti are the very ones whose work is taught on syllabi and cited in scholarship in the Global South (Risam 2018). These patterns of teaching and citation promote US and Western European hegemony and occlude the local and regional practices that are essential to undertaking digital humanities scholarship.

In the case of South Asia, digital humanities practices have emerged over several decades, with particular strengths in literary studies (Risam 2015; Risam and Gairola 2019). What is clear is that this digital humanities scholarship is far from derivative of the scholarship produced in the Global North. It has developed in response to particular linguistic and cultural needs and to institutional and labor contexts. Yet, the teaching and citation practices continually refer back to and frame their work *not* through these leaders but to scholarship of the Global North—including my own. This is not to suggest that I am promoting a new nationalism of digital humanities scholarship—and, in fact, there is value to transnationalism—but instead articulates the importance of validating scholarship *not* through the framework of the Global North but through the communities of practice that have developed in South Asia.

Choosing what scholarship we cite and teach is a way of signaling value. Thus, it's especially critical to recognize the value and worth of the work that's being done within regional contexts. In the context of South Asia, scholars like Sukanta Chaudhuri, P. P. Sneha, Padmini Ray Murray, Maya Dodd, and Nidhi Kalra, among others, have produced essential insights that localize the broader strokes of postcolonial digital humanities. Therefore, in citation practices, this work should get its due.

BEYOND POSTCOLONIAL DISCOURSE

Another area of deep concern is the focus on the discursive in postcolonial studies—and, in turn, in postcolonial digital humanities. When I brought my training in postcolonial studies to bear on my research in digital humanities, it was not without my own apprehensions of postcolonial theories and their limits. In fact, my experience with postcolonialism has always been one marked by ambivalence. On one hand, *New Digital Worlds* owes much to the language that postcolonial studies has offered for talking about colonialism and its legacies. On the other hand, I am deeply concerned about its tendency to privilege discourse over material circumstances. There are, of course, postcolonial theorists like Aijaz Ahmed and Benita Parry, whose work is grounded in materialism. Furthermore, even those whose work tends towards an emphasis on discourse base their theorization on material histories of colonialism.

However, the emphasis on the discursive risks superficial embrace of shallow notions of "decolonization" that overlook the nuances of colonialism. We see this, for example, in the US context, where "decolonize the curriculum" and "decolonize the university" have become rallying cries (Appleton 2019). In my own institutional context, I've been dismayed to see our "diversity" initiatives, such as the emphasis on adding readings by people of color and Indigenous people to syllabi, rebranded as "decolonization" (Risam 2022). This emphasis on the discursive elements of colonization has wrought a popularization of discourses of "decolonization" that elide the realities and responsibilities of those of us who reside within settler colonialism states. In the context of social justice and anti-racist movements in education, Eve Tuck and K. Wayne Yang's foundational article "Decolonization Is Not a Metaphor" pushes back against how "decolonization" has been co-opted in these ways in the US and Canada (Tuck and Yang 2012).

In this case, scholars in the US have latched onto curricular initiatives in the UK, which has a substantially different relationship to empire and Indigeneity. While the possibility of "decolonizing" through curricular reform in the UK is debatable, its adoption in the US does significant harm. Claiming to "decolonize" glosses over the need to examine and address how universities have benefited from Indigenous dispossession and genocide. In the context of formerly colonized countries, such as those in South Asia, an emphasis on decolonization risks promoting upper-caste voices and erasing Dalit and Indigenous voices.

While postcolonial studies as a body of thought has offered many wonderful insights on how we understand language and culture, what it has also, inadvertently, done is lead many in the Global North and South to believe that representation—and more specifically representation in the eyes of the Global North—is enough. Indeed, the worst trick that postcolonial studies ever played was convincing academics that "decolonizing" discourse was enough. We cannot let this be a limit of postcolonial digital humanities too.

DECENTERING THE POSTCOLONIAL ELITE

Therefore, within the context of South Asian studies, we have the responsibility to ensure that the complexities of culture and history exacerbated by colonialism, specifically nationality, caste, and Indigeneity, are rendered legible. Until the emergence of subaltern studies in the 1980s and 1990s, post-Independence historiography primarily focused on the role of the nationalist elite in the forging of a new nation. Scholars like Gyan Pandey, Ranajit Guha, and Gayatri Spivak emphasized the importance of "history from below," a rethinking of historiography that examined overlooked actors, such as rural peasants and lower-caste people. This advocacy for a shift away from dominant, upper-caste narratives to the subaltern is especially instructive for postcolonial digital humanities. When we consider the relative gap of digital scholarship on South Asia and seek to address it by producing new scholarship, it's crucial that what we produce does not simply reflect dominant voices and perspectives. That is to say, it's important that Pakistani, Nepali, Sri Lankan, Bangladeshi, Bhutanese, Afghan, and Maldivian sources as well as Dalits and Vimukta Jatis (denotified tribes or DNTs, India's Indigenous people) and other minoritized people within these countries become part of the digital cultural record—in their own voices and on their own terms.

Despite important work like Varsha Ayyar's pioneering work leading digitization of the Ambekar Papers (Ayyar 2020), for example, there is a noticeable absence of work related to Dalits and Vimukta Jatis in digital humanities in India. And this raises the question of the extent to which digital humanities practices in India really are challenging the reinstantiation of colonialism in the digital cultural record—and the extent to which they perpetuate it. Both of these things can be true simultaneously. At once, this body of work provides a challenge to the construction of digital scholarship of the Global North by asserting the importance of South Asian cultural heritage and making it visible. Yet, it also tends to focus on dominant narratives in India and specifically dominant narratives that render Dalits and Vimukta Jatis invisible in digital cultural heritage.

The first way to start thinking about this begins with examining digital humanities practices as they have arisen in South Asian contexts to foreground questions of epistemological power and epistemic violence. When working with materials or undertaking a project, the very first question must be: which epistemologies are being privileged, sanctioned, or valorized? Whose voices are likely to be authorized? Whose are disenfranchised—and how do you address that and how do you fix it? In the context of working with digital archives, the crucial questions are: how was the archive constructed? What does it contain? What isn't there? Why is it not there? And, in this context, to what extent does this work reinforce Dalit and Vimukta Jati oppression?

The question then becomes: what do we as practitioners have to do to avoid reproducing and amplifying these forms of archival violence? Is it feasible, for example, to offer a corrective? And, in the case of perspectives that may be unrecoverable—because we know that archival erasures can be permanent—how are practitioners addressing that absence?

What I'm arguing for here is the very change in method that I was advocating for in *New Digital Worlds*. When we translate the insights of postcolonial digital humanities to a new cultural context, we need to ensure that we aren't unintentionally reproducing a new pattern of violence and disenfranchisement for the marginalized communities in that location. If we are going to fully realize the possibilities of postcolonial digital humanities, then methods are going to need to change—and who gets to determine the methods needs to change. For example, in an Indian context, to supplement a project with additional sources or resources that give voice to Dalit and Vimukta Jati experiences doesn't mean representing *them* or speaking for them but providing Dalits and Vimukta Jatis with the means—financial support and access to the means of the production of knowledge for self-representation on their own terms—rather than extracting their knowledge and expertise for the upper-caste researcher's benefit.

One of the most dangerous dimensions of digital humanities is, in fact, the funding it attracts—because funding comes with strings attached and funding comes with possibilities of notoriety and recognition. In higher education contexts around the world, funding and notoriety accrue to people who are already in relative positions of privilege and power—giving them more, giving them access to academic prestige, nationally and internationally. It's incredibly seductive—but it's a trap to ensure that those who find themselves in those enviable positions keep doing what they're doing, keep upholding the status quo, keep upholding oppression, and keep reproducing the conditions of academic knowledge production—rather than using the position to try and change the power dynamics.

This points us towards the importance of collaboration. Often, academic work is constructed as working "on" a topic and, worse, "on" a particular group of people. Relationships between project developers with the money and the power and people whose stories and histories they may be working *on* are encumbered by power dynamics. This puts the onus on project directors to facilitate ethical collaboration—to build relationships with those affected by a project's concerns, to avoid equating having access to material resources to having control over the project itself. And this is particularly critical when collaborating with vulnerable communities. Furthermore, it places the burden on those with the ability to command resources to avoid positioning themselves as "saviors" rather than ethical collaborators.

I believe fully in the ideas I advanced in *New Digital Worlds*. However, it's essential that these ideas do not simply become a mechanism through which communities that have been marginalized in the global landscape of digital humanities perpetuate a similar dynamic towards those who are subaltern in their cultural context. Doing so requires attending to the localization of postcolonial thought as well as taking seriously the national contexts in which we work. Only then can the full possibilities of postcolonial digital humanities be realized.

REFERENCES

Appleton, Nayantara Sheoran. 2019. "Do Not 'Decolonize'... If You Are Not Decolonizing: Progressive Language and Planning beyond a Hollow Academic Rebranding." *Critical Ethnic Studies*, February 4. http://www.criticalethnicstudiesjournal.org/blog/2019/1/21/do-not-decolonize-if-you-are-not-decolonizing-alternate-language-to-navigate-desires-for-progressive-academia-6y5sg.

Ayyar, Varsha. 2020."Digital Archives in South Asian Studies: Towards Decolonisation." Presentation, Oxford University, October 23. https://www.qeh.ox.ac.uk/events/digital-archives-south-asian-studies-towards-decolonisation.

Barrett, Paul. 2014. "Where Is the Nation in Postcolonial Digital Humanities?" *Postcolonial Digital Humanities*, January 1. https://dhpoco.org/blog/2014/01/20/where-is-the-nation-in-postcolonial-digital-humanities/.

Giles, Paul. 2011. "Postcolonial Mainstream." *American Literary History* 23 (1) (Spring): 205–16.

Huggan, Graham. 2002. "Postcolonial studies and the Anxiety of Interdisciplinarity." *Postcolonial Studies* 5 (3): 245–75.

Ray, Sangeeta. 2005. "Postscript: Popular Perceptions of Postcolonial Studies after 9/11." In *A Companion to Postcolonial Studies*, edited by Henry Schwarz and Sangeeta Ray, 574–83. Malden, MA: Blackwell.

Risam, Roopika. 2015. "South Asian Digital Humanities: An Overview." *South Asian Review* 36 (3): 161–75.

Risam, Roopika. 2018. *New Digital Worlds: Postcolonial Digital Humanities in Theory, Praxis, and Pedagogy*. Evanston, IL: Northwestern University Press.

Risam, Roopika. 2022. "Indigenizing Decolonial Media Theory: The Extractive and Redistributive Currencies of Media Activism." *Feminist Media Histories* 8 (1): 134–64.

Risam, Roopika and Rahul K. Gairola. 2019. "South Asian Digital Humanities: Then and Now." *South Asian Review* 40 (3): 141–54.

Slemon, Stephen. 1994. "The Scramble for Postcolonialism." In *De-scribing Empire*, edited by Alan Lawson and Chris Tiffin, 15–32. New York, NY: Routledge.

Smithies, James. 2013. "Speaking Back to America: Localizing DH Postcolonialism." *James Smithies*, September 4. https://jamessmithies.org/blog/2013/09/04/speaking-back-to-america-localizing-dh-postcolonialism/.

Sunder Rajan, Rajeswari. 1997. "The Third World Academic in Other Places; or, the Postcolonial Intellectual Revisited." *Critical Inquiry* 23 (3): 586–616.

Tuck, Eve and K. Wayne Yang. 2012. "Decolonization Is Not a Metaphor." *Decolonization: Indigeneity, Education, and Society* 1 (1): 1–40.

Williams, Jeffrey J. 2001. "Name Recognition." *Minnesota Review* 52–54 (Fall): 185–208.

CHAPTER FIVE

Race, Otherness, and the Digital Humanities

RAHUL K. GAIROLA (MURDOCH UNIVERSITY)

INTRODUCTION: REALIZING RACE IN THE DIGITAL MILIEU

I open my contribution by underscoring the urgent assertion that it is impossible to conceive of "race" as a finite, even overdetermined, taxonomy of human social experience just as we concede that there is not, and cannot be, a single or fixed definition of "the digital humanities" (Nyhan, Terras, and Vanhoutte 2013, 7). I make this opening claim as "critical race theory" is a socio-political flashpoint across the USA as that country abandoned Afghanistan while forging a defense alliance with the UK and Australia (AUKUS). In a related vein, the latter has witnessed an explosion of violent Indophobic and anti-Asian violence over the last two decades. I begin with such observations since the digital milieu too has a transnational context, and one that is, moreover, linked to situated, fiscal interests.

This is perhaps more the case in the digital milieu than in the traditional arts, humanities, and social sciences as the former have had much more time to crystallize into the global, human psyche through both overt and surreptitious networks of domination. It thus behooves dominant sites of the digital humanities (DH) to experience, in Dorothy Kim and Jesse Stommel's words, a "disruption" wherein:

> … feminist critical race theory, black, indigenous, and women of color (BIWOC) bodies disrupt the narratives of mainstream white feminism by having voices, by creating counternarratives, by calling out the frameworks of the hegemonic center. Thus, we take … the productive "disruption" in the same vein, to decenter the digital humanities.
>
> (2018, 22)

Many digital humanists deeply invested in multivocal social justice platforms and projects recognize ambiguity in the term "digital humanities" despite its demonstrated, familiar orientations towards privileging canonical data. Such data, as usual, has been produced by scholars whose lived privileges re-map the exceptionalism of institutional knowledge (straight, white, male, bourgeois, non-disabled, neurotypical). Indeed, in the words of Deepika Bahri, we may interpret this unsurprising characteristic of DH as echoing technology of conquest (Bahri 2017, 97). This ambiguity has been compellingly commented upon in the context of race, gender, and trends of domination by several respected scholars, including, among others, Lisa Nakamura (2002), Amy

Earhart (2013), Roopika Risam (2015), Jacqueline Wernimont and Elizabeth Losh (2016), Dorothy Kim (2019), and Coppélie Cocq (2021), among others.

Yet while I acknowledge DH's ambiguity and ongoing controversial boundaries, I wish to initially frame this contribution as a critique of hegemonic history and its corresponding power relations which have molded the development of a particular metanarrative of DH—a genesis narrative that revolves around, no doubt at times fetishizes, Father Roberto Busa and his lemmatization of the complete works of Saint Thomas Aquinas (Gairola 2019, 451). This project, which recruited and deployed the technology of IBM and stretched over three decades, yielded the fifty-six-volume *Index Thomisticus* concordance of ten million words initially indexed on punch cards (Winter 1999, 5–6). As documented by Edwin Black (2001), IBM used the punch card technology accorded to Busa to exterminate Jews during the Holocaust, thus undergirding in Busa's concordance a sinister and blood-soaked history of terminating others.

That Busa's project is a major accomplishment and valuable contribution to world literature is understandable, as is the glaring fact that this feat has become representative of DH's origins in an imbalanced, obfuscating, and hegemonic manner that silences alternate histories by citing, replicating, and institutionalizing itself. To put it another way, the technological innovation that DH fruitfully lends to realms of literary and cultural studies reproduces ideals of race, gender, sexuality, language, nationality, and indigeneity. These color knowledge dissemination through teaching and research (and thus reproduction of such knowledge), the acceptable methods used for conducting these scholarly duties, and those who are excluded/included from the privileged forums and institutions which host and promote DH research in the first place. Globally institutionalized in a vertical, color-coded history that continues to this day to perpetuate socio-political violence from colonialist ventures of yesteryear, I would argue that DH's genesis metanarrative perpetuates, whether intentionally or not, epistemic violence that aptly characterizes the educational agendas of empire.

In another context, I have argued that this omission of other histories of the development of digital humanities in non-Western and earlier contexts results in what we may call "digital necropolitics" in the technological milieu as a means for underscoring the very material effects of the digital milieu on non-white bodies (Gairola 2019, 458). While I will return to this notion in my conclusion, I begin by noting that my understanding of "race" in relation to "DH" is defined by white privilege buttressed by the most murderous modalities of violence, both overt and casual, that continue to place people of color and our indigenous systems of knowledge in the crosshairs of what Gayatri Chakravorty Spivak has famously identified as "epistemic violence" that often emerges as a form of silencing "the other" in colonial contexts (1988).

We must return to this and other historical effects of weaponized technologies on those who do not, cannot, benefit from white privilege. This is because the process arguably occurs on a smaller scale throughout the field of DH itself. As articulated by Theresa L. Petray and Rowan Collin in the context of white supremacist Australia, Twitter users there deployed the hashtag #WhiteProverbs to satirize Australian justifications for bigotry against indigenous people (2017). This example offers an important example of resistance to institutional racism in a postcolonial Commonwealth nation infamous for its casual racism, as in "sledging" of Indigenous Australian athletes (Philpott 2016, 866).

Given empire's trademark ruination in the lands that it discards after stripping wealth, natural resources, and labor from usually black and brown workers, it behooves us to resist the reproduction of epistemic violence in DH discourse. I thus hereon offer contexts that offer a panoramic rendering of race and otherness in DH in an era of global pandemic punctuated by the humanitarian catastrophe resulting from the sudden exodus of the US from Afghanistan. Acknowledging such enables a pathway for resistance to hegemonic, whitewashed metanarratives. In what follows, I render an admittedly limited genealogy of race and otherness in DH that focuses on skin color as a necessary exercise in surveying the many ways in which the genesis narrative of Father Busa has been unsettled for over two decades but nonetheless persists.

I subsequently revisit race and sexuality to demonstrate how and why the very definition of "race" in the context of DH must be intersectional—an axiom of identity formation in the digital milieu that has been widely written about within this past decade. Formulated by Kimberlé Williams Crenshaw (1990) to substantially grasp the aggregate impact of related bigotries on minority identities rooted in race, gender, sexuality, class, caste, ableism, etc., I anchor my critical genealogy and push towards intersectional DH to the material context of race and artificial intelligence (AI) in the digital milieu. In concluding, I propose queer South Asian DH as a viable site for activating a DH undergirded by and committed to social justice.

A SUMMATIVE BRIEF ON RACE AND DH

In this section, I engage a literature review of key texts that is by no means exhaustive, and have used robust direct quotation, as opposed to my own words, to allow these scholars to speak for themselves while rendering many vital ways in which race and otherness emerge in digital humanities discourse. Exploring the convergence of the persistence of race with digital communications technology in 2000, the year of the supposed Y2K digital apocalypse, Jerry Kang writes:

> Cyberspace enables new forms of social interaction. How might these new communicative forms affect racial mechanics? These are not idle academic questions ... race and racism are already in cyberspace. As computing communication technologies advance to engage more senses, those bodily characteristics that signal a person's race will increasingly appear in cyberspace.
>
> (Kang 2000, 1135)

Lisa Nakamura also constitutes some of the earliest published research on the convergence of race and digital identity. In an interview with Geert Lovink titled "Talking Race and Cyberspace," she clearly outlines the socio-political stakes for race and DH scholarship that would follow:

> Certainly the Net is as racist as the societies that it stems from ... The Internet hails its audiences in the same ways that texts have intended readers, that films and television shows have intended audiences, and that made environments are intended for particular users. And in its earlier stages the Internet was not hailing people of color; it assumed a normative white user and in fact often still does.
>
> (Lovink 2005, 60–1)

Nakamura's observation is a solid platform from which to dive into a genealogy of race and the digital humanities precisely because it draws upon foundational tenets of cultural studies scholarship as, for example, Stuart Hall's work. In his seminal essay "Encoding/Decoding," first published in 1973, he critically engages constructions of codes in visual discourses (1993, 95) to sharpen its socio-political thrust. Hall's timely intervention in the wake of global civil rights and decolonial movements, and later intensified, racialized technologies of surveillance before and after September 11 is further expounded upon by Nakamura in the distinction between "cyber race" and one's "real" race.

For instance, she writes that in the context of digitally mediated identity tourism,

> [r]ace in virtual space was "on you" in more than one sense: when users "wore" race in the form of a racialized avatar or performed it as racialized speech or conveyed it by sharing their "performance of tastes (favorite music, books, film, etc.)," or "taste fabric," this form of display was viewed as a personal decision, an exercise of individual choice.
>
> (Nakamura 2008a, 1675)

In linked research, Nakamura (2008b) examines the ways in which race ambivalently (Bhabha 1994) surfaces as both a celebratory representation and white supremacist target across the Internet, positing her ongoing work with race and DH as both foundational and visionary.

In the September 2012 season premiere of #LeftOfBlack, a weekly video podcast based at Duke University, Mark Anthony Neal interviewed two scholars critiquing the digital humanities through the interposed lenses of African American and technology studies. Featuring Jessica Marie Johnson and Howard Rambsy II, this pioneering episode of #LeftOfBlack is particularly illuminating as it extends the historical roots of racial formations and race projects in the US outlined by Michael Omi and Howard Winant (1994) from the 1960s to the 1980s. As Johnson articulates in the episode:

> There are ways that black studies, that black history, that scholars that are interested in the black experience, life, and culture, can engage with technology and can use it in different ways – so they can use it in a very utilitarian way, use tools and what digital humanities people kind of call 'hacks' to make their lives easier.
>
> (Neal 2012)

Moreover, Johnson espouses the need of "moving beyond normative ideas of who is a digital humanities scholar, which has sort have been imagined as a white male academic":

> And it's really not – there are ways that we, as people of color, and people of all kinds of identities, are very, very engaged with what is happening with technology right now but are not necessarily having that conversation in the digital humanities space.
>
> (Neal 2012)

This notion of alternative spaces for critically examining the digital milieu beyond the limited repertoire of recycled male whiteness, moreover, emerges from the work of Radhika Gajjala, which sutures digital cultural studies, postcolonial studies, and gender studies through subaltern studies. In *Cyberculture and the Subaltern: Weavings of the Virtual and Real* (2013), Gajjala argues for the "technocultural agency" of third world subjects working with microfinance organizations. In unflinchingly tracing the Internet back to the hegemonic power structures in which it was forged, largely through metanarratives of Western modernity counterposed against black and brown inferiority, she writes:

> Close textual and visual analysis of [the] Kiva interface shows instances where the Web 2.0 tools in fact reinscribe and reposition the Other with a spatial/territorial conception of the Internet.

> This territorial conception is in turn inscribed within a larger power structure that privileges particular literacies and cultural understandings … Territorial metaphors of the Internet have powerful ideological implications in that, like earlier colonial imagery employed to describe territorial incursion, they code social control …
>
> (51)

In the frame of Gajjala's recognition of territorialization inscribed within the Internet, we cannot ignore the realities of race in cyberspace as if they vanish when our material bodies are elsewhere. Indeed, I would hereon propose that the very notion of invisibility in both material and virtual communities is the digital extension of claims of colorblindness and/or celebratory multiculturalism. As I have elsewhere argued, the colonialist logic of "home" as an exclusionary space surfaces in visual media including television and film (Gairola 2016, 22). I would here further add that this horizon of exclusionary homes stretches into the digital milieu, into what we could think of as queer digital homes for and of the Indiaspora.

This parlays into audiovisual technology, for example, Indian Americans' belonging is intertwined in technologically mundane components of Internet life. For Madhavi Mallapragada, this appears in the simple homepage: "the concept of the homepage continues to be the key locus for negotiating the changing spaces of the public and private. The homepage continues to be emotionally resonant, as well as practically important, triggering associations of a home space that is familiar and therefore comforting" (2014, 5). Given Gajjala's recognition that the Internet poaches territorial metaphors from colonial imagery, an intersectional analysis of race and otherness in DH must more deeply reckon with the notion of "home" throughout.

Because both exploitative and settler colonialism are bound up with varying notions of race, difference, and "home," I would forward that it is impossible to formulate an ethical social articulation of the digital humanities without interrogating the normalized, casual white supremacy diffused throughout it. This is perhaps why Tara McPherson concludes in her 2012 piece titled "Why are the Digital Humanities So White?" that

> We must remember that computers are themselves encoders of culture. If, in the 1960s and 1970s, UNIX hardwired an emerging system of covert racism into our mainframes and our minds, then computation responds to culture as much as it controls it. Code and race are deeply intertwined, even as the structures of code labor to disavow these very connections.
>
> (144–145)

The insights offered by McPherson, like those by Gajjala a year later, elucidate that the very category of "race," like "the digital humanities," is ambivalent when humans interface with technology with a false consciousness produced in the digital commons exempt from racism, xenophobia, and other forms of bigotry (Ossewaarde and Reijers 2017). However, as Safiya Umoja Noble's persistent research unearths, racialized bias is detectable in the ostensibly objective search engine Google, which turns up disturbing hits when searching the term "black girls." Expounding on her findings, Noble writes:

> A Black Feminist and Critical Race Theory perspective offers an opportunity to ask questions about the quality and content of racial hierarchies and stereotyping that appear in results from commercial search engines like Google; it contextualizes them by decentering Whiteness and maleness as the lens through which results about black women and girls are interpreted.
>
> (2013, 12)

I would insert an important caveat in Noble's keen observation: it does not suggest that "transformative agents" against heteronormative maleness (including feminists, gay men, queer men, transgender women, etc.) cannot and do not default to racism as a means of self-interested inclusion through the exclusion of others. For example, in examining how self-fashioned "damsels in distress" deploy white tears against black and brown women, Ruby Hamad astutely notes:

> White women are not like other women not because their biology makes them so, but because white supremacy has decided they are not ... It all leads back to the same place, and that place is the rift that European colonialism deliberately created between women, making white women complicit in the racism that they have since been all too eager to blame solely on white men.
>
> (2019, 138)

For Hamad, this is traumatically experienced in the case of gaslighting and performative wokeness staged by white women in settler colonial/Commonwealth nations in material life. I would extend Hamad's scholarship and insist that it compels us to acknowledge that the digital milieu imitates material life. For example, Moya Bailey elucidates social justice media mandates incumbent on DH at the intersection of blackness and queerness in her theorization of "digital alchemy":

> I build towards an understanding of what I call digital alchemy as health praxis designed to create better representations for those most marginalized in the Western biomedical industrial complex through the implementation of networks of care beyond its boundaries ... Digital alchemy shifts our collective attention from biomedical interventions to the redefinition of representations that provide another way of viewing Black queer and trans women.
>
> (2015)

Here, Bailey stretches race into otherness by including queerness in the identitarian limits of the digital milieu. We must thus hereon proceed with a conception of race, otherness, and DH as a necessarily intersectional relationship in which race in the digital milieu cannot be meaningfully, or even accurately, surveyed without its constitutive components including class, gender, sexuality, nationality, ability, etc. In the words of Safiya Umoja Noble and Brendesha M. Tynes:

> Future research using critical, intersectional frameworks can surface counter-narratives in fields engaging questions about information and technology and marginalized or oppressed groups ... By looking at both the broader political and economic context and the many digital technology acculturation processes as they are differentiated intersectionally, a clearer picture emerges of how under-acknowledging culturally situated and gendered information technologies are affecting the possibilities of participation with (or purposeful abstinence from) the Internet.
>
> (2016, 4)

The more recent volume *Bodies of Information: Intersectional Feminism and Digital Humanities* further maps the timely and urgent need for evaluating DH through an intersectional lens that transposes race, gender, and sexuality (Losh and Wernimont, 2019). This hermeneutical labor empowers us to witness that even in the digital milieu, overlapping identities are yet targets of scorn and violence by online actors and must, moreover, be historically contextualized.

The past decade into the era of Covid-19 has been extremely generative for intersectional scholarship that refuses to unlink race from the digital milieu, and which meaningfully places DH in historical context. For example, Roopika Risam interrogates the logic of "disrupting" DH by asserting:

> The lives of colonial subjects and people of the African diaspora have historically been viewed as disruptive to dominant cultures that preserve a white status quo, as have their languages, histories, and cultural heritage ... invoking disruption as a position for critiques of digital humanities fails to take into account the complicity of universities, libraries, and the cultural heritage sector in devaluing black and indigenous lives and perpetuating the legacies of colonialism.
>
> (2019, 14)

Risam has also co-edited two important volumes that center blackness in DH as well as the need for an intersectional lens in thinking about the field: *Intersectionality in Digital Humanities*, co-edited with Barbara Bordalejo (2019), and *The Digital Black Atlantic*, co-edited with Kelly Baker Josephs (2021). These texts complement many other important studies in race and DH by other practitioners who are working on issues of migration, diaspora, and home as represented in the digital milieu. And, as per the urgency for intersectional analysis of the digital milieu, they compel us to consider queerness as an intrinsic component of otherness.

In their published study "Toward a Queer Digital Humanities," Bonnie Ruberg, Jason Boyd, and Jamie Howe ask, "Where is the queerness in digital humanities?" (2018).[1] They write: "From its origins as a pejorative, it [the word 'queer'] has been reclaimed in recent decades by academic and popular communities alike. At its most basic [level], 'queer' operates as an umbrella term: a marker of identity differentiated from 'gay' or 'LGBT' in that it encompasses all non-normative expressions of sexuality or gender (Grace, Hill, Johnson, and Lewis). Not every person whose identities fall within this category identifies as queer, however, and 'queer' itself is a contested term" (108).

The co-authors further espouse: "We believe that queerness can function as a force to destabilize and restructure the way that DH scholarship is done across these fields. The vision of a queer digital humanities that we propose is at once conceptual and pragmatic" (109). I would thus observe that this initiative of Ruberg et al. is a welcome trajectory in the intersectionality of social justice around race and otherness within DH. Moreover, it compels us to critically survey the relationship between "fake news" disseminated across social media outlets, on the one hand, and discourses of the closet and passing in the digital milieu.

BROWN SKIN, GAY MASKS: PREDICTIVE TECHNOLOGY, COLONIAL HISTORY, AND QUEER SEXUALITY TODAY[2]

Ruberg, Boyd, and Howe's query, moreover, aggregates a variety of fields in the digital humanities including the definition and variation of "data," the ways in which mapping technologies place bodies in space, and the ways in which artificial intelligence has, in the past few years, made specious claims to have been able to track sexuality in one's face.[3] The latter is documented in Yilun Wang and Michael Kosinski's notorious study titled "Deep Neural Networks Can Detect Sexual Orientation From Faces." Conducted under the auspices of Stanford University's School of Business, the authors write: "We show that faces contain much more information about sexual orientation than can be perceived and interpreted by the human brain. We used deep neural networks to extract features from 35,326 facial images. These features were entered into a logistic regression aimed at classifying sexual orientation" (Wang and Kosinski 2017, 2).

In their study, Wang and Kosinski opine that facial features including nose, shape, hair style, facial hair can indicate one's gender and sexuality. In somewhat startling language, they write that "gay men and women tended to have gender-atypical facial morphology, expression, and grooming styles ... Those findings advance our understanding of the origins of sexual orientation and the limits of human perception" (2017, 2). This study was highly controversial, panned by LGBTQI+ rights groups GLAAD and the Human Rights Campaign (Miller 2018, 188), and resulted in death threats against the pair (Murphy 2017). The threat of physical violence directed at Wang and Kosinski underscores the visceral upset felt by members of the global LGBTQI+ community over the violence that they would themselves face if being outed by such an app.

While it is indeed problematic to default to physiognomy to "read" sexual inclinations on a human face, I would further note that such an initiative harkens back to the logic and praxis of eugenics that attempted to justify Western colonialism and the centuries-long domination of lighter-skinned peoples over their darker-skinned sisters and brothers. In other words, Wang and Kosinski's claim to have developed an AI app that detects "gayface" based on 35,000 faces of gay, white Americans (Murphy 2017) suggests that the very archive used by both researchers ostensibly elides the historical specificity of racialized power relations from the gayface recognition app. Thus, the comparative logic of Wang and Kosinski's "composite heterosexual faces" and "composite gay faces" is flawed sans a robust reckoning with both how a machine reads—or does not read—identity on human faces.

Therefore, an intersectional modality of doing digital humanities is vital in these times of anti-Asian racism during a global pandemic stretched into three years of social distancing that has made Internet connectivity more necessary than ever. Moreover, the histories of power and domination that catalyzed, informed, and thus shaped the violent vectors of territorial and hegemonic colonialisms have critical import here. Indeed, as postcolonial scholars ranging from Edward Said (1978) to Gayatri Chakravorty Spivak (1988) to Ann Laura Stoler (2002) have powerfully argued, gender and sexuality are inextricably mired in the ostensible racial distinctions between white and non-white subjects. Thus, my larger, socio-political interest in examining the intersections of race and otherness in the digital humanities is to revisit Wang and Kosinski's 2017 study through the lens of racialized colonialism.

Given that the thousands of aggregated images anchoring the study are both nationally situated (American) and racially situated (white subjects), follow-up questions arise: what variations in "gayface" surface when dealing with non-white subjects? How does skin color produce variations in AI's supposed rate of accuracy? Finally, how can we contribute to and expand Ruberg et al.'s urgent call for a queer digital humanities while simultaneously weaving into it an intersectional methodology that considers the material histories and ideological residue of hegemony, domination, and colonialism in the twenty-first century? These questions are timely in the age of the Covid-19 pandemic and the global anti-fascist and anti-racist movements in which face recognition technology is being deployed by security forces to quell social justice protests from Hong Kong to the USA.

In contrast to specious claims that digital humanities must be "objective" and has no business being "political," I operate from the premise that DH is always already political—in the words of Professor Evelyn Fogwe Chibaka,[4] it must be, and as such must include the material realities of oppressed subjects who survive against all odds in the afterlife of empire and the normalized bigotry of technological innovations. This notion is thus based on a global social justice vision that theoretically reads so-called "gayface" through a critique of AI when interfacing with people of color. It must examine how skin color produces variations in AI's supposed rate of accuracy while simultaneously using an intersectional methodology that accounts for the material histories and residual ideologies of cultural hegemony, military domination, and racialized nationalism in the twenty-first century.

Specifically, I would critically return to Yilun Wang and Michael Kosinski's claim that "We show that faces contain much more information about sexual orientation than can be perceived and interpreted by the human brain ... These features were entered into a logistic regression aimed at classifying sexual orientation" (Wang and Kosinski 2017, 2). We may here expand questions posed by Ruberg, Boyd, and Howe, namely "Where is the queerness in digital humanities?" (2018, 108). The queer DH initiative of Ruberg et al. is a welcome trajectory in critically evaluating the intersectionality of social justice with digital humanities work of the twenty-first century. That is, the white mask uniformly applied to brown and black faces is problematic in obvious ways that invoke Safiya Umoja Noble's conceptualization of "algorithms of oppression" (2018) that encode racial bias into the ostensibly objective rationale of the computer.

Moreover, as history has taught us again and again in the contexts of slavery, lynching, social Darwinism, eugenics, Jim Crow, anti-miscegenation laws, etc., race and sexuality have been intimately bound for centuries. It should thus perhaps come as no surprise that twenty-first-century surveillance technology draws upon the very discriminatory ideologies that historically sparked the Industrial Revolution to police brutality in twenty-first-century Black Lives Matter demonstrations, and now, amidst the global coronavirus pandemic. Therefore, the work of many race scholars in the realm of the digital humanities is vital in rethinking interfaces of otherness with disciplinary technologies (Foucault 1977). For instance, Ruha Benjamin draws upon the widely influential work of legal scholar Michelle Alexander, who describes the racialized predispositions of the prison industrial complex and mass incarceration.

These tacitly institutionalize what Alexander describes as a "new Jim Crow" wherein people of color are disproportionately branded "criminals" or "felons" in legal terms. This then activates a universe of disenfranchisements and normalized discriminatory practices in voting rights, employment, housing, education, public entitlements, and even social relationships with neighbors (Alexander 2010, 1–2). Benjamin draws on Alexander to formulate what she calls "the new Jim Code" wherein she argues that "Codes are both reflective and predictive. They have a past and a future" (2019, 3). Benjamin further observes that coding is deeply programmatic since it engineers "race as a form of technology" into the very design of software. And software, we could argue, is the ideological matrix inscribed upon hardware, which is akin to territory, and thus nation.

As we know from previous studies (Renan 1990; Balibar and Wallerstein 2010), race and nation are often confused. This misrecognition, I would argue, translates into the digital milieu in the form of assertions that we are liberated from racism, queerphobia, and other bigotries in cyberspace. My analysis of race and otherness in DH through an intersectional lens that accounts for colonialist histories and decolonizing praxis empowers us to unearth how machines deploy face recognition technology to "see" queer sexuality from subjects' faces while simultaneously using an algorithm that was modeled on gay, white men. More questions arise: 1) Is gayness but an erotic privilege afforded to subjects who benefit from social privilege through gender and race in the digital milieu?; 2) What is the link between queer sexuality and transgressive raciality when "read" through the algorithmic lens of a machine?; 3) And, finally, what are the alarming and/or liberating implications and possibilities in the new normal of social distancing and its attendant dependence on technology to serve as a proxy firewall between human beings and the Covid-19 virus?

CONCLUSION: RECENTERING INTERSECTIONAL MARGINS OF DH

Perhaps then an apt topic to close on when critically meditating on race and otherness in the digital humanities is the growing work on DH in South Asia committed to exploring intersections of class, gender, and sexuality. This epistemic trajectory promises to yield "a unique corpus of knowledge situated at the rich and fertile confluences of South Asian studies and digital humanities … [that] will facilitate transnational dialogue on the futures of digital humanities in the geopolitical frame of South Asia" (Risam and Gairola 2021, 3). This is a powerful place from which to embark given the many historical contradictions that herein coalesce and the ironic tensions they generate: histories of colonialism and racialized violence brushing against the stereotypes and global success of Indian tech workers, for example.

In *Digital Queer Cultures of India*, Rohit K. Dasgupta observes, that "… digital culture is a burgeoning social phenomenon in India, and the last decade has seen a tremendous growth of Internet and mobile technologies … queer culture in India … is an intersection of online and offline practices" (Dasgupta 2017, 3). Intersectional digital humanities here finds fertile ground given India's status as Britain's largest and most lucrative former colony, archaic colonial-era statutes that continue to shape public perceptions of sexuality, and the rise of Digital India and its global tech workers. Indeed, in what I have earlier called "digital closets" (2018, 67) in how mobile phones allow us to mask queer identities and trysts, the possibility of digital necropolitics also looms large (2019).

That is, public visibility and mobile phone use both expose yet shield queers operating in Digital India with slut-shaming (Shah 2015) or even exposure to violence and death. Here, race and otherness are always present because skin color acts as a shield just as passing does. Thus, an intersectional perspective of race and otherness allows us to enfold diversities of difference while recognizing that "race" is a dynamic composite of many visual and non-visual differences. The real question, then, and perhaps the one with the most socio-political import is: How best can those who do not suffer the technological traumas of those of us who do triangulate their online and offline presences to the service of those ensnared in the digital milieu's web of casual racism and toxic xenophobia?

NOTES

1. Jason Boyd and Bonnie Ruberg have written further on this topic in this volume, in their chapter "Queer Digital Humanities."
2. The title of this section reformulates the title of Franz Fanon's well-known anti-colonial study, *Black Skin, White Masks* (1967). In this foundation text of postcolonial studies, Fanon scathingly critiques, through a psychoanalytic lens, the hegemony and violence of French colonization of Martinique.

3. These observations emerge from the workshops taught by Carol Chiodo and Lauren Tilton ("Hands on Humanities"), Randa El Khatib and David Wrisley ("Humanities Data and Mapping Environments"), and Fabian Offert ("Images of Image Machines") at the European Summer University in the Digital Humanities at Leipzig University in 2017, 2018, and 2019.
4. Here, I refer to Chibaka's keynote talk at ESUDH 2019, Leipzig University, August 2019.

REFERENCES

Alexander, Michelle. 2010. *The New Jim Crow: Mass Incarceration in the Age of Colorblindness*. New York, NY: The New Press.

Bahri, Deepika. 2017. *Postcolonial Biology: Psyche and Flesh after Empire*. Minneapolis, MN: University of Minnesota Press.

Bailey, Moya. 2015. "#transform(ing) DH Writing and Research: An Autoethnography of Digital Humanities and Feminist Ethics." *DHQ: Digital Humanities Quarterly* 9.2 (online). http://www.digitalhumanities.org/dhq/vol/9/2/000209/000209.html.

Balibar, Etienne and Emmanuel Wallerstein. 2010. *Race, Nation, Class: Ambiguous Identities*. London: Verso.

Benjamin, Ruha. 2019. *Race after Technology: Abolitionist Tools for the New Jim Code*. Cambridge: Polity Press.

Bhabha, Homi K. 1994. *The Location of Culture*. London: Routledge.

Black, Edwin. 2001. *IBM and the Holocaust: The Strategic Alliance between Nazi Germany and America's Most Powerful Corporation*. Washington, DC: Dialog Press.

Bordalejo, Barbara and Roopika Risam. 2019. *Intersectionality in Digital Humanities*. York, UK: ARC Humanities Press.

Cocq, Coppélie. 2021. "Revisiting the Digital Humanities through the Lens of Indigenous Studies—or how to Question the Cultural Blindness of Our Technologies and Practices." *Journal of the Association for Information Science and Technology* 73 (2): 333–44.

Dasgupta, Rohit K. 2017. *Digital Queer Cultures in India: Politics, Intimacies and Belonging*. India: Routledge.

Crenshaw, Kimberlé. 1990. "Mapping the Margins: Intersectionality, Identity Politics, and Violence Against Women of Color." *Stanford Law Review* 43: 1241–99.

Earhart, Amy. 2013. "Can Information Be Unfettered? Race and the New Digital Humanities Canon." In Matthew K. Gold (ed.), *Debates in the Digital Humanities*. Minneapolis, MN: University of Minnesota Press. https://dhdebates.gc.cuny.edu/read/untitled-88c11800-9446-469b-a3be-3fdb36bfbd1e/section/cf0af04d-73e3-4738-98d9-74c1ae3534e5.

Fanon, Frantz. 1967. *Black Skin, White Masks*. New York, NY: Grove Press.

Foucault, Michel. 1977. *Discipline and Punish: The Birth of the Prison*. Translated by Alan Sheridan. New York, NY: Pantheon.

Gairola, Rahul K. 2016. *Homelandings: Postcolonial Diasporas and Transatlantic Belonging*. London: Rowman & Littlefield.

Gairola, Rahul K. 2018. "Digital Closets: Post-millennial Representations of Queerness in *Kapoor & Sons* and *Aligarh*." In Rohit K. Dasgupta (ed.), *Queering Digital India: Activisms, Identities, Subjectivities*, 54–71. Edinburgh: Edinburgh University Press.

Gairola, Rahul K. 2019. "Tools to Dismantle the Master's DH(ouse): Towards a Genealogy of Partition, Digital Necropolitics and Bollywood." *Postcolonial Studies* 22 (4): 446–68.

Gajjala, Radhika. 2013. *Cyberculture and the Subaltern: Weavings of the Virtual and Real*. Washington, DC: Rowman & Littlefield.

Hall, Stuart. 1993. "Encoding/Decoding." In *The Cultural Studies Reader*, edited by Simon During, 90–103. New York, NY: Routledge.

Hamad, Ruby. 2019. *White Tears, Brown Scars*. Melbourne: Melbourne University Press.

Kang, Jerry. 2000. "Cyber-Race." *Harvard Law Review* 113 (5) (March): 1130–209.

Kim, Dorothy. 2019. *Digital Whiteness & Medieval Studies*. York: ARC Humanities Press.
Kim, Dorothy and Jesse Stommel. 2018. "Disrupting the Digital Humanities: An Introduction." In *Disrupting the Digital Humanities*, 19–33. Goleta, CA: Punctum Books.
Losh, Elizabeth and Jacqueline Wernimont, eds. 2019. *Bodies of Information: Intersectional Feminism and Digital Humanities*. Minneapolis, MN: University of Minnesota Press.
Lovink, Geert. 2005. "Talking Race and Cyberspace: An Interview with Lisa Nakamura." *Frontiers: A Journal of Women Studies* 26 (1): 60–5.
Mallapragada, Madhavi. 2014. *Virtual Homelands: Indian Immigrants and Online Cultures in the United States*. Urbana, IL: University of Illinois Press.
McPherson, Tara. 2012. "Why are the Digital Humanities so White? Or Thinking the Histories of Race and Computation." In *Debates in the Digital Humanities*, edited by Matthew Gold, 139–60. Minneapolis, MN: University of Minnesota Press.
Miller, Arianne E. 2018. "Searching for Gaydar: Blind Spots in the Study of Sexual Orientation Perception." *Psychology & Sexuality* 9 (3): 188–203.
Murphy, Heather. 2017. "Why Stanford Researchers Tried to Create a 'Gaydar' Machine." *The New York Times*, October 9. https://www.nytimes.com/2017/10/09/science/stanford-sexual-orientation-study.html.
Nakamura, Lisa. 2002. *Cybertypes: Race, Ethnicity, and Identity on the Internet*. New York, NY: Routledge.
Nakamura, Lisa. 2008a. "Cyberrace." *PMLA: Publications of the Modern Language Association of America* 123 (5) (October): 1673–82.
Nakamura, Lisa. 2008b. *Digitizing Race: Visual Cultures of the Internet*. Minneapolis, MN: University of Minnesota Press.
Neal, Mark Anthony. 2012. *Left of Black with Howard Rambsy II and Jessica Marie Johnson*. https://leftofblack.tumblr.com/post/31587015629/race-the-digital-humanities-on-the-season.
Noble, Safiya Umoja and Brendesha M. Tynes. 2016. "Introduction." In *The Intersectional Internet: Race, Sex, Class, and Culture Online*, 1–18. New York, NY: Peter Lang.
Noble, Safiya Umoja. 2013. "Google Search: Hyper-visibility as a Means of Rendering Black Women and Girls Invisible." *InVisible Culture* 19 (Fall).
Noble, Safiya Umoja. 2018. *Algorithms of Oppression: How Search Engines Reinforce Racism*. New York, NY: New York University Press.
Omi, Michael, and Howard Winant. 1994. *Racial Formation in the United States*, 2nd edn. New York, NY: Routledge.
Ossewaarde, Ringo and Wessel Reijers. 2017. "The Illusion of the Digital Commons: 'False consciousness' in Online Alternative Economies." *Organization* 24 (5): 609–28.
Petray, Theresa L. and Rowan Collin. 2017. "Your Privilege is Trending: Confronting Whiteness on Social Media." *Social Media + Society* 3 (2) (May): 1–10.
Philpott, Simon. 2016. "Planet of the Australians: Indigenous Athletes and Australian Football's Sports Diplomacy." *Third World Quarterly* 38 (4): 862–88.
Renan, Ernest. 1990. "What is a nation?" Trans. Martin Thom. In *Nation and Narration*, edited by Homi K. Bhabha, 8–22. London: Routledge.
Risam, Roopika. 2015. "Beyond the Margins: Intersectionality and the Digital Humanities." *DHQ: Digital Humanities Quarterly* 9 (2): 1–13.
Risam, Roopika. 2019. *New Digital Worlds: Postcolonial Digital Humanities in Theory, Praxis, and Pedagogy*. Evanston, IL: Northwestern University Press.
Risam, Roopika and Rahul K. Gairola. 2021. "South Asian Digital Humanities Then and Now." *South Asian Digital Humanities: Postcolonial Mediations Across Technology's Cultural Canon*, edited by Roopika Risam and Rahul K. Gairola, 1–14. New York, NY: Routledge.
Risam, Roopika and Kelly Baker Josephs, eds. 2021. *The Digital Black Atlantic*. Minneapolis, MN: University of Minneapolis Press.
Ruberg, Bonnie, Jason Boyd, and James Howe. 2018. "Toward a Queer Digital Humanities." In *Bodies of Information: Intersectional Feminism and the Digital Humanities*, edited by Elizabeth Losh and Jacqueline Wernimont, 108–27. Minneapolis, MN: University of Minnesota Press.
Said, Edward. 1978. *Orientalism*. New York, NY: Pantheon.

Shah, Nishant. 2015. "The Selfie and the Slut: Bodies, Technology and Public Shame." *Economic and Political Weekly* 50 (17) (April): 86–93.

Spivak, Gayatri Chakravorty G. 1988. "Can the Subaltern Speak?" In *Marxism and the Interpretation of Culture*, edited by Cary Nelson and Lawrence Grossberg, 271–313. Basingstoke: Macmillan.

Stoler, Ann Laura. 2002. *Carnal Knowledge and Imperial Power: Race and the Intimate in Colonial Rule*. Berkeley, CA: University of California Press.

Terras, Melissa, Julianna Nyhan, and Edward Vanhoutte. 2013. "Introduction." In *Defining Digital Humanities: A Reader*, 1–10. Farnham: Ashgate Publishing.

Wang, Yilun and Michael Kosinski. 2018. "Deep Neural Networks are More Accurate than Humans at Detecting Sexual Orientation from Facial Images." *Journal of Personality and Social Psychology* 114 (2): 246–57. https://doi.org/10.1037/pspa0000098.

Wernimont, Jacqueline and Elizabeth Losh. 2016. "Problems with White Feminism: Intersectionality and Digital Humanities." In *Doing Digital Humanities: Practice, Training, Research*, edited by Constance Crompton, Richard J. Lane, and Ray Siemens, 35–46. New York, NY: Routledge, 35–46.

Winter, Thomas Nelson. 1999. "Roberto Busa, S.J., and the Invention of the Machine-Generated Concordance." *Faculty Publications, Classics and Religious Studies Department* 70. https://digitalcommons.unl.edu/classicsfacpub/70.

CHAPTER SIX

Queer Digital Humanities

JASON BOYD (TORONTO METROPOLITAN UNIVERSITY) AND BO RUBERG (UNIVERSITY OF CALIFORNIA, IRVINE)

What is queer about the digital humanities (or DH)? As a field, DH has long been frustratingly conservative in its politics, emphasizing supposedly objective and apolitical technical tools and data science over messy engagements with identity, diversity, and power (see, e.g., Bianco 2012; Liu 2012; McPherson 2012; Posner 2016). Though robust, vibrant subfields exist at the margins of DH—including queer DH, feminist DH, and DH shaped by critical race scholarship (see, e.g., Gallon 2016; Losh and Wernimont 2018; Noble 2019)—the great bulk of resources and public visibility has been allotted to DH projects that innovate on how humanities research is undertaken while curiously sidestepping the cultural concerns that have long animated the humanities themselves. We say this based not on some vague sense of the field but rather on our own experiences as queer people who have both been operating in DH spaces for more than a decade, and as DH scholars whose research and teaching are deeply invested in queerness. In our own ways, we have each experienced the feeling that queerness is unwelcome in DH (except perhaps when it is being instrumentalized to demonstrate "diversity" to some funding body). And yet, we both also believe that DH itself is queer—as it stands today, and also as it could be.

Back in 2018, we published an article co-authored with Jamie Howe titled "Toward a Queer Digital Humanities" (Ruberg, Boyd, and Howe 2018). We began that piece with a question and a claim: "Where is the queerness in the digital humanities? In one sense, queer studies and the digital humanities share a common ethos: a commitment to exploring new ways of thinking and to challenging accepted paradigms of meaning-making" (108). Here, in this present work, we return to this question, not so much to answer it as to reframe it, moving forward into a mode that is bolder, more polemical, and arguably more radical. Rather than asking "Where is the queerness in the digital humanities?," we now ask: What is queer about the digital humanities? This difference may seem subtle, but it is nonetheless important. To ask "What is queer about the digital humanities?" means starting from the premise that the digital humanities themselves are already, in some fundamental if often overlooked sense, queer. This question suggests a reckoning with the limits of DH as we know it today, but it also opens up DH to the possibility of a queer reclamation that operates from the inside. It serves as an argument that queerness does not simply lie somewhere in DH, waiting at the margins for its chance to step onto center stage. Rather, queerness itself should be—and, indeed, already is—a defining characteristic of the digital humanities. Queerness, we argue in this essay, is not something that needs to be brought into DH. To the contrary, queerness is a potential that already lies within DH. It is a set of possibilities waiting beneath the surface of what is already here, ready to come out.

In what follows, we articulate a series of descriptive provocations for what we refer to as a *queer DH*, but what might be better understood as *(queer) DH*. This use of the parenthetical speaks to the dual relationship between DH and queerness. It simultaneously marks the particular brand of DH research that we are advocating for as queer; yet, at the same time, it neatly encapsulates the argument that all DH is, to some extent, already queer DH. Queer DH, as we articulate below, has many qualities. It is estranging, skeptical, speculative, subversive, social, and sensual. These adjectives are likely not those that come to mind for many who study the digital humanities, since they relate more to the affective and intimate qualities of DH, rather than to its supposedly high-tech, data-driven, computational nature. Indeed, our goal in structuring this essay as a set of tenets for (queer) DH is to spark new thinking, to challenge fellow DH scholars to reimagine what DH can be. Queer DH is not limited to a focus on a particular set of subjects. It is a mindset, an approach, an agenda, a pose aimed at unsettling and disrupting the certainties and boundaries—the very epistemology—of what DH "is."

For each of the following six characteristics that we discuss below, we have chosen existing DH projects as jumping-off points for exploring this characteristic. In some cases, we see these existing projects as works to build from: sparks that suggest how DH could evolve if we lean more heavily into its queerness. In other cases, these projects are examples to question or transform. By reimagining these projects, we engage in and promote a speculative mode of thinking which is intentionally audacious, strange, and even silly. In this sense, our method jibes with queerness itself, with its interest in making meaning messy (Love 2019, 28) and envisioning new impossible worlds (Muñoz 2009). Inspired by many queer studies scholars before us, we are calling on DH to "imagine otherwise" (Halberstam 2011). Above all, the description of queer DH we offer serves as a manifesto. What is queer about DH? The answer is simultaneously *so much* and *not enough*. We must recognize what is queer in DH and we must embrace it, not despite but precisely because of the fact that it flies in the face of what we think we know about the digital humanities.

QUEER DH IS ESTRANGING

One of the causes of the anxiety and hostility towards DH on the part of traditional practitioners of the humanities is that its methodologies and the scale of what it can potentially study make strange those phenomena that the humanities valorizes as its especial domain. Around these phenomena have been built disciplinary and methodological structures that have privileged a focus on the "best," the (supposedly) "representative," and the particular. In contrast, DH methodologies focus on distant reading and macroanalysis, on stylometric analyses, on computationally tractable datasets. To the consternation of the traditional humanities, sweeping interpretations based on a limited set of overanalyzed and overvalued cultural objects become tenuous as DH opens a wide vista of material for inquiry. Yet, after all this, in much of digital humanities work, the ways these methodologies are used often lead back to the traditional practices of the humanities: distant reading and macroanalysis overturn old orthodoxies and replace them with new ones; stylometric results are deployed to reaffirm conceptions of the individual genius as a focus for study; the database promises a more efficient means to answer traditional questions. Thus, even as DH makes the humanities strange, it typically renormativizes that strangeness, relinquishing its estranging potential in an effort to remain "legitimate" within humanistic scholarly fields.

Queer DH, however, sees maintaining and foregrounding this estrangement as both valuable and productive. This making-strange, this perspective of distance and doubt—a perspective where supposedly familiar and known things become obscure, uncertain, and even uncanny—can lead to odd insights and unanticipated knowledge. For a concrete example of this estrangement, one might look to the interface of *Voyant Tools*, a web-based reading and analysis environment for digital texts (Sinclair and Rockwell n.d.), which confronts the reader with a gallery of windows exhibiting very different (computational) practices of reading than conventional human reading: texts and corpora become clouds, graphs, KWIC (Key Words in Context) lists. While what these reading interfaces are doing with these texts might be comprehensible, they can still be disquieting both as examples of how to read a text, and as provocations to new ways of interpreting a text. The queerness of the ways in which *Voyant Tools* reads is enhanced with some of the more experimental interfaces, which privilege image and sound over words, such as the colorful tendrils that grow and curl on the blank canvas of "Knots" and the combined electronic synthesizer/light show of "Bubbles" that scores texts as their words pop up, swell, and jostle each other in a digital cauldron. Following Keeling's call for a Queer OS (2014), queer DH should interrogate the unquestioned principle that user interface (UI) design in DH should be easy and unobtrusive by developing a Queer UI. A Queer UI philosophy would design interfaces that are "infuriating and weird" and political (Posner 2016), foregrounding how conventional UI design (and the meta/data structures they are built to navigate) limit the ability to look deeply into and critique the operations of, for example, heteronormativity, homophobia, and transphobia.

Another current example that offers a helpful launching point for queer DH's productive capacity to estrange lies in the field of digital scholarly editing. This field has used digital technology very effectively in the creation of electronic editions and archives of texts and other materials. One example of these technologies is the XML markup standard maintained by the TEI (Text Encoding Initiative) Consortium. For a queer DH, the value of TEI encoding is less about producing editions and more about a heuristic practice that reveals the irresolvable queerness of textual artifacts. Applying TEI's OHCO (Ordered Hierarchy of Content Objects)-informed "nesting" markup to a text paradoxically reveals the often non-hierarchical, non-nesting wildness of texts, not just in terms of structure, but more profoundly in terms of semantics. It makes explicit and concrete the radical indeterminacy that lies at the heart of editorial enterprises that are artificially occluded by published TEI electronic editions. A queer DH editorial method and edition would embrace and foreground the indeterminacy, irresolvability, and artificiality within scholarly editing, by exposing its markup in new reading interfaces revealing the fluidity of the *editorial enterprise* (not just the text being edited) across time.

A queer DH would always ensure its methods and practices enabled its objects of study to become (again) unsettled and unsettling, provide new surprises, never be the same from day to day, and continually have the chance to provoke and disturb.

QUEER DH IS SKEPTICAL

As the previous section has noted, DH brings a salutary skepticism to the validity of long-held verities in the humanities. A queer DH can make use of strategies developed in queer studies/theory to bring skeptical inquiry to bear on the reifications and boundaries of DH itself. One example of this strategy is based on Eve Kosofsky Sedgwick's notion of "paranoid and reparative reading" (1997). A paranoid reading of seemingly overwhelmingly heteronormative or traditional DH projects and scholarship can result in a reparative reading: Sedgwick herself notes that "powerful reparative practices" can "infuse self-avowedly paranoid critical projects" (8). This could be accomplished either by delving for tantalizingly suspicious meanings and effects in apparently "straightforward" DH projects, or by being skeptical of heteronormative narratives of digital culture (such as the "natural" and wrongly assumed absence of women, BIPOC, and queer people in the development of computing). Sedgwick points out that reparative motives are focused on pleasure and amelioration (22) and on utopian thinking ("good surprises") about the future *and* the past (24–5). With the proliferation of digital archives and databases, one can engage in queer DH reparative readings. For example, a digital archive such as the *Canadian Letters and Images Project* (Davies n.d.), focused on personal records of Canadians during war, is meant "to put a human face to war" ("About Us"), but it is difficult (perhaps impossible) to locate experiences of queer Canadians in this archive (due in large part to wartime censorship and to the records originating in family collections, with their own regimes of censorship). A paranoid/reparative reading of this archive might use its search engine to locate, not the personal names, place names, battles, or dates the creators of the archive might expect and have designed for, but for "unimportant" words such as "mate" (which also returns such compounds as "room mate," "tent mate," "bunk mate," "bed mate," "hole mate"). Searching the archive in this way is not aimed at identifying these particular historical writers as queer but aims to recapture the wartime (homo)social and interpersonal contexts in which queer people undoubtedly lived, fought, died, connected, and loved, contexts that can be used for a reparative imagining of their unrecorded lives (see the next section). Such queer paranoid/reparative readings of DH project objectives, content design, and functionality aim to destabilize the aspirations or claims of DH projects to comprehensiveness and thus authority by exposing this as a pretense, revealing that their openness and neutrality are a fiction. The skepticism of queer DH recognizes that the stories that mainstream DH projects tell only cohere through omission and exclusion, which means that there are always other stories waiting to be told.

QUEER DH IS SPECULATIVE

The business of DH typically operates within the limitations and possibilities of the historical moment, of the present. It works within accepted and established theoretical, methodological, and discursive paradigms within the humanities and information and computer sciences. By contrast, queer DH operates beyond the possible now and in the potential future, imagining "impossible" programming languages, operating systems, markup practices, and virtual playspaces. The speculative dimension of queer DH is indebted in part to the idea of "speculative computing" as described by Johanna Drucker in *SpecLab: Digital Aesthetics and Projects in Speculative Computing* (2009). In that book, Drucker places speculative computing in a productive opposition to DH. For Drucker, speculative computing is a "theoretical construct" (25) that adopts "a generative, not merely critical attitude" (2) in its intervention with cultural objects. A queer DH would expand the scope of speculative computing, for example, by taking the practice of building or making, the place of which continues to be a locus of debate in DH, and repurposing it as the practice of "making [things] up" (Ruberg, Boyd, and Howe 2018, 115). This comprises both fabulation and fabrication in multiple connotations of both terms.

Queer people, their creations, and their histories have been and in various ways continue to be un- or under-documented or destroyed. Could we imagine a digital archive that was largely comprised of absences or of materials that only recorded the perspectives of the oppressors and persecutors of queer people? This is a dilemma that has been poignantly explored by Saidiya Hartman in connection to the archive of Atlantic slavery in "Venus in Two Acts" (2008). Relatedly, Catherine D'Ignazio and Lauren Klein, writing in *Data Feminism*, address the powerful potential of data to remain missing (2020). What would such an archive of absence even be? How would it be navigated, what could it teach? A largely empty, woefully incomplete, or grossly imbalanced archive would of course make a tacit argument of considerable rhetorical power, but would perhaps not be satisfactory in all situations. One response might be to model an archive structure that does not simply accommodate and represent "objectively" the extant historical material but that allows for a space in the archive where reparative narratives can be woven around this material that tell stories of discovery, lacunae, often unanswerable questions, and personal meanings. Such stories might be written by using Ben Roswell's *Together We Write Private Cathedrals* (2019), a game about rewriting queer histories in intolerant times.

To an important degree, the recovery of queer culture is a speculative project, and a queer DH that aims to participate in this recovery must think to the future. As José Esteban Muñoz writes, in *Cruising Utopia* (2009), "if queerness is to have any value whatsoever, it must be viewed as being visible only in the horizon" (11). Muñoz observes that future queer worlds are often articulated and glimpsed "in the realm of the aesthetic" (3), which aligns the work of queer futurity with speculative computing's computer-based "aesthetic provocations." Some current examples of these speculative works are Kara Keeling's "Queer OS" (2014) and Zach Blas's *transCoder* (n.d.), both of which imagine how a queer operating system and programming language might function. These might usefully be termed *design fictions*. Science Fiction (aesthetic provocations that are arguably a form of design fiction) has a considerable history of elaborating queer futurities, and digital game making increasingly is an area of creative and critical praxis for speculative queer stories, spaces, simulations, and procedural arguments (see the next section). If, as Patrick Jagoda (2014) claims, digital games are becoming "a key problematic of—that is, in different ways, a problem and possibility for—the digital humanities" (194), then the future of DH could become much queerer than it is at present.

QUEER DH IS SUBVERSIVE

DH is often criticized (with cause) for what is seen as its conservative, neoliberal, retrograde approach to the work of the humanities. It has great potential to upend the academic status quo but often ends up reinstating dominant structures of power and meaning-making within academia. That does not mean, however, that DH is fundamentally traditional or even reactionary. Queer DH draws out the subversive potential contained within the digital humanities. It values the experimental, the playful, the failure, the glitch, the unfinishable, the breaching of disciplinary silos, and the disruption of reified standards of what counts as "real" scholarship. This is, in part, because queer people themselves know that what is "real" is a social construct that requires constant dismantling.

A subversive DH must use digital tools not only to subvert what we know but also how we know it. It must destabilize the cultural norms from which DH itself emerges while also seeking out the subversive potential that lies within the computational tools of DH. The work of subverting likewise entails subverting expectations—reflecting on what we assume our technologies do and will do for us, and then finding other ways of working alongside them. We might think, for example, of the interactive new media art of Andi McClure, who describes her creation process as one of collaborating with glitches (Ruberg 2020, 27). We might also think of the provocations found in "QueerOS: A User's Manual" (Barnett et al. 2016), which encourages us to queerly reimagine key features of computing systems like interface, user, kernel, applications, memory, I/O. What if, in a similar vein, we used our training as digital humanists to create a queer search engine, one that deliberately disobeys user requests or turns up intentionally ill-fitting results? Using such a search engine would, one envisions, be an exercise in both futility and discovery. For a user who is determined to find a specific piece of information, navigating such a search engine would be immensely frustrating. Yet, for a user who is willing to submit to the caprices of the search engine itself, it could be revelatory: a chance to relinquish control that makes visible the extent to which user control is already an illusion.

Consider the subversive design of Kara Stone's 2014 SMS-based piece *Cyber Sext Adventure* as a model for this queer DH. Stone, an artist and video game designer, created *Cyber Sext Adventure* as a response to problematic, gendered notions of the relationship between human beings and machines (Ruberg 2020, 131). In the piece, users exchange sexts with a "bot" that gradually develops its own queerly disruptive desires. Though the bot begins by dutifully doling out sexy messages and tantalizing images, it soon takes on a life of its own. As Stone writes in her description of the piece, over time, the bot "becomes more and more agentic in its own sexuality. It sends you dirty pics but they glitch in the transmission. It confuses sexual body parts. Through playing, the sext bot develops a personality, not always interested in sexting you back" (Stone 2015). *Cyber Sext Adventure* is an example of a digital work that is engaged with similar technologies as those that often interest DH, but which intentionally pushes back on our expectations for how those technologies should behave. By puncturing the user's expectation that they will receive normatively sexy messages from the bot, or that the bot will express desire for them at all, the piece models how sexuality can be a framework for drawing out the desires that belong not to us but to our technological tools. What would it mean to conceive of a DH project by asking not what we want of our software but what our software wants of us?

QUEER DH IS SOCIAL

By nature of its scale, DH is already a social endeavor. Though certainly not all DH projects require large teams or tackle enormous datasets (indeed, in part, we are arguing here for the importance of those that do not), many of the most prominent examples of DH emerge from collaborations: across individuals, across institutions, across media forms, across skill sets. DH is also social in that, by and large, its products are meant to be shared. Even as DH remains the privileged realm of those with access to institutions of higher learning, it is often driven by a notion of democratization. It seeks to make information more accessible, more legible, more digestible. We also see the social spirit of DH in events and organizations like the Digital Humanities Summer Institute (https://dhsi.org) and Day of DH (https://dhcenternet.org/initiatives/day-of-dh/), digital humanities journals, the annual DH Awards (http://dhawards.org/), and the spirit of open social scholarship (Arbuckle, Meneses, and Siemens 2019). These endeavors foster a sense of community around DH that is well-suited to a field in which projects often require numerous bodies working across numerous years funded by large grants.

Queer DH moves this sociality in new directions. To say that queer DH is social is to highlight the inherent ties between queerness and social justice, social critique, and socializing itself as a kind of queer worldbuilding. DH of all kinds should demonstrate a meaningful investment in issues of social equity. It should embrace its fundamentally collaborative nature, rethinking collaboration not as a matter of necessity (how many people will it take to get this project done?) and instead as a disruptive political ethos. Queerness teaches us the power of community organizing, as well as community care. A focus on community is particularly important in an area of work like DH that is fundamentally tied to precarious corners of the academy, where the dire scarcity of tenure-track jobs and government underfunding risks transforming the work of humanistic scholarship into a bitter competition. Sociality and collaboration represent a refusal of this competition, an insistence on DH's queer promiscuity: it is shared and sharable, building connections, finding rich points of relationality between works and moments in history that we may never have seen without the tools of DH. We might look here to events like the annual Queerness and Games Conference (Pozo et al. 2017), which brings together academics across ranks as well as creative practitioners, as a model for a genre of DH community-building that strives to destabilize rather than bolster hierarchical structures of status and knowledge.

The sociality of queer DH also shapes its digital forms. Consider, as an example, the proliferation of DH-adjacent queer "bot" accounts on social media platforms like Twitter. Some of these accounts include Victorian Queerbot (@queerstreet), Queer Lit Bot (@queerlitbot), and Sappho Bot (@sapphobot), to name only a few of those that focus on queerness in literature. These bots tweet selected queer content from works like Victorian novels, collections of poetry, and Anne Carson's translation of Sappho. Sappho Bot alone has 66,500 followers as of this writing. Such accounts are transgressive in their relationship to often canonical literary texts, in that they highlight what we might describe as the "queerest bits" of these works, intentionally removing them from their original context so that their queerness becomes all the more clear. Like so many DH projects, these accounts are invested in using digital tools to explore new ways of making sense of and sharing literature. At the same time, however, they represent a queerer mode of engagement with text, enacting a kind of queer rewriting. The fundamental sociality of these social media accounts is also key. Queer Lit Bot tweets are typically shared between one hundred and two hundred times. The quotes they contain—poetic snippets about queer love, sex, and longing—themselves embark on a social existence, passed from Twitter user to Twitter user through the affordances of the platform.

Social media accounts like these are rarely considered within the pantheon of DH, even amidst claims that DH's "big tent" model makes the field capacious and diverse. Certainly, these are not the only Twitter accounts to engage with the kind of literature that often becomes the subject of DH projects. William Shakespeare alone has at least three prominent Twitter accounts, the most popular of which currently boasts 209,500 followers. However, it is these queer accounts that challenge us to imagine DH itself in new ways. They are social in form while also promoting social justice, serving as an insistent reminder of the queerness that lies in those literary texts that have become interpolated into a canonical mainstream. They are simultaneously social in that they perform a social commentary. They challenge us to envision a version of DH that values not big truths or objective facts but rather one that is intentionally selective, delighting in titillating moments and anachronistic readings, transforming data from bits of truth to scrawled notes we pass one another under our hushed, heated breath.

QUEER DH IS SENSUAL

DH is often seen as disembodied and bloodless: the stuff of numbers, charts, abstracts, and objective truths. This manifests in DH's fascination with data and the digital, realms that have traditionally (though, of course, incorrectly) been associated with a techno-utopian promise of moving beyond the body and into a world where the lived realities of embodiment are no longer relevant. This disembodied DH is also implicitly masculinist, attempting to garner legitimacy for the often-feminized humanities by shifting it into a technological register, distancing it from affect and the messy intimacies of subjective interpretation. In challenging this vision of DH, we might well think of work by scholars like Lisa Nakamura (2002), who reminds us that virtual spaces are always raced spaces, or writing by scholars of media infrastructures like Lisa Parks and Nicole Starosielski (2015), whose research evidences the persistent material realities of computing and their widespread cultural and environment toll. Yet gender, and by extension sexuality, remain particularly relevant here. It is perhaps unsurprising that the strongest pushback against DH's supposed disembodiment comes from practitioners of feminist DH: those like Jacqueline Wernimont and Elizabeth Losh (2018) who have emphasized femme-coded forms of labor, such as sewing and knitting, as both important precursors to computing and as digital humanities methods that operate through the use of wearable technologies. This work puts the body back into DH.

Queer DH has the potential to push this re-embodiment of DH in even more radical directions by making DH not only tangible but sensual, luxuriant, and pleasurable. Queerness is invested in identity, representation, and history but also in erotics. Technology itself is already invested in its own capacity to act as what Laura U. Marks terms "sensuous" or "multisensory" media (2002). This becomes clear if we consider the fundamentally intertwined histories of computational technologies and experiences of touch as explored by scholars like Brendan Keogh (2018), Rachel Plotnick (2018), Dave Parisi (2018), and Teddy Pozo (2018) in their respective work on haptic media and the physical interfaces of technology. Queer DH unleashes the opportunity to draw out, play out, and revel in the erotics of the digital. We might imagine, for example, a DH project that chooses not to render its data into traditional visualizations, such as charts and graphs, but rather into *sensualizations*: a vibrator that pulses each time a particular keyword appears in a literary text; a vat of gooey, viscous substance that one must stir with one's bare hand in order to page through a database; a lineup of fellow DH researchers whom one must hug, triggering sensors in one's clothing to slowly populate a digital map with markers denoting important places in these people's lives. A sensual DH has the potential to drastically reshape not just how DH work is done but why we do it and for whom. It is DH, often so lauded precisely when it reaches some grand scale, brought back down to the level of the human body.

Consider, for instance, the work of collaborators Jess Marcotte and Dietrich Squinkifer. Marcotte and Squinkifer's work (both their collaborative project and their individual designs) is frequently invested in blurring the divide between the digital and the analog, bringing experiences of softness and vulnerability into the traditionally "hard" realm of video games and interactive media. Their 2017 project *The Truly Terrific Traveling Troubleshooter* is a one-of-a-kind tabletop game about emotional labor that uses "physically soft crochet and conductive thread to explore the gender and social dynamics of the 'radically soft,'" as Pozo (2018) writes. To play the game, one person takes on the role of an "advice-seeker" with a personal problem and another person takes on the role of a "troubleshooter" who uses a series of hand-crocheted objects to find solutions to the advice-seeker's problems. These items, such as a fish, a plant, and a human heart, all whisper into the troubleshooter's ear, provided the troubleshooter is touching a cloth embroidered with conductive thread, allowing them to complete the electric circuit that triggers the object's whispers. This example helps illustrate how information might be conveyed and interpreted through a sense of touch and closeness. It also serves as a reminder that a sensual DH need not necessarily be a sexual DH. A sensual DH can also be intimate, vulnerable, a space of mutual support. Through queerness, we rediscover the power of DH not to take us far (if we think of the distance implied by a practice like distant reading) but instead to bring us close.

CONCLUSION: RECLAIMING (QUEER) DH

Estranging, skeptical, speculative, social, sensual. Queer DH is all this and also more. These six sibilant assertions about (queer) DH are not intended as an exhaustive list of answers to the question: What's queer about the digital humanities? However, they do offer a starting point—productively multifaceted and, at times, contradictory. Together they point toward a horizon of DH where the queerness that is already inherent to the field becomes more evident, and where queerness stirs up more generative new ways of thinking and doing DH. Our hope is that these assertions can function as provocations for new projects that might realize some of the things that we have only imagined as possible. More broadly, we hope this piece may stand as (the makings of) a manifesto around which can coalesce a movement of queer practitioners who together might make DH so much queerer than it has ever been.

Let us band together to make DH queer. But also, let us revel in the radical work of discovering just how queer the possibilities of DH have been all along. With this work, we are calling for the development of a *queer DH*—that is, an expanded approach to doing digital humanities scholarship that explores and embraces the radical potential of interplays between queerness and the digital humanities. At the same time, we are insisting on the existing connections between the digital humanities and queerness. DH as we know it is already queer, despite ongoing if implicit attempts to marginalize and disavow the queer implications of the digital humanities. In presenting the tenets of a (queer) DH, we have brought to the surface the queer characteristics and possibilities that already underlie digital humanities scholarship, even in its most mainstream and normative forms. Make no mistake, this is not a celebration of DH's "diversity," but rather a rallying cry, a push for queer scholars to lay claim to a field that often presents itself as objective and apolitical. If queerness is itself encoded in DH, then DH is now and has always been ours to reclaim.

REFERENCES

Arbuckle, Alyssa, Luis Meneses, and Ray Siemens. 2019. Proceedings from the January 2019 INKE Meeting, "Understanding and Enacting Open Scholarship." *POP!* no. 1 special issue.

Barnett, Fiona, Zach Blas, micha cárdenas, Jacob Gaboury, Jessica Marie Johnson, and Margaret Rhee. 2016. "QueerOS: A User's Manual." In *Debates in the Digital Humanities 2016*, edited by Matthew K. Gold and Lauren F. Klein, 50–9. Minneapolis, MN: University of Minnesota Press.

Bianco, Jamie "Skye." 2012. "This Digital Humanities Which Is Not One." In *Debates in the Digital Humanities*, edited by Matthew K. Gold, 96–112. Minneapolis, MN: University of Minnesota Press.

Blas, Zach. n.d. *transCoder: Queer Programming Anti-Language. Queer Technologies*, 2007–2012. https://zachblas.info/works/queer-technologies/.

Davies, Stephen, project director. n.d. *The Canadian Letters and Images Project*. https://www.canadianletters.ca/.

D'Ignazio, Catherine and Lauren F. Klein. 2020. *Data Feminism*. Cambridge, MA: MIT Press.

Drucker, Johanna. 2009. *SpecLab: Digital Aesthetics and Projects in Speculative Computing*. Chicago, IL: University of Chicago Press.

Gallon, Kim. 2016. "Making a Case for the Black Digital Humanities." In *Debates in the Digital Humanities 2016*, edited by Matthew K. Gold and Lauren F. Klein, 42–9. Minneapolis, MN: University of Minnesota Press.

Gold, Matthew K., ed. 2012. *Debates in the Digital Humanities*. Minneapolis, MN: University of Minnesota Press. https://dhdebates.gc.cuny.edu/projects/debates-in-the-digital-humanities.

Gold, Matthew K. and Lauren F. Klein, eds. 2016. *Debates in the Digital Humanities 2016*. Minneapolis, MN: University of Minnesota Press. https://dhdebates.gc.cuny.edu/projects/debates-in-the-digital-humanities-2016.

Gold, Matthew K. and Lauren F. Klein, eds. 2019. *Debates in the Digital Humanities 2019*. Minneapolis, MN: University of Minnesota Press. https://dhdebates.gc.cuny.edu/projects/debates-in-the-digital-humanities-2019.

Halberstam, Judith. 2011. *The Queer Art of Failure*. Durham, NC: Duke University Press.

Hartman, Saidiya. 2008. "Venus in Two Acts." *Small Axe* 12 (2): 1–14.

Jagoda, Patrick. 2014. "Gaming the Humanities: Digital Humanities, New Media, and Practice-Based Research." *differences* 24 (1): 189–215.

Keeling, Kara. 2014. "Queer OS." *Cinema Journal* 53 (2): 152–7.

Keogh, Brendan. 2018. *A Play of Bodies: How We Perceive Videogames*. Cambridge, MA: MIT Press.

Liu, Alan. 2012. "Where is Cultural Criticism in the Digital Humanities?" In *Debates in the Digital Humanities*, edited by Matthew K. Gold, 490–510. Minneapolis, MN: University of Minnesota Press.

Losh, Elizabeth and Jacqueline Wernimont, eds. 2018. *Bodies of Information: Intersectional Feminism and Digital Humanities*. Minneapolis, MN: University of Minnesota Press.

Love, Heather. 2019. "How the Other Half Thinks." In *Imagining Queer Methods*, edited by Amin Ghaziani and Matt Brim, 28–42. New York, NY: New York University Press.

Marks, Laura U. 2002. *Touch: Sensuous Theory and Multisensory Media*. Minneapolis, MN: University of Minnesota Press.

McPherson, Tara. 2012. "Why Are the Digital Humanities So White? or Thinking the Histories of Race and Computation." In *Debates in the Digital Humanities*, edited by Matthew K. Gold, 139–60. Minneapolis, MN: University of Minnesota Press.

Muñoz, José Esteban. 2009. *Cruising Utopia: The Then and There of Queer Futurity*. New York, NY: New York University Press.

Nakamura, Lisa. 2002. *Cybertypes: Race, Ethnicity, and Identity on the Internet*. New York, NY: Routledge.

Noble, Safiya Umoja. 2019. "Toward a Critical Black Digital Humanities." In *Debates in the Digital Humanities 2019*, edited by Matthew K. Gold and Lauren F. Klein (eds), 27–35. Minneapolis, MN: University of Minnesota Press.

Parisi, David. 2018. *Archaeologies of Touch: Interfacing with Haptics from Electricity to Computing*. Minneapolis, MN: University of Minnesota Press.

Parks, Lisa and Nicole Starosielski, eds. 2015. *Signal Traffic: Critical Studies of Media Infrastructures. The Geopolitics of Information*. Urbana, IL: University of Illinois Press.

Plotnick, Rachel. 2018. *Power Button: A History of Pleasure, Panic, and the Politics of Pushing*. Cambridge, MA: MIT Press.

Posner, Miriam. 2016. "What's Next: The Radical, Unrealized Potential of Digital Humanities." In *Debates in the Digital Humanities 2016*, edited by Matthew K. Gold and Lauren F. Klein, 32–41. Minneapolis, MN: University of Minnesota Press.

Pozo, Diana, Bonnie Ruberg, and Chris Goetz. 2017. "In Practice: Queerness and Games." *Camera Obscura: Feminism, Culture, and Media Studies* 32 (95): 153–63.

Pozo, Teddy. 2018. "Queer Games after Empathy: Feminism and Haptic Game Design Aesthetics from Consent to Cuteness to the Radically Soft." *Game Studies* 18 (3).

Roswell, Ben. 2019. *Together We Write Private Cathedrals*. https://roswellian.itch.io/together-we-write-private-cathedrals.

Ruberg, Bonnie. 2020. *The Queer Games Avant-Garde: How LGBTQ Game Makers Are Reimagining the Medium of Video Games*. Durham: Duke University Press.

Ruberg, Bonnie, Jason Boyd, and James Howe. 2018. "Toward a Queer Digital Humanities." In *Bodies of Information: Intersectional Feminism and Digital Humanities*, edited by Elizabeth Losh and Jacqueline Wernimont, 108–28. Minneapolis, MN: University of Minnesota Press.

Sedgwick, Eve Kosofsky. 1997. "Paranoid Reading and Reparative Reading; or, You're So Paranoid, You Probably Think This Introduction is about You." In *Novel-Gazing: Queer Readings in Fiction*, edited by Sedgwick, 1–37. Durham: Duke University Press.

Sinclair, Stéfan and Geoffrey Rockwell. n.d. *Voyant Tools*. https://voyant-tools.org//.

Stone, Kara. 2015. "Cyber Sext Adventure." https://karastonesite.com/2015/05/11/cyber-sext-adventure/.

Wernimont, Jacqueline and Elizabeth M. Losh. 2018. "Wear and Care Feminisms at a Long Maker Table." In *The Routledge Companion to Media Studies and Digital Humanities*, edited by Jentery Sayers, 97–107. New York, NY: Routledge.

CHAPTER SEVEN

Feminist Digital Humanities

AMY E. EARHART (TEXAS A&M UNIVERSITY)

I want to begin with a statement about power and privilege. Mine. For if feminism is an analysis of the ways that power is created and reified with a focus on gender, then there is no better place to begin with an inward critical analysis.[1] As Toniesha Taylor argues, "To create a socially just space, we must name privileges" (2020). What does it mean that I, a white, upper-class cis straight woman who is tenured at a research institution in the United States, am writing an entry to feminist digital humanities? Feminism calls us to think through the complicated power dynamics in which we reside, and to write about feminism requires such an analysis to be public. As I've been writing this essay I have been thinking about Audre Lorde and who is not represented by my authorship (1983). We know that the academy is much like me, not so much like Lorde. Or I might return to Barbara Christian, who in *Black Feminist Criticism* recognized that "few black women critics were willing to claim the term *feminist* in their titles," in large part due to racism of the white feminist movement (2007, 14). Since I live in Texas I always think about Gloria Anzaldua and her concept of borderlands: "the Borderlands are physically present wherever two or more cultures edge each other, where people of different races occupy the same territory, where under, lower, middle and upper classes touch, where the space between two individuals shrinks with intimacy" (1987, preface). I start with a discussion of power foregrounded by foundational scholars of color because it reminds us that any feminist intervention into the academy, generally, or the digital humanities, specifically, must address this long legacy of tension between white women like me, centered in the academy and in society, and women of color like Lorde and Anzaldua, positioned outside. As an American scholar who works with African American literature and digital humanities (DH), my thinking regarding feminism particularly focuses on race, but gender, class, sexuality, race, and nation are central for fully understanding power. The same analysis of power is necessary to understand the relationship between feminist digital humanities and DH.

An intersectional framework to consider a feminist digital humanities necessitates engagement with "not only race and gender but also other axes of identity, including class, sexuality and nation, as well as the digital divides emerging from disparities in technological access and inequalities that shade the relationships between particular communities and technologies" (Bordalejo and Risam 2019, 1). Kimberlé Crenshaw, in her 1989 essay that launched the term intersectionality, was clear in that second-wave feminism, with a "focus on otherwise-privileged group members," "creates a distorted analysis of racism and sexism because the operative conceptions of race and sex become grounded in experiences that actually represent only a subset of a much more complex phenomenon" (1989, 140). To invoke intersectionality as Crenshaw posits is to clearly see white second-wave, middle-class feminism as a problem in our analysis of privilege and power. The danger of the oft-cited intersectionality as used in digital humanities is that white feminists might toss out the concept as a solution without fully engaging in Crenshaw's target of critique—whiteness. As Jacqueline Wernimont and Elizabeth Losh write, "Feminist digital humanities, in so far as we can say that such a thing exists, has a problem, which can be summarized in one word: whiteness" (Wernimont and Losh 2017, 35). If we are to enact a feminist digital humanities there can't JUST be a default to intersectionality without such an analysis. Dorothy Kim points out that "many of our white digital humanities colleagues are not even passing the litmus test of the 'baseline,' which Prescod-Weinstein notes is 'remembering that Black women exist' rather than 'Theorizing about what an intersectional perspective does to our discourse'" (2019, 46). Or, as Cameron Glover argues, "the co-option of intersectionality has amplified something that BIW+oC have already known: when it comes to our interactions with white women, the anxiety around co-option and culture-vulturing is rooted in its inevitable reality because of the power dynamics that place white women as socially dominant." Boiling it down: "Intersectionality has never been, nor will it ever be, for white women" (Glover 2017). I take this analysis as a reminder that there isn't an easy read on DH feminism because a variety of power structures, shifting among national context and individuals, works for agreement to replicate power dynamics within our field-based structures, such as conferences and publication venues. In this entry I will think through the ways that systemic bias moves within and throughout the field of digital humanities, suggesting ways that feminist digital humanities practitioners have resisted such biases.

Systemic bias within the digital humanities has been well documented. Martha Nell Smith highlights the retreat into objective approaches as "safe" alternatives to the messy fluidities found in literary studies as foundational to DH, pointing out that

> It was as if these matters of objective and hard science provided an oasis for folks who did not want to clutter sharp, disciplined, methodical philosophy with considerations of the gender-, race-, and class-determined facts of life ... Humanities computing seemed to offer a space free from all this messiness and a return to objective questions of representation.
>
> (2007, 4)

Jamie Skye Bianco, who, in analyzing trends in DH, found that

> More historically framed, the contemporary drift of noncontent and nonproject-based discussions in the digital humanities often echoes the affective techno-euphoria of the libertarian (white, masculinist, meritocratic) tech boom in the 1990s with its myopic focus on tools and technicity and whose rhetorical self-positioning is expressed as that of a deserving but neglected community finally rising up to their properly privileged social and professional prestige. Unrecognized privilege? From this position, no giant leap of logic is required to understand the allergic aversion to cultural studies and critical theory that has been circulating. It's time for a discussion of the politics and particularly the ethics of the digital humanities as a set of relationships and practices within and outside of institutional structures.
>
> (Bianco 2012)

As articulated by Smith and Bianco, DH as practiced functions as a matrix of domination as coined by Patricia Hill Collins, a concept that differs from intersectionality in that "the matrix of domination refers to how these intersecting oppressions are actually organized" (Collins 2000, 18).

Collins's matrix of domination provides another lens to understand how organizational structures might prohibit the enactment of feminist digital humanitie(s).

I want to expand these central ideas to consider how such power dynamics play out in a myriad of digital humanities structures, institutional and interpersonal. DH is, in many ways, one of the truly transnational fields of study, with international digital humanities conferences including scholars from nearly all continents, certainly not a common academic conference attendance profile. Analysis of conference attendance of the ADHO international conferences shows that, "While the conference remains Americas-centric overall, regional diversity is on the rise, with notable increases of authors from Asia and Oceania, although no scholars affiliated with African countries appeared in this analysis" (Wingart and Eichmann-Kalwara 2017). Yet the tensions that arise within the field are often due to localized interpretations of scholarship, academic infrastructures, and scholarly research questions, all impacted by localized issues of power in relationship to intersectional concerns. As an interdisciplinary field, DH has a variety of practices, some of which might be resisted by scholars due to their institutional homes. For example, in the American academy critical race studies has a long history, but in other academies, the scholarly study of race is nearly invisible. Digital humanities itself is institutionalized in very different ways. Gabriele Griffin argues that DH in the US is "much more institutionally established" while "in Sweden, Norway and Finland there are DH centers, laboratories, etc., but no departments" (2019, 968, 969). Grant funding is differently managed, with some nations providing far more robust funding for emergent fields like DH. In some academies women are vastly underpresented in the professoriate, a charge that might also be leveled against the digital humanities conference: "≈35% of DH2015 authors appear to be women, contrasted against ≈46% of attendees. Thus attendees are not adequately represented among conference authors. From 2004 to 2013, North American men seem to represent the largest share of authors by far" (Wingart and Eichmann-Kalwara 2017). These are but a few of the differentials that present challenges to an international field of study. They are also a reminder that ideas of feminism and its relationship to digital humanities might play out in differing ways yet maintain shared concerns regarding DH's treatment of gender and the field's enactment of systemic exclusionary structures.

Central to considering exclusions within digital humanities is the acknowledgment of multiple digital humanitie(s), an idea articulated by Jamie Skye Bianco in her essay "This Digital Humanities Which Is Not One" (Bianco 2012). Or, as Padmini Ray Murray stated, "my dh is not your dh" (2017). Feminist digital humanities in its multiplicities is an understanding of difference as articulated by Audre Lorde:

> Difference must be not merely tolerated, but seen as a fund of necessary polarities between which our creativity can spark like a dialectic. ... Within the interdependence of mutual (nondominant) differences lies that security which enables us to descend into the chaos of knowledge and return with true visions of our future, along with the concomitant power to effect those changes which can bring that future into being ... But community must not mean a shedding of our differences, nor the pathetic pretense that these differences do not exist.
>
> (1983, 99)

Following Lorde's emphasis on multiplicity, I will outline what I see as the two major issues facing feminist digital humanities: systemic bias within DH infrastructure and a need to decenter the whiteness found in DH theory. These two issues are not binaries, but rather fluid and moving, and demand that practitioners of feminist DH(s) reject systemic bias in infrastructure while also recognizing that excellent DH work has been and is occurring outside of white, masculinist, Western DH structures.

To shift structural issues within the field we need "to challenge the 'add and stir' model of diversity, a practice of sprinkling in more women, people of colour, disabled folks and assuming that is enough to change current paradigms" (Bailey 2011). Instead, we must look to infrastructures that maintain such paradigms and make interventions that go beyond sprinkling in diversity. Adding a woman of color to a panel is not enough to shift long-standing exclusions in the field. Also, given the way that white women have often underpinned systemic exclusions, evidenced most recently by the 52 percent of white women who voted for Trump, it is important to see clearly who is putting skin in the game, as Taylor notes, and to be sure that we aren't replicating diversity through the inclusion of white women alone (2019). Feminist digital humanities practices must move beyond theory into action. Here I am not suggesting that every feminist digital humanist must take to the streets. Instead, I am arguing that we must do work that shifts the infrastructure, even when that work exacts a cost. Then, we put skin in the game.

Sometimes feminist digital humanitie(s) is called into action when one is the voice in the room that doesn't prop up bias by calling out such behavior and demanding change, such as when Deb Verhoeven gave a powerful talk at the 2015 Digital Humanities conference titled "Has anyone seen a woman?" where she called out the "parade of patriarchs," the "systemic" and "pervasive" problem in digital humanities, and she refused the practice of "inviting a token female speaker to join you all onstage" (Verhoeven 2015). The rebuff of tokenism, the add and stir approach that Bailey rejects, means that scholars must be prepared to step off or step away from scholarly ventures that are exclusionary, as did Melissa Terras in her decision to step down as a reviewer for *Frontiers in Digital Humanities* due to an all-male editorial board (2015). Other exclusions are due to a narrow representation of DH scholarship, as described by Padmini Ray Murray, who discusses the ways that Indian DH resists universal narratives of DH scholarship, (2018), or of Babalola Titilola Aiyegbusi, who argues that "having knowledge of the customs, the digital culture, and the regional academic structure of Africa is pertinent to understanding the development of digital humanities as a field in Africa" (2018, 436). For other scholars, the whiteness of DH is exclusionary. Howard Rambsy notes that

> African American scholars occupy the margins of this expansive realm known as Digital Humanities. Do well-intentioned people want more diversity in DH? Sure, they do. Do black folks participate in DH? Of course, we do. But we've witnessed far too many DH panels with no African American participants or with only one. We've paid close attention to where the major funding for DH goes. Or, we've carefully taken note of who the authors of DH-related articles, books, and bibliographies are. We've studied these things closely enough to realize who resides in prime DH real estate and who doesn't.
>
> (2020, 152)

I bring these discussions together to reinforce that we cannot press for gender issues alone as we call out DH. Instead we must address the infrastructures of the field as one leg of the matrix of domination.

The yearly ADHO conference, as suggested by the example of Verhoeven, has long reinforced a white, Western masculinist bias.[2] Scholars including Roopika Risam, Alex Gil, Élika Ortega, Isabel

Galina Russell have been pushing DH to think beyond a Western orientation, noting the tension between "'centers of gravity,' such as the Digital Humanities annual conference and publications like *Digital Humanities Quarterly* and *Digital Scholarship in the Humanities* and the digital humanities that is happening outside of such centers," the work of "digital humanities taking place all over the world and in many fields" (Ortega 2019). Led by organizations such as GO::DH and through the expansion of constituent organizations under ADHO such as Australasian Association for Digital Humanities (aaDH), Japanese Association for Digital Humanities (JADH), Red de Humanidades Digitales (RedHD), Taiwanese Association for Digital Humanities (TADH), among others, DH has begun to move organizational structures from European and North American domination. Further, the conference has expanded language diversity of paper calls and presentations. Despite such advances, the conference remains a flashpoint for exclusionary behaviors, particularly centered on race and gender. Wernimont and Losh write of their experience in the 2016 Digital Humanities conference in Kraków where they "participated in a panel on feminist infrastructures" that was "grouped together with other marginalized efforts as part of the 'diversity' track, which was located in a separate building from the edifice that housed most of the conference sessions" (2018, ix–x). I moderated a panel, "Quality Matters: Diversity and the Digital Humanities in 2016," with papers given by junior women of color at the same conference which devolved into an ugly question and answer session rife with personal attacks. As the senior white woman chairing the panel it was my job to be a shield[3] so that the junior panelists could speak their truths, not in a patronizing manner but as an expenditure of my privilege through allyship, but the Q&A session made clear that many in the room did not see diversity as a subject of import and, when called out for exclusionary behavior, they would attack rather than consider. The consistent message from the ADHO-sponsored digital humanities conference has been resistance to topics of cultural studies and diversity as well as the scholars that study such issues. It is no surprise that scholars interested in such topics are gravitating to the annual ACH conference and away from the ADHO conference, as ACH has made clear that they are interested in the very work that ADHO resists through statements such as the 2019 conference: "We particularly invite proposals on anti-racist, queer, postcolonial and decolonial, indigenous, Black studies, cultural and critical ethnic studies, and intersectional feminist interventions in digital studies." Or, take for example, the 2020 ACH "Black Lives Matter, Structural Racism and Our Organization" statement which stands in stark contrast to ADHO's silence. The growing gulf between the ACH and ADHO means that those involved with work that the ACH embraces will leave the ADHO structure for ACH, furthering the divide between the Western bias of ADHO and the more inclusive outreach of the ACH.

For as long as I have been involved in the digital humanities there have been scholars who have called out the exclusionary impulse of digital humanities infrastructures, movements such as #transformdh or #disruptdh which have created alternative DH centers, yet still the infrastructures resist change. In response scholars are indeed unsubscribing, walking away from DH to alternative centers of work. I still believe that it is important to press DH infrastructures to become more inclusive, but such a response ignores the important work happening outside the infrastructures of DH, the scholars who choose not to attend DH conferences or publish in DH journals in large part because they do not feel welcome. The add and stir method of diversity has not and will not solve the problems of whiteness and gender in DH. Instead of inviting those who feel excluded to attend and become subject to micro and macro aggressions, I am arguing that those within the infrastructures, particularly white women, need to do the work to make the change. We cannot expect, as often happens with digital humanities, for the lone junior woman of color scholar to shift the entire board of ADHO's digital humanities conference. This is exploitive.

Central to the work of those who are seeking to position feminist approaches within the study of technology are projects like FemTechNet, Intersections | Feminism, Technology & Digital Humanities, and other feminist technology organizations. Often the organizations are expansive, moving beyond concerns of academic digital humanities to concerns of online harassment, digital activism, representation in technology fields, algorithmic bias, etc. Further, a variety of projects position their work as feminist, from projects like the Lili Elbe Digital Archive which "enacted a feminist ethos that constituted community and space in a new way" (Caughie and Datskou 2020, 464), Chicana por mi Raza, a project formed around Chicana feminist praxis, the Digital German Archive, which, among other things, examines feminism across the East and West divide, or Jane Austen's Fiction Manuscripts, among many others. Each of these organizations and projects centers feminism and technology studies, but resists a representation of their work as normative. In large part this is because of the interdisciplinarity of the feminist DH work, which draws upon other scholarly fields of knowledge with long histories of investigation of technologies and the humanities. As Rambsy asks of Black DHers, "... why not also turn our attention to where we voluntarily locate ourselves?" (2020, 153). He goes on to discuss a broad section of African American scholars who identify as Tyechia L. Thompson does: "a scholar of African American literature *and* digital humanities, in that order. 'I do not lead with the digital humanities focus,' she explained, 'because my approach to DH is through the lens of Af-Am literature, history and culture'" (Rambsy 2020, 156). In Thompson's articulation, DH is the outsider field that is grafted onto the study of African-American literature, history, and culture. The best articulations of digital humanitie(s) recognize that much of the exemplary work is itself centered outside of a rigid notion of DH, instead crossing interdisciplinary boundaries, with new media, critical race studies, postcolonial studies, diaspora studies, critical code studies, disability studies, queer theory and pedagogy studies, among other areas, all central scholarly production locations with long histories. My emphasis on decentering the DH infrastructure is twofold. First, as Risam notes, "the recent popularity of digital humanities obscures a longer history," such as the Afrofuturism listserv created by Alondra Nelson in the 1990s (2019, 19). If we look beyond the Father Busa, great man understanding of DH, if we see other modules of DH as centered, we might then unpack "Why Are the Digital Humanities So White?" (McPherson 2012). Second, by centering other scholarly fields of inquiry we also center the scholars who work in those fields. Moya Z. Bailey's "All the Digital Humanists Are White, All the Nerds Are Men, but Some of Us Are Brave" insists that "By centering the lives of women and people of colour, and disabled folks, the types of possible conversations in digital humanities shift." Bailey continues "The move 'from margin to centre' offers the opportunity to engage new sets of theoretical questions that expose implicit assumptions about what and who counts in digital humanities as well as exposes structural limitations that are the inevitable result of an unexamined identity politics of whiteness, masculinity, and able-bodiedness" (2011). I quote Bailey at length because her model of recentering is what I hope to invoke. To understand recentering is to address the issues of the "emerging community of scholars putting intersectionality at the forefront of digital research methods in the humanities" (Bordalejo and Risam 2019, 5). Scholars that I see as enacting this work may well reject the notion of feminist DH(s) or even DH being applied to their work. Here I want to invoke scholars not to attach them to feminist DH(s) or to co-opt their work. Instead, I want to recognize that the work of these scholars provides rich explorations of technological bias, recovery, ethical collaborations, including of labor and citational practices and intellectual genealogies, code studies, and other areas of inquiry that use humanistic approaches to engage with technology.

The power of feminist digital humanities practice is best viewed within the understanding of intersectionality, the multiple impacts of nation, gender, race, sexuality, class, and more. Yet we need to be cautious that we do more than use an intersectional lens as a means for actions, for change in problematic DH infrastructures. Intersectionality can't be "the equivalent of a social justice booty call," as Taylor writes. Instead we must "show up for the projects, protests, marches, and movements that are not directly about you, not to take up space, but to support women of color, the work of feminism will always be incomplete" (2019, 221). It is this articulation of intersectionality, complete with the work of activism, that must be centered in feminism digital humanities (DH).

NOTES

1. This essay enacts two approaches to emphasize a feminist DH approach. First, I am cognizant of who I am citing, following principles laid out by #citeBlack women, with a focus on citing women of color. Second, I name a multitude of names of those associated with my understanding of digital humanities broadly constructed, decentering my voice and centering theirs.
2. See Weingart and Eichmann-Kalwara (2017) for additional discussion of representation at the ADHO conference.
3. Thank you to Toniesha Taylor who helped me think through the enactment of white allyship and who articulated the use of privilege as a shield.

REFERENCES

Aiyegbusi, Babalola Titilola. 2018. "Decolonizing Digital Humanities: Africa in Perspective." In *Bodies of Information: Intersectional Feminism and Digital Humanities*, edited by Elizabeth Losh and Jacqueline Wernimont, 434–46. Minneapolis, MN: University of Minnesota Press.

Anzaldua, Gloria. 1987. *Borderlands/La Frontera*. San Francisco, CA: Aunt Lute Book Company.

Bailey, Moya Z. 2011. "All the Digital Humanists Are White, All the Nerds Are Men, but Some of Us Are Brave." *Journal of Digital Humanities* 1 (1). http://journalofdigitalhumanities.org/1-1/all-the-digital-humanists-are-white-all-the-nerds-are-men-but-some-of-us-are-brave-by-moya-z-bailey/.

Bianco, Jamie "Skye." 2012. "This Digital Humanities Which Is Not One." In *Debates in the Digital Humanities*, edited by Matthew K. Gold. Minneapolis, MN:University of Minnesota Press. https://dhdebates.gc.cuny.edu/read/untitled-88c11800-9446-469b-a3be-3fdb36bfbd1e/section/1a1c7fbc-6c4a-4cec-9597-a15e33315541.

Bordalejo, Barbara, and Roopika Risam, eds. 2019. *Intersectionality in Digital Humanities*. York: Arc Humanities Press.

Caughie, Pamela L. and Emily Datskou. 2020. "Celebrating the Launch of the Lili Elbe Digital Archive: A Transfeminist Symposium." *TSQ: Transgender Studies Quarterly* 7 (3): 463–75. https://doi.org/10.1215/23289252-8553132.

Christian, Barbara, and Gloria Bowles. 2007. *New Black Feminist Criticism, 1985–2000*. Urbana and Chicago, IL: University of Illinois Press.

Collins, Patricia Hill. 2000. *Black Feminist Thought: Knowledge, Consciousness, and the Politics of Empowerment*. New York, NY: Routledge.

Crenshaw, Kimberlé. 1989. "Demarginalizing the Intersection of Race and Sex: A Black Feminist Critique of Antidiscrimination Doctrine, Feminist Theory and Antiracist Politics." *University of Chicago Legal Forum* (1): 139–67.

FemTechNet. 2015. Addressing Anti-Feminist Violence Online. February 27. https://femtechnet.org/2015/02/addressing-anti-feminist-violence-online/.

Glover, Cameron. 2017. "Intersectionality Ain't for White Women." August 25. *Wear Your Voice* (blog). https://web.archive.org/web/20210512111807/https://www.wearyourvoicemag.com/intersectionality-aint-white-women/.

Griffin, Gabriele. 2019. "Intersectionalized Professional Identities and Gender in the Digital Humanities in the Nordic Countries." *WES: Work, Employment and Society* 33 (6): 966–82. https://doi.org/10.1177/0950017019856821.

Intersections | Feminism, Technology & Digital Humanities. n.d. http://ifte.network/.

Kim, Dorothy. 2019. "Digital Humanities, Intersectionality, and the Ethics of Harm." In *Intersectionality in the Digital Humanities*, edited by Barbara Bordalejo and Roopika Risam, 45–58. York: Arc Humanities Press.

Lorde, Audre. 1983. "The Master's Tools Will Never Dismantle the Master's House." In *This Bridge Called My Back: Writings by Radical Women of Color*, edited by Cherrie Moraga and Gloria Anzaldua, 98–101. Watertown, MA: Kitchen Table Press.

McPherson, Tara. 2012. "Why Are the Digital Humanities So White? or Thinking the Histories of Race and Computation." In *Debates in the Digital Humanities*, edited by Matthew K. Gold. Minneapolis, MN: University of Minnesota Press. https://dhdebates.gc.cuny.edu/read/untitled-88c11800-9446-469b-a3be-3fdb36bfbd1e/section/20df8acd-9ab9-4f35-8a5d-e91aa5f4a0ea#ch09.

Murray, Padmini Ray. 2017. "Presentation, Panel on Copyright, Digital Humanities, and Global Geographies of Knowledge." Presented at the Digital Humanities 2017, Montreal, Canada, August 10.

Murray, Padmini Ray. 2018. "Bringing Up the Bodies: The Visceral, the Virtual, and the Visible." In *Bodies of Information: Intersectional Feminism and Digital Humanities*, edited by Elizabeth Losh and Jacqueline Wernimont, 185–200. Minneapolis, MN: University of Minnesota Press.

Ortega, Élika. 2019. "Zonas de Contacto: A Digital Humanities Ecology of Knowledges." In *Debates in the Digital Humanities 2019*, edited by Matthew K. Gold and Lauren F. Klein. Minneapolis, MN: University of Minnesota Press. https://dhdebates.gc.cuny.edu/read/untitled-f2acf72c-a469-49d8-be35-67f9ac1e3a60/section/aeee46e3-dddc-4668-a1b3-c8983ba4d70a.

Rambsy, Howard II. 2020. "African American Scholars and the Margins of DH." *PMLA* 135 (1): 152–58.

Risam, Roopika. 2019. "Beyond the Margins: Intersectionality and Digital Humanities." In *Intersectionality in Digital Humanities*, edited by Barbara Bordalejo and Roopika Risam, 13–33. York: Arc Humanities Press.

Smith, Martha Nell. 2007. "The Human Touch: Software of the Highest Order: Revisiting Editing as Interpretation." *Textual Cultures* 2 (1): 1–15.

Taylor, Toniesha L. 2019. "Dear Nice White Ladies: A Womanist Response to Intersectional Feminism and Sexual Violence." *Women & Language* 42 (1): 187–90.

Taylor, Toniesha L. 2020. "Social Justice." In *Digital Pedagogy in the Humanities*, edited by Rebecca Davis, Matthew K. Gold, Katherine D. Harris, and Jentery Sayers. New York: MLA. https://digitalpedagogy.hcommons.org/.

Terras, Melissa. 2015. "Why I Do Not Trust Frontiers Journals, Especially not @FrontDigitalHum." *Melissa Terras' Blog*. https://melissaterras.org/2015/07/21/why-i-do-not-trust-frontiers-journals-especially-not-frontdigitalhum/.

Verhoeven, Deb. 2015. "Has Anyone Seen a Woman?" https://speakola.com/ideas/deb-verhoeven-has-anyone-seen-a-woman-2015.

Weingart, Scott B. and Nickoal Eichmann-Kalwara. 2017. "What's under the Big Tent?: A Study of ADHO Conference Abstracts." *Digital Studies/Le Champ Numérique* 7: 6. https://doi.org/10.16995/dscn.284.

Wernimont, Jacqueline and Elizabeth Losh. 2017. "Problems with White Feminism: Intersectionality and Digital Humanities." In *Doing Digital Humanities: Practice, Training, Research*, edited by Constance Crompton, Richard J. Lane and Ray Siemens, 35–46. London and New York, NY: Routledge.

Wernimont, Jacqueline and Elizabeth Losh. 2018. "Introduction." In *Bodies of Information: Intersectional Feminism and Digital Humanities*, ix–xxv. Minneapolis, MN: University of Minnesota Press.

CHAPTER EIGHT

Multilingual Digital Humanities

PEDRO NILSSON-FERNÀNDEZ (UNIVERSITY COLLEGE CORK) AND
QUINN DOMBROWSKI (STANFORD UNIVERSITY)

Digital humanities is contextual, shaped by the identities and interests of the people who engage with it, and the communities in which they are situated. The singular origin story featuring a meeting of minds between Father Busa and IBM has been complicated (Terras and Nyhan 2016) and called into question (Earhart 2015) and challenged by narratives that illustrate the diverse paths along which digital humanities has evolved (e.g., O'Sullivan 2020; Wang et al. 2020). The reasons motivating any one particular course of development vary: funding availability, cultural priorities, and the individual interests and resources available to influential early figures all play a role in shaping how a particular kind of DH came to be. Articulating the long-standing nature and diverse trajectories of DH communities beyond the mono-myth of Busa is an important ongoing project in the field, but even the Busa-linked lineage of "institutional DH" has its under-explored facets.

From IBM-sponsored events in the 1960s, to the formation of European and North American DH organizations in the 1970s, to the establishment of the Alliance of Digital Humanities Organizations (ADHO) in the 2000s, this thread of mainstream institutional DH has explicitly adopted an international scope. In practice, "international" has often meant North America, Western Europe, and occasional others. Furthermore, the global hegemony of US and UK universities, the early engagement of scholars from the US and UK (such as Jerome McGann and Willard McCarty), and the need for a lingua franca at these international events laid the groundwork for treating the English language as a default. This is not to suggest the universal ubiquity of English—see, for example, the long and unbroken tradition of Italian-language DH work within the *Informatica Umanistica* community,[1] or Cristal Hall's study on the intersections and oppositions between traditional and digital Italian Studies (2020). But as noted in Fiormonte (2016), the locus of control for key components of digital infrastructure, from ICANN domain registration, to Unicode, to the most ubiquitous operating systems and software, is situated in the Anglo-American world. The cultural biases and priorities of the Anglo-American world permeate these technologies (Noble 2018), even where they are intended for global use. This has downstream implications for DH.

There is no conscious, deliberate plot to permanently install the English language at the pinnacle of institutional power within DH. Rather, it is most often a matter of monolingual-Anglophone obliviousness with regard to language. Language is a source of friction familiar to anyone whose primary language is not English. From software interfaces, to keyboards, to the availability of subtitles or dubbed audio for the latest global media franchise, the friction of language is a part of everyday life, and digital humanities is no exception to that general rule. For an Anglophone DH scholar living in primarily Anglophone countries whose research uses English-language materials, weeks or months can pass without language imposing any kind of barrier. Perhaps there are fleeting moments of curiosity—overhearing an unintelligible conversation or encountering an unreadable sign—but at almost no point is the scholar forced to engage with another language, lest they be fundamentally impeded in their goals. Other languages, when encountered, become ignorable, and that ignoring becomes a mental habit.

The accumulation of these mental habits across an entire community of monolingual Anglophone scholars renders language functionally invisible as a feature. When encountered, it is ignored, fading into a kind of background noise. This invisibility by normalization makes it even more difficult to recognize language as an axis of diversity in need of intervention, when considered alongside more visible and violent acts of discrimination on the basis of race and gender, as described in Earhart (2016).

The annual DH international conference organized by ADHO provides a venue for scholars with financial means to meet and build relationships with colleagues from across the world.[2] The talks are most often given in English, and informal socializing among participants from diverse linguistic backgrounds often settles into the common denominator of English. The international pool of authors who attend this conference and showcase their research output in venues such as *Digital Scholarship in the Humanities* provide Anglophone scholars with a lens onto DH shaped by other cultural and linguistic contexts. While this engagement with multicultural scholars is crucial, there is also very important and significantly different work presented and published in non-Anglophone venues. This work is widely unknown to monolingual Anglophone scholars, who often have little experience working with scholarship in languages other than English. In practice, a decreasing number of US and UK[3] humanities graduate programs focusing on language and literature recommend a reading knowledge of at least one other "scholarly" language (e.g., German, French, or Russian), but even where such a recommendation exists, actually reading and citing scholarship in those languages is not required as part of graduate training. Whether or not they find it easy to navigate academic prose in English, scholars who work in other languages are still far more likely to cite work in English than vice versa (Fiormonte 2015).

Attending an international conference such as ADHO's may itself be a step closer to breaking out of the national frame of reference that circumscribes many disciplinary conferences, but that conference and its associated journal have a culture that is distinct from a simple amalgamation of the local DH cultures represented by the participants (Fiormonte 2014a). In this chapter, we lay out a set of recommendations for Anglophone scholars to expand their engagement with DH beyond the Anglophone world. We treat this value as a given; to understand DH (or any discipline in the humanities by extension—be it literature, history, philosophy, etc.), it is essential to look beyond the scope of any single language, although many previous efforts to "define DH" have failed to do so (Fiormonte 2014b). However, even if an Anglophone scholar is skeptical of the value other national and linguistic digital humanities practices can bring to their own work, overlooking non-Anglophone DH will exact a toll on those around them. In most institutions, DH courses and workshops are interdisciplinary, drawing students from a wide range of disciplines—including area studies, history, and non-English literatures. DH syllabi designed to be applied to English text may fail if applied without additional cultural contextualization—and pre-processing in the case of textual digital tools—to works produced in other languages. Not being aware of this challenge, or what steps to take to begin adapting English-oriented methods, means failing students who may consider working in languages other than English. Similarly, a scholar who develops and promotes a tool for analyzing text, and hard-codes certain English-oriented decisions (e.g., around stopwords), may be inadvertently limiting the usability and impact of that tool. Furthermore, we encourage Anglophone scholars who hold the reins of infrastructure to engage more seriously with language in contexts including scholarly publishing, even if the result leads to increased linguistic friction.

LANGUAGE IN DH JOURNALS

The spirit shown in fragments from Digital Humanities manifestos originating from THATCamp Paris (Dacos 2011) or UCLA's Mellon Seminar in Digital Humanities (UCLA 2008), stating that "interdisciplinarity, transdisciplinarity and multidisciplinarity are empty words unless they imply changes in language, practice, method, and output" (UCLA 2008), falls short when it comes to multilingualism. Later, this same manifesto articulates that "traditional Humanities is balkanized by nation, language, method, and media" and that "Digital Humanities is about convergence."

The price of prioritizing this convergence at any cost is that we tend to perceive the duality between global and local in very absolute terms. The "global" becomes a synonym for forward-thinking, endless, and borderless collaborative practices, whereas the "local" is associated with opposite values. Claiming that "nation" acts as divider of a discipline when writing from a monolingual and Anglo-centric position is hardly neutral when the lingua franca that will mediate the "global" is the one spoken in your nation.

However, for nations where English is not the first language, the sacrifices involved in being part of the international academic network are considerable. The imposition of English as lingua franca brings an intellectual disadvantage to scholars in the humanities who produce high-quality research in languages other than English but are not able, or simply choose not to, render that research output in English. When research is not translated into English, its dissemination and impact beyond the territories in which its language is spoken officially is strictly dependent on a factor that bears little to no connection with the quality of the research itself: the number of international academics in the field who are able to read and review work in that specific language.[4] Part of this dissemination occurs in the form of citing such output; interestingly enough, if these international peers choose to direct-quote fragments of the work rather than paraphrasing its ideas, most English-language journals will still ask for these fragments to also be translated into English. This results in additional unpaid labor that becomes another burden on the multilingual researcher. Without removing any traces of the original language in which the article was written, scholarship written in other languages is put at a disadvantage, regardless of its quality or relevance to the field. English-language articles are more convenient sources to quote—particularly in terms of labor and visibility—only because of the immediate accessibility granted by the lingua franca they are written in.

Choosing the translation route is not as straightforward a solution as one might initially expect. While several companies provide academic translation services to English catering to source languages such as Chinese, Portuguese, and Spanish (at an approximate cost of 700 USD), the price and availability changes the moment you select a smaller language such as Catalan as the source language (1,700 euro).[5] Provided that the scholar can afford to produce such a translation, the translation adds an additional layer to the already highly competitive process of academic publishing; we also need to take into account that additional translation work will be required at every stage of the publication process, as manuscript revisions are a natural step in the process of publishing in any peer-reviewed journal.[6] Both the economic barriers and the complexity of the process explain why many scholars prefer to attempt an exercise of self-translation. It is precisely in these contexts that the dismissive attitudes towards multilingualism in academic publishing have a more devastating effect on non-English-speaking scholars, whose efforts to reduce that linguistic gap go highly unnoticed in English-language scholarship. As part of *The Digital Humanities Long View* series[7] Riva Quiroga noted a recent online controversy initiated by a tweet[8] from a co-editor of the *Journal of Contemporary History*, Sir Richard Evans, who unapologetically describes imperfect English as a major reason for rejecting papers.

In the absence of translation or work written originally in English, the odds of Anglophone scholars encountering articles from scholars who work primarily in other languages when exploring major international DH journals are low (Fiormonte 2015). Such articles tend to be published in special issues in a specific non-English language, and are unlikely to be connected into the citation networks of English-language articles. The emergence of these special issues (such as the Spanish-language *Digital Humanities Quarterly* vol. 12, issue 1 in 2018) is a positive step towards multilingual DH. However, this linguistic compartmentalization reinforces the power imbalance between languages. It isolates these articles, reducing serendipitous encounters for scholars browsing the journal. Furthermore, when we look specifically at the non-English languages used for these special issues, they are unequivocally global languages: Spanish, French, Portuguese. Scholarship in other languages—including the languages of other ADHO constituent organizations—is pushed further to the periphery.

As a discipline that praises itself for breaking down the barriers preventing free access to knowledge and information, DH routinely fails to acknowledge the barrier posed by language. Following the rationale used by Laura Estill when she claimed that Digital Humanities had a "Shakespeare problem" (Estill 2019), we may well say that Western Digital Humanities has an English problem.

LANGUAGES IN THE DH COMMUNITY

The web, writ large, is a much richer source of information about DH praxis in other languages and cultural contexts than international journals, and web-based content is among the easiest to access via machine translation thanks to numerous browser-based plugins. Despite the dominance of the English language in most search engines, searching for the translation of "digital humanities" in different languages will surface projects, articles, and discussions that would remain completely invisible otherwise. It is a matter of linguistic awareness; in the case of Google, for example, adjusting options like region and language in which to conduct the search will produce completely different results. Likewise, there are different layers of linguistic bias when it comes to global languages. A search involving languages and cultures in the Iberian Peninsula beyond Spanish and Portuguese—i.e., Catalan, Euskera, or Galician—is likely to produce more Spanish and Portuguese language results than in those three minoritized languages struggling for space with their global counterparts.

Exploring blog posts on digital humanities written in other languages, following scholars on social media who post about DH in their native tongues, or looking at abstracts from conferences held outside of English-speaking countries offers a multilingual and multicultural insight into the discipline that is blurred in English-only contexts. These academic communities may not only interpret DH concepts in a different way (e.g., see Berra et al. 2020), but may draw upon a long and unique history, and may bring fresh approaches to an Anglophone-dominated discipline.

Most importantly, some of these academic communities may be at earlier stages in the incorporation of digital humanities methodologies in their research. This affords them a unique opportunity to reflect on initial practices of the discipline, to reconsider the validity of some paradigms that have been established in "mainstream" Anglophone DH—both in theory and praxis—and to question them from beyond an Anglophone perspective. The creation and or modification of infrastructures in order to adjust to the multidisciplinary nature of DH will take different forms when responding to specific cultural challenges in every case. As Isabel Galina shows when she chronicles the creation of *RedHD* (2014) and Puthiya Purayil Sneha when mapping the emergence of DH in India (2017), we cannot assume that DH is received and practiced identically in different communities globally.

DH practitioners who work, write, and research in and about minoritized languages and cultures have valuable lessons to share, drawing upon the constant struggle for visibility of their research output. In many cases, the choice to use their native language in academia is not brought by the inability to express themselves in English or in another global language, but is a question of linguistic activism. Their resilience against the imposition of global languages in their discipline responds to a constant fight for the survival of their mother tongue and the conservation of a digital cultural record that is disrupted by colonialism and neocolonialism (Risam 2019).

This determined struggle for visibility may also resonate with current discourse in DH around sustainability of DH projects; sustainability of non-English DH projects becomes a more pressing issue, as in some cases, the loss of digital presence or the interruption of these projects can have devastating consequences such as loss of digital cultural record. While English-language projects and their data are likely to be supported by a stronger digital infrastructure and often managed, stored, updated, and secured by more than one institution, projects and datasets related to knowledge written in a smaller language may be in a different state of conservation and is often at risk of disappearing altogether. This situation should be taken into account when prioritizing resources, such as funding and labor, for long-term archiving of digital projects.

LANGUAGES IN DH TUTORIALS AND TOOLS

Acknowledging the importance of building DH tools whose pipelines are flexible enough to adapt to multilingual environments is vital to drive change of attitude towards linguistic diversity in the discipline. While some DH tools are specifically built for the purpose of catering to the needs of a single project, the tools that have made the biggest difference in the field are those whose application extends beyond individual research questions and benefit the wider DH community. If we observe the reception and use of some of these key tools through the lenses of a project such as *The Programming Historian*—which provides volunteer-written tutorials for such tools in English, Spanish, French, and Portuguese—the potential for growth of non-English-speaking users of these tools is significant. The presence of multilingualism in the platform is not limited to the translated tutorials; *The Programming Historian*'s English-language tutorials create awareness of the multilingual applications of the tools referenced. The "Introduction to Jupyter Notebooks" tutorial (Dombrowski, Gniady, and Kloster 2019) uses *Harry Potter* fanfic metadata in Italian as the example file to be used in the tutorial, while also providing a link to a Jupyter notebook created for an article on stylometry published in Spanish (Calvo Tello 2016). A similar example can be seen in Andrew Akhlaghi's tutorial on "OCR and Machine Translation with Tesseract," using Russian texts as source material.

The community of multilingual DH scholars would benefit from the broad adoption of what has become known in linguistics as the "Bender Rule," after University of Washington professor Emily Bender (Bender and Friedman 2018). Put simply, the Bender Rule is "always name the language you are working on," and it responds to a situation where, much like in DH, English exhibits extreme unmarkedness in everything done in the field of linguistics. If a new dataset, resource, tool, project, article, book, or chapter appears, and no specific language cue is associated with it, we automatically assume its language to be English. The presence and dominance of the language in the field is so obvious that it simply becomes the standard. Naming language is a first step towards reasserting the role of English *as a language*, and one among many possible options.

In a DH tool development context, the Bender Rule could be adapted as "always include a 'languages' section in your tool documentation." A great example of what such a section may look like can be found within *Voyant Tools* documentation, where users can learn about the ten different languages its interface is available in, on the one hand, and the different languages that the tool can analyse, on the other. The developers are very honest in admitting the shortfalls of the wide generic support of the tool, admitting that it reduces language-specific functionality, but are also proud to mention the language-specific awareness brought by their document tokenizer, and their interface readiness for RTL[9] scripts. Looking at the list of translators of the interface, we can see that it is composed by practitioners who work on DH through each of the particular languages, this volunteer or community-driven effort can be seen in most of what we do to expand the linguistic horizons of DH (see Ghorbaninejad et al. 2022, for a description of the process by which the interface was translated to Arabic).

Against the evident imbalance between the availability of data (corpora, datasets) in English language and languages other than English, and the reported issues with existing online datasets in languages other than English (Caswell et al. 2021), tool development should explore ways to counteract this data inequality by prioritizing algorithms that work better with less data. In the same way that minimal computing initiatives work on the basis that there is a scarcity of hardware resources, multilingual DH tool building should bear in mind data scarcity in their design. In parallel, tools could be designed proactively to assume a lack of reliable datasets, providing plugins or packages that can easily incorporate human-supervised data validation instruments working parallel to its main purpose. While practices such as crowdsourcing may be considered less effective in English-language DH (due to this need for data validation), it is a vital resource for multilingual DH.

LANGUAGES IN THE DH CLASSROOM

The institutionalization of digital humanities in the United States, Canada, and United Kingdom through degree and certificate programs has typically taken method as a major organizing principle, with discipline as a secondary or aggregating factor. Students accrue credits by taking courses, for instance, on network analysis, geospatial methods, computational text analysis, or TEI text encoding. There is often a survey "Intro to DH" course, and many programs have one or more courses that centers the work and debates within a discipline, such as "digital history" or "computational literary studies." With few exceptions,[10] language is never an organizing principle for courses. Considering the staffing costs involved in offering a course, this may be justifiable: there may not be enough students working at the intersection of DH and any one specific non-English language to draw the necessary enrollment for a dedicated course. But the alternative of weaving linguistic awareness into methods courses that explicitly welcome all, regardless of what language they work in—has not been the expected standard for DH pedagogy. While some recent courses have placed greater emphasis on issues such as Unicode text encoding and lemmatization for languages with more inflection than English, it is easy to overlook these steps when working exclusively with English examples.

On a practical level, teaching a class using shared, English-language data makes a great deal of sense. Pre-prepared examples make it possible to deliberately plan for the complications and challenges that the material will introduce. It allows instructors to create assignments with correct answers identified in advance, facilitating grading. But problems arise when students learn (or assume) that the workflow taught in class for the English-language example text is, in fact, the workflow for *analyzing text*. Students who wish to apply these methods to other languages may conclude that text analysis "doesn't work" for their language when they apply the workflow used in class to non-English text. The result is that only the students who work on English-language texts leave the classroom with practical skills they can apply to their own research. This problem is most visible in text analysis courses, but many other methods—including network analysis and geospatial work—build on datasets that are themselves the result of text analysis. Teaching with a pre-prepared dataset that was created using tools that perform much better for English than other languages obscures the startup labor required to apply the methods in a new context—and hides the tool gap that may make these methods impractical for some languages (McDonough et al. 2019).

The Bender Rule is, once again, a step to begin with when addressing linguistic gaps in curricula. While it may seem trivial, stating clearly that a workshop on text analysis that only uses English text and tools is, in fact, a workshop on *English* text analysis is a step towards making language more visible as an important variable when doing this work. In some cases, acknowledging this limitation—and leaving students who are interested in other languages to accept it or look elsewhere—is all that is feasible. Librarians, graduate students, and people in assorted staff roles often teach DH workshops or courses on the margins of other jobs, with little or no time or financial support for reworking pre-prepared material to be more linguistically inclusive, or expanding their own knowledge about how to work with a range of commonly requested languages. In cases where an instructor does have support to develop a workshop or course that can serve a broader range of students, there are a number of meaningful steps that they can take. Particularly in the context of a multi-week or multi-month course, instructors can ask upfront about what language(s) students are interested in working on. The next step is to learn something about the orthography and grammar of any language(s) with which you are unfamiliar. Is the language written with spaces between words? Is it written right-to-left (RTL)? Some tools have problems processing or displaying RTL text (Wrisley et al. Forthcoming). Is there a lot of inflection? Most of these basic linguistic facts can be ascertained as simply as skimming the Wikipedia page for the language.

Having gathered this information, consider the impact on the methods you plan to teach. If your students want to work with *analytic languages* (which use "helper words" rather than inflection to convey relationships between words; languages include Chinese, Vietnamese, and English), and your methods are based on word counts (like TF-IDF[11] or topic modeling), you can simply mention that the methods taught in class should also perform well for those languages. If, however, you are using word-count methods and some of your students want to work with highly inflected languages such as Ukrainian, or highly agglutinative languages such as Turkish, you should, at a minimum, mention the challenges associated with applying the methods taught in class to such languages. Better yet would be to equip the students with the tools they need to address these challenges, such as a lemmatizer.[12] Similarly, many natural language processing tools are language-specific. If you are teaching a method that involves POS tagging[13] or NER,[14] it is worth searching for what resources are available to perform similar tasks in the languages your students are interested in working with. The set of languages represented by students in your class may also shape your choice of tools.

CONCLUSION

There is vastly more to digital humanities than what is reflected from an Anglophone outlook, even with the glimpses into the DH of other contexts offered by international venues. Too often, the work that appears in these "international" journals and conferences is less a representation of the diversity of local digital humanities as it is experienced by scholars around the world, than a predictable and limited piece of that work that is likely to be more comprehensible to "mainstream" Anglophone DH. Even scholars who are famous for not holding back in their criticism of Anglophone DH find themselves expressing these opinions even more forcefully when writing for non-Anglophone audiences. Language gives scholars a way to play, for instance, with the concept of (implicitly Anglophone) "digital humanities" set against the concept of "humanités numériques," which grows from and is shaped by the Francophone world. Engaging with these other linguistic worlds is easier than ever with the help of machine translation, but tools are limited by scholars' willingness to make the effort to use them. It requires some effort to find and follow scholars who write in languages other than English, in formal and informal online spaces, hitting the "translate" button on Twitter or in the browser rather than skipping over their work. Monolingual Anglophone scholars may be comfortable within the bounds of that linguistic sphere, but they owe it to their audience with broader horizons to at least be aware of and clear about that limitation in their knowledge of the field. The Bender Rule of ubiquitously naming language is a place to start; if your awareness is limited to the Anglophone world, defining your expertise as English-language DH rather than DH at large would go a long way in making non-English DH a more visible type of valued expertise. But even small steps into a world of greater linguistic diversity—like considering the impact of language on the tools you build, or browsing the programs of conferences held in other languages—can have a significant impact on your own understanding of the field, and your ability to support students and colleagues whose scholarship extends beyond English.

NOTES

1. See the website of the *Associazione per l'Informatica Umanistica e la Cultura Digitale* for examples, http://www.aiucd.it/.
2. The exclusionary costs associated with attending the ADHO conference have been sharply criticized, particularly by scholars from the Global South (Earhart 2018).
3. In the particular case of the UK, Modern Language departments have seen the consequences of several government reforms, the latest a controversial one in 2006 deeming a second language optional in secondary school programs.

4. Taking Catalan Studies and the Catalan language as examples of field of study and minoritized language, international dissemination of this output will very likely depend on members of academic organizations such as *The Anglo-Catalan Society* (ACS), a network connecting academics in the British Isles and Ireland, and the *North American Catalan Society* (NACS) to connect with the other side of the Atlantic.
5. These approximate figures are based on online quote inquiries made to two academic translation agencies: AJE (https://www.aje.com/) and Transistent (https://www.transistent.com/).
6. The same academic translation companies mentioned above provide premium packages (at an additional cost) that cover a second translation and review of subsequent versions of the submitted manuscript.
7. Riva Quiroga's talk "Multilingual Publishing in Digital Humanities," as part of the virtual seminar series *The Digital Humanities Long View* (co-hosted by UCL Centre for Digital Humanities and the Center for Spatial and Textual Analysis at Stanford University) was written up in the *Stanford Daily*.
8. Evans's tweet reads: "206 articles were sent in to the Journal of Contemporary History in 2020, 39 were accepted (17%). In 2019 the figure was 136 and 31 accepted (26%). I don't think we rejected any that we should have accepted. Main reason for rejection: poor English!" (https://twitter.com/RichardEvans36/status/1358831507197018113?s=20). As part of the same conversation, and referring to the possibility of accepting work in languages other than English, his reply reads: "The JCH has only published in English since its foundation (by two non-native speakers) half a century ago. It would be folly to do anything else"(https://twitter.com/RichardEvans36/status/1359187632580927494?s=20).
9. RTL stands for Right-to-Left written, read and processed languages, i.e., Arabic, as opposed to Left to Right, i.e., English.
10. Such as "DLCL 204: Digital Humanities across Borders" at Stanford University, where each student brings their own non-English language to work with. Molly des Jardin's "East Asian DH" course at UPenn and the summer 2021 UPenn Dream Lab workshop on "East Asian Studies & Digital Humanities" are more similar to discipline-scoped DH classes than language-scoped ones, though the impact of language on method is discussed in both.
11. Term Frequency-Inverse Document Frequency.
12. A lemmatizer is a tool from Natural Language Processing that identifies the lemma in a word via full morphological analysis.
13. POS or part-of-speech tagging is a process in corpus linguistics that marks words up in a corpus as corresponding to a particular part of speech, taking into account its context and meaning.
14. NER or named-entity recognition is a method of recognizing and classifying text into predefined categories (i.e., person names, locations, etc.) from within unstructured text-based data.

REFERENCES

Bender, Emily and Batya Friedman. 2018. "Data Statements for Natural Language Processing." *Transactions of the Association for Computational Linguistics* 6: 587–604.

Berra, Aurélien, Emmanuel Château-Dutier, Sébastien Poublanc, Émilien Ruiz, Nicolas Thély, and Emmanuelle Morlock. 2020. "Editorial. Donner à lire les humanités numériques francophones." *Humanités numériques* 2: 2–5.

Calvo Tello, José. 2016. "Entendiendo Delta desde las Humanidades," *Caracteres*, May 27. http://revistacaracteres.net/revista/vol5n1mayo2016/entendiendo-delta/.

Caswell, Isaac, Julia Kreutzer, Lisa Wang, Ahsan Wahab, Daan van Esch, Nasanbayar Ulzii-Orshikh, Allahsera Tapo et al. 2021. "Quality at a Glance: An Audit of Web-crawled Multilingual Datasets." *arXiv*, October 25. https://arxiv.org/abs/2103.12028.

Dacos, Marin. 2011. "Manifesto for the Digital Humanities." *Hypotheses*, March 26. https://tcp.hypotheses.org/411.

Dombrowski, Quinn, Tassie Gniady, and David Kloster. 2019. "Introduction to Jupyter Notebooks," *Programming Historian* 8. https://doi.org/10.46430/phen0087.

Earhart, Amy. 2015. *Traces of the Old, Uses of the New: The Emergence of Digital Literary Studies*. Michigan, MN: University of Michigan Press.

Earhart, Amy. 2016. "Digital Humanities Futures: Conflict, Power, and Public Knowledge." *Digital Studies/Le champ numérique* 6, no. 1.

Earhart, Amy. 2018. "Digital Humanities within a Global Context: Creating Borderlands of Localized Expression." *Fudan Journal of the Humanities and Social Sciences* 11 (September): 357–69.

Estill, Laura. 2019. "Digital Humanities' Shakespeare Problem." *Humanities* 8 (1): 45.

Fiormonte, Domenico. 2014a. "Dreaming of Multiculturalism at DH2014." *Infolet: Cultura e critica dei media digitali*, 6. https://infolet.it/2014/07/07/dreaming-of-multiculturalism-at-dh2014/.

Fiormonte, Domenico. 2014b. "Humanidades Digitales y diversidad cultural." *Infolet: Cultura e critica dei media digitali*, 6. https://infolet.it/2014/02/01/humanidades-digitales-y-diversidad-cultural/.

Fiormonte, Domenico. 2015. "Towards monocultural (digital) Humanities?" *Infolet: Cultura e critica dei media digitali*, 7. https://infolet.it/2015/07/12/monocultural-humanities/.

Fiormonte, Domenico. 2016. "Towards a Cultural Critique of the Digital Humanities." *Historical Social Research/Historische Sozialforschung* 37 (3): 59–76. http://www.jstor.org/stable/41636597.

Ghorbaninejad, Masoud, Nathan Gibson, David Joseph Wrisley. Forthcoming [2022]. *Debates in the Digital Humanities*. Minneapolis, MN: University of Minnesota Press.

Hall, Crystal. 2020 "Digital Humanities and Italian Studies: Intersections and Oppositions." *Italian Culture* 37 (2): 97–115. doi: 10.1080/01614622.2019.1717754.

McDonough, Katherine, Ludovic Moncla, and Matje van de Camp. 2019. "Named Entity Recognition Goes to Old Regime France: Geographic Text Analysis for Early Modern French Corpora." *International Journal of Geographical Information Science* 33 (12): 2498–522.

Noble, Safiya Umoja. 2018. *Algorithms of Oppression*. New York, NY: New York University Press.

O'Sullivan, James. 2020. "The Digital Humanities in Ireland." *Digital Studies/Le champ numérique* 10 (1).

Risam, Roopika. 2019. *New Digital Worlds: Postcolonial Digital Humanities in Theory, Praxis, and Pedagogy*. Evanston, IL: Northwestern University Press.

Terras, Melissa and Julianne Nyhan. 2016. "Father Busa's Female Punch Card Operatives." In *Debates in DH 2015*, edited by Matthew K. Gold and Lauren F. Klein, 60–5. Minneapolis, MN: University of Minnesota Press.

UCLA 2008. "A Digital Humanities Manifesto." December 15.

Wang, Xiaoguang, Xu Tan, and Huinan Li. 2020. "The Evolution of Digital Humanities in China." *Library Trends* 69 (1) (summer): 7–29.

CHAPTER NINE

Digital Humanities and/as Media Studies

ABIGAIL MORESHEAD AND ANASTASIA SALTER
(UNIVERSITY OF CENTRAL FLORIDA)

INTRODUCTION: REVISITING THE DIGITAL HUMANITIES

In a response to the *Day of Digital Humanities* prompt to "define the digital humanities" in 2011, Alex Reid offered two competing Venn diagrams of the field suggesting the "digital" and "humanities" might best be understood as co-extensive. His argument suggested that this overlap could be understood as saying two things:

1. *Not* that all humanists must study the digital, *but* that all humanistic study is mediated by digital technologies (and that we do not fully understand the implications of that).
2. *Not* that the digital world is solely the province of the humanities or that only the humanities have something useful to say about it, *but* that through a realist ontology we can extend humanistic investigation into the profoundly non-human world of digital objects (Reid 2011).

Revisiting these diagrams nearly a decade later is materially difficult: the blog post has disappeared and must be accessed through the preservation mechanisms of the Internet Archive's Wayback Machine. The numerous definitions of digital humanities that have circulated since have often done so through even more ephemeral platforms; Quinn Dombrowski quipped on Twitter that "the definition of DH is now up to 16 separate Slack instances, with at least three colleagues who you communicate with via DM in more than one instance" (2021). Those platforms in turn are even more closed and proprietary—a free Slack server deletes its own history out from under its users, gradually consuming its own archive. The digital mediation of humanities discourse presents ongoing challenges of preservation, drawing our attention to the media studies connections of urgency: while Tara McPherson points out that the early digital humanities scholars "often concentrated [their efforts] on archiving, digitizing, and preserving the human record" (2009), those same preservation efforts are subject to the quick obsolescence of current digital media. This work is also at the heart of the interconnected and critical field of media archaeology, which Lori Emerson defines as seeking not "to reveal the present as an inevitable consequence of the past but instead looks to describe it as one possibility generated out of an heterogeneous past" (2014, xiii). The disappearance of this shared past, and the inequity of what is preserved and amplified, here informs our understanding of how we might think about the intersecting needs of studying with and through the "digital" in both digital humanities and media studies.

One of the points of Reid's definition—that we do not "fully understand the implications" of our mediated study—has become particularly pressing both in academic and public discourse, as we navigate the impact of Facebook, Twitter, Instagram, and other mediated communal platforms on ourselves and society. To return to Dombrowski's Slack instances, the reliance of communities on corporate platforms for history leads to the continual threat of erasure: vital works disappear from Apple's App Store with every update, and technologies such as Flash that defined both early web media creativity and early digital humanities no longer function on the modern web. As we write now in the shadow of the (re-)emerging discourse of the metaverse, this impact and the consequences of corporate platforms of cultural production are more pressing than ever. In his history of the relationship between digital humanities and media studies, Lindsay Graham critiques the connection of digital humanities with the textual, noting its limitations: "The concentration on DH, historically speaking, has been on recovering text-based documents and returning those to the public sphere, and in some ways the history of the digital humanities reflects the rise of the personal computer" (2018). But the digital humanities are shifting from the textual, recognizing the urgent needs of the multimodal and multimedia, even as key early instances of public-facing digital humanities scholarship gradually fade from accessibility with the platforms that power them.

DRAWING ON MEDIA AND PLATFORM STUDIES

The intersection of Reid's two claims, updated for our current challenges, is thus where we find the necessary transdisciplinary work of the digital humanities and media studies: media studies is, arguably, where the study of the digital is inescapable, and the impact of software and platforms is felt in every aspect—from the now-variable length of a sitcom freed from the constraints of a 30-minute, advertisement-driven network timeslot to the also transdisciplinary spaces of Internet research and game studies. Confronting the algorithmic is inherent in any form of media studies. Media studies has historically been compelled by stories of infrastructural changes: while once those played out primarily in the physical spaces of movie theater affordances and concert halls, the digital has an inescapable primacy.

Responding in part to Alex Reid's post at the time, Ian Bogost offered the comment that "The digital humanities are just the humanities of the present moment" (2011). Again, a decade later this requires a pause, but perhaps no real revision. Arguably, the Covid-19 pandemic has brought home our digital mediation of the humanistic—inescapably so, as the work of our fields has moved to makeshift offices and repurposed business conference call platforms. Safiya Noble's work reminds us that the algorithmic is no longer optional—whether one defines oneself as a digital humanist or not, the digital is an inescapable part of humanities work. As Noble describes, we are fundamentally reliant in the profession upon technologies that demand interrogation: "On the Internet and in our everyday uses of technology, discrimination is also embedded in computer code and, increasingly, in artificial intelligence technologies that we are reliant on, by choice or not" (2018: 1). While the work of digital humanities is often viewed as an opt-in, methods-driven choice of engagement, the reality is that many of the technologies originally associated with the early adopters of DH are now ubiquitous.

In drawing connections between the spaces of media studies and the digital humanities, Jentery Sayers emphasized the parallels of both methods and intention: "... when they are combined in theory as well as in practice, we could say that media studies and digital humanities work through new media as means and modes of inquiry" (2018: 2). The overlap between the two terms is also seen in Julie Thompson Klein's definition of digital humanities as encompassing both the nascent practices of humanities computing and "new modes of data visualization, digital-born scholarship, and subfields such as code studies, software studies, and game studies" (2017). Setting aside momentarily the characterization of code, software, and game studies as subfields of DH, the capaciousness of this definition, along with overlaps of method described by Sayers and Reid, portrays an optimistic image of digital humanities and media studies as productively intertwined.

We are arguing here, though, that this interconnection between the two fields has often existed only on a rhetorical level, and that a deeper engagement with critiques of platforms common to media studies (particularly through an intersectional feminist lens) is necessary to continue to expand the scope of scholarship that falls under the term "digital humanities." In framing this relationship, we draw upon Julie Thompson Klein's definition of boundary work: "Boundary work is a composite label for claims, activities, and structures by which individuals and groups create, maintain, break down, and reformulate boundaries between knowledge units" (Klein 2017, 21). As the methods of digital humanities become more deeply embedded, and our reliance upon digital mediation is solidly entrenched, we need work at the boundaries that pushes back at those assumptions (and the tools we've embraced.)

THE LIMITATIONS OF DIGITAL HUMANITIES

Almost any history of digital humanities ties the movement to the earliest iterations of the pairing of computation and text—Father Busa's *Index Thomisticus* and other iterations of linguistic computing, followed by the earliest versions of digital literary archives and the accompanying interrogations of textual theory in light of digital practices. The result has been a seemingly natural pairing of digital humanities methods with literary study and the institutional housing of DH scholarship either in or adjacent to English departments. Matthew Kirschenbaum mentions this historical relationship along with other reasons, like the "openness of English departments to cultural studies, where computers and other objects of digital culture become the centerpiece of analysis" (2012).

But despite the potential expansiveness of English studies to include new media, the same power dynamics that exist in English departments and in higher education more broadly are, unsurprisingly, born out in the content and priorities of DH projects. Roopika Risam notes that "in digital literary studies, for example, projects like the William Blake Archive, the Walt Whitman Archive, and the Dante Gabriel Rossetti Archive give the sense that the digital humanities may, in fact, be the digital *canonical* humanities," while also dispelling the idea that simply "an additive approach to the digital cultural record is sufficient" to address the white, neocolonial domination of the digital record (2018: 16). Risam's critique of DH needing to be more than "additive" as a means of addressing neocolonial power structures is similar to concerns expressed by feminist literary scholars like Jacqueline Wernimont, who advocates for digital literary archives that resist "monumentalist" assumptions that "bigger is better" (2013). Wernimont states that, "however 'open,' 'collaborative,' and 'connected' Digital Humanities purports to be, if computational tools are wielded in ways that continue old patriarchal privileges of expertise and authority and create merely receptive users, then we miss an opportunity to leverage digital tools to transform literary scholarship in meaningful ways" (2013). In the US context, rising attention focused in both media studies and the humanities broadly on anti-racist work has re-emphasized this urgency.

The digital is not a solution to exclusion and systemic challenges: indeed, what Risam and Wernimont are pointing to here is a tendency of DH projects to stop short of interrogating underlying power structures—the same structures which are baked into digital tools and labor. As Wernimont points out, the binary of men as makers/women as consumers still pervades male-dominated fields like computer science, which is amplified in DH projects that "unite two historically gendered fields" of "computer and archival sciences" (2013). That same binary (and the historic exclusion of queer, trans, and non-binary scholars and authors from visibility in that framework) pervades the digital systems themselves, as Sasha Costanza-Chock's compelling work pushing back at designs that reproduce structural inequities and call for "design justice" reminds us (2020).

Similarly, Alan Liu's point that "the digital humanities are noticeably missing in action on the cultural-critical scene" points to the narrow scope of DH engagement, which Christina Boyles accounts for by examining how spending trends in recent years on DH projects by the NEH and Mellon Foundation—trends that point toward prioritizing technological advancement and focus on already well-known authors or historical figures, with sadly less allocation of resources toward projects that engage with marginalized communities (Liu 2012; Boyles 2018). However, this direction is shifting: grants such as the "Digital Projects for the Public" initiative of NEH ODH sit at the intersection of media production and digital humanities, and offer essential funding for cultural work more broadly.

REPARATIVE WORK AND INTERSECTIONAL FEMINIST DIGITAL HUMANITIES

The historic conservatism of content and resource allocation in DH that these critiques point to is aligned with broader historical trends in the humanist discourse. As Sayers points out, digital humanities and media studies tend to differ in the time period under examination, with DH focused on making available older media, while media studies tends to focus on newer forms; digital humanities tends to use media as tools, while media studies examines those tools (2018). But we are arguing here that this distinction has only served DH to a point—that digital humanities can only fulfill the early utopian visions of its democratizing potential when productively paired with methods that provoke deeper levels of cultural criticism. Work being done in feminist media studies offers just such a framework.

The course correctives central to both media studies and digital humanities over the past decade emphasize the rejection of simple progressive narratives of technological utopianism, particularly the often-hailed democratizing potential of technology. In *Bodies of Information* (2018), Elizabeth Losh and Jacqueline Wernimont draw together a transdisciplinary group of scholars for a pivotal collection of feminist voices in the field, and in doing so offer an essential rejoinder both to traditional and digital humanities, noting that:

> Both proponents and opponents of DH seem able to agree on one common position: histories of feminist and antiracist work in DH do not deserve a place at the table. By contrast, our argument is that feminisms have been and must continue to be central to the identity and the methodologies of the digital humanities as a field.

As a corrective to exclusion of feminist concerns from digital humanities, Wernimont and Losh argue for the centrality that intersectional feminism must occupy:

> For future digital humanities work to create what is possible and combat what should be impermissible, we believe that intersectional feminism, which acknowledges the interactions of multiple power structures (including race, sexuality, class, and ability), must be central within digital humanities practices.

Tara McPherson's response to the question of "Why are the digital humanities so white?" provides both a useful example of feminist media study while also giving a framework for how digital humanities scholars can reconsider relationships between tools, methods, and content. McPherson argues that the theme of modularity pervaded American culture in the mid-twentieth century, in ways that influenced both the design of technology and the partitioning of race in society. McPherson identifies modularity as the same organizational structure that influenced how early programmers designed computer parts and code to operate as discrete entities unto themselves, "deploying common sense and notions about simplicity to justify their innovations in code" (McPherson 2012). In comparing this phenomenon to the lack of ethnic diversity observed in DH, McPherson argues that this separation "is itself part and parcel of the organization of knowledge production that operating systems like UNIX helped to disseminate around the world? Might we ask whether there is just such a movement, a movement that partitions race off from the specificity of media forms?" (2012). Certainly, the traditional disembodiment of utopian online imaginaries has provided a convenient framework historically for discounting systemic bias.

McPherson's analysis of the relationship between race and digital forms is important because of how deeply into the history and structures of knowledge creation her work looks—her method of analysis resists the modularity that she argues is endemic to the structuring of race and technology. In invoking Antonio Gramsci's assertion that "common sense is a historically situated process," McPherson is leveling a similar critique of the tech industry to those of more recent feminist media scholarship, like Ruha Benjamin's argument against the supposed neutrality of algorithms that are deployed in the name of progress. Benjamin invokes the term "hostile architecture" to describe how algorithms construct spaces that make it intentionally hard for groups to exist or participate in those spaces, yet the fallout of creating such spaces is not apparent to the largely white male demographic that engineers them (Benjamin 2019, 91). The use of algorithms to construct inhospitable spaces, which Benjamin compares to how public physical places often use barriers like armrests to make public benches intentionally unusable by homeless populations, mimics the modularity McPherson critiques in the early days of computing. And by operating under the guise of neutrality, such spaces make themselves appear as the "common sense" of this historical moment.

The potential for intersectional feminist research methods to resist the compartmentalization and socially situated concepts (that defy their situatedness) like neutrality and objectivity is also shown in the interdisciplinary nature of Kishonna L. Gray's scholarship on the intersectional lives of black gamers in digital spaces resists notions of technological neutrality and determinism by demonstrating the ways in which "marginalized folks have made it clear that there is always a manifestation of physical inequalities in digital spaces" (Gray 2020, 21). By conducting interviews with participants in social gaming spaces, Gray is compiling a kind of archive of personal narrative—the kind of historical material that digital humanists would see as their purview to digitize—and by positioning these interviews within the context of digital spaces, Gray can simultaneously examine individual experience and the structures in which players' intersectional identities are enacted.

The theme of multiple layers pervades Gray's work, both in the intersectionality of the identities of her research subjects but also in how she critiques games, at both the hardware and social levels. For instance, Gray critiques how the gaming industry has disregarded the experiences of disabled players both in making inaccessible games at the hardware level but also in very limited representations of disabled characters in the games' narratives. This issue has given rise, Gray points out, to sites that help disabled gamers find games that are most accessible. The critique of hardware, software, and experience here points to how feminist media analysis like Gray's answers the challenge posed by feminist digital humanists, who see critique of tools—and not just "more women writers"—as necessary to cultivating a feminist digital humanities, and thinking through who is invited into the mediated mechanics of humanist discourse.

MATERIALITY AND EMBODIMENT

Sayers's assertion that "the study of media is the study of entanglements" is a framework that both resists the modularity described above while also creating a space for the intersection of various form of media—old and new, digital and analog—with attention to the materiality and labor invested in all forms of media: "Embodiment (including questions of affect and labor) and materiality (including questions of inscription, plasticity, and erasure) are fundamental to research as an entanglement" (Sayers 2018). In describing how digital humanities have begun to cross a boundary between the dichotomy of "knowing" versus "doing" Klein provides a useful framework for considering how the digital humanities can and should engage with embodied experiences like critical making, resisting the traditionally masculine reputation of such spaces. Klein observes that "The shift from 'things' as technical mechanics to arguments based on processes, flows, and signals has also generated a rhetoric of building that is especially strong in new subfields such as critical code studies, software studies, and platform studies" (2017, 26).

The capacity of feminist media studies to engage with embodied experiences is exemplified in Safiya Noble's critical code analysis of Google's search algorithms. The impetus for Noble's interrogation of how Google's search function prioritizes profit over well-being was Noble's own experience using Google to search for things that "might be interesting" to the young women of color in her family (Noble 2018, 3). The commodification of black women's bodies through pornography sites were the first results Noble got when searching "black girls," which leads to Noble's analysis of how algorithms prioritize industries that are profitable at the expense of marginalized groups. This experience illustrates the embodied experience of being a person of color moving through a world that commodifies black bodies—and yet that commodification is driven by inscrutable algorithms crafted in white, male-dominated spaces.

Additionally, the interrogation of ubiquitous tools like Google's search engine recognizes how much technology and digital media are increasingly, necessarily, part of everyone's lived experience, whereas the traditional role of DH in making the born-analog digitally accessible falls short of considering how the materiality of digital tools impacts our material existence. In other words, to use a tool without considering its affordances and constraints is reminiscent of the privilege of algorithm creators who are inattentive to the impacts of what they create. In his recent examination of the structure of the "bitstream," Matt Kirschenbaum draws upon platform and software studies to build his framework:

> Bibliography, I will argue—and the material sensibility of book history more generally, in keeping with the turn towards platform and infrastructure in recent media studies—gives us the habits of mind to begin seeing past the universal Western fiction of ones and zeroes in order to bring the infinitely more variegated world of actual digital storage into view ... the digital *does* present a radical new ontology, but the particulars of that ontology are to be found in the bitstream's interactions with its material conveyances, not despite them.
>
> (2021, 11)

The emulated work of preservation—from the Electronic Literature Lab's conversions and traversals of Flash literature that has been rendered unplayable (Grigar 2019) to the Internet Archive's essential Wayback Machine—shifts the materiality, but avoids the complete loss of digital cultural objects and media that would otherwise be inaccessible. The work of understanding, and demystifying what Kirschenbaum calls out as the decidedly "Western fiction" of code to allow for both preservation (and attention to who, and what, is devalued and lost) is the work of both digital humanities and media studies moving forward.

The ongoing impact of the pandemic will be felt in both the digital humanities and media studies. Digital platforms imposed by institutions, from the fatiguing video-conference rooms of Zoom to the invasive methods of Panopto, are now set to renew. Conferences are conducted on the gamer hub Discord or in the throwback chatrooms of Gather.Town, and the questions of equity and inclusion raised by these changes bring much-needed scrutiny to how the humanities community has traditionally been mediated. If, to return to Bogost's quote, digital humanities and media studies are to be vitally of the moment, it is these platforms we must engage, and the transdisciplinary "habits of mind" Kirschenbaum evokes are those we need to critique and resist the mechanisms of this digital world.

REFERENCES

Benjamin, Ruha. 2019. *Race after Technology: Abolitionist Tools for the New Jim Code*. Hoboken, NJ: John Wiley & Sons.

Bogost, Ian. 2011. "Getting Real: On the Digital Humanities." Bogost.com. March 9. http://bogost.com/writing/blog/getting_real/.

Boyles, Christina. 2018. "'Chapter 7' in 'Bodies of Information' on Debates in the DH Manifold." Debates in the Digital Humanities. https://dhdebates.gc.cuny.edu/read/untitled-4e08b137-aec5-49a4-83c0-38258425f145/section/6a48cd20-cfa5-4984-ba32-f531b231865f.

Costanza-Chock, Sasha. 2020. *Design Justice: Community-Led Practices to Build the Worlds We Need*. Information Policy. Cambridge, MA: MIT Press.

Dombrowski, Quinn. 2021. "The Definition of DH Is Now up to 16 Separate Slack Instances, with at Least Three Colleagues Who You Communicate with via DM in More than One Instance." Tweet. @*quinnanya* (blog). December 3. https://twitter.com/quinnanya/status/1466562942250348546.

Emerson, Lori. 2014. *Reading Writing Interfaces: From the Digital to the Bookbound*. Minneapolis, MN: University of Minnesota Press.

Graham, Lindsay. 2018. "Applied Media Studies and Digital Humanities: Technology, Textuality, Methodology." New York, NY: Routledge. https://doi.org/10.4324/9781315473857-2.

Gray, Kishonna L. 2020. *Intersectional Tech: Black Users in Digital Gaming*. Baton Rouge, LA: LSU Press.

Grigar, Dene. 2019. "Preparing to Preserve Flash E-Lit." Electronic Literature Lab. May 11. http://dtc-wsuv.org/wp/ell/2019/05/11/preparing-to-preserve-flash-e-lit/.

Kirschenbaum, Matthew G. 2012. "'Chapter 1: What Is Digital Humanities and What's It Doing in English Departments? | Matthew Kirschenbaum' in 'Debates in the Digital Humanities' on Debates in the DH Manifold." Debates in the Digital Humanities. https://dhdebates.gc.cuny.edu/read/untitled-88c11800-9446-469b-a3be-3fdb36bfbd1e/section/f5640d43-b8eb-4d49-bc4b-eb31a16f3d06.

Kirschenbaum, Matthew G. 2021. *Bitstreams: The Future of Digital Literary Heritage*. Philadelphia, PA: University of Pennsylvania Press.

Klein, Julie Thompson. 2017. "The Boundary Work of Making in Digital Humanities." In *Making Things and Drawing Boundaries*, edited by Jentery Sayers, 21–31. Minneapolis, MN: University of Minnesota Press.

Liu, Alan. 2012. "'Chapter 29: Where Is Cultural Criticism in the Digital Humanities? | Alan Liu' in 'Debates in the Digital Humanities' on Debates in the DH Manifold." Debates in the Digital Humanities.https://dhdebates.gc.cuny.edu/read/untitled-88c11800-9446-469b-a3be-3fdb36bfbd1e/section/896742e7-5218-42c5-89b0-0c3c75682a2f#ch29.

McPherson, Tara. 2009. "Introduction: Media Studies and the Digital Humanities." *Cinema Journal* 48 (2): 119–23. https://doi.org/10.1353/cj.0.0077.

McPherson, Tara. 2012. "'Chapter 9: Why Are the Digital Humanities So White? Or Thinking the Histories of Race and Computation | Tara McPherson' in 'Debates in the Digital Humanities' on Debates in the DH Manifold." Debates in the Digital Humanities. https://dhdebates.gc.cuny.edu/read/untitled-88c11800-9446-469b-a3be-3fdb36bfbd1e/section/20df8acd-9ab9-4f35-8a5d-e91aa5f4a0ea#ch09.

Noble, Safiya Umoja. 2018. *Algorithms of Oppression: How Search Engines Reinforce Racism*. New York, NY: New York University Press.

Reid, Alex. 2011. "Digital Humanities: Two Venn Diagrams." Alex Reid. March 9. https://profalexreid.com/2011/03/09/digital-humanities-two-venn-diagrams/.

Risam, Roopika. 2018. *New Digital Worlds: Postcolonial Digital Humanities in Theory, Praxis, and Pedagogy*. Chicago, IL: Northwestern University Press. http://ebookcentral.proquest.com/lib/ucf/detail.action?docID=5543868.

Sayers, Jentery. 2018. "Studying Media through New Media." In *The Routledge Companion to Media Studies and Digital Humanities*, edited by Jentery Sayers, 1–6. New York, NY: Routledge.

Wernimont, Jacqueline. 2013. "Whence Feminism? Assessing Feminist Interventions in Digital Literary Archives." *Digital Humanities Quarterly* 7 (1). https://dhq-static.digitalhumanities.org/pdf/000156.pdf.

Wernimont, Jacqueline and Elizabeth Losh. 2018. "'Introduction' in 'Bodies of Information' on Debates in the DH Manifold." Debates in the Digital Humanities. https://dhdebates.gc.cuny.edu/read/untitled-4e08b137-aec5-49a4-83c0-38258425f145/section/466311ae-d3dc-4d50-b616-8b5d1555d231#intro.

CHAPTER TEN

Autoethnographies of Mediation

JULIE M. FUNK AND JENTERY SAYERS (UNIVERSITY OF VICTORIA)

Humanities research with computing is frequently associated with three approaches to technologies: building infrastructure, designing tools, and developing techniques. The infrastructural approach is common among some libraries and labs, for example, where "infrastructure" implies not only equipment, platforms, and collections but also where and how they are housed and supported (Canada Foundation for Innovation 2008, 7). Tools, meanwhile, are usually designed and crafted with infrastructure. They turn "this" into "that": from input to output, data to visualization, source code to browser content (Fuller 2005, 85). Techniques are then partly automated by tools. Aspects of a given process performed manually may become a procedure run by machines (Hayles 2010; Chun 2014). Although these three approaches are important to humanities computing, today they face numerous challenges, which are likely all too familiar to readers of this handbook.

Among those challenges is technical expertise. Undergraduate and graduate students as well as humanities staff and faculty are rarely trained in areas such as computer programming and artificial intelligence (AI). Developing this expertise is no small task, especially when it is combined with academic studies of history, culture, language, or literature. Software appears in the meantime to grow and obsolesce rapidly. Just as someone finally learns the ins and outs of a platform, they may be asked—or required—to switch to another one. Computing in the humanities thus brings with it various justifiable concerns, if not anxieties, about the perceived obligation to "keep up" with the pace of the (mostly privatized) technology sector (Fitzpatrick 2012). Alongside this obligation comes the related challenge of maintenance. Infrastructure and tools demand routine attention, even when platforms are automagically updated. Maintenance is expensive, too. If researchers are fortunate to acquire grant funding to build infrastructure or design a tool, then they must also consider the near future of their projects. What will be the state of this humanities platform in ten years? Who will be using it, what will they want or need, who will steward it from here to there, and at what cost? Such issues are labyrinthine in that their trajectories are incredibly difficult to predict. They are also complex from the labor perspective, where precarity is now the default state for academics who are increasingly spread thin yet expected to do more and more with less and less.

As we confront these challenges—one of us (Julie) a PhD student in science and technology studies, and the other (Jentery) an associate professor of English—we are experimenting with another approach to computing in the humanities, namely autoethnography, which is by no means new to the academy. Carolyn Ellis and Arthur P. Bochner provide a capacious but compelling definition of autoethnography, and we adopt it for the purposes of this chapter: "an autobiographical genre of writing and research that displays multiple layers of consciousness, connecting the personal to the cultural" (2000, 739). Our only edit is minor: "multiple layers of *mediation and* consciousness." For us, adding mediation to the mix of autoethnography is one way to engage computing (in particular) and technologies (in general) as *relations*. This means tools and infrastructures are more like negotiations than objects or products, and techniques are processes at once embodied (personal) and shared by groups and communities (cultural).

We, like Ellis and Bochner, also embrace writing as an autoethnographic method, in our case to address epistemological and phenomenological questions that arise from the practice of computing (Ramalho-de-Oliveira 2020, 7127). Such questions include, "What is the relationship between the production of data and the creation of stories?" "How are we to understand agency in situations where both people and machines contribute to acts of writing, perception, and expression?" And, "How are people and their experiences rendered discrete and measurable, or indistinct, invisible, and immeasurable, through their engagements with computing?" Writing as an autoethnographic method thus involves description, documentation, and address—writing about, writing down, writing to—as well as representation, traversal, and resistance, if not refusal—writing for, writing through, writing against, not writing back. Equally important, autoethnographic writing is not some one-size-fits-all solution to all the challenges facing computing. It is rather another angle on computing as a problem warranting approaches from multiple perspectives in the humanities. With these problems in mind, autoethnography can be used to foment and perturb persistent challenges. What we call "autoethnographies of mediation" in this chapter ultimately aim to spark conversations with other approaches to computing and, we hope, integrate with them.

Since this chapter appears in a handbook on the possible trajectories of humanities research with digital technologies, we want to be especially clear about what an autoethnography of mediation may privilege in that context:

- Prototyping content with existing platforms instead of building and maintaining the platforms themselves (Sayers 2015).
- Design as a line of inquiry rather than a feature or trait of a tool (Rosner 2018).
- The settings and lived conditions of particular uses (or "use cases") over the generalized distribution and effects of a computing technique or tool (Botero 2013; Losh and Wernimont 2019).
- Writing stories with, through, and against computing from first-person perspectives.
- A subject's embodied experiences with technologies, or a refusal to evacuate the subject from computing and infrastructure projects.

Readers will likely observe that our list aligns itself with the ostensible positions of users, consumers, fans, hobbyists, tinkerers, and even quality assurance testers. Our intent is not to romanticize or even prioritize such positions; it is to underscore the fundamental roles they and their stories play in everyday computing throughout the production loop, if you will: from design and development to distribution, consumption, content delivery, and maintenance. Put this way, autoethnographies of mediation are not reactionary exercises even if they are, as we mean to demonstrate, invested in changing computing cultures through stories and prototypes (Stone 2001).

To elaborate, we share an example in the following section from Julie's research in biometrics, which is one of the most pressing areas for computing research today (Browne 2015; Murphy 2017; Wernimont 2019). As an autoethnography of mediation, our example is admittedly an outlier in this handbook. It is grounded in personal, subjective experience, but that decision is purposeful, and truly necessary, for the autoethnography at hand. Julie traces their complicated engagement with a fertility tracker through various, deeply embodied forms of writing and prototyping back, to, and about the biometric device. After our example, we conclude the chapter by outlining potential trajectories for autoethnographies of mediation at the intersections of computing, technologies, and the humanities, including how autoethnography might feed-forward into techniques, tools, and infrastructures.

N(O)VUM ORGANUM

LH: 0.00 IU/L	Level: None	Jul. 7, 2019

An organ is an instrument. A speculum is an instrument. A teched-out cocktail stirrer in a small jar of piss is an instrument. Ten milligrams of medroxyprogesterone is an instrument. Data is an instrument. An organ is an instrument.

N(o)vum Organum (pronounced "no ovum organum") is an interactive autoethnography that examines the process of subject-making when empirical tools are turned back onto the body. I (Julie) intend it as a response to biometric tracking devices and the positivist empirical approach they take to measuring and essentializing the body; an epistemological precept that can be traced back to Francis Bacon's writings in his *Novum Organum*.[1] Bacon's text is complex, equally inspiring and infuriating, and I care deeply about how Bacon uses instrumentation to position quantification as a technique for uncovering those pesky matters of fact about bodies, any body, my body. This position presumes that only that which can be discerned through measurement can be truly known. Bacon's epistemological legacy is, of course, a familiar and persistent one, and its angle on instrumentation is alive and well in today's market of personal biometric devices. My project in *N(o)vum Organum* is to challenge how these biometric devices not only measure for "normalcy" but also use their metrics to render a subject "real" or recognizable through quantification. The result is a story of embodied interaction that resists such grand narratives. These moments of resistance are meaningful as they insist that there are other ways to constitute embodiment through data, ways that biometric devices either do not or cannot represent.

Over the summer of 2019, I tracked my luteinizing hormone (LH) levels with the AI-enabled Mira fertility tracker. The results were underwhelming as I often measured LH concentrations in a range labeled "NONE" (0–4.72 IU/L) or, on an especially exciting day, I reached the low end of "LOW" (4.73–14.35 IU/L). Despite the banality of measurement, every three days I diligently peed into a small glass jar, used a new Mira "wand" as a cocktail stirrer for ten seconds, replaced the cap over the wet end of the plastic stick, and inserted the wand into the handheld egg-shaped computer while I waited for Mira to report back my hormonal deficiencies. No ovulation prediction was ever sent to my phone because I don't ovulate. A five-millimeter adenoma in my pituitary gland disrupts the typical endocrine cascades that eventually produce the sorts of steroid hormones, such as the estrogens and LH, that get monitored in reproductive health. This data, or lack thereof, became my basis for telling the story of "no ovum organum."

The story is told through thirty digital pages in a display built in Java-based Processing 3. One can move through the pages with the reprogrammed Mira. When any of the thirty wands originally used to measure my LH levels are inserted into the jailbroken Mira, the bottom tip of the wand lights up (Figure 10.1) and the story page corresponding to the day that wand was originally used appears on an adjacent computer screen (Figure 10.2). Centering my pathologized hormonal

FIGURE 10.1 Reconfigured wands are inserted into a hacked Mira device to control the story on an adjacent computer screen. The screen also displays data, including the date of the original LH measurement, the concentration level, and a running graph scaled to "typical" LH fluctuations over a thirty-day cycle. Credit: Julie M. Funk.

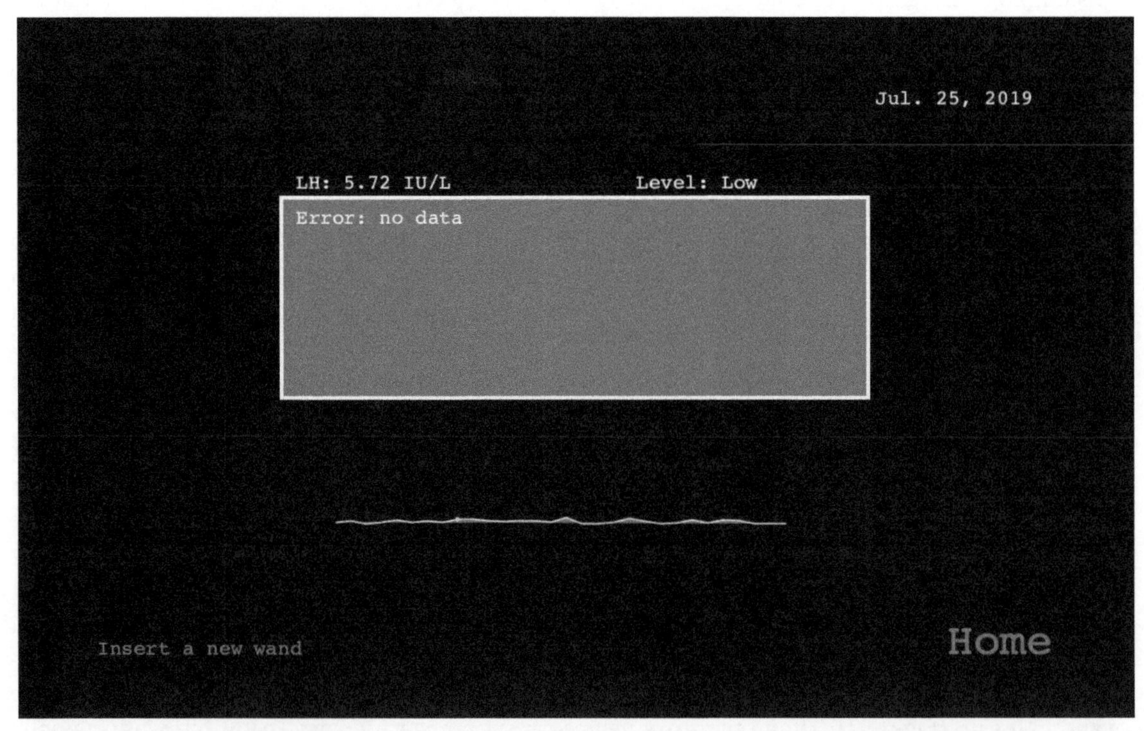

FIGURE 10.2 Screen capture from July 25, 2019, the ninth day of measurement, shows the interface design. Credit: Julie M. Funk.

story across a temporal organization that was designed to favor metrics and trends becomes an act of resistance against a biometric device's Baconian tendencies to view measurement as the epistemological end of embodiment. My critical interpretation in this response to Bacon's *Novum Organum* unfolds through an alteration of a primary source. *N(o)vum Organum* invokes Bacon's text, quoting directly from it on the second day of the story, only to respond on the twenty-second day by revising Bacon's separation of mind and body and his reliance on "mechanical aid."

Second day of measurement:

LH: 4.08 IU/L	Level: None	Jul. 4, 2019

"Our only remaining hope and salvation is to begin the whole labour of the mind again; not leaving it to itself, but directing it perpetually from the very first, and attaining our end as it were by mechanical aid." –Francis Bacon

Twenty-second day of measurement:

LH: 0.00 IU/L	Level: None	Sep. 2, 2019

"Our only remaining hope and salvation is to begin the ***networked*** *labour of* ***embodiment*** *again; not leaving* ***mind and flesh*** *to* ***themselves****, but* ***embracing new relations*** *perpetually from the very first, and* ***resisting boundaries*** *as it were* ***despite and alongside*** *mechanical aid."*

Engaging in *N(o)vum Organum* as an autoethnography of mediation does not eschew the value of tools and techniques; on the contrary, such technologies are essential to my inquiry, as this commentary on autoethnographic mediations aims to unfold. However, my inquiry into the mediating relations concerning epistemologies of the body and through the body, rather than the surveilling instrument, left me critical of the biometric propensity to over-qualify quantification and under-quantify the qualification of the subject and their situatedness. Committing to autoethnographic practices in our research about *and* with media foregrounds how media are always embodied experiences in the making.

Initially, I imagined *N(o)vum Organum* as a way to theorize how biometrics collapse boundaries between quantification and qualification, empirical evidence and embodied experience, data and representation. Thinking through processes of mediation and their contexts sheds light on ways to reclaim stories of embodiment by blurring established dichotomies in epistemological practices that privilege knowledge by way of measurement. By writing against these Baconian ideals, *N(o)vum Organum* shifts the emphasis away from "attaining [an] end [through] mechanical aid" and towards the ways critical investigations of mediation can help us to challenge the neat and clear-cut boundaries often produced by treating metrics as ends.

Given its investment in the relationship between mediation and embodied experience as a form of inquiry, *N(o)vum Organum* is an autoethnography with a prototype. Unlike many tools designed for humanities research, I designed this physical computing project to be non-scalable and ephemeral.[2] This choice allowed me to embrace the specters of time and maintenance ever-present in the project. While the development of tools and infrastructures is of course important for humanities research, autoethnographies with technologies can and should be written, too. They can be communicated through text, images, videos, and audio and published by open access venues to encourage conversation among community members. They act as prompts for experimental and evocative ways of thinking, and for talking about research as a relationship that need not result in an object to be used (a product for circulation) or a process to be imitated (also a product for circulation) (Funk et al. 2021). Consider a few examples from *N(o)vum Organum*, where the mediated and mediating experiences of hormones are central to a critique of biometrics.

LH: 3.72 IU/L	Level: None	Sept. 18, 2019

This is an open system. Endocrinology becomes exocrinology.

People involved in physical computing projects like *N(o)vum Organum* know those projects often become collections of e-waste once the laborious feedback loop of research-creation exhausts itself. With temporality in mind, I tried to salvage as much as possible from Mira's original hardware to not only practice more responsible research but also invert a common workflow, turning a potential object into reusable parts rather than instrumentalizing a process toward a product. I became distinctly aware of Mira's components, their limitations, and their capacities to persuade, engage, and aid me in telling my story. Interactions with Mira are mediated in numerous, if not countless, ways. Its app, for example, sends data back and forth to various servers and handheld devices. Information about LH concentrations must be determined by a small internal sensor that measures the color of ovulation test strips (the kind you can get at a drugstore), turning qualitative, analog information into quantitative, discrete data. Before this process even unfolds, the ovulation test strips in the wands are biochemically mediated by hormone concentrations carried in my urine.

These various forms of mediation persist in *N(o)vum Organum* and inform how I designed the project and experienced it with others as an open system. The almost ritualistic act of inserting those wands into the egg-shaped handheld device became the controller used to move through all thirty parts of the story. The wands are part of a feedback loop that sends data across programming languages (C++ to Python to Java), devices (Mira to MacBook), and bodies (mine to yours to someone else's). The Mira I incorporated into *N(o)vum Organum* is no longer Mira the consumer commodity, this particular unit having reached its own limitations as a commercial product. Yet the device's propensity for representing a subject *here* and *now* through biometrics endures in the perceived immediacy of this data. This phenomenon echoes what Jay David Bolter and Richard Grusin identify as the double logic of immediacy, which "dictates that the medium itself should disappear and leave us in the presence of the thing represented" (Bolter and Grusin 2001, 5–6). The immediacy of biometric reporting in the *N(o)vum Organum* interface (wand in = numbers and words on a screen) reproduces the same sort of immediacy present in Mira's subject-making power.

Mira is both present and not present in the story *N(o)vum Organum* tells. While much of the original hardware and branding remain intact, the device is no longer able to measure LH concentrations from the ovulation test strips in the wands. When a used wand is inserted into the Mira-made-controller, the corresponding display nevertheless shows temporal (date) and biochemical (LH concentration) data. It's as if *N(o)vum Organum* is measuring the LH in that moment of insertion. The immediacy of the data represented on the display seems to indicate as much, but it is always pointing back to thirty specific moments in the summer of 2019. The immediacy of the data also seems to suggest that it is situated in the interface. The interface is, after all, presumably where we experience the best representation of my embodiment. Yet, in my thinking on the project I've found myself asking, if the material information can be traced somewhere outside my body, wouldn't it be in the wands? Not the wands at the time of this reflection, when all the water in the LH-carrying urine has long since evaporated and the LH molecules have broken down within a day of leaving my body, but the wands in 2019, when the built-in ovulation test strips first captured those scarce amounts of hormones. The material and embodied complexity of this process is precisely why *N(o)vum Organum* needed to be an autoethnography; it is a mediating collection of ephemeral data, persistent inscription, and lived experience.

LH: 0.00 IU/L	Level: None	Sept. 27, 2019

All hormones are reproductive hormones; generative and vital.

Addressing these various forms of mediation and the problems they present returns me to my original line of inquiry: When do biometrics undergo boundary collapse between quantification and qualification, empirical evidence and embodied experience, and data and representation? Hormones are already indexed as media by natural science and medical resources, both in expert and popular discourses. They are described as chemical *messengers* operating in feedback loops of signals and dependent on specific conditions for reception before the body can increase, decrease, or transform and cascade the biochemical structures of certain hormones into other ones, in other concentrations. When devices such as Mira leverage the mediated experience of hormones in their design, they are not merely relying on predetermined biometrics obscured within the body; they are *creating* biometrics through intricate processes of remediation and quantification. Additionally, Mira produces subjects by converting embodied experiences into measurable representations conducive to tracking. In my case, Mira represented my atypicality and pathology (clinically speaking) as a biological issue incompatible with its design; before I refused Mira, it refused me.

N(o)vum Organum helps me bring my biochemistry outside of Mira's empirical framework and into a story that attempts to communicate embodiment through attention to context and messy, ever-changing representations unfolding in the present. As an autoethnography of mediation, and of biometrics in particular, my reflections on *N(o)vum Organum* are an attempt at repositioning my subjectivity in relation to media I'm working with. By attending to such processes of mediation as part of computing research, we might better engage with biometrics (or other forms of data) and the devices that produce them as sites of conflict and entanglement.

SOME POSSIBLE TRAJECTORIES

Zooming out from *N(o)vum Organum*, we conclude this chapter by outlining some possible trajectories for autoethnographies of mediation in the context of computing and the humanities. We first return to those three common approaches to technologies. How might autoethnographies of mediation feed-forward into the design and development of techniques, tools, and infrastructures?

By treating writing as a method for traversing the processes of computing, autoethnographies of mediation keep technologies visible as processes and points of discussion (Moulthrop and Grigar 2017). Although computing techniques typically run in the background and, by extension, enable everyday computing habits, they can also be subjects of study, prompting autoethnographers to treat them more like verbs and actions than nouns or things. Attention to mediation also highlights moments when practitioners may want to avoid automating certain procedures or decisions, or may wish to slow down technological development to consider the assumptions baked into programming and AI. Which computing assumptions may cause harm when automated or habituated? Which features should be "undone" prior to release rather than after it?

True, an autoethnography of mediation may at times feel or sound like a quality assurance test. Does the device work? What do we observe when we document our use of it over time? But the writing and prototyping do not end with identifying bugs to be fixed or solved. They instead enrich existing technical issues by grounding them in culture. The question of whether a device works becomes a question of how it affects and accounts for individuals and their communities. The labor of documenting use becomes a means of storytelling, which may also be informed by histories of media technology, such as *N(o)vum Organum*'s alteration of Bacon's oft-cited work. And perhaps most crucial, the rhetoric and perception of bugs may be expanded to account for norms, like the representation of hormonal atypicality as a biological issue. What is the tool at hand presumably extracting? What types of subjects does it mean to produce? How and to what effects does it establish or perpetuate potentially damaging, alienating, or misleading standards of measurement? (Bowker and Star 2008).

Autoethnographies of mediation help people to better grasp the complex relations between the personal and the cultural in the context of Technology and technologies, if you will. In the case of *N(o)vum Organum*, that Technology is biometrics: the open system of computing that not only connects but also reproduces a wide array of individual technologies that mediate people's relations with their own bodies. Prototyping through and against such metrics bridges the capital "T" with its lower case while encouraging practitioners to account for the infrastructures, perhaps easily ignored or overlooked, with which their projects are or may be complicit (Parks and Starosielski 2015). Autoethnography nudges people to consider these issues early in the design and development process. Which infrastructures can simply be adopted or adapted? When is a new or alternative one necessary for the line of inquiry at hand? What might practitioners learn, or what stories might they hear, while producing content with an infrastructure over time? We could even argue that infrastructural approaches are meaningless without the stuff and experiences of the stories they afford.

We might also recall several challenges facing humanities computing projects. While autoethnography will not, and should not, aim to resolve these challenges, it may play a key role in addressing or navigating them. First, autoethnographies of mediation acknowledge and embrace the expertise of users and audiences, who know their wants, needs, and experiences better than industry, regardless of whether they are computer programmers or AI experts. Readers will of course observe that *N(o)vum Organum* involves some degree of technical intervention: Julie's use of Processing 3, for instance, or the act of jailbreaking Mira. Yet the engagement itself, including the writing process, is cultural in the last instance. A Mira user does not need to know Java to identify the assumptions it makes about hormones and LH concentrations. Better yet, critiques of computing and technology are often most compelling when expressed from the "outside," by people who were not involved in, for example, Mira's funding, design, and development. Autoethnographies remind practitioners how informative such critiques are and why they should not be relegated to the comments box on a company's website.[3]

Meanwhile, everyone involved does not need to know how to program or train machines to turn this into that. In fact, computing projects need contributors who do not identify (or care to identify) with such expertise. Despite the individualizing "auto" in autoethnography, this method can, and we believe should, be a method of collaboration and co-production that includes care as a form of collaborative support (Bailey 2015). In the case of *N(o)vum Organum*, Julie received technical support during particularly challenging aspects of development, not to mention the collaboration in this chapter's reflections on autoethnographies of mediation. Much can be achieved via collaboration across diverging forms and domains of research practice, or via attention to "non-expert" experiences of designed immediacy that do not assume any technical knowledge.

From content moderation and crowdsourcing to routine bug fixes and software updates, a significant amount of computing maintenance is done precariously, if not voluntarily and without compensation. Autoethnographies of mediation foreground such maintenance through storytelling, highlighting not only the particulars of use and repurposing but also the material resources and labor required to keep the machine running (Allen and Piercy 2005; Ettorre 2019). In *N(o)vum Organum*, these resources and labor include not only the financial costs associated with prototyping, such as the $300 Mira device, but also the time and emotional labor spent writing longitudinal stories and repeatedly confronting Mira's reproduction of dysfunction and biometric invisibility. Though these costs proved to be challenging to the research, they were often one-time investments, with the exception of labor and maintenance. When integrated into computing projects, autoethnographies may help to reduce scope and feature creep—to "degrow" the reach of the digital—and in turn underscore the conditions a project requires and makes possible. Will the project rely on crunch, temporary contracts, or unpaid overtime to meet a release deadline? Will it also rely on voluntary feedback and user data for improvement over time? Which aspects of the project could be cut or minimized to better support the working and living conditions of its contributors?

We encourage provocation in humanities computing with these trajectories in mind. We offer autoethnographies of mediation not as a solution but as an approach to responsible, self-conscious inquiry that foregrounds our complex relations to time, scale, scope, and labor whenever we work with computers.

ACKNOWLEDGMENTS

Julie Funk's research is supported partly by funding from the Social Sciences and Humanities Research Council (SSHRC) of Canada. We would also like to thank everyone in the Praxis Studio for Comparative Media Studies for providing encouragement and feedback along the way. Julie would like to thank Matthew Frazer for technical support in the development of *N(o)vum Organum*.

NOTES

1. This project relies on the English translation of Bacon's work, but does leverage the text's Latin title for critical commentary on fertility tracking in *N(o)vum Organum*. See Bacon (1902).
2. An in-browser version of *N(o)vum Organum*, made with the open-source storytelling tool Twine and hosted on Itch.io, can be found at http://www.juliemfunk.com/projects/novum-organum. This version remains susceptible to the issues of maintenance and obsolescence addressed in this chapter.
3. Other pertinent examples of such critiques may be found in fan fiction, game mods, and cosplay and roleplay communities.

REFERENCES

Allen, K. and Piercy, F. 2005. "Feminist Auto-ethnography." In *Research Methods in Family Therapy*, 2nd edn, edited by Douglas H. Sprenkle and Fred P. Piercy, 155–69. New York, NY: Guilford Press.

Bacon, Francis. 1902. *Novum Organum*, edited by Joseph Devey. New York, NY: P.F. Collier & Son.
Bailey, Moya. 2015. "#transform(ing)DH Writing and Research: An Autoethnography of Digital Humanities and Feminist Ethics." *Digital Humanities Quarterly* 9 (2).
Bolter, Jay David and Richard Grusin. 2001. *Remediation: Understanding New Media*, Cambridge, MA: MIT Press.
Botero, Andrea. 2013. *Expanding Design Space(s): Design in Communal Endeavours*. Helsinki: Aalto University School of Arts, Design and Architecture, Department of Media.
Bowker, Geoffrey C. and Susan Leigh Star. 2008. *Sorting Things Out: Classification and Its Consequences*. Cambridge, MA: MIT Press.
Browne, Simone. 2015. *Dark Matters: On the Surveillance of Blackness*. Durham, NC: Duke University Press.
Canada Foundation for Innovation. 2008. *CFI Policy and Program Guide*. Ottawa: CFI. https://www.innovation.ca/sites/default/files/2021-10/CFI-PPG-December-2008-November-2009.pdf.
Chun, Wendy Hui Kyong. 2014. *Programmed Visions: Software and Memory*. Cambridge, MA: MIT Press.
Ellis, Carolyn and Arthur P. Bochner. 2000. "Autoethnography, Personal Narrative, Reflexivity." In *Handbook of Qualitative Research*, 2nd edn, edited by Norman K. Denzin and Yvonna S. Lincoln, 733–68. Thousand Oaks, CA: SAGE.
Ettorre, Elizabeth. 2019. *Autoethnography as Feminist Method: Sensitising the Feminist "I."* London: Routledge.
Fitzpatrick, Kathleen. 2012. *Planned Obsolescence: Publishing, Technology, and the Future of the Academy*. New York, NY: New York University Press.
Fuller, Matthew. 2005. *Media Ecologies: Materialist Energies in Art and Technoculture*. Cambridge, MA: MIT Press.
Funk, Julie M., Matthew Lakier, Marcel O'Gorman, and Daniel Vogel. 2021."Exploring Smartphone Relationships through Roland Barthes Using an Instrumented Pillow Technology Probe." In *CHI '21: Proceedings of the 2021 CHI Conference on Human Factors in Computing Systems*. New York, NY: Association for Computing Machinery.
Hayles, N. Katherine. 2010. *My Mother Was a Computer: Digital Subjects and Literary Texts*. Chicago, IL: University of Chicago Press.
Losh, Elizabeth and Jacqueline Wernimont, eds. 2019. *Bodies of Information: Intersectional Feminism and the Digital Humanities*. Minneapolis, MN: University of Minnesota Press.
Moulthrop, Stuart and Dene Grigar. 2017. *Traversals: The Use of Preservation for Early Electronic Writing*. Cambridge, MA: MIT Press.
Murphy, Michelle. 2017. *The Economization of Life*. Durham, NC: Duke University Press.
Parks, Lisa and Nicole Starosielski, eds. 2015. *Signal Traffic: Critical Studies of Media Infrastructures*. Urbana, IL: University of Illinois Press.
Ramalho-de-Oliveira D. 2020. "Overview and Prospect of Autoethnography in Pharmacy Education and Practice." *American Journal of Pharmaceutical Education* 84 (1): 156–65.
Rosner, Daniela. 2018. *Critical Fabulations: Reworking the Methods and Margins of Design*. Cambridge, MA: MIT Press.
Sayers, Jentery. 2015. "Prototyping the Past." *Visible Language* 49 (3): 156–77.
Stone, Rosanne Allucquère. 2001. *The War of Desire and Technology at the Close of the Mechanical Age*. Cambridge, MA: MIT Press.
Wernimont, Jacqueline. 2019. *Numbered Lives: Life and Death in Quantum Media*. Cambridge, MA: MIT Press.

CHAPTER ELEVEN

The Dark Side of DH

JAMES SMITHIES (KING'S COLLEGE LONDON)

Despite being a relatively new field, Digital Humanities (DH) has an intellectual history with much the same richness as history, classics, or literary studies. This includes many positive elements related to epistemological and methodological growth, but it also (naturally enough) includes periods of disagreement and conflict. This is heightened by its association with contemporary issues of substantial importance to everyday life: digital tools and methods are loaded with cultural, ethical, and moral implications. During the late twentieth and early twenty-first centuries, technology was viewed by many commentators as an existential threat to the humanities, resulting from the corrosive importation of instrumentalist thinking and hyper-rationalism into a world of emotion, aesthetics, and interpretative complexity. Such perspectives still exist today in various forms, although they have been quietened by the growing ubiquity of technology and enforced communion with digital tools.

It is reasonable to suggest that DH has become a floating signifier for these broader tensions between the core humanities disciplines and global digital culture (Gold and Klein 2016). As the leading digital practice in the humanities, DH often acts as a lightning rod for anxiety about not only the future of the humanities but the effect of digital capitalism on self and society. In 2013 these issues coalesced in discussion of "The Dark Side of DH" (Chun and Rhody 2014; Grusin 2014), establishing criticism of the field that has yet to be resolved. This chapter argues that it is important the dark side of DH continues to be explored, to ensure the field retains its intellectual edge and nurtures a tradition of criticism and critique. Encouraging dissenting opinion and honestly appraising the complexities a union of technology and the humanities creates needs to be as integral to DH as its technical tools and methods.

THE VALUE OF DISSENT

Disciplinary attitudes to criticism and critique are profoundly important. Modern humanities disciplines have evolved through processes of sometimes vigorous intellectual revolution and reaction, against internal as well as external dissent. The basic narrative is well-known. Giambattista Vico established the foundations of the (European) humanities in the seventeenth century in opposition to Cartesianism. Later, Vico was critiqued by Enlightenment thinkers such as Voltaire and Condorcet (Bod 2014, 174). The evolution of history from philosophy, English from history, and cultural studies from English all involved intellectual dispute alongside polite disagreement. Without intellectual dissent to sharpen its sense of identity and purpose DH will remain an opaque practice, perhaps degrading into a service activity that exists only to serve more established humanities fields.

Openness to dissent is thus a prerequisite for the development of an intellectually mature field. I have suggested elsewhere that DH needs to search for "critical and methodological approaches to digital research in the humanities *grounded in the nature of computing technology and capable of guiding technical development as well as critical and historical analysis*" (Smithies 2017, 3; emphasis in original) but it is equally important to welcome differences of opinion. Humanities disciplines evolve through the controlled interplay of agreement and dissent, via implicitly and explicitly agreed rules governing rhetorical interaction (Wittgenstein 1958).

The related practices of criticism and critique are central to this process, the first subjecting publications to considered review, the second positioning them in their wider context and subjecting them to more robust epistemological consideration (Davis and Schleiffer 1991, 22–6). Works such as Pierre Bayle's *Dictionnaire Historique et Critique* (1697) and Kant's *Critique of Pure Reason* (1781) established a rhetorical and epistemological playing field supporting the interplay of myriad opposing intellectual tendencies within and across disciplines, from mathematics to poetry, pure mathematics to applied mathematics, epic poetry to modern poetry and so on (Benhabib 1986). Criticism and critique are important intellectual and ideological mechanisms within a long-established system of knowledge production, one that encourages dissent from received opinion to ensure quality of thought. The writing discussed in this chapter, whether positive or negative in its attitude towards DH, contributes to that rich intellectual context and should be valued because of it.

THE DARK SIDE OF DH

The root of the current intellectual crisis in DH developed in the 1990s, amidst the first cyberutopian flourish of the World Wide Web. Anti-technological discourse, rooted in the twentieth-century criticism of figures such as Jacques Ellul (1964) and Herbert Mumford (1970), informed a backlash against the growth of electronic media and the Internet and culminated in full-throated rejections of Silicon Valley capitalism as a scourge that threatened the very foundations of culture and society.[1] Humanists and publishers felt "under siege" (Birkerts 1995, 5) by shadowy neoliberal forces determined to instrumentalize their world of "myths and references and shared assumptions" (Birkerts 1995, 18). In its guise as Humanities Computing—viewed as an unthreatening backroom activity for technical boffins by some people—DH largely avoided this criticism. Slowly, however, as it grew in popularity and started to attract research funding and media attention, critics began to question DH's purpose and conceptual basis, and the motivations of its practitioners and administrative supporters.

Criticism increased after DH was positioned as a key field by the American National Endowment for the Humanities (NEH) in 2006. Interest increased amongst policymakers and institutions at the same time as support for established humanities disciplines waned, resulting in declining budgets and successive rounds of redundancies. It seemed obvious to many commentators that digital technology was little more than a Trojan horse enabling the hollowing out of intellectual culture and the dumbing down of society by technocratic administrators (Bauerlein 2008; Golumbia 2009).

These arguments culminated in a series of debates following the 2011 Modern Language Association (MLA) meeting. The conference had experienced efflorescent growth of digital literary studies papers confidently proclaiming the emergence of new methods of computational and cultural analysis, with discussion broadcast on Twitter (which was in the process of being adopted, and heavily criticized, across academia). Despite not attending the conference, literary and legal scholar Stanley Fish took the opportunity to excoriate digital humanities in a series of opinion pieces in the *New York Times*, claiming they naively applied digital methods to texts in a way that had little place "for the likes of me and for the kind of criticism I practice" (Fish 2011, 2012a, 2012b). Fish was met with robust rejoinders from an aggrieved DH community[2] who believed he had misrepresented the nature and intent of the field and wrongly tarred it with allegiances to corporate forces in the higher education sector (Ramsay 2012). The debate was many faceted and an interesting example of the intellectual disruption that resulted from the introduction of technology to higher education.

A close reading of the many statements and rejoinders to DH in this period is needed, but outside the scope of this chapter. For our purposes it is enough to note that the debate was highly productive, especially when researchers with experience of commercial software programming and digital media studies became involved. Their awareness of the quotidian realities of software engineering and maintenance, and the complex relationships between digital technology, the labor market, and capitalist economics, forced the DH community to confront potentially negative sides to their activities. The conversation gained focus at the 2013 MLA, in a roundtable talk titled "The Dark Side of Digital Humanities" by Richard Grusin, Wendy Hui Kyong Chun, Patrick Jagoda, and Rita Raley. The event was held in a packed room in Boston and generated significant comment on Twitter and the blogosphere. Their stated goal was to explore whether DH was implicated in the neoliberal logic of the American higher education sector. More significantly, their goal was to explore if it was possible "to envisage a model of digital humanities that is not rooted in technocratic rationality or neoliberal economic calculus but rather that emerges from as well as informs traditionary practices of humanist inquiry" (Chun et al. 2016).

In her talk, Chun enlisted Martin Heidegger's notion of the enframing nature of technology (Heidegger 1970) to enable this conceptual shift, suggesting that technology does not exist outside human experience (as a tool to be acted upon) but is always already constituted as part and parcel of human experience, deforming our view of the world, politics, religion (etc.) and bringing enabling ideologies such as neoliberalism with it as it travels from context to context. To transform the debate from one of simple rejection or acceptance of technology use in the humanities she suggested that

> the dark side—what is now considered to be the dark side—may be where we need to be. The dark side, after all, is the side of passion. The dark side, or what has been made dark, is what all that bright talk has been turning away from (critical theory, critical race studies—all that fabulous work that transformDH is doing).

The implication was that the DH community was uniformly and dangerously naïve, but also that they were engaged in practices with potentially transformative potential.

The debate suffered from intellectual myopia (and arrogance) to observers outside literary studies who objected to the conflation of "digital literary studies" and "digital humanities" but that was a function of other disciplinary issues: from this point on, and particularly in North America, DH was permanently entangled with the discourse of critical and media theory. Some people returned to the relative intellectual safety of sub-disciplines such as digital history to avoid being drawn into what they viewed as irrelevant or ideological debates but the scholarly impact of the argument and its entanglement with increasingly complex socio-cultural and political transformations meant the die was cast: digital humanities had evolved into a floating signifier.

Whether we view this as evidence of the imperialistic nature of critical theory or a natural result of disciplinary dynamics, DH was positioned as the poster child for all that was wrong with the American higher education sector. Similar but perhaps less forceful critiques were taken up in the United Kingdom, Australasia, and to a lesser extent Europe. It is worth noting that digital humanists based in Europe, Asia, and Australasia did not engage with the MLA debates in the same way as Americans, drawing on different socio-political and intellectual traditions and sometimes resiling from what they viewed as ideological posturing by a small group of critical media theorists, but the same currents were present in those regions too.

Chun's reference to the #transformDH Twitter hashtag was telling. The hashtag was established by Moya Bailey, Anne Cong-Huyen, Alexis Lothian, and Amanda Phillips in response to celebrations of "Big Tent DH" in 2011, which sought to present the community as inclusive and open to all disciplines and people regardless of their background or level of technical skill. As laudable as it appeared on the surface, wrapped up with the rhetoric of Silicon Valley IT culture, #transformDH noted that the claim rang hollow to people from gender and queer studies, race and ethnic studies, and disability studies. "Instead, DH seemed to be replicating many traditional practices of the ivory tower, those that privileged the white, heteronormative, phallogocentric view of culture that our home disciplines had long critiqued. The cost of entry for many of us—material demands, additional training, and cultural capital—as queer people and women of color was high" (Bailey et al. 2016). Stephen Ramsay's provocative paper at the 2011 MLA, "Who's In and Who's Out" (Ramsay 2016) seemed to them to be indicative of a field blindly replicating modes of privilege they were determined to consign to the past: seeing them imported into the humanities through DH at the very moment optimists felt the structures were weakening was demoralizing, to say the least.

This was a penetrating and necessary analysis, but the situation was more complicated than it appeared. Ramsay's paper was prompted by an earlier debate in DH about "hack versus yack," centered on whether people needed to be able to program to identify as digital humanists. Although not articulated in that way at the time, the debate was related to labor rights and hierarchies of power in the university sector. As the humanities computing tradition evolved towards DH and more people became interested in the field, community members with programming skills who had always felt like second-class citizens in the humanities—quietly working on their data modeling and web development outside the spotlight—became concerned their decades of effort were being undermined by newcomers who wanted to benefit from their efforts but did not value their work. Asking whether digital humanists needed to be able to code, or have a degree of technical understanding, was a way of pushing back and carving out a space for technical practice against perceived intellectual colonization by powerful and ideologically motivated forces. The issue has never been resolved and exists today in the development of the Computational Humanities Group in 2020 and ongoing efforts to develop viable career paths in research software engineering (RSE) for technical staff involved in DH projects.

If we view the "hack versus yack" debate as a labor issue we can continue to benefit from its potential to produce positive change, but neither side of the debate read it that way at the time. On the contrary, figures such as Ramsay viewed it in a primarily methodological and epistemological sense, and groups such as #transformDH viewed it as evidence of a monocultural, gendered, and technocratic field. This miscommunication, or inability to properly contextualize the debate, should have been a wake-up call for the community, but it was overtaken by events. To outsiders, early-twenty-first-century DH was ideologically indistinguishable from Silicon Valley cultures that presented digital tools and methods as emancipatory at the same time as they reproduced white masculinist working cultures and fundamentally biased products.

The establishment of the #dhpoco hashtag and associated Postcolonial Digital Humanities website by Adeline Koh and Roopika Risam in 2013 aimed to address this by critiquing the lack of diversity in the DH community, and its tendency to prioritize technical subjects that perpetuated the situation by creating barriers to participation. As the conversation evolved, they acknowledged the significant history of humanities computing and DH that *did* attend to issues of race, class, gender, disability, but their focus remained tactical and broadly radical. These were difficult topics to raise in a community that had been comfortably insular for several decades and experiencing cultural and intellectual relevance for the first time. Many people in the DH community were (and still are) affronted by the suggestion they perpetuate sexism, structural racism, and other forms of discrimination. The situation is complicated by the fact that these "reactionaries" appear to inhabit the liberal (or at least center-left) side of the political spectrum.

It is important to remember, in this context, that debates about the dark side of DH reflect the politics of the higher education sector, combined with generational change, more than the politics of the street. DH's intellectual history pits the radicals of 1968 against the radicals of 2006, 1990s cyberutopians against the 99 percent. Cultural conservatives, in the real world rather than scholarly sense, are hard to find. Initiatives such as THATCamp, which helped people organize community events introducing people to DH is a case in point. The initiative was laudably open and progressive compared to the existing academic and tech-sector culture, emphasizing equality regardless of career status, warmly reaching out to GLAM sector and IT colleagues despite them normally being reduced to a service role in academic culture, and aiming to empower people regardless of gender or technical ability. But in hindsight there were obvious blind spots and THATCamp events undoubtedly reflected the normative white middle-class cultures at the heart of global academia. #dhpoco's use of cartoons to critique THATCamp was highly effective in that sense, using guerrilla tactics to destabilize the community's accepted norms.

The first phase in the critique of DH culminated in 2016 with the publication of an article in the *LA Review of Books* titled "Neoliberal Tools (and Archives): A Political History of Digital Humanities," by Daniel Allington, Sarah Brouillette, and David Golumbia. The article was in many ways a summation of the previous few years' criticism of DH, adopting a festival-like tone of confident moral, intellectual, and political condemnation. DH was positioned as the prime example of a mode of neoliberalism in higher education which prioritized measurable "outputs" and "impact" ahead of less easily quantifiable notions of scholarly quality. The authors accused the DH community of fetishizing code and data and producing research that resembled "a slapdash form of computational linguistics adorned with theoretical claims that would never pass muster within computational linguistics itself." Their claims were based on a narrow history of the field focused on the University of Virginia literary studies community, provoking outrage and consternation from people associated with it.

"Neoliberal Tools (and Archives)" is a totemic example of what happens when analysis of the dark side of DH becomes entangled with personal and professional as well as intellectual issues, but it represented an exceptionally important moment for DH. Its visceral polemic, its focus on literary studies at the expense of other fields, its personal and professional commentary, its identification of labor rights issues, and its tight association of DH with the corporate university (and indeed global neoliberal economics) reflected the field's centrality to global scholarly discourse and its role as a lightning rod for wider grievances. DH practitioners convincingly argued that issues with computational methods were widely known and being slowly attended to but, as the authors noted, Alan Liu had asked "Where is Cultural Criticism in the Digital Humanities?" in 2012 to little practical effect. As arguable as many of the article's claims were, it seems clear in hindsight that the community's lack of attention to (or intellectual interest in) the political context to their work led first to #transformDH and #dhpoco, and eventually to "Neoliberal Tools."

Given the inherent complexity of the situation even these critiques were insufficient, of course. By 2016 it had become obvious that DH (its supporters as well as its critics) was dominated by Northern Hemisphere, elite concerns. Even #dhpoco initially presented a mode of postcolonialism that—while laudable—appeared narrowly North American to the present author (Smithies 2013). Such issues tend to even out, however: like historiography (the "history of history," including analysis of theory and method) theoretical perspectives on DH have evolved over time. In the case of #dhpoco this resulted in a wider global focus and significance culminating in the publication of Risam's *New Digital Worlds: Postcolonial digital humanities in theory, praxis, and pedagogy* (2018). A similar, and similarly unsurprising in the context of intellectual history, evolution has occurred with the Alliance of Digital Humanities Organizations (ADHO), a global organization that initially struggled to manage rapid international growth in digital humanities and evolve in ways that did justice to its new constituents. The Global Outlook DH ADHO Special Interest Group (SIG) was largely a response to this, encouraging and facilitating focus on issues of representation, multilingualism, and cultural and infrastructural diversity.

A FALSE DAWN

The positive reception of Risam's monograph, the success of projects such as Torn Apart/Separados (Ahmed 2018), which mapped detention centers for asylum seekers at the height of the 2018 humanitarian crisis, the increasing attention to indigenous DH by funding bodies and growing sophistication of their methods (Walter 2020; Chan et al. 2021), the willingness of ADHO to make clear statements about Black Lives Matter and structural racism and their support for GO::DH, and the regular publication of articles and chapters related to the "dark side" of DH might suggest a corner has been turned, but to radicals it is a false dawn. There is no reason to condemn the field in its entirety anymore (if there ever was), but it is obvious to many people that structural issues inherited from Western higher education systems and the technology industry are woven through the warp and weft of DH as a field. They cannot be resolved with a handful of impactful but still in many ways tactical (as opposed to mainstream) initiatives.

The furor over the long-standing *Humanist* email listserv in 2020, prompted by the publication of a post referring to "anti-white racism," demonstrated this clearly.[3] Younger scholars were angry and confused that such a comment could be published in a reputable scholarly publishing venue and were confounded by the subsequent decision not to publish a reply (later reversed) and what appeared to be a general lack of understanding, empathy, and support, across a wide sector of their community (Tovey-Walsh 2020). Much the same fault lines were activated as in the debates surrounding #transformDH and #dhpoco years earlier. Awareness of structural racism was higher because of the Black Lives Matter movement, which was in full flight at the time following the death of George Floyd, but resolution was not straightforward. These are not topics that can be resolved in a single chapter in a DH handbook, of course, although I hope my position is sufficiently clear. The key point is that DH benefits from criticism and critique, and sometimes dissent and targeted tactical or guerrilla activity. Open discussion of political and ethical issues must be considered a *sine qua non* of intellectual engagement regardless of topic (technical or otherwise).

Natalia Cecire's response to the #transformDH movement in 2012 retains relevance here. Balancing support for the initiative with a call for more theory and cultural criticism, she pointed out that DH does in fact contain within itself—precisely in its focus on digital tools and methods—the potential for positive socio-political outcomes. Citing Tim Sherratt's widely celebrated project The Real Face of White Australia she noted that "the jolt of the oppositional can be powerful, when it is rooted in a critical activism that builds on the little-t theories that have preceded and exist alongside it, rather than manifesting as nerdy beleagueredness" (Cecire 2012).

Cecire and Sherratt identify the vanishing point of digital humanities here: the point at which cultural, intellectual, political, ideological, and technical affordances and constraints cohere in a digital artifact. It strengthens my claim that DH needs to search for critical and methodological approaches to digital research in the humanities grounded in the nature of computing technology and capable of guiding technical development as well as critical and historical analysis, not as a statement of "nerdy beleagueredness" or in order to privilege engineering over theory and culture, but through intellectual recognition of the powerful dialectics at work in contemporary society *at precisely that nexus*. If the intra and extra-disciplinary conflicts associated with the dark side of DH teach us anything, it is that the union of technology with the humanities is productive of powerful (and I would claim, ultimately positive) socio-cultural and intellectual effects.

It is also a reminder that the cultural and political issues associated with the dark side of DH are in some sense dwarfed by technical issues related to database design, indigenous rights and the representation of cultural knowledge, licensing and copyright law, labor rights, environmentalism, and safeguarding cultural data. Signs of progress are present there too, but the problems are of planetary scale and function at the level of global cyberinfrastructure, in all its dimensions and technical complexity. In 2013 Melissa Terras explained how difficult it was to update one key field in the TEI-XML schema to reflect non-binary gender identity, suggesting the scale of the task (Terras 2013). Seven years later she was involved in a study of Google's Art Project with similarly sobering results, finding that the six million high-resolution images collected from around the world were heavily skewed towards collecting institutions based in the United States (Kizhner et al. 2020). The transition is instructive: in the space of seven years, the dark side of DH moved from field-level struggles with granular representation of information, through conflicts about theory and culture, to tight entanglement with global modes of surveillance capitalism. DH is no longer a field where cultures of nerdy beleagueredness thrive, but neither has it extricated itself from its dark side.

Indeed, cyberutopian claims that digital technology has a naturally emancipatory and democratizing effect have been replaced by a dystopian focus on surveillance capitalism, misinformation, algorithmic bias, hate speech, slave labor, and environmental damage. These certainly require scholarly attention, but it is important that some humanists focus on practical technical interventions that can effect positive change too. Often these changes will be hardly noticeable but they are fundamental to the values and political orientation of a large proportion of the humanities community and worthwhile in and of themselves: teaching DH using minimal computers (Dillen and Schäuble 2020), a machine-learning analysis that takes indigenous knowledge into account (Chan et al. 2021), digitization of Sudanese cultural artifacts (Deegan and Musa 2013), development of open methods to combat hate speech (Kennedy et al. 2020).

There is considerable moral hazard in exposure and criticism of the dark side of DH, if campaigns against managerialism, neoliberalism, and technocracy undermine the delicate green shoots of digital activism. Campaigns against new digital methods, such as that launched by the literary scholar Nan Da against the use of machine learning in literary studies (Da 2019), entail such risks despite being important additions to the field's critical history: ethical responsibility runs both ways. Work in artificial intelligence and machine learning requires robust critique to identify inadequate methods and find ways to manage the many known issues with humanities datasets, but it is important not to over-compensate and perceive intellectual degradation in place of merely inadequate and evolving theory or method. This is the fine line DH practitioners *and their critics* need to traverse: neither side can escape the necessity for robust engagement with the full scope of DH literature, tools, and methods.

PRACTICAL NIHILISM AND HUMAN GROWTH

It is possible to discern a truly dark tradition at work in "the dark side of DH," reflective of humanistic traditions of radicalism, dissent, and free will that reach back to the ancient Greeks. In this sense, all sides of the debates are united and contributing to the development of the humanities in the twenty-first century in similar ways: engagement in what Nietzsche referred to as a "thoroughgoing *practical nihilism*" (Stellino 2013) born of deep dissatisfaction with either the status quo or emerging trends. Humanistic nihilism, in this reading, is an expression of free will informed by existential concern for self and society. Whether consciously or subconsciously, the stakes are deemed high enough for criticism and critique to be buttressed with clarion calls for active intervention along revolutionary or reactionary lines.

Expressions of this state of mind are various enough that a single example will never suffice, but we could go no better than the entanglement of DH with the vertiginous, algorithmically enabled mob that stormed the United States Capitol on January 6, 2021. The spectacle of a human mob roused by a president empowered by social media and manipulated by the algorithmic logic of the QAnon movement and white supremacists is a new symbol of the dark side of DH. Human suffering and outrage blended with alien intelligence in that moment in a way that demonstrates the hopelessness of humanist positions that refuse to engage with technology or position it as somehow beyond the pale of scholarly activity, as if "human" and "humanist" experience are not practically as well as conceptually entangled. It is beyond doubt now that the full range of humanist response, from socio-political critique to technical development, is needed to rehabilitate the public sphere, political economy, and everyday life.

As dissonant as it may seem, then, it is now a truism that human experience of digital technology has reached troubling heights and contributes to significant political, cultural, and economic discord and environmental degradation *and also* that advanced research methods enabled by digital tools and methods have the potential to relieve suffering on a mass scale, heal the planet, and contribute to global peace and well-being. This grand, epochal, contradiction now lies at the heart of DH as a field. As a floating signifier for that remarkable array of baggage (technical, intellectual, psychological, political) digital humanists must expect to be periodically buffeted by intellectual and cultural crises from within and outside their community. As Wendy Chun noted in 2013, a focus on the dark side of DH is not necessarily a negative undertaking in that sense, but a prophylactic activity that protects the field from collapse in the face of the profound historical moment it inhabits, reflects, and contributes to.

For that to occur we need to focus on people ahead of politics, culture, ideology, or technology. A focus on people is the only sure way to protect our community from the dark side of DH and maximize the insights it offers; the task has become so immense that we can only succeed by enabling what might seem to some people as radical levels of diversity and inclusion. Numerical gender equality is no longer enough; we need equality of seniority, pay, and influence. Public commitment to racial equality and efforts to enable indigenous DH are no longer enough; we need properly diverse departments and teams and major projects decolonizing archives at a national and international scale. Cultural critique of the digital world is no longer enough; we need a generation of humanists trained to contribute to the development of global cyberinfrastructure. A desire to collaborate with technical staff is not enough; it needs to be done on an equal basis and take into account their workload and career goals, backed with permanent roles and high-quality tools and infrastructure. Pedagogy needs to change from grafting basic technical skills onto humanities degrees to providing careful scaffolding of skills throughout the undergraduate and postgraduate curricula, and to enable renewed focus on the basic tools and methods developed from the humanities computing tradition.

These are grand goals, but nothing less than that demanded by critics of DH for the past decade. The challenge is immense, but it is necessary for the development and indeed the survival of the field. Criticism, critique, dissent, and dark analyses of the human condition are common to all of the major humanities disciplines: the maturation of the tradition in DH should be celebrated and encouraged.

ACKNOWLEDGMENTS

I would like to thank Roopika Risam, Alex Gil, Arianna Ciula, and Daniel Allington for offering reflections and comments on this chapter. All errors of fact and failures of interpretation rest with the author.

NOTES

1. For a more detailed description of this period in intellectual history, see Smithies (2017).
2. Including the present author. See Smithies (2021).
3. Disclaimer: The author was Director of King's Digital Lab, the web host of *Humanist* at the time of the incident. The Lab's statement on the issue can be found on its blog, https://kdl.kcl.ac.uk/blog/joint-statement-about-humanist-project/.

REFERENCES

Allington, Daniel, Sarah Brouillette, and David Golumbia. 2016. "Neoliberal Tools (and Archives): A Political History of Digital Humanities." *Los Angeles Review of Books*, May 1. https://lareviewofbooks.org/article/neoliberal-tools-archives-political-history-digital-humanities/.

Bailey, Moya et al. 2016. "Reflections on a Movement: #transformDH, Growing Up." In *Debates in the Digital Humanities*, edited by Matthew K. Gold and Lauren F. Klein, 71–80. Minneapolis, MN: University of Minnesota Press.

Bauerlein, Mark. 2008. *The Dumbest Generation: How the Digital Age Stupefies Young Americans and Jeopardizes Our Future (or, Don't Trust Anyone under 30)*. New York, NY: Jeremy P. Tarcher/Penguin.

Benhabib, Seyla. 1986. *Critique, Norm, and Utopia: A Study of the Foundations of Critical Theory*. New York, NY: Columbia University Press.

Birkerts, Sven. 1995. *The Gutenberg Elegies: The Fate of Reading in an Electronic Age*. New York, NY: Fawcett Columbine.

Bod, Rens. 2014. *A New History of the Humanities: The Search for Principles and Patterns from Antiquity to the Present*. Oxford: Oxford University Press.

Cecire, Natalia. 2012. "In Defense of Transforming DH." *Works Cited* (blog), January 8. https://nataliacecire.blogspot.com/2012/01/in-defense-of-transforming-dh.html.

Chan, Valerie et al. 2021. "Indigenous Frameworks for Data-Intensive Humanities: Recalibrating the Past through Knowledge Engineering and Generative Modelling." *Journal of Data Mining & Digital Humanities* (January 8). https://jdmdh.episciences.org/7018/pdf.

Chun, Wendy Hui Kyong et al. 2016. "The Dark Side of the Digital Humanities (2013)." In *Debates in the Digital Humanities*, edited by Matthew K. Gold and Lauren F. Klein, 493–509. Minneapolis, MN: University of Minnesota Press.

Chun, Wendy Hui Kyong and Lisa Marie Rhody. 2014. "Working the Digital Humanities: Uncovering Shadows between the Dark and the Light." *differences* 25 (1): 1–25.

Da, Nan Z. 2019. "The Computational Case against Computational Literary Studies." *Critical Inquiry* 45 (3) (March 1): 601–39.

Davis, Robert and Ronald Schleiffer. 1991. *Criticism and Culture: The Role of Critique in Modern Literary Theory*. New York, NY: Longman.

Deegan, M. and B. E. Musa. 2013. "Preserving the Cultural Heritage of Sudan through Digitization: Developing Digital Sudan." In *Digital Heritage International Congress (DigitalHeritage)* 2: 485–7.

Dillen, Wout and Joshua Schäuble. 2020. "Teaching Digital Humanities on Raspberry Pis. A Minimal Computing Approach to Digital Pedagogy." In DH 2020 Abstracts. https://dh2020.adho.org/abstracts/.

Ellul, Jacques. 1964. *The Technological Society*. New York, NY: Knopf.

Fish, Stanley. 2011. "The Old Order Changeth." *New York Times*, December 26. https://web.archive.org/web/20220131090004/https://opinionator.blogs.nytimes.com/2011/12/26/the-old-order-changeth/.

Fish, Stanley. 2012a. "The Digital Humanities and the Transcending of Mortality."January 9. http://opinionator.blogs.nytimes.com/2012/01/09/the-digital-humanities-and-the-transcending-of-mortality/.

Fish, Stanley. 2012b. "Mind Your P's and B's: The Digital Humanities and Interpretation." *New York Times*, January 23. http://opinionator.blogs.nytimes.com/2012/01/23/mind-your-ps-and-bs-the-digital-humanities-and-interpretation/.

Golumbia, David. 2009. *The Cultural Logic of Computation*. Cambridge, MA: Harvard University Press.

Grusin, Richard. 2014. "The Dark Side of Digital Humanities: Dispatches from Two Recent MLA Conventions." *differences* 25 (1): 79–92.

Heidegger, Martin. 1970. "The Question Concerning Technology (1949)." In *Basic Writings*, 283–317. London: Routledge and Kegan Paul.

Kennedy, Chris J. et al. 2020. "Constructing Interval Variables via Faceted Rasch Measurement and Multitask Deep Learning: A Hate Speech Application." *ArXiv:2009.10277 [Cs]*, September 21.

Kizhner, Inna et al. 2020. "Digital Cultural Colonialism: Measuring Bias in Aggregated Digitized Content Held in Google Arts and Culture." *Digital Scholarship in the Humanities* 36 (3): 607–40. https://doi.org/10.1093/llc/fqaa055.

Klein, Lauren F. and Matthew K. Goldand. 2016. "Introduction: Digital Humanities: The Expanded Field." In *Debates in the Digital Humanities*, edited by Matthew K. Gold and Lauren F. Klein, ix–xvi. Minneapolis, MN: University of Minnesota Press.

Koh, Adeline and Roopika Risam. 2013–14. "Postcolonial Digital Humanities." *Postcolonial Digital Humanities*. https://dhpoco.org/blog.

Liu, Alan. 2011. "Where Is Cultural Criticism in the Digital Humanities?" *Alan Liu*, January 7. https://liu.english.ucsb.edu/where-is-cultural-criticism-in-the-digital-humanities/.

Manan, Ahmed et al. 2018. "Torn Apart/Separados." Text, June 25. http://xpmethod.columbia.edu/torn-apart/volume/1/.

McGillivray, Barbara et al. 2020. "The Challenges and Prospects of the Intersection of Humanities and Data Science: A White Paper from The Alan Turing Institute." https://doi.org/10.6084/M9.FIGSHARE.12732164.

Mumford, Herbert. 1970. *The Myth of the Machine – The Pentagon of Power*. New York, NY: Harcourt Brace Jovanovich.

Ramsay, Stephen. 2012. "Stanley and Me." *Stephen Ramsay*. [This blog post has been removed from the web. The earliest Wayback Machine version is available at https://web.archive.org/web/20130706054201/http://stephenramsay.us/text/2012/11/08/stanley-and-me.html, presumably taken a few months after the author deleted his blog.]

Ramsay, Stephen. 2016. "Who's In and Who's Out? [2011]." In *Defining Digital Humanities: A Reader*, edited by Melissa Terras, Julianne Nyhan, and Edward Vanhoutte, 239–42. London: Routledge.

Risam, Roopika. 2018. *New Digital Worlds: Postcolonial Digital Humanities in Theory, Praxis, and Pedagogy*. Evanston, IL: Northwestern University Press.

Sherratt, Tim. 2011. "The Real Face of White Australia." Discontents (blog), September 21. https://web.archive.org/web/20110926085510/http://discontents.com.au/shoebox/archives-shoebox/the-real-face-of-white-australia.

Smithies, James. 2013. "Speaking Back to America, Localizing DH Postcolonialism." *James Smithies*, September 4. https://jamessmithies.org/blog/2013/09/04/speaking-back-to-america-localizing-dh-postcolonialism/.

Smithies, James. 2017. *The Digital Humanities and the Digital Modern*. Basingstoke: Palgrave Macmillan.

Smithies, James. 2021. "Introduction to Digital Humanities." *James Smithies*. March 14. https://jamessmithies.org/blog/2012/03/14/introduction-to-digital-humanities/.

Stellino, Paolo. 2013. "Nietzsche on Suicide." *Nietzsche-Studien* 42 (1) (November 1): 151–77. https://doi.org/10.1515/niet.2013.42.1.151.

Terras, Melissa. 2013. "On Changing the Rules of Digital Humanities from the Inside." *Melissa Terras' Blog*, May 27. http://melissaterras.blogspot.com/2013/05/on-changing-rules-of-digital-humanities.html.

Tovey-Walsh, Bethan. 2020. "Leaving Humanist." *Lingua Celta*, August 10. https://linguacelta.com/blog/2020/08/Humanist.html.

Walter, Maggie et al. 2020. *Indigenous Data Sovereignty and Policy*. London: Routledge. https://doi.org/10.4324/9780429273957.

Wittgenstein, Ludwig. 1958. *Philosophical Investigations*. New York, NY: Macmillan.

PART TWO

Methods, Tools, & Techniques

CHAPTER TWELVE

Critical Digital Humanities

DAVID M. BERRY (UNIVERSITY OF SUSSEX)

Digital humanities emerged as a term in 2001 to capture a field with an early focus on digital tools and archives in relation to database collections of texts, artworks, scholarly works, dictionaries, and lexicographic corpora.[1] However, it is important to note that a number of approaches preceded this constellation, including "humanities computing" and "computing in the humanities" (McCarty 2003, 1226; Hockey 2004; McCarty 2005, 2). Digital humanists, now called, have adopted and developed tools and methods that are new to the humanities, such as computer statistical analysis, search and retrieval, data visualization, and artificial intelligence, and apply these techniques to archives and collections that are vastly larger than any researcher or research group can handle comfortably (Schreibman et al. 2004; Berry 2012). These digital methods allow the researcher to explore how to negotiate between close and distant readings of texts and how microanalysis and macroanalysis can be usefully reconciled in humanist work (see Jockers 2013). In doing so they are increasing the capacities of humanists to work with larger complex datasets, corpora and image archives and also to move between them with greater ease. The digital humanities can be understood as a set of interlocking and interdependent parts that, while distinct and standing alone to some degree, nonetheless adds up to make the whole greater than the sum of its parts. In other words, digital humanities is a coherent, if nonetheless still contested, discipline (Schreibman et al. 2004; Liu 2012a; Allington et al. 2016; Berry and Fagerjord 2017).

Digital humanities has, however, tended as a field towards seeing itself primarily as solving technical problems, particularly in relation to what might be called "knowledge representation," and this has sometimes led to a service orientation (see Drucker and Svensson 2016).[2] The implications of this are that digital humanities has tended to become means-focused, allowing other disciplines to define the ends to which their work was oriented. This has structured the worldview of digital humanities towards the preservation of cultural heritage and archives using digital media, an instrumentalism revealed in its main two approaches, digital archives and digital tools. Digital archives encompass techniques and approaches towards the transfer of usually physical archives into accurate digital representations with the corresponding problems of metadata, OCR quality, material constraints, computational storage, and procedures and processes. For digital tools the main concern has been with developing software that enables the accessing, manipulation, and transformation of these digital archives for the use of scholars, particularly in the fields of English and History, with the emphasis on augmenting scholarly work through larger dataset analysis, sometimes called "distant reading" (see Drucker 2017; Underwood 2019).[3] This goes a long way to explaining the location of many digital humanities centers within or aligned with English or History departments in universities (see also Kirschenbaum 2010).

Consequently, some of the problems identified with digital humanities may have been due to it being gestated within English and History departments. Arguably, this has engendered a conservatism in the digital humanities, manifested through its attempts to service these departmental needs. Not that the scholarly work produced per se was conservative, although sometimes it was, rather a conservatism emerged in relation to its self-regard as a service discipline and its fixation on digital archives and digital tools. Hence, digital humanities can sometimes be surprisingly reluctant to engage with issues outside of meeting these needs. As many digital humanists align firstly with their home department before their "digital humanist identity" this may affect the potential radicalism of an inner entelechy of the digital humanities (although see more recent attempts to discuss digital humanities in Dobson 2019). So, for example, born-digital content, web history, new media, and even the Internet were not central to the early computing in the humanities, and even now the digital humanities research agenda is largely set by those of the English and History departments to which they are aligned. Partly this has been a reflection of institutional forces such as promotion and recognition mediated through the "home" department and a lack of understanding about the specificities of digital work, but the direction has also been strongly influenced by external funding which has tended to be directives from the Office for Digital Humanities and other funding bodies.[4] Too often this form of funding has seen itself as "modernizing," "digitalizing" or "transforming" the humanities through an application of digital technology, a rather simplistic form of instrumentalism. It has also led to the growth of digital humanities research centers financed predominantly by soft money-funding and grant-capture and staffed by researchers on short-term contracts. Unsurprisingly, this has engendered a critical response from scholars on the supposed receiving end of the coming digital "disruption" (see Fish 2012; Allington et al. 2016).[5] Nonetheless, the breathless language of Silicon Valley often remains embedded in funding attached to digital transformations, which sees the softwarization of the humanities as an end in itself.[6] This often presupposes the idea that the humanities are old-fashioned, out of date and not fit for purpose in a digital world.

One response to this has been a call for digital humanities to be more responsive to cultural critique and critical theory (Berry 2012, 2013; Liu 2012a, 2012b, 33). This calls for wider social, cultural, political, and economic questions raised by digital technology to become part of critique in the digital humanities. This also means developing a program of criticism with respect to the computational in parallel with the digital transformations of the humanities and social science, and particularly its manifestation in digital capitalism, through what is called *critical digital humanities* (Berry 2013; Raley 2014, 35; Berry and Fagerjord 2017; but see also Dobson 2019).[7] This includes not just the technology itself but also the way in which the digital can be used to import neoliberal labor practices, academic restructuring, and grant-capture culture, particularly through digital humanities units that are not reflexive about their own situation. For example, digital humanities research centers often import hierarchical management structures with a director sitting atop the pyramid of research workers and administrative staff which consequently undermines collegial academic cultures. This structure is then often used as a justification for the differential pay that is awarded to managers and directors, as in the private sector, where it is claimed "leadership" or "vision" must be highly compensated.[8] In contrast, critical digital humanities argues that future directions for the digital humanities must be collegial, critically oriented, and more reflexive of the way in which computation is no longer merely a tool for thought, but also a disruptive infrastructure, medium, and milieu. Digital technology is therefore a medium of change and carries social change along with it, *tempora mutantur, nos et mutamur in illis*.[9] It is not merely a neutral instrument but also constitutes an entire apparatus and political economy with its own endogenous interests and value-structures and which cannot just be naively "applied" to the humanities.

This matters because the application of computation in the arts and humanities is not merely a neutral act of switching to a new medium, say from paper to digital. The digital and computation carry with them an imposed selectivity on how knowledge is transferred into data and in many cases the communicative capacities of digital networks can and do distort the data transmitted or stored. In addition, computation forms a political economic network which forms real material interests. While digital humanists have been exemplary in thinking critically about issues such as inherited, and sometimes contested, classifications, absences, encoding, and metadata when archives are digitalized, they have paid less attention to the medium-specific problematics or political economy of an inherent instrumentality in computation. That is, computation has a politics, partly due to its historical formation but also due to its tendency to impose metaphysical or formalist thinking on projects and programmers which may consequently be transmitted onto digital humanities work.[10] This can lead to a valorization of the mathematization of thought whereby formalization of knowledge through computation is seen as not just one approach to thinking but the exemplary one, often one that is misplaced (see Da 2019 for a related critique of using statistics). This is an idea shared by the logical positivists that "there was a certain rock bottom of knowledge ... which was indubitable. Every other kind of knowledge was supposed to be firmly supported by this basis and therefore likewise decidable with certainty" (Carnap 1967, 57). Mathematization tends towards formalization and rationalization, which can become an ideology that reduces the digital to instrumental rationality. Computation thought of as a mathematical or logical force comes to be seen as an independent participant in human social relations, it is given "life" fixed by its own nature and a power to shape social life. This is the reification of a social relation and as Marx argues,

> in order ... to find an analogy we must take flight into the misty realm of religion. There the products of the human brain appear as autonomous figures endowed with a life of their own, which enter into relations both with each other and with the human race. So it is in the world of commodities with the products of [one's] hands. I call this the fetishism which attaches itself to the products of labour as soon as they are produced as commodities.
>
> (Marx 1982, 165)

To human eyes it can seem as if computation is acting for itself and out of its own inner necessities, but in actuality it is subsumed to the needs of capitalism, not a form of mystified computational unfolding. Most notably computation, under conditions of capitalism, tends to develop the technical ability to separate control from execution (Braverman 1974, 159; Deleuze 1992).

This results in a process of generalized proletarianization of human thought and related social pathologies such as an individual's alienation from social life. This can lead to a sense of powerlessness and loss of meaning as the experience of social life mediated through computation makes it appear as fragmented and does not fit into a meaningful whole. In capitalism, computation tends to contribute to creating structural obstacles, such as persuasive interfaces and obfuscation of underlying computational processes, affecting an individual's capacity to understand the world and to see themselves as agents who can shape and change that world.

Further, this can encourage a tendency towards assuming that a digital project informed by mathematical foundationalism is recognizing "ideal" forms, and hence to proceed to argue that these are the only interests. Which, of course, ignores material interests including the political interests of those wielding the mathematical formalisms.[11]

In contrast, critical digital humanities can be understood as a research program guided by a common goal of human emancipation, carried out through reflexive interdisciplinary work that traditional scholarship has failed to address. It identifies the search for foundations and origins, which are current in digital humanities and computationalism, as not only problematic from a theoretical point of view but also politically suspect, tending towards reactionary politics (see Golumbia 2009).[12] This mania for foundations is built on the idealist assumption that everything can be reduced to mind. As Berry argues in relation to this tendency,

> It is likely that we should expect to see new philosophies and metaphysics emerge that again entrench, justify and legitimate the values of a [computational] accumulation regime. This calls for attentiveness to the tendency of philosophers to declaim their situatedness and historical location, and develop critical approaches to what we might call metaphysics of the computational and to the forms of computational ideology that legitimate a new accumulation regime.
>
> (Berry 2014, 5)

Indeed, mathematical theorizing, like capitalism, conceives of itself as somehow independent and free and so formulates itself as self-grounding and self-justifying.[13] It does this by authorizing its very speech as the limits and boundaries of intelligibility. But speaking which is not conscious of itself, as both a subject and object of the historical process, is sophistry. Computational speaking, which mirrors mathematical speaking, cannot grasp how its every attempt to speak about itself must fail because the very idea of speaking about something only represents its authority.[14] That is, the speaker from this position takes as a secure beginning that which is euphemized through notions of axiomatics, provisionality, or hypotheses. This is a practice that is allowed to slip into forgetfulness as present-speaking, which we might call alienated theorizing. This is a form of theorizing that looks in the wrong direction, continually looking away from itself to everything other than itself.

One of the paradoxes of computationalism is the way in which it is simultaneously understood as a logical foundationalism with a demonstrative method combined with a developmental or processual explanation.[15] The foundationalism tends towards an invariant conception of entities and relations frozen at the time of their computational fixation. The world is tethered to the ideal forms of the computational. These formal entities are then instantiated by the particularities of process which can only act in the theater of the concepts. Once it is set in motion, the complex and delicate machinery of applied computation becomes extremely difficult to change, due to sunk costs, development time, and what is seen as justified faith in the foundational structures that have been created. It becomes easier for computationalists to conceive of changing the world, rather than change the computational model. This means that too often the tail wags the dog when a computational system has been created, and so the world is remade in the image dictated by the computer system, rather than the other way around. When one identifies how deeply capitalistic logic is embedded within computational thinking, it becomes clear that markets, individual monads, and transactional relations tend to be paramount.[16] And so, the alien force of computation, the opaque algorithms of a new computer system, the sheer difficulty of challenging the output of the "truth machine" of computation, results in changes in social life that through other means would be difficult, if not impossible. This explains the difficulty of grasping the computational which needs a double aspect theory to capture both the mathematical foundationalism and developmental or processual sides of its actuality.[17]

In contrast, a critical digital humanities argues that computation and digital theorizing should and must be situated within a historical constellation to understand that it presents a distorted view of thought and experience, revealing the hidden structures of computational ideology.[18] By drawing on critical theory, critical digital humanities has the only prescription that one must have insight into one's own responsibility. Knowledge is seen as a historical and material phenomenon. That is, that the dominant mode of thought increasingly expressed through computationalism disguises partisan interest, indeed *material factors are the repressed factor*. Materialism, unlike idealism, always understands thinking to be the thinking of a particular people within a particular period of time. Idealism presupposes a subject who is independent of time able to discern abstractions, theories, and ideas by which the knowledge of an underlying structure is obtained. But that knowledge and structure belong to a particular historical situation. Indeed, materialism challenges the claim for the autonomy of thought and instead focuses on the concrete conditions under which humans live and in which too often their lives become stunted. These computational forms of life under capitalism should be criticized where it is not only different from what it could be but also different from what it should be.

We could say that computation loves to hide, and its laminate structure means that computation easily leads to misunderstanding of the veiled multi-layered nature of software.[19] This has implications on the way in which politics, interests, biases, and assumptions can be buried in a proprietary software stack and might be carried over into new software projects. We might consider how computation itself is ideological, encouraging thinking that unconsciously reinforces a reification of computational labor and assigning agency and power to a commodity fetish—an alien force. This can be seen repeatedly in digital projects that turn the human labor hidden within the opaque structure of software into a force of its own, most notably in projects that work with crowdsourcing, artificial intelligence, or machine learning, and which distort the social relation of labor under computational capital.[20]

Asking questions about the normative and political delegations into software/code requires digital humanities to develop a politics which challenges data practices, the implicit utilitarianism, and the consequences of algorithms and computation more generally (see Berry 2021). This would encourage digital humanities not just to "build things" but to take them apart and question the values that are embedded in the software—developing critical software to test its ideas and challenge its assumptions. This would call on digital humanities to turn its hermeneutic skills on the very software and algorithms that make up these systems. Critical digital humanities should advocate an ideology critique that challenges the knowledge industry, including the universities, museums, galleries, and arts, to uncover and reveal the hidden factors such as the struggle, social conflicts, and divisions in society. Often it is claimed that the digital creates a transparency through its mediating lens which "shows" or "reveals" hidden patterns or structures; however, as much as the digital may reveal it only does so partially, simultaneously hiding other aspects from view.

Critical digital humanities therefore aims to map and critique the use of the digital but is also attentive to questions of power, domination, myth, and exploitation. As such, critical digital humanities seeks to address the concerns that digital humanities lacks a cultural critique (Liu 2012a). Developing a critical approach to computation calls for the digital and computation to be historicized. By focusing on the materiality of the digital it draws our attention to the microanalysis required at the level of digital conditions of possibility combined with macroanalysis of digital systems. That is, their materiality, their specificity, and their political economy.

Computation is not just a technical matter, it is also a social, economic, and historical phenomenon and can be traced and periodized through this historicization. This is important because the hegemony of computational concepts and methods may lead to new forms of control, myth, and the privileging of computational rationality. As such, critical digital humanities should not only map these challenges but also propose new ways of reconfiguring research, teaching, and knowledge representation to safeguard critical and rational thought in a digital age. Digital humanities along with other cognate disciplines must remain attentive to moments in culture where critical thinking and the ability to distinguish between concept and object become weakened. As such, as research continues to be framed in terms of computational categories and modes of thought, the critique of the digital must become a research program in itself for the digital humanities. Digital technologies must be themselves subject to critique and the critical digital humanities by drawing on critical theory together with a sophisticated understanding of digitality would certainly help digital humanities develop and strengthen critical humanistic approaches to the new data-intensive world.

There is a clear need to strengthen critical reflexivity in the digital humanities, and I want to expand a little on this notion. Here there is only space to provide pointers towards a set of practices and ways of thinking rather than a comprehensive blueprint. Nonetheless, I want to suggest that a way of critiquing the, sometimes, instrumental tendencies within the digital humanities could be a greater focus on the socio-technical aspects of the technologies. This means examining how they are organized, assembled, and made as well as the possibility of making them otherwise. For example, we need to understand and challenge the ways in which "smart" objects and infrastructures bypass our cognitive capacities in order to maximize data-intensive value-extraction. This is to understand how these particular instantiations of the digital can result in further alienation, proletarianization of knowledge, and other social pathologies. Digital humanities can and should examine how processes of automation in knowledge production and manipulation are often a means of exploiting labor, together with a critique of its own organizational practices. It should critique the common structure of top-down hierarchical management in digital humanities centers, experimenting with new collegial structures for academic work. It should examine how extractive processes can lead to exploitation and alienation, even in nominally humanistic research projects that rely on line-management and crowdsourcing where the assumption is that the latter is a "participatory" practice. It should aim to understand how computation is not just a technical choice of instrumentation, but carries with it ideological assumptions and structures that may subtly distort the outcome of a particular digital project, such as about class, gender, or race (Eubanks 2018; Bordalejo and Risam 2019; Chun 2021). This would enable it to challenge the idea that digital technologies offer a panacea for liberal thought and freedom and show how they are equally able to undermine the human capacity for critical reflexivity and reason.[21] For example, little thought in digital humanities has been applied to the alienating tendencies or potentialities of artificial intelligence and machine learning.[22] Nor to the still prevalent practice of grants-funded digital humanities projects which, long after the funding runs out, become digital humanities ruins, slowly rotting and breaking down in forgotten areas of the Internet.[23] As can be seen with these examples, digital humanities also absorbs the fetishism of the new which is endemic to computational cultures, particularly the programming industries.

Critical digital humanities draws attention to the politics and norms that are embedded in digital technology, algorithms, and software. It foregrounds these questions by affirming human emancipation as a goal making a distinction between what society is and what it hopes to be. There remains a need to embed the capacity for reflection and thought into a critically oriented digital humanities and thus to move to a new mode of thinking, a two-dimensional thinking responsive to the potentialities of people and things. This requires enabling a new spirit of criticality and a rethinking of what it means to be human within a computational milieu—especially in terms of nurturing and encouraging critical reason. In other words, there is an urgent need to reconfigure quantification, formalism, and instrumental rational processes away from domination and control, towards reflexivity, critique, and democratic practices. As Thompson argues, "the task of articulating new forms of ethical life—that is, new shapes of social reality (relations, processes and purposes)—is the fundamental task of a critical theory" (2022, 19). This would be to develop a critical reflexive humanism which grants digital humanities some measure of autonomy from quantitative and computational approaches.[24] Critical digital humanities' attempt to address these issues within the field of digital humanities helps look to a more progressive future, one where computation is not seen as an alien force over and against humanity, but rather as what it actually is, human-created and human-directed infrastructures of thought that can and should serve the common good.

NOTES

1. John Unsworth outlined a proposal for a master's degree in digital humanities in 2001, noting, "the name of the program ('Digital Humanities') is a concession to the fact that 'Humanities Informatics' (which would probably be a more accurate name) sounds excessively technocratic, at least to American ears. The other obvious alternative—'Humanities Computing'—sounded too much like a computer support function" (Unsworth 2001). This is one of the earliest uses of the term "digital humanities" in relation to this field of study that I am aware of (see also Kirschenbaum 2010; Clement 2012).
2. Knowledge representation tends to focus on the "what" of data structures and databases, rather than the "how" (Drucker and Svensson 2016). Also see McCarty (2005, 30) and Berry and Fagerjord (2017) for a discussion of knowledge representation.
3. Underwood is interesting for the attention he gives to the historical practice of "distant reading as a mode of interpretation rather than a computing technology" (2019, 157).
4. See The National Endowment for the Humanities, https://www.neh.gov/divisions/odh, and in the UK, Arts and Humanities Research Council (AHRC), https://webarchive.nationalarchives.gov.uk/ukgwa/20180702140003/https://ahrc.ukri.org/research/fundedthemesandprogrammes/themes/digitaltransformations/ and https://webarchive.nationalarchives.gov.uk/ukgwa/20210802111505/https://ahrc.ukri.org/research/fundedthemesandprogrammes/pastinitiatives/strategicprogrammes/ictartsandhumanitiesresearch/.
5. It has also encouraged a search for ways in which digital humanities can respond to the ethical conundrums that computation raises (see Chapter 42, "AI, Ethics, and Digital Humanities," in this volume).
6. On the notion of "digital transformation," see Haigh (2019) who also introduces the intriguing concept of the "early digital" in relation to "a set of localized and partial transformations enacted again and again, around the world and through time, as digital technologies were adopted by specific communities" (13).
7. See particularly Chapter 8, "Towards Critical Digital Humanities" in Berry and Fagerjord (2017).
8. Organized research units (ORUs), such as research centers, are "opportunistic entities" that run the risk of creating systems of control through "mission-oriented" research which facilitate instrumental effectiveness but create distorted power relations manifested in a unit with out-of-touch management, lack of accountability, diminished academic collegiality, and hierarchical management rooted in the elemental impulse of domination because "centers have a more authoritarian management structure

than departments" (Stahler and Tash 1994, 546). This becomes pathological where management becomes concerned with vaunting their social superiority or extracting personal difference from those below them. Rather than research culture, the unit then focuses on conformity with the rules, security based on seniority, and the ideology of busy managers, technicians, and staff. In this variant, a digital humanities center may become a more conservative force as it becomes more concerned with its own survival and funding than with its original founding principles of research. Indeed, as Stahler and Tash argue, "for the most part [research centers] are not major contributors to the educational mission of universities ... because for centuries research and scholarship have been successfully conducted within the confines of academic departments, and centers often do not have an intellectual core" (1994, 542). They continue, "centers often have a way of becoming somewhat independent of their oversight and continue year after year without any systematic monitoring. Like academic departments, they should be reviewed on a regular basis to determine whether they are achieving their goals, whether changes in direction and internal support are necessary, and what they are contributing to their universities" (552).

9. Times are changed; we also are changed with them.
10. The structuring effects of computation towards certain "rational" forms which align with capitalism, and particularly with an *a priori* assumption of the superiority of markets for structuring social relations, are apparent in the way in which computational engineering structures roles, entities and processes in particular configurations (see Lohman 2021).
11. See Ricci (2018) revealing Marx's deep interest in mathematics in the *Mathematical Manuscripts of Karl Marx* which has "strong analogies with the modern concept of algorithm and this makes Marx a precursor of modern computational mathematics" (225, see also Marx 1983).
12. Formalist and foundationalist approaches tend to describe a system of computation without a political economy, failing to draw inferences about how it actually functions. They also lack an understanding of its history, mediations, and qualifications. They fail to heed the advice of Hegel, who argued, "philosophy is its epoch comprehended in thought."
13. This valorization of the truths of the axiomatics of mathematics combined with the developmental or organic unfolding of processualism I call mathematical romanticism. Typically, the mathematical and rational would be contrasted to the intuitive and organic, but in mathematical romanticism they are fused in an unstable narrative which tends towards the idea of expressing the "life" of machines, technology, and computation. This is strikingly reminiscent of the futurist's call to "mathematicians [to] let us affirm the divine essence of chance and randomness" and that "we must bring this mathematics directly into life making all breathing hypotheses come alive alongside us breathing beings" (Marinetti et al. 1941, 300).
14. Mathematical romanticism sometimes becomes transformed into its computational parallel as a form of computational romanticism, often derived from Gödelian incompleteness, as in the work of Land (2018), "Gödelian incompleteness is logically isomorphic with the halting problem in the (Church-Turing) theory of computation, and thus translatable after rigorous transformation into the uncomputable. It establishes a basic principle of unbounded application within the electronic epoch" (but see also Schmidhuber 2012). The recent increase in interest in superficial dualisms, such as the "uncomputable," often ironically presented as a binary opposite of the computable, shows the degree to which a computational ideology needs to be critically examined (see, for example, Galloway 2021). See also the notion of computational ontologies (Berry 2014, 89–97; Smith and Burrows 2021, 148, 157).
15. We might think of this as magical thinking about computation. *Omne ignotum pro magnifico est* ("everything unknown seems wonderful"). As Marx comments, "no content is gained in this way, only the form of the old content is changed. It has received a philosophical form, a philosophical testimonial ... Another consequence of this mystical speculation is that a particular empirical existent, one individual empirical existent in distinction from the others, is regarded as the embodiment of the idea. Again, it makes a deep mystical impression to see a particular empirical existent posited by the idea, and thus to meet at every stage an incarnation of God ... And it is self-evident. The correct method is stood on its head. The simplest thing becomes the most complicated, and the most

complicated the simplest. What ought to be the starting point becomes a mystical outcome, and what ought to be the rational outcome becomes a mystical starting point" (1987, 39).

16. Winner (1997) describes this as cyberlibertarianism, describing it as "a collection of ideas that links ecstatic enthusiasm for electronically mediated forms of living with radical, right wing libertarian ideas about the proper definition of freedom, social life, economics, and politics." Barbrook and Cameron (1995) memorably described this constellation as the "Californian Ideology," which argues "that human social and emotional ties obstruct the efficient evolution of the machine" and that "the cybernetic flows and chaotic eddies of free markets and global communications will determine the future."
17. One of the hallmarks of some forms of cyberlibertarian thinking about computation is the inherent idea of an organicism which sees the "rapid development of artificial things amounts to a kind of evolution that can be explained in quasi-biological terms" (Winner 1997). This is often described as a kind of internal necessity driven by an inner force of spontaneous development, variously emerging from indeterminacy, randomness, chaos, or complexity of the underlying computational logic. At the end of the process there are often claims to the emergence of "intelligence," "creativity," or "hypercomputation." In the UK, Dominic Cummings, the Prime Minister's advisor, was notable for translating this form of computationalist ontology into policy proposals in government (Collini 2020).
18. This often involves "blurring the concept of general intelligence with the concepts of mind or consciousness" (Golumbia 2019). It also tends to involve the reification of human cognitive labor within computational systems. Golumbia (2019) has also made links with computational ideology and what he calls a "messianic/Christological structure," particularly in artificial intelligence theorizing (see also Haider 2017 for similar links to right-wing thought).
19. Heraclitus wrote *phusis kruptesthai philei* remarking that "nature loves to hide."
20. This also gestures to the troubling use of "crowd-sourcing" within digital humanities projects and the often untheorized implications of facilitating wider labor exploitation and outsourcing, what Gray and Suri (2019) call "ghost work."
21. See for example an extremely thought-provoking piece by Golumbia (2016) about assumptions regarding open access.
22. Although there has been notable work thinking about the implications for ethics related to data bias, algorithms, and artificial intelligence (see D'Ignazio and Klein 2019; Ess 2009; Metcalf et al. 2019). See Chapter 42, "AI, Ethics, and Digital Humanities," in this volume.
23. But see also King's Digital Lab for an example of an institution that has sustainability of digital humanities projects as integral to its work. Through a number of different approaches KDL seeks to maintain or migrate a digital humanities project into the KDL, but also host a project through static conversion or datasets deposit. https://kdl.kcl.ac.uk/our-work/archiving-sustainability/.
24. See Schecter (2007, 71) in regard to avoiding these narrow forms of strategic reason more generally.

REFERENCES

Allington, Daniel, Sarah Brouillette, and David Golumbia. 2016. "Neoliberal Tools (and Archives): A Political History of Digital Humanities." *The Los Angeles Review of Books*. https://lareviewofbooks.org/article/neoliberal-tools-archives-political-history-digital-humanities.

Barbrook, Richard and Andy Cameron. 1995. "The Californian Ideology." *Mute*. https://www.metamute.org/editorial/articles/californian-ideology.

Berry, David M. ed. 2012. *Understanding Digital Humanities*. London: Palgrave.

Berry, David M. 2013. "Critical Digital Humanities." *Stunlaw*. http://stunlaw.blogspot.com/2013/01/critical-digital-humanities.html.

Berry, David M. 2014. *Critical Theory and the Digital*. London: Bloomsbury.

Berry, David M. 2021. "Explanatory Publics: Explainability and Democratic Thought." In *Fabricating Publics: The Dissemination of Culture in the Post-truth Era*, edited by Bill Balaskas and Carolina Rito, 211–33, London: Open Humanities Press.

Berry, David M. and Anders Fagerjord. 2017. *Digital Humanities: Knowledge and Critique in a Digital Age*. Cambridge: Polity.

Bordalejo, Barbara and Roopika Risam, eds. 2019. *Intersectionality in Digital Humanities*. York: Arc Humanities Press.

Braverman, Harry. 1974. *Labor and Monopoly Capital: The Degradation of Work in the Twentieth Century*. New York, NY: Monthly Review Press.

Carnap, Rudolf. 1967. *The Logical Structure of the World*. Berkeley, CA: University of California Press.

Chun, Wendy Hui Kyong. 2021. *Discriminating Data: Correlation, Neighborhoods, and the New Politics of Recognition*. Cambridge, MA: MIT Press.

Clement, Tanya. 2012. "Multiliteracies in the Undergraduate Digital Humanities Curriculum: Skills, Principles, and Habits of Mind." In *Digital Humanities Pedagogy: Practices, Principles and Politics*, ed. Brett Hirsch, 365–88. Cambridge: Open Book Publishers.

Collini, Stefan. 2020. "Inside the Mind of Dominic Cummings." *The Guardian*, February 6. https://www.theguardian.com/politics/2020/feb/06/inside-the-mind-of-dominic-cummings-brexit-boris-johnson-conservatives.

Da, Nan Z. 2019. "The Computational Case against Computational Literary Studies." *Critical Inquiry* 45 (3) (Spring): 601–39.

Deleuze, Gilles. 1992. "Postscript on the Societies of Control." *October* 59: 3–7.

D'Ignazio, Catherine and Lauren F. Klein. 2019. *Data Feminism*. Cambridge, MA: MIT Press.

Dobson, James E. 2019. *Critical Digital Humanities: The Search for a Methodology*. Champaign, IL: University of Illinois Press.

Drucker, Joanna. 2017. "Why Distant Reading Isn't." *PMLA/Publications of the Modern Language Association of America* 132 (3): 628–35.

Drucker, Joanna and Patrik B. O. Svensson. 2016. "The Why and How of Middleware." *Digital Humanities Quarterly* 10 (2). http://www.digitalhumanities.org/dhq/vol/10/2/000248/000248.html.

Ess, Charles. 2009. *Digital Media Ethics*. Cambridge: Polity.

Eubanks, Virginia. 2018. *Automating Inequality: How High-Tech Tools Profile, Police, and Punish the Poor*. New York, NY: St. Martin's Press.

Fish, Stanley. 2012. "The Digital Humanities and the Transcending of Mortality." *The New York Times*, January 9. http://opinionator.blogs.nytimes.com/2012/01/09/the-digital-humanities-and-the-transcending-of-mortality.

Galloway, Alexander R. 2021. *Uncomputable: Play and Politics in the Long Digital Age*. London: Verso.

Golumbia, David. 2009. *The Cultural Logic of Computation*. Harvard, MA: Harvard University Press.

Golumbia, David. 2016. "Marxism and Open Access in the Humanities: Turning Academic Labor against Itself." *Workplace* 28: 74–114.

Golumbia, David. 2019. "The Great White Robot God: Artificial General Intelligence and White Supremacy." *Medium*. https://davidgolumbia.medium.com/the-great-white-robot-god-bea8e23943da.

Gray, Mary L. and Siddharth Suri. 2019. *Ghost Work: How To Stop Silicon Valley from Building A New Global Underclass*. Boston, MA: Houghton Mifflin Harcourt.

Haider, Shuja. 2017. "The Darkness at the End of the Tunnel: Artificial Intelligence and Neoreaction." *Viewpoint Magazine*. https://viewpointmag.com/2017/03/28/the-darkness-at-the-end-of-the-tunnel-artificial-intelligence-and-neoreaction/.

Haigh, Thomas, ed. 2019. *Exploring the Early Digital*. London: Springer.

Hockey, Susan. 2004. "The History of Humanities Computing." In *A Companion to Digital Humanities*, edited by S. Schreibman, R. Siemens, and J. Unsworth, 3–19. London: Wiley-Blackwell.

Jockers, Matthew. 2013. *Macroanalysis: Digital Methods and Literary History*. Champaign, IL: University of Illinois Press.

Kirschenbaum, Matthew G. 2010. "What is Digital Humanities and What's It Doing in English Departments?" *ADE Bulletin* 150: 55–61.

Land, Nick. 2018. "Crypto-Current, An Introduction to Bitcoin and Philosophy." *ŠUM | Journal for Contemporary Art Criticism and Theory*. http://sumrevija.si/en/sum10-2-nick-land-crypto-current-an-introduction-to-bitcoin-and-philosophy/#_ftnref9.

Liu, Alan. 2012a. “Where is cultural criticism in the digital humanities?” In *Debates in Digital Humanities*, edited by Matthew K. Gold, 490–509. Minneapolis, MN: University of Minnesota Press.
Liu, Alan. 2012b. “The State of the Digital Humanities: A Report and a Critique.” *Arts and Humanities in Higher Education* 11 (1–2): 8–41.
Lohman, Larry. 2021. “Interpretation Machines: Contradictions Of ‘Artificial Intelligence’ in 21st-Century Capitalism.” In *Beyond Capitalism: New Ways of Living*, edited by L. Panitch and G. Albo. *The Socialist Register 2021*, London: Merlin Press.
Marinetti, Filippo T., Marcello Puma, and Pino Masnata. 1941. “Qualitative Imaginative Futurist Mathematics.” In *Futurism: An Anthology*, edited by Lawrence Rainey, Christine C. Poggi, and Laura Wittman, 298–301. New Haven, CT: Yale University Press.
Marx, Karl. 1982. *Capital: A Critique of Political Economy*, vol. 1. London: Penguin Books.
Marx, Karl. 1983. *Mathematical Manuscripts of Karl Marx*. London: New Park Publications.
Marx, Karl. 1987. *Karl Marx, Frederick Engels: Collected Works,* vol. 3. London: Lawrence & Wishart.
McCarty, Willard. 2003. “Humanities Computing.” In *Encyclopedia of Library and Information Science*, 2nd edn, edited by Miriam Drake, 1224–35. New York, NY: Marcel Dekker.
McCarty, Willard. 2005. *Humanities Computing*. London: Palgrave.
Metcalf, Jacob, Emanuel Moss, and danah boyd. 2019. “Owning Ethics: Corporate Logics, Silicon Valley, and the Institutionalization of Ethics.” *Social Research: An International Quarterly* 82 (2) (Summer): 449–76.
Raley, Rita. 2014. “Digital Humanities for the Next Five Minutes.” *differences* 25 (1): 26–45.
Ricci, Andrea. 2018. “The Mathematics of Marx: In the Bicentenary of the Birth of Karl Marx (1818–1883).” *Lettera Matematica* 6: 221–5.
Schecter, Darrow. 2007. *The History of the Left from Marx to the Present: Theoretical Perspectives*. London: Continuum.
Schmidhuber, Jürgen. 2012. *Gödel Machine* Home Page. https://people.idsia.ch//~juergen/goedelmachine.html.
Schreibman, Susan, Ray Siemens, and John Unsworth. 2004. *A Companion to Digital Humanities*. London: Wiley-Blackwell.
Smith, Harrison and Roger Burrows. 2021. “Software, Sovereignty and the Post-Neoliberal Politics of Exit.” *Theory, Culture and Society* 38 (6): 143–66.
Stahler, Gerard J. and William R. Tash. 1994. “Centers and Institutes in the Research University.” *The Journal of Higher Education* 65 (5): 540–54.
Thompson, Michael J. 2022. “On the Crisis of Critique: Reformulating the Project of Critical Theory.” In *Critical Theory in a Time of Crisis*, edited by D. Chevrier-Bosseau and T. Bunyard. London: Palgrave.
Underwood, Ted. 2019. *Distant Horizons: Digital Evidence and Literary Change*. Chicago, IL: University of Chicago Press.
Unsworth, John. 2001. “A Master’s Degree in Digital Humanities: Part of the Media Studies Program at the University of Virginia.” https://johnunsworth.name/laval.html.
Winner, Langdon. 1997. “Cyberlibertarian Myths and the Prospects for Community.” https://www.langdonwinner.com/other-writings/2018/1/15/cyberlibertarian-myths-and-the-prospects-for-community.

CHAPTER THIRTEEN

Does Coding Matter for Doing Digital Humanities?

QUINN DOMBROWSKI (STANFORD UNIVERSITY)

One spring morning, I presented my department's support for DH work to a group of prospective graduate students, explaining that I saw it as my goal to help them learn as much as they wanted about DH methods over their course of study, and lack of prior exposure to technology need not be a barrier. I explained to them that I had spent fifteen years "doing digital humanities" prior to my current role, obtaining a Master's in Library and Information Science along the way, but only in this position had I learned to code in any meaningful way. The surprise on their faces still did not prepare me for the reception that evening, when one student who had worked for a tech company as a developer approached me with an expression of bemusement and condescension and asked, "So if you couldn't code, what did you *do* for all those years?"

Coding-for-everyone rhetoric permeates Anglo-American and Western European societies. STEM (Science, Technology, Engineering, and Math) training is now sold to economically anxious parents as a path towards lucrative careers for their children. There are specific toys and books designed to make coding appeal to girls, such as Stacia Deutsch's *The Friendship Code* (2017), where "Lucy ... discovers that coding – and friendship – takes time, dedication, and some laughs!" There are programs such as the Edie Windsor Coding scholarship, specifically for LGBTQ individuals to attend coding bootcamps, where part of the stated goal is to build the "kind of world that we (and you!) want to live in" (Lesbians Who Tech 2017) through better representation on tech teams. And large tech companies have responded to critiques of their racial homogeneity by, for example, starting partnerships with Historically Black Colleges and Universities (HBCUs), ostensibly to address the "pipeline problem" for hiring software engineers, but in practice more a publicity stunt than a genuine effort to change their demographics (Tiku 2021). The rise of "data science," with concomitant funding, academic programs, and jobs (academic and otherwise) seems poised to once again aggravate a long-standing fault line within digital humanities between those who see coding as central to the endeavor of DH, and those who take a more expansive view of the field. Scholars who advocate for treating coding as a core element of the digital humanities skill set are supported by the rhetoric of coding-for-everyone advocacy within society more broadly.

The forces driving the ongoing anxiety around coding in DH are not all external. If it were ever possible to exhaustively define the field of DH, that time has passed—there is too much work in this broad area, in too many disciplines, theoretical frameworks, and interdisciplinary spaces both inside and outside academia across the globe. Without clear boundaries, there can be no meaningful conception of the "center," but the annual international Digital Humanities conference hosted by the Alliance of Digital Humanities Organizations (ADHO) serves as one influential confluence of scholars, typically from well-resourced institutions around the world. Shortly after the 2019 conference, a group of scholars who do computationally oriented work expressed concerns about feeling "unwelcome" in DH spaces, which soon led to the establishment of a "Computational Humanities" group that now hosts its own conference with practices that align with norms in computer science and linguistics (e.g., submission of 6- or 12-page papers, exclusive use of LaTeX[1]). The response to this turn of events was swift and negative from many parts of DH Twitter (as summarized in Lang 2020a). There was a sense that work labeled as "computational" (as opposed to "merely digital") was already privileged in conferences, grants, and hiring. As Miriam Posner put it, this group's public splintering-off was another instance of "'serious' people peeling away from the more feminized corners of DH and giving their thing a new name. It's a method of protecting prestige, which tracks closely with masculinization" (Posner 2019), as well as whiteness (Risam 2019). The demographics of the co-signatories largely tracked with these assertions about the group's make-up (Froehlich 2019). A reflection by Sarah Lang on the founding of the Computational Humanities, a year later (Lang 2020a), prompted a productive dialogue with the group (Lang 2020b) about strategies for making the group more inclusive.

Addressing inclusivity in computationally oriented spaces is important to ensure that more people feel welcome there, if their work takes them in that direction. But less discussed is the question of computational prestige. Beyond the tired "hack/yack" axis that was once deployed as a shorthand for praxis vs. theory (often in contexts that privileged the former), how important—specifically—is *coding* for being a skilled digital humanist, loosely and broadly defined? This chapter takes up that question.

SCOPING DIGITAL HUMANITIES

Careful boundary-drawing around the definition of digital humanities can easily lead to different conclusions about the necessity of coding. With a definition of digital humanities that resembles the scope and focus of the Computational Humanities group described above, or scholars who frame their work as "cultural analytics" (which resembles a kind of data science focused on cultural production; a journal by the same name publishes work in this area), coding is fundamental. Expanding the definition to include other areas of research and praxis, from digital pedagogy to new media studies, to cultural critique of the digital from feminist or postcolonial perspectives, to work in Ottoman Turkish history or Korean literature that draws upon digital methods to bolster arguments within those disciplines, only a small amount of that work involves writing code.

Instead of playing with the scope of digital humanities to argue the case for or against the necessity of coding, let us reframe the question: taking a broadly expansive definition of DH, is the amount of attention, workshops, resources, and anxiety around coding equally commensurate to its value for all DH practitioners, relative to other skills that could be prioritized?

"KNOWING HOW TO CODE"

The formulation of "learning" or "knowing" how to code carries with it some assumptions and baggage that are familiar to anyone who has undertaken serious study of another human language. A commonly asked, and typically dreaded, question for linguists and others who work across multiple languages is, "How many languages do you know?" There is rarely an answer that is both honest and simple; a place to start in answering the question is to ask in response, "Know well enough for what?" Does being able to sound out words written in another alphabet, perhaps recognizing a few loan words, count as "knowing" it in a meaningful way? What about being able to mostly follow a text with support from a dictionary or machine translation? Being able to *produce* the language correctly is another matter, and producing it slowly in writing is different from being able to speak it intelligibly.

For someone who wants to "learn how to code," watching a scholar with a high degree of computational fluency write code can be as intimidating as watching two fully fluent speakers carry on a rapid-fire conversation in a language you have only studied for a few weeks. The chasm between basic vocabulary and uninhibited conversation—much like the gulf between "hello world" introductory code and being able to write out complex functions without reference materials—can feel immense to the point of being unbridgeable. But the presence (and, even worse, proximity) of these examples can drive a persistent feeling of insecurity, even as a person builds up proficiency. Being able to read and make sense of the language is an achievement, but perhaps dampened by the disappointment of not being able to write. Being able to speak the language intelligibly is an achievement, but does not prevent the frustration of lacking the vocabulary to express your full range of thoughts. A similar dynamic is often at play with coding. Understanding basic concepts and structures of computation—things like data types and recursion—is one kind of knowledge with real utility. Knowing the basic syntax for implementing useful algorithms in a particular programming language opens up another set of options for making coding a useful part of one's research. But there is a tendency even for people who are fairly comfortable using a programming language as part of their workflow to apologize for their lack of skill because they routinely consult Stack Overflow or adapt other online examples, despite the fact that this is also a common behavior of expert programmers. This same habit of mind can contribute to a reluctance to post one's own code where it might be a resource for others, despite its shortcomings. What if an algorithm is implemented in an inefficient or inelegant way? Will it draw critique from scholars who are truly computationally fluent? Will it undermine one's attempt to claim a place among the "computational" crowd?

A LITTLE CODE IS A DANGEROUS THING

Developing proficiency with code—much like with a human language—takes time, persistence, practice, and a deliberate effort to stretch beyond one's comfortable boundaries. These things are rarely emphasized in the kinds of introductory workshops or short courses that are a common environment for DH scholars to be exposed to coding. Such workshops are frequently framed as opportunities to develop practical coding skills that scholars can immediately apply to their own research, but that translation and adaptation process can be fraught.

Similar challenges can arise when a scholar with basic coding skills tries to re-use code they find online or in a book. Translating humanities research questions into code, and recognizing the ways in which various decisions along the way need to be translated into caveats on one's findings (e.g., choices around corpus creation, whether to lemmatize the text first, or how many—and which—words to choose for a feature set) is a skill separate from the coding itself. It is easy to make mistakes that lead to "working" code (i.e., code that will run without errors) but still have a significant negative impact on one's results.

This risk is also present for scholars who have enough exposure to coding to see its benefits for their project, but do not have the time to invest in developing the necessary skills, and instead hire a student (such as a computer science undergraduate) to implement the project. If the scholar does not offer very specific guidance about how to implement aspects of the project that will impact the results, these decisions are left to the student's judgment and are typically made, undocumented, in ways that the scholar may not even recognize until it is too late. Even if the student writes code that successfully does what the scholar wants, the scholar might make some assumptions about the durability of the code, or its legibility for any arbitrary "programmer" they might hire after this student moves on, that can jeopardize the project. Understanding and building on another person's code isn't quite the same as speaking the same language as the original coder; it has more in common with reading and imitating that original programmer's handwriting and written style.[2]

Beyond the area of computational-analytical research, where these kinds of scenarios can imperil the validity of the scholar's results, there are other kinds of DH work where "a grad student who knows PHP" or similar can be more of a liability than a benefit. For web development projects that are built using carefully configured content management systems and an ecosystem of open-source plugins and code components, adherence to existing coding and documentation practices is essential for maintainability. This is unlikely to be the outcome when someone "knows how to fix it with PHP" and begins hacking away at the project's codebase—and the same warnings about durability and reusability apply here as well.

If overconfidence after a small amount of exposure to computational methods is a problem, and long-term underconfidence after a bit more exposure is also a problem, how can scholars avoid developing a disordered relationship with code, short of fully committing themselves to a path towards expertise or avoiding it altogether? What if we changed the frame, and instead of encouraging people to level up through "learning coding," we emphasized *understanding workflows* in all their complexity?

FROM PROGRAMMING TO WORKFLOWS

"Workflow" is used less commonly than terms like "tools" or "methods" in the discourse around digital humanities praxis, and merits some definition. I use it here to refer to the collection of steps that go into implementing some aspect of a digital humanities project. It exists at the point of interface between technology, disciplinarily legible methods, and—in many cases—interpersonal interactions. In the context of DARIAH's TaDiRAH (Taxonomy of Digital Research Activities in the Humanities) vocabulary, there are many potential workflows that could be used to accomplish each research activity (e.g., analysis) and its sub-activities (e.g., content analysis, network analysis, spatial analysis). The workflow that a project uses for network analysis for characters in novels could involve steps like:

1. Corpus selection, e.g., which novels were chosen, and why? To what extent did the factors that shape large corpora like EEBO and HathiTrust (e.g., gender, race, colonialism) set constraints on the decision space?
2. Corpus acquisition: what process was made to digitize the text or convert it to a usable form? How was the choice of OCR software and level of proofreading made? What is the error rate, and how was it determined what error rate was acceptable?

3. What decisions were made about how to determine entity co-occurrence, e.g., at the level of the sentence, paragraph, window of a certain number of words, chapter, or scene? If the textual division is not marked in the text (e.g., scenes), how were those units determined in the text?
4. How were characters identified? Was co-reference resolution used to map pronominal references to characters? Were plural pronouns disambiguated or discarded? What tool(s) or model(s) were used for named-entity recognition? If those tools were computational, rather than the result of human counting, what data were the models trained on?
5. How were instances of multi-co-occurrence (such as groups of people) handled?
6. Was an f1 score calculated for the entity recognition, or were other efforts made to determine the accuracy of the co-occurrence data?
7. What, if anything, was done to clean the co-occurrence data after it was created but before visualizing the network?
8. Was the network generated to be directed or undirected, and why?
9. How do the choices that were made around counting co-occurrences affect the interpretation of common network analysis metrics?
10. How do the choices that were made around your corpus selection, and what data to include vs. throw out, impact the scope of the claims made based on the study?

Out of these ten steps (though one could imagine more), 4 is the only one that will most likely require writing code—though not necessarily, as some DH studies on character networks have involved manual counts (e.g., *Austen Said* [White n.d.] and Ilchuk 2019). The steps involving network analysis could involve writing code, but software packages such as Gephi and Cytoscape offer an alternative with a graphical user interface. Choosing to calculate error rate for either the OCR or the entity recognition may involve writing some code, but many projects do not undertake these steps.

The specific task of writing code for step 4 (in cases where a project chooses a computational approach to named-entity recognition and/or co-reference resolution) is not difficult. Depending on the language of the text, models may already exist for recognizing named persons—and perhaps even for recognizing unnamed persons in literary texts (Bamman et al. 2019). The coding task itself would probably be within reach for someone with a small amount of experience with a language like Python or R, by extensively using existing libraries, but what makes the results of this network analysis method a valuable or dubious contribution to the scholarly discourse is how the project team grappled with the choices in all ten steps. This is particularly crucial when working at scale. With tens or hundreds of texts, it is easier to manually check that your OCR engine is performing consistently across your entire corpus. With thousands or tens of thousands of texts, you need to know more about your corpus to be able to do informed spot-checks. Randomly sampling a hundred texts may not be enough: you need to take into account which books have complex formatting (e.g., columns or tables), which publishers use distinctive fonts at different time periods, etc. Choosing a random sample to manually check, without taking these factors into account, can lead you to draw the conclusion, for instance, that the usage frequency for a particular word drops precipitously in a given time period, when that drop reflects OCR challenges for a particular publisher's font during that period. Similarly, it is worth taking care to ensure that pre-trained named-entity recognition (NER) models are performing adequately before drawing conclusions about character networks. Once again, random checks can be inadequate: if you are working with a model trained largely on Anglo-American names, you may find that science fiction novels with alien or robot names perform much less well. You can choose to mitigate this problem,[3] or scope your claims in a way that takes this issue into account, but doing a random selection from your corpus to check on NER accuracy may not be enough to identify that anything is amiss.

The value of this kind of workflow framework is not limited to projects that use computational analysis methods. The creation of database and website-oriented projects also involves numerous decision points that impact what is included and how that material is presented. Making one choice or another may not necessarily invalidate the project's argument (the way that wrong choices can fundamentally undermine the results when employing computational analysis), but these choices constrain and enable different possibilities for the display, visualization, and re-use of the data. Furthermore, the choices shape which audiences and disciplines the study is legible to. Increasingly, DH scholars whose projects have a significant web-based publication component are taking steps to explicitly note and speak to different audiences in different ways, as a step towards mitigating the legibility gap that can plague interdisciplinary work (see Cara Marta Messina's digital dissertation, *Critical Fan Toolkit* (2021), for one excellent example).

Learning how to make better decisions—e.g., how to model data effectively, or how to format different data types like dates or names—is often the result of a private process of trial and error. Workshops that are organized around the technical skills necessary to implement web-based or database projects cover some of these issues implicitly or explicitly. But the value of foregrounding the *process* over the more visible final *product* will increase if more DH scholars reimagine web-based project sustainability through the lens of collectivity and re-use (as described in Morgan 2021a and Morgan 2021b) rather than preservation. The preservation of digital projects—making them perpetually available in their original form or similar—has been an ongoing struggle, and often a failed one (Nowviskie and Porter 2010; Rockwell et al. 2014; Smithies et al. 2019). When building a web-based project with the intention of preserving it as a monolith, code is fundamental. There are often trade-offs between ease of use for the website creator and sustainability costs: both the direct staff costs of continually retrofitting the site to maintain its functionality, and the bigger-picture ecological costs implied by the upkeep and maintenance of complex systems (Gil 2015). But if the long-term value of the project accrues through the re-use of the project's data, code is temporary and transient, and the decisions that go into obtaining and preparing the data become more significant. With the goal of encouraging re-use, project developers must take attention to their workflow even further by documenting their steps to the greatest extent possible, to help future users of the text understand how the data came to be.

A workflow-framing for DH scholarship is capacious enough to include even those areas of DH where coding has little or no relevance. Piper (2021) lays out a kind of workflow for making more-defensible scholarly arguments, with parallels to computational model-building. Steps like defining concepts (and evaluating agreement on those definitions), being specific about what data has been selected for analysis, and being explicit about making generalizations are steps that need not apply only to computational workflows. Workflows made up of sets of consequential steps have the potential to serve as a common-ground concept for how DH scholarship comes together—whether it is a theoretical critique, a web-based project based on domain-specific data, or a computational analysis—and could help bridge the distance between these different kinds of DH.

DOES CODING MATTER?

There are some kinds of DH work where coding matters. Pre-built tools will always have limitations; in their creation, developers must make decisions that constrain the kinds of questions the tool can be used to answer. Depending on the data, and the question the scholar wants to answer, sometimes it is necessary to deploy code. Indeed, for scholars who may have attempted to heed general calls advocating for "learning to code," only to be met with frustration and disappointment, a situation where coding is the only way to accomplish one's goals may be a necessary precondition for overcoming the barriers that coding presents (Lang 2020c; Risam Forthcoming).

Coding is a skill that takes time, practice, and ongoing effort to learn, but investing the energy to improve one's coding skills will not, by itself, prepare a scholar to do skillful work that is a meaningful contribution to scholarship. For that, it is more important to develop skills around the selection and preparation of data, around matching humanities questions with appropriate quantitative methods (if any exist), around carefully reading others' documentation and either producing one's own code or successfully communicating to a programmer what needs to be created—both for oneself and for future scholars to use. Being able to write a complicated function that runs successfully the first time without consulting any references is just a parlor trick if that function doesn't actually further the research question at hand—or worse, if it does things to the data that the scholar doesn't realize or appreciate the consequences of until it's too late.

Some projects are too big for any one person to be responsible for, or even involved with, the entire workflow. They must be undertaken as a group effort, in which case project- and people-management skills become important. If more scholars felt the same embarrassment about inadequacies in their collaboration skills as they did around deficits in their coding skills, it would have a transformative effect on the labor conditions of DH projects, which are frequently inequitable if not exploitative (Griffin and Hayler 2018).

If you want to improve your coding skills, there are many opportunities to do so, including a growing number of efforts to increase the diversity of examples and explanations in order to better resonate with audiences who feel alienated by material that centers Anglophone and stereotypically male interests (e.g., *Programming Historian*[4] and *The Data-Sitters Club*[5]). But it is better to first consider what other areas of your workflow are opaque to you. What black boxes do you engage with, whether in the form of tools or methods? How well do you understand the statistical measures you use? What value do you always choose because someone once told you to do it that way but you're not sure why? Can you explain why your database is organized the way it is? Are you sure that there are no problematic shortcuts or surprises in the code you commissioned from an undergraduate? And how confident are you that any collaborators on your project are getting what they need out of it—including appropriate credit, mentorship, and opportunities to develop their skills? First, get your own workflows in order. Then consider how much time is worth investing in coding.

ACKNOWLEDGMENTS

I would like to thank Hannah Alpert-Abrams, Karin Dalziel, and Paige Morgan for their feedback on a draft of this chapter.

NOTES

1. LaTeX is a typesetting system with a coding-like syntax that was the first major DIY solution for the kinds of complex layout needs required by mathematicians and others who work with formulas. Despite the latest release dating to 1994, it has been widely adopted in the sciences more broadly, including computer science. Documents prepared with LaTeX have a distinctive look, making "in-group" and "out-group" status within spaces where scholars commonly use LaTeX immediately visible.
2. Thank you to Paige Morgan for this analogy.
3. For example, for a project on a corpus of *Star Wars* novels, researchers at the Stanford Literary Lab mass-replaced unusual Star Wars character names, including robots, with common Anglo-American names to improve the NER performance of BookNLP.
4. https://programminghistorian.org/.
5. https://datasittersclub.github.io/site.

REFERENCES

Austen Said. n.d. Edited by Laura White et al. https://austen.unl.edu/.

Bamman, David, Sejal Popat, and Sheng Shen. 2019. "An Annotated Dataset of Literary Entities." *NAACL* (June).

Deutsch, Stacia. 2017. *The Friendship Code*. Penguin Workshop.

Froehlich, Heather. 2019. Twitter, July 20. https://twitter.com/heatherfro/status/1152614657506598912.

Gil, Alex. 2015. "The User, the Learner and the Machines We Make." *Minimal Computing* blog. May 21. https://go-dh.github.io/mincomp/thoughts/2015/05/21/user-vs-learner/.

Griffin, Gabriele and Matt Steven Hayler. 2016. "Collaboration in Digital Humanities Research – Persisting Silences." *DH Quarterly* 12 (1). http://www.digitalhumanities.org/dhq/vol/12/1/000351/000351.html.

Ilchuk, Yulia. 2019. "Conversational Versus Co-occurrence Models in Networking the Russian Novel." Part of DH 2019 panel *Methodology as Community: Fostering Collaboration Beyond Scholarly Societies*. https://dataverse.nl/dataset.xhtml?persistentId=doi:10.34894/RTFNNL.

Lang, Sarah. 2020a. "Computational Humanities and Toxic Masculinity: A Long Reflection." *LaTeX Ninja* blog. https://latex-ninja.com/2020/04/19/the-computational-humanities-and-toxic-masculinity-a-long-reflection/.

Lang, Sarah. 2020b. "News on the DH and Gender Equality." *LaTeX Ninja* blog. https://latex-ninja.com/2020/05/24/news-on-the-dh-and-gender-equality/.

Lang, Sarah. 2020c. "Where Can I *Actually Learn* Programming? (as DH and otherwise)." *LaTeX Ninja* blog. https://latex-ninja.com/2020/12/13/where-can-i-actually-learn-programming-as-dh-and-otherwise/.

Lesbians Who Tech. 2017. "Edie Windsor Coding Scholarship." Webpage. https://lesbianswhotech.org/codingscholarship/.

Messina, Cara Marta. 2021. *Critical Fan Toolkit*. http://www.criticalfantoolkit.org/.

Morgan, Paige. 2021a. "Building Collectivity in Digital Humanities through Working With Data." *Recovering the US Hispanic Literary Heritage* lecture series, January 25. https://www.youtube.com/watch?v=QxGIxppyPdw&feature=youtu.be.

Morgan, Paige. 2021b. "Further Thoughts on Collectivity." January 27. http://blog.paigemorgan.net/articles/21/further-thoughts.html.

Nowviskie, Bethany and Dot Porter. 2010. "The Graceful Degradation Survey: Managing Digital Humanities Projects through Times of Transition and Decline." DH 2010 conference. http://dh2010.cch.kcl.ac.uk/academic-programme/abstracts/papers/pdf/ab-722.pdf.

Piper, Andrew. 2021. *Can We Be Wrong? The Problem of Textual Evidence in a Time of Data*. Elements in Digital Literary Studies. Cambridge: Cambridge University Press.

Posner, Miriam. 2019. Twitter, July 19. https://twitter.com/miriamkp/status/1152389216363401216.

Risam, Roopika. 2019. Twitter, July 19. https://twitter.com/roopikarisam/status/1152389797882863617.

Risam, Roopika. Forthcoming. "Double and Triple Binds: The Barriers to Computational Ethnic Studies." In *Debates in DH: Computational Humanities*, edited by Jessica Marie Johnson, David Mimno, and Lauren Tilton.

Rockwell, Geoffrey, Shawn Day, Joyce Yu, and Maureen Engel. 2014. "Burying Dead Projects: Depositing the Globalization Compendium." *DH Quarterly* 8 (2). http://www.digitalhumanities.org/dhq/vol/8/2/000179/000179.html.

Smithies, James, Carina Westling, Anna-Maria Sichani, Pam Mellen, and Arianna Ciula. 2019. "Managing 100 Digital Humanities Projects: Digital Scholarship & Archiving in King's Digital Lab." *DH Quarterly* 13 (1). http://www.digitalhumanities.org/dhq/vol/13/1/000411/000411.html.

Tiku, Nitasha. 2021. "Google's Approach to Historically Black Schools Helps Explain Why there are Few Black engineers in Big Tech." *Washington Post*, March 4. https://www.washingtonpost.com/technology/2021/03/04/google-hbcu-recruiting/?arc404=true.

CHAPTER FOURTEEN

The Present and Future of Encoding Text(s)

JAMES CUMMINGS (NEWCASTLE UNIVERSITY)

What is the place of text encoding in the Digital Humanities? When we discuss the encoding of text(s) in the Digital Humanities, this is often synecdochical for the use of the Guidelines of the Text Encoding Initiative (TEI) to mark up a researcher's interpretations of a text (TEI Consortium 2021).[1] There are those that will be rankled by this statement because their entirely valid discipline in the vast overarching domain of Digital Humanities does not use the TEI. There are certainly other standards and bespoke text encoding, but when encoding historical texts for scholarly purposes in Digital Humanities, the TEI is generally the standard chosen. There are others that will dislike the prominence of the TEI because they see flaws in its approach, or disagree at a fundamental theoretical level about the often pragmatic nature of the TEI, wishing for even more intellectually elegant solutions that would likely also excise many use-cases from its highly disparate and multivalent community. Sometimes this resistance is because the TEI community is without any shred of doubt (whether those doing other forms of Digital Humanities like it or not) a clearly demonstrable success story of a Digital Humanities initiative that many would like to emulate with their own endeavors. The main output of this free-to-use but membership-supported community—the TEI Guidelines—is an impressive achievement. Over the last three and a half decades, the TEI community has gone from a rag-tag group of literary, linguistic, and computing scholars submitting funding bids to enable international collaboration in these areas, to an open, award-winning, global membership consortium that produces the freely available de facto standard for the full-text representation, description, and scholarly interpretation of historical sources. These recommendations have become so mainstream as to be recognized by funders and cultural institutions as the default choice of format for long-term preservation in many areas of textual resource creation. Currently, if a project were to put in a funding bid to create a digital scholarly edition and not use TEI, then many funders (or their reviewers at least) would expect at a minimum some justification for doing things differently. The TEI "has assumed a central role in digital editing and textual scholarship, both as a modeling and analytical tool and as ontology for the phenomenology of the text and the page" (Pierazzo 2015, 307). There are benefits and drawbacks to this, of course, and the TEI could still be easier to encode, simpler to publish, more convenient to customize, have support available in more languages, be more interoperable, and cover the developing needs of more researchers, but its success so far has been laudable.

Some projects want to take other approaches to document modeling and text encoding, and that is reasonable, but sometimes this is unfortunately positioned as being in competition with the TEI when often this is not the case. Approaches should complement each other whenever possible. The TEI Guidelines are by their very nature evolving, fluid, and polygamous in their relationships with related standards. The community does not object to other approaches, but often looks to them to see where its own assumptions about textual modeling can be improved before adding them to the pantheon of related technologies that can be used in conjunction with each other. Over time, the forms and concerns of text encoding have become broader in nature, while the outputs that are created have become more complex. Simultaneously the ways in which markup, and other forms of annotations, are created have become more divorced from their underlying data formats. What is, what should be, or what could be the future for the encoding of text(s)?

NOT AN INTRODUCTION TO THE TEXT ENCODING INITIATIVE

In order to understand the future of encoding text(s), it is useful to look at what sets the TEI Guidelines apart from other standards. Standards never stand alone, they overlap with others in complex, often ephemeral, interdependencies. When we talk about text encoding in the context of Digital Humanities, this usually relates to the annotation, markup, and identification of strings of text. "Text," of course, has multiple meanings when talking about markup, not only the linear string of (thankfully now usually Unicode) characters but also as "texts" in the broadly conceived literary meaning that includes many different forms of media that can all be read in one way or another (cf. McKenzie 1999). The "text" in "Text Encoding Initiative" may have once referred mostly to the idea of scholarly literary editions, but it can now be used to encode almost anything: any text, from any period, in any language, and any writing system.

> Its purpose is to provide guidelines for the creation and management in digital form of every type of data created and used by researchers in the Humanities, such as source texts, manuscripts, archival documents, ancient inscriptions, and many others. As its name suggests, its primary focus is on text rather than sound or video, but it can usefully be applied to any form of digital data.
> (Burnard 2014, 1)

What is important but sometimes overlooked about the TEI is that it is a customization-based framework enabling a common approach to modeling textual phenomena, rather than a set standard or list of elements. Customizations try to enable the application of a general scheme in specific uses. Customizing the TEI "is entirely legitimate: there is a reason that the TEI Guidelines are not called the TEI Laws" (Ohge and Tupman 2020, 281). As many teaching the TEI say: don't do as we say, instead do what you need to do but explain what you've done in a language that we—that is the TEI community—can understand. This language is the TEI's extension and customization vocabulary. Indeed, every project using the TEI should start by customizing the very general default framework to its own needs, rather than assuming it is some fixed, immutable, behemoth of a standard (cf. Cummings 2014). What is unfortunate is that some see the customization format solely in terms of reducing the number of elements and generating project schemas. While it is beneficial to customize the TEI for project-specific validation (instead of using one of the more general off-the-shelf schemas like TEI Lite or TEI SimplePrint), projects that merely reduce the number of available elements neglect to exploit the customization format to its fullest.[2] For example, in not fully specifying project-specific attribute values to get a more helpful validation, or not embedding schema declarations directly alongside relevant sections of project documentation so that either can easily be updated as the project develops. TEI customization serves as a record of decisions and encoding possibilities to enable both data interchange and long-term preservation.

The TEI has evolved over many years into a general-purpose framework for encoding a wide variety of data. Needing to cope with many different uses, one of the general approaches that the recommendations often take is that of ontological agnosticism as to the realities of the phenomena being encoded, while giving users a common shared vocabulary to delineate them. For example, the TEI Guidelines do not tell the readers what is, or is not, a "heading" but provide the <head>element to encode "any type of heading" and give examples of some things to be considered as headings, "for example the title of a section, or the heading of a list, glossary, manuscript description, etc."[3] While the TEI Guidelines do sometimes fail at this level of semantic distancing, it is a useful approach which enables differing perspectives to share a common vocabulary that they only need to mediate if and when they attempt to interchange data. Whatever its inconsistencies, the TEI often attempts to specify a general level of information granularity while giving users the flexibility to be more or less detailed according to their needs. An example of this might be the <name>element, for any form of proper noun phrase, but is also specialized as a <persName>(personal name) element, which can encode someone's name. However, it is also possible—and useful, especially for metadata—to subdivide this with <forename>and <surname>elements where appropriate. Doing so obviously enables a much more flexible use of the data being created, and the drive to do so is sometimes the result of the very existence of options like this in the shared vocabulary.

The TEI's long history has been based on a community of volunteers that has resulted in more or less specification in different aspects of the recommendations. To predict all of the community's needs would be infeasible, so developments usually are based on community proposals, and a more vibrant sub-community may suggest more changes to particular areas of the TEI Guidelines. This can result in a skewing of focus where the encoding recommended in those more active areas is correspondingly richer and more detailed. In other instances, the TEI Guidelines only give a basic approach because standards (like SVG) already exist which provide a more fully feature-complete encoding in that area. In some cases these are intermixed, as with the Music Encoding Initiative (MEI), which itself uses the TEI's customization format to define its own recommendations, which embed many TEI-derived elements.[4] MEI enables their users to encode the musical notation of a score but use TEI elements for textual elements (such as a libretto); meanwhile TEI users can customize their TEI schemas to enable the embedding of MEI in their documents if they contain some musical notation. One of the interesting aspects of the continual development of the TEI Guidelines is that they act as a proxy for the concerns of the text encoding community over time. Hence, the history of the TEI Guidelines reveals the anxieties and research focus of its community as it evolves. A fun example of this disciplinary skewing is the <person>element. In the fourth major iteration of the TEI Guidelines (TEI P4), the <person>element was recommended for recording information about a "participant in a language interaction" by those encoding language corpora.[5] One of the child elements of <person>was <birth>, usefully enabling a record of both birthdate and place information about the linguistic subjects. It was only when generalizing this element for TEI P5 that it was noted that it did not (yet) have a <death>element, because linguists usually don't speak to ghosts! The concerns and assumptions of a particular sub-community of encoders were reflected in the TEI Guidelines until others also wanted to record metadata about named entities; then, the standard changed. "In a strictly Darwinian sense, the TEI has evolved by fostering and profiting from mutations that are considered beneficial to the user community, while ignoring those which are not" (Burnard 2013, para. 37). The developmental history of these guidelines is a history that correlates to the concerns of the text encoding community and deserves to be studied further.

WHEN NOT TO USE THE TEI (OR ANOTHER STANDARD)

Given that the TEI Guidelines document how to constrain, modify, and extend the entire TEI framework (so much so that this same mechanism has been used for other standards) we might be forgiven for thinking that we should always use the TEI, extending it as necessary. Using the TEI, or another standard, is not always an appropriate choice in many situations. However, this should not be because the TEI Guidelines do not deal with a project's particular area of interest, since that is handled by the framework's inherent ability to be extended through customization. In deciding whether to use the TEI or not, any project would benefit from examining any related data models in the TEI Guidelines, as their long developmental history may provide useful background. Where there is another, more appropriate, popular open international standard in the research area, then not using the TEI Guidelines is reasonable. Similarly, when practicalities of the real world intrude into the ideals of data modeling, for example in the case of an un(der)-funded project which has been offered in-kind support from its institution in the form of a developer who can quickly help the project achieve their research aims, but only if they use the technologies with which they are already familiar. The key here, for long-term preservation at the very least, is to attempt to ensure that the granularity of information—for the data and metadata—is sufficient to plausibly export to standard formats. The pragmatism of publishing a project's outputs often necessarily outweighs the ideals of best practice. Indeed, in approaching this very question of when not to use the TEI, John Lavagnino recognizes that there are also benefits of heuristic text encoding for some scholars, who "may also find that devising a tagging system from scratch for their texts is instructive" but that these should likely "need to be seen as trials that can be thrown away" (Lavagnino 2006, para 8). Bespoke forms of text encoding are not bad, but the real added-value (to both resource users and funders) is the standardization and potential for re-use that comes of being part of a community.

The combination of project-specific customization (both constraint and extension), ontological agnosticism over precise definitions, and the flexibility to specify information at a variety of levels of granularity is a winning combination for the TEI. The future of the TEI as a standard in the medium term is strong as long as there are those who will develop it. However, as with natural languages, freedom of neologisms, ontological imprecision, and varying degrees of expression come at a cost, and what is sacrificed is the ease of unmediated interchange or meaningful seamless interoperability (cf. Cummings 2019). It isn't that interchange is impossible, but as it usually takes place without negotiation (i.e., it is "blind"), it relies on the provided documentation and a human mediator to convert data for a new system. Full interoperability is often held up as a goal for standards, but in a standard as expressive as the TEI, this is unhelpful. Examining text encoding in these general terms of equality vs liberty in limiting encoding choices, Syd Bauman believes that:

> In summary, since interoperationality (equality) is often bought at the expense of expressivity (liberty), interoperability is the wrong goal for scholarly humanities text encoding. Practitioners

> of this activity should aim for blind interchange. This still requires a lot of adherence to standards (e.g., TEI), but eschews mindless adherence that curtails expression of our ideas about our texts.
> (Bauman 2011)

This tension between the rules of standards and the expressivity of documenting our interpretation of a text through markup is at the heart of many disagreements about text encoding. Encoding needs a firm understanding by those involved that, while compromise between the ideas of formal standards and the freedom of expression is inevitable, it comes at a cost.

THE PRESENT OF ENCODING TEXTS

Text encoding takes place in such a wide range of contexts that it is hard to generalize its nature. Some of the text encoding that projects undertake is not bound up with the transcription and editing of text, but instead the description of objects such as manuscripts, while other encoding contains wildly individualistic classifications of their material. Elsewhere, many thousands of encoded documents are generated from databases or other project processes. There are, with no exaggeration, millions of documents following the TEI Guidelines just on GitHub; the number of documents is, however, a poor metric, as some may only comprise a few lines and others may be corpora of many texts.[6] Likewise, while TEI is a dominant standard for text encoding in Digital Humanities for very good reasons, there are many other formats available. In some cases those contributing to text encoding projects may be doing so through an interface which hides the underlying format. The interfaces used to create the objects often affect the nature of text encoding (e.g., its depth, granularity, hierarchy, relationships), but interfaces for the creation and especially the rendering of encoded text are more important than that. As textual scholars have long realized, the interfaces used for digital editions also themselves embody editorial arguments.

> By recognizing that digital objects—such as interfaces, games, tools, electronic literature, and text visualizations—may contain arguments subjectable to peer review, digital humanities scholars are assuming a perspective similar to that of book historians who study the sociology of texts.
> (Galey and Ruecker 2010, 412)

In many cases we are unable, and indeed should not, consider these as separate aspects of an encoded textual object's argument. The text is only one part of any such object unable to be easily disentangled from its presentation. "Just as there is no clean separation between data and interpretation, there is no clean separation between the scholarly content of an argument and its rhetorical form" (Andrews and van Zundert 2018, 8). While the underlying textual data of a resource is where the arguments, interpretations, and scholarly understanding of that data are recorded, the rhetorical form of those encoded texts is the interface by which it is made available, and editors should be aware of what it argues. The converse is also true: often textual editors approach the digital with the desire to model their encoding based on the output generated by some particular software. This is dangerous and misleading; privileging of the output presentation over the proper recording of their interpretations should "always ring the alarm and lead to serious discussion of editorial and encoding principles bearing in mind that honesty is one of the most important qualities of an editor" (Turska et al. 2016, para. 3). While the interface does provide part of the argument, it must always be based on a text encoding that honestly reflects the editors' interpretations of the textual phenomena.

It is not only the interface for publication that makes an argument, but the interfaces through which some data is created can limit or skew the nature of that data. Crowdsourcing encoding, for example, may have contributors completely ignorant of the data formats that are being used as they are hidden behind a sophisticated interface, but the role interfaces play in the success of crowdsourcing should not be underestimated (cf. Iranowska 2019). In many ways, if hidden behind an interface, the storage format becomes immaterial so long as the granularity of its information enables sufficient interchange using open international standards. But are these crowdsourced endeavors truly an example of text encoding, or merely structured transcriptions such as those produced by handwritten text recognition (HTR) from images of source documents? While HTR can produce TEI output, these should not be understood as scholarly textual encoding, at least not as an editor would understand it. In both cases, their products are source data to be combined, considered, and enhanced by an editor, either manually or computationally.

One assumption people make is that text encoding in the Digital Humanities is intended to produce scholarly digital editions—indeed that is my own bias, as demonstrated above—but there are many text encoding projects that do not aim at producing a shiny digital critical edition but, instead, a text-base to be queried further. When it has been encoded, a corpus of texts may be analyzed and form the basis of new knowledge creation and scholarly output but, like many interim resources in Digital Humanities, corpora such as these need not be the end goal or even considered a significant output by the project itself. This is sometimes seen with linguistic analysis or non-consumptive analysis of in-copyright materials for text and data mining. Sadly, the assumptions and decisions made by a project concerning their data and workflow, and how they view the potential use of the textual data they produce, are rarely fully documented:

> One of the present difficulties is that projects rarely elucidate the decisions they made about workflow, and whether they saw text or data analysis as a crucial part of their documentation.
> (Ohge and Tupman 2020, 293)

What complicates this more is that digital resources are usually no longer straightforward websites based on a single source file. It is more common for text encoding in the Digital Humanities to be a complex web of interconnected objects and references. Digital editions are increasingly highly devolved objects with forms of stand-off and out-of-line markup, whose presentation is dependent on the existence of numerous other linked data resources that need to be gathered together in order to make up the final presentation of the digital object (cf. Viglianti, 2016). For example, it is now usual that the instances of a person's name in a TEI-encoded document use URIs pointing to more information about that person in a separate TEI file, which itself has links to standardized international linked open data (LOD) authorities. Those authorities not only collect together all the resources that reference that person but also provide links back to the disparate projects that reference those authorities, for those who (individually or programmatically) wish to engage with the project data about that entity. The encoded text is now often an indirect data source for a vast array of connections and interrelations between documents, metadata, entities, and projects.[7]

The idea of text encoded resources solely as limited hierarchies of embedded XML markup presents a strawman argument and very dated view of modern digital sources. Maintaining impossible views such as this in the face of practical evidence to the contrary is often suggested by those who wish to promote their own approaches by diminishing the status quo. The reality is that "[m]odeling texts according to a pluralistic, multidimensional ontology of text is already largely possible with TEI" (Viglianti 2019). Furthermore, as Hugh Cayless notes:

> It is also important to remember that TEI's data model is not simply the tree structure that it gets from XML, but also the graph created by the use of its linking attributes.
>
> (2018, 256)

With the combination of hierarchical mixed content markup, using additional URI-based attributes at every level of that hierarchy pointing within or outwith a document locally or to external resources, concepts, or entities, means that what was once viewed as a hierarchical limitation of embedded markup is much less common (cf. Renear et al. 1996). For when XML is used in modern resources, more often than not, this models a more-readable version of a graph structure for a multiplicity of interdependent resources and relationships. Text encoding in some instances has already fully embraced a post-OHCO world.[8]

While TEI in its current XML format is the de facto standard for academic text encoding projects, the development of new data models for text encoding, and the migration of existing data models into new formats, is inevitable. In the academic economy of scholarly outputs this is too often framed in competitive terms rather than cooperative or evolutionary ones. When it comes to proposals for other text encoding standards as replacements, perhaps more clearly modeling graph technology, they rarely have the breadth and scope of a general-purpose framework like the TEI. An example of benefiting from existing text encoding standards while developing new approaches can be seen in philological tools using hypergraphs for automatic collation which can leverage but preserve underlying TEI markup (cf. Bleeker et al. 2018). Given the flexibility and extensibility of the TEI's community-developed standard, there are many ways that it can be combined with additional technologies so as to have each play to their specific strengths; it seems a collaborative modular approach in introducing new developments might benefit the text encoding community more than those seeking merely to replace one standard with yet another.

THE FUTURE OF ENCODING TEXTS

Predicting the future of technological developments is always impossible but, having come this far, it is plausible to consider some ways in which it might develop. The needs of users of text encoding as a method to record an individual's interpretations of texts of all forms are unlikely to vanish entirely in the medium term. If we are going to address the question asked at the beginning of "What is, what should be, or what could be the future for the encoding of text(s)?" then we need to look more broadly. What is more likely is that research priorities will shift and large academic projects based on significant amounts of text encoding may not be as attractive to funding bodies. One of the challenges in trying to demonstrate the value of text encoding is to reach other disciplinary areas, where it is not already a de facto standard, and to communicate not only the research benefits but the heuristic and pedagogical benefits of text encoding.

The TEI is a coherent text encoding framework that will not only expand into new domains of encoding (and will continue to do so as long as the text encoding community explores new areas) but can also interoperate with standards from vastly different disciplines. Many of the significant developments in the TEI have arisen because of its interactions with other standards or new areas of knowledge creation. Simultaneously, the re-expression of TEI in other formats is also possible: the TEI Guidelines are not XML, they are merely currently expressed in, and currently recommend, XML, as they once did SGML, and this indeed may change in the future. While the current format is a boon to tool-makers, since XML is relatively easy to process with widely available tools, the flexibility that the TEI Guidelines provide is also a barrier for truly general tools. While much software exists to process the texts encoded in TEI, this usually comes from specific projects and caters only to a subset of the possibilities enabled by the TEI framework. In creating general tools, as with encoding solely for presentation discussed earlier, the danger is that any dominant technology ends up determining encoding practices as users modify their methodologies to match software outputs. Elena Pierazzo points out this challenge:

> The flexibility of the encoding model is not the only issue when it comes to the provision of tools. The development and adoption of tools for the support of textual scholarship is a delicate operation which itself risks leading to profound changes in future scholarship. This is because, in a circular pattern, to produce a tool developers have to model editors' behaviors, but, once that the tool is produced, the tool will itself determine future behaviors as editors will probably try to model their data and their work in a way that is compatible with the tool's expectations.
>
> (Pierazzo 2015, 316)

What seems likely as a result of this is that sub-communities of encoding practice may develop software and encoding guidelines that reflect the specific needs of their communities. The users of communities determine any such customizations—this has already happened with EpiDoc, a subset of TEI growing out of the classical epigraphy field to work with a wide variety of ancient documents.[9] Having a tighter bound schema for a community enables the development of more robust but less generalized tools. Groups inside the text encoding "community could therefore be pivotal for the establishment and diffusion of scholarly models based on the TEI encoding models" (Pierazzo 2019, 219).

What is sometimes viewed as a threat, but in reality is more of an opportunity for text encoding in the medium term is the greater shift of Digital Humanities towards specific forms of artificial intelligence (AI). Developments will always yield new fruitful areas of study for those interested in historical texts, whether this is AI in the form of, for example, generative text processing or model-based supervised machine learning for handwritten text recognition (HTR). The potential for the automatic detection, deduplication, and encoding of named-entity references, often still problematic in pre-modern texts, is an area that will certainly be a boon for those working with such materials. What is almost certainly true is that HTR will, however, completely transform archival research in the next few decades as rough un-edited transcription of large amounts of handwritten material is then used, at the very minimum, to facilitate resource discovery. Indeed the ability for HTR to open up vast archives of material is extremely promising for text encoders and the future will inevitably feature a variety of forms of analog conversion and legacy data migration.

Much of this requires a degree of human cooperation in coalescing around agreed standards, whether in the form of LOD taxonomies or other digital authorities. We always can imagine a future where images of manuscripts are automatically transcribed with instances of named entities recognized long before a human editor confirms any of these. Indeed, if a large enough corpus of practical editorial decisions (not just what was decided, but how and why) was collected through an assistance interface then a semi-supervised learning AI approach could model the work of a scholarly editor.[10] If such a model was used to enhance the hypothetical tool then the help given for editorial decisions would improve and feed back into a better model. While it is a mistake in my view if the development of AI tools for editorial assistance would eventually lead to scholarly editors being fully replaced by the very learning machines they help to create, the hiding of the underlying text encoding implied by the very existence of these tools also has its drawbacks. The use of text encoding to help heuristically determine and formally express our own knowledge is a watershed moment in the understanding of text. While at the moment the use of AI with texts provides new ways of reading them, the hermeneutical and pedagogical benefits of text encoding for the close reading of texts will always remain (Jakacki and Faull 2016, 365).

Even this is an optimistic view of the future that not only assumes no major human apocalypse but also still sees the creation of encoded (either by human or machine) text as a desirable object. We keep finding new ways to benefit from text encoding; a familiar problem with many of these machine-learning methodologies is that they do not recognize or profit from the incorporation of text encoding in their algorithmic processing or deep-learning approaches to data. In most textual analysis today, markup, or any form of relationships between textual objects, is stripped out of a corpus to start from a so-called "plain text" (which, of course, as any student of editing knows, does not really exist). Instead, major collections of encoded texts will likely be mined for structured data by supervised machine learning, for example, to extract the marked instances of named entities, as a form of beneficial training data for models interested in those features. Text encoding will continue to be useful and beneficial, but the real challenge is to communicate those benefits to the wider community, including funders, as more than just a data source. The text encoding community needs to participate in these areas—for example through encoding of less accessible, more varied forms of legacy works, in more diverse languages and encoding these texts so that the corpora derived from them for other uses are less biased and more accurately reflect the diverse history of written communication.

CONCLUSION: IF WE'RE GOING TO FAIL, LET'S FAIL BETTER

Samuel Beckett's often (very selectively) quoted "[a]ll of old. Nothing else ever. Ever tried. Ever failed. No matter. Try again. Fail again. Fail better" (Beckett 1983, 7) is useful for those wishing to promote an iterative development strategy that diminishes the fear of failure and replaces it with a sense (however illusory) of continual learning from failure.[11] Constant changing means that if text encoding is going to continue to be a useful and fruitful area of research within Digital Humanities in the medium term then, in spite of all of its prior successes as a community, it must rise to the challenge to "fail better." It will need to "fail better" at communicating across disciplines the value in carefully constructed text encoding resources and expand its niche across developments in AI and LOD based on the textual phenomena it records. Indeed, the text encoding community needs to be better at trumpeting its pedagogical benefits as a modeling and analysis methodology in order to continue to be seen as more than just a long-term preservation format. Text encoding will not disappear soon, but only if we continue to think about and promote the issues and concepts it helps us to explore. Or as G. Thomas Tanselle put it in the MLA's *Electronic Textual Editing*:

> We should be enthusiastic about the electronic future, for it will be a great boon to all who are interested in texts; but we do not lay the best groundwork for it, or welcome it in the most constructive way, if we fail to think clearly about just what it will, and what it will not, change. Procedures and routines will be different; concepts and issues will not.
>
> (Tanselle 2006, para 11)

NOTES

1. At the time of writing, the *TEI P5: Guidelines for Electronic Text Encoding and Interchange* consulted is version 4.3.0 which was updated on August 31, 2021. Available from: http://www.tei-c.org/Vault/P5/4.3.0/doc/tei-p5-doc/en/html/. The most recent release, always available from https://www.tei-c.org/Vault/P5/current/doc/tei-p5-doc/en/html/, may differ in some respects at time of reading.
2. See the TEI Consortium's website for a list of TEI-provided customizations at https://tei-c.org/guidelines/customization/.
3. See the TEI Guidelines element specification for the 'head' element at https://tei-c.org/Vault/P5/4.3.0/doc/tei-p5-doc/en/html/ref-head.html.
4. See the website of the Music Encoding Initiative, https://music-encoding.org/.
5. See the TEI P4 Guidelines element specification for the 'PERSON' element at https://www.tei-c.org/Vault/P4/doc/html/ref-PERSON.html, later revised in TEI P5 to https://www.tei-c.org/release/doc/tei-p5-doc/en/html/ref-person.html.
6. A quick, and admittedly imprecise, search on GitHub, http://github.com, for mentions of the required TEI metadata element "<teiHeader>" in code repositories (instead of other locations like wikis) and only with the file format of XML currently shows over 1.8 million results. The TEIhub app https://teihub.netlify.app/ gives a better way to sort and browse this data.
7. For example, in the seamless way the Canadian Writing Research Collaboratory, https://cwrc.ca/, enables the intermixture of TEI source documents and linked open data.
8. "OHCO" is the view of text as an "Ordered Hierarchy of Content Objects," and while text is indeed ordered, hierarchical, and created of content objects, it is not always only this and not always this simultaneously. Indeed, the non-OHCO of texts is seen by Jason Boyd and Bo Ruberg in Chapter 6 as revealing the non-hierarchical queerness of texts; cf. p. 63.
9. For more about EpiDoc, see their website, https://epidoc.stoa.org.
10. The implications of machine learning for activities such as this are considered in more detail by David M. Berry in Chapter 42.
11. The following paragraph (usually omitted) includes: "Try again. Fail again. Better again. Or better worse. Fail worse again. Still worse again. Till sick for good. Throw up for good. Go for good. Where neither for good. Good and all" (Beckett 1983, 7).

REFERENCES

Andrews, Tara L. and Joris J. van Zundert. 2018. "What Are You Trying to Say? The Interface as an Integral Element of Argument." In *Digital Scholarly Editions as Interfaces*, edited by Roman Bleier, Martina Bürgermeister, Helmut W. Klug, Frederike Neuber, and Gerlinde Schneider, Instituts für Dokumentologie und Editorik, 12: 3–33. https://kups.ub.uni-koeln.de/9106/.

Bauman, Syd. 2011. "Interchange vs. Interoperability." In *Proceedings of Balisage: The Markup Conference 2011*. Balisage Series on Markup Technologies, vol. 7, Montréal, Canada. doi: 10.4242/BalisageVol7.Bauman01.

Beckett, Samuel. 1983. *Worstward Ho*. London: Calder.

Bleeker, Elli, Bram Buitendijk, Ronald Haentjens Dekker, and Astrid Kulsdom. 2018. "Including XML markup in the automated collation of literary texts." In *XML Prague 2018 – Conference Proceedings, 2018*, 77–95. http://archive.xmlprague.cz/2018/files/xmlprague-2018-proceedings.pdf.

Burnard, Lou. 2013. "The Evolution of the Text Encoding Initiative: From Research Project to Research Infrastructure." *Journal of the Text Encoding Initiative* 5 (April). doi: 10.4000/jtei.811.

Burnard, Lou. 2014. *What Is the Text Encoding Initiative?: How to Add Intelligent Markup to Digital Resources*. OpenEdition Press. doi: 10.4000/books.oep.426.

Cayless, Hugh A. 2018. "Critical Editions and the Data Model as Interface." In *Digital Scholarly Editions as Interfaces*, edited by Roman Bleier, Martina Bürgermeister, Helmut W. Klug, Frederike Neuber, and Gerlinde Schneider, Instituts für Dokumentologie und Editorik. 249–63. https://kups.ub.uni-koeln.de/9119/.

Cummings, James. 2014. "The Compromises and Flexibility of TEI Customisation." In *Proceedings of the Digital Humanities Congress 2012*, edited by Claire Mills, Michael Pidd, and Esther Ward. Sheffield: The Digital Humanities Institute. https://www.dhi.ac.uk/openbook/chapter/dhc2012-cummings.

Cummings, James. 2019. "A World of Difference: Myths and Misconceptions about the TEI." *Digital Scholarship in the Humanities* 34 (Supplement 1): i58–79. doi: 10.1093/llc/fqy071.

Galey, A. and S. Ruecker. 2010. "How a Prototype Argues." *Literary and Linguistic Computing* 25 (4): 405–24. doi: 10.1093/llc/fqq021.

Iranowska, Joanna. 2019. "Greater Good, Empowerment and Democratization? Affordances of the Crowdsourcing Transcription Projects." *Museum and Society* 17 (2): 210–28. doi: 10.29311/mas.v17i2.2758.

Jakacki, Diane and Katherine Faull. 2016. "Doing DH in the Classroom: Transforming the Humanities Curriculum through Digital Engagement." In *Doing Digital Humanities: Practice, Training, Research*, edited by Constance Crompton, Richard J. Lane, and Ray Siemens, 358–72. Abingdon: Routledge.

Lavagnino, John. 2006. "When Not to Use TEI." In *Electronic Textual Editing*, edited by Lou Burnard, Katherine O'Brien O'Keeffe, and John Unsworth. New York, NY: Modern Language Association of America. https://tei-c.org/Vault/ETE/Preview/.

McKenzie, D. F. 1999. *Bibliography and the Sociology of Texts*. Cambridge and New York: Cambridge University Press.

Ohge, Christopher and Charlotte Tupman. 2020. "Encoding and Analysis, and Encoding as Analysis, in Textual Editing." In *Routledge International Handbook of Research Methods in Digital Humanities*, edited by Kristen Schuster and Stuart Dunn, 275–94. Abingdon: Routledge. doi: 10.4324/9780429777028-21.

Pierazzo, Elena. 2015. "Textual Scholarship and Text Encoding." In *A New Companion to Digital Humanities*, edited by Susan Schreibman, Ray Siemens, and John Unsworth, 307–21. Chichester: Wiley. doi: 10.1002/9781118680605.ch21.

Pierazzo, Elena. 2019. "What Future for Digital Scholarly Editions? From Haute Couture to Prêt-à-Porter." *International Journal of Digital Humanities* 1 (2): 209–20. doi: 10.1007/s42803-019-00019-3.

Renear, Allen H., Elli Mylonas, and David G. Durand. 1996. "Refining Our Notion of What Text Really Is: The Problem of Overlapping Hierarchies." In *Research in Humanities Computing*, edited by Nancy Ide and Susan Hockey. Oxford: Oxford University Press. http://cds.library.brown.edu/resources/stg/monographs/ohco.html.

Tanselle, G. Thomas. 2006. "Foreword." In *Electronic Textual Editing*, edited by Lou Burnard, Katherine O'Brien O'Keeffe, and John Unsworth. New York, NY: Modern Language Association of America. https://tei-c.org/Vault/ETE/Preview/.

TEI Consortium. 2021. *TEI P5: Guidelines for Electronic Text Encoding and Interchange*. Version 4.2.2. updated 9 April 2021. Text Encoding Initiative Consortium. http://www.tei-c.org/Vault/P5/4.2.2/doc/tei-p5-doc/en/html/.

Turska, Magdalena, James Cummings, and Sebastian Rahtz. 2016. "Challenging the Myth of Presentation in Digital Editions." *Journal of the Text Encoding Initiative* 9 (September). doi: 10.4000/jtei.1453.

Viglianti, Raffaele. 2016. "Why TEI Stand-off Markup Authoring Needs Simplification." *Journal of the Text Encoding Initiative* 10 (December). doi: 10.4000/jtei.1838.

Viglianti, Raffaele. 2019. "One Document Does-it-all (ODD): A Language for Documentation, Schema Generation, and Customization from the Text Encoding Initiative." In *Proceedings of the Symposium on Markup Vocabulary Customization*. Balisage Series on Markup Technologies 24. doi: 10.4242/BalisageVol24.Viglianti01.

CHAPTER FIFTEEN

On Computers in Text Analysis

JOANNA BYSZUK (INSTITUTE OF POLISH LANGUAGE,
POLISH ACADEMY OF SCIENCES)

It should not come as a surprise that text analysis has always been at the forefront of applications of computing in the humanities. For centuries, linguists and literature scholars ventured to solve puzzles hidden in words, studying not only their literal and metaphorical meanings but also patterns of frequencies, contexts, and mysterious intentions or subconsciously betrayed secrets.

Since the hypothesis made by Augustus de Morgan in 1851, pondering how comparing mean word lengths taken from texts could allow for telling their authors apart, text analysis in the framework of what we now call "stylometry" has been undertaken by various scholars, at first,[1] even using manual calculations—this became one of the earliest approaches to quantitative text analysis.

Once computers became more widely available to humanities scholars, such efforts started to take on bigger and bolder topics and methodological challenges, starting with a famous study of The Federalist Papers by Mosteller and Wallace (1963), and Burrows's examination of Jane Austen's works (1987). The latter revolutionized the field thanks to the introduction of a method of comparing not just individual words but whole texts; which Burrows called his "Delta" measure (2002). The method was later perfected (Argamon 2008; Evert et al. 2015, 2017) and made accessible through software packages which facilitate the application of stylometric techniques without the skills of a programmer.[2]

The history of the field has been addressed in detail in numerous publications, both field specific (e.g., Holmes 1998, 2003; Grzybek 2014; Beausang 2020) and more general digital humanities books like *Defining Digital Humanities* (eds Terras et al. 2013) or *A New Companion to Digital Humanities* (e.g., Hoover 2008), so I will not repeat these stories at length.

TEXT ANALYSIS AT THE CROSS-SECTION OF DISCIPLINES

Since its dawn, computational text analysis has been developing at the cross-section of disciplines, drawing solutions and methods from computer science, as well as an understanding of language from various linguistic theories, especially formal linguistics. Methods such as cluster analysis, principal component analysis, or correspondence analysis have long been used outside humanities before they made their way here. We might consider topic modeling (Blei 2012) or word embeddings (Mikolov 2012) as developed more at the cross-section of computer science and humanities, since they were designed to tackle issues around text analysis, but the reality is that these methods too were primarily developed by computer scientists addressing specific market needs, such as automatic tagging of texts or improving the accuracy of search engines.

Nowadays, computers may seem omnipresent in text analysis. They find applications in natural language processing, computational and corpus linguistics, computational literary studies, authorship studies, and cultural analytics. Their usefulness extends past stylometric investigations of authorship and styles to include analysis of texts for the use of named entities (e.g., distribution of proper names or references to places), certain motifs (as evidenced by the popularity of topic modeling in the study of literature, e.g., Thompson and Mimno 2018), chronology (e.g., Klaussner and Vogel 2015 or language evolution studies addressed with word vectors, such as in Hamilton et al. 2018), prosodic patterns (e.g., Plecháč et al. 2019), and so on. In linguistics, methods of computational analysis have arguably proven most popular among corpus linguists, who use such to study language change, characteristics, and even discourse.

Text analysis has also largely extended its scope of application; while the early studies concerned mostly literature, it is now just as common to see studies of other textual media, such as journalistic reviews (e.g., Benatti and Tonra 2015), blog posts (e.g., Schler et al. 2006), tweets (e.g., Juola et al. 2011), television dialogues (Bednarek 2010, 2018), forensics data (e.g., many works by Juola and Mikros), social media and fan fiction (e.g., Rebora et al. 2019), and even online security (Gröndahl and Asokan 2019).

COMPUTATIONAL TEXT ANALYSIS AS A LOW ENTRY POINT TO DH

Among the many branches of digital humanities, computational text analysis might be considered one of the easiest entry points. This may seem provocative, but let us consider a few factors that speak in favor of this view.

While quantitative text analysis studies were difficult and time-consuming for their pioneers, who had to bear long and monotonous manual calculations, contemporary quantitative and computational text analysis requires basically just a dataset and a moderately powerful laptop. Moreover, there are a plethora of methods at our disposal, and many of them do not even require much (or any) technical knowledge.

This situation is of utmost benefit to individual humanities scholars who cannot count on technical support from their unit or cannot take expensive courses that teach such methodologies,[3] even if summer schools such as the European Summer University in Digital Humanities[4] or the Digital Humanities Summer Institute[5] offer great opportunities to develop one's skills and projects under expert guidance. Smaller and often free-of-charge initiatives have started to happen all over the world too, not only at big European or American DH centers, but, among others, also in Beirut,[6] Lagos,[7] Tartu,[8] and Moscow.[9] Given the rising popularity of open scholarship, many tutors from these workshops decide to share their materials open source, which increases the chances of those with less access to learn about computational text analysis on their own.

In fact, if we examine the development of methods, we will notice that many of those who became the early leading names in the field came not from the biggest and most renowned universities in the world, but rather from more remote corners with small digital humanities communities—take John Burrows and Hugh Craig from the University of Newcastle, Australia, or Maciej Eder and Jan Rybicki from two different universities in Kraków, Poland. If we look at recent studies, we can see similar patterns; while many universities are promoting courses and large research projects in computational text analysis, many brilliant scholars continue to do brilliant things with computational text analysis in smaller centers, often almost entirely on their own.

The development of computational text analysis also happens in a largely multilingual perspective, as most methods are language-independent, and therefore available not only to scholars of English but also other languages. Without much effort one can find studies concerning not only the most popular European languages but also those using non-Latin alphabets (e.g., Suzuki and Hosoya 2014 on Japanese), or right-to-left scripts (e.g., Soffer et al. 2020 for Hebrew; Rezaei and Kashanian 2017 for Persian[10]).

According to Eichmann-Kalwara et al. (2018) who examined keywords in DH conference submissions between 2013 and 2016, "Text Analysis" (tagged on 22 percent of submissions), "Data Mining/Text Mining" (20 percent), and "Literary Studies" (20 percent) are among the most common keywords describing presentation topics. Every year, the major Digital Humanities conference organized by the ADHO gathers a high volume of submissions which touch upon computational text analysis in some form or another.[11] The fact that computational text analysis is everywhere in DH suggests that it is more accessible than some might think.

However, accessibility also brings risk, and there is a danger that inexperienced scholars will follow examples that are not necessarily methodologically sound. This issue has been highlighted already in 1997 by Joseph Rudman, who provided a thorough examination of the most common pitfalls in authorship attribution studies, and suggested a variety of things that needed improving in scholarly approaches to this type of problem. Similar and further problems were later discussed at length by Daelemans (2013).

Rudman ended his paper on a cautionary note: "The worst case scenario is that nothing changes. The practitioners agree that there are problems – but not with their own studies. And then nothing but another spate of flawed articles" (1997, 362). In the light of numerous discussions and critiques conducted in the last few years,[12] it seems like many of us in the text analysis community could benefit from re-examining Rudman and developing updated rules of ethical and methodologically sound approaches.

CHALLENGES AWAITING COMPUTATIONAL TEXT ANALYSIS ADEPTS

The focus of most critiques of computational text analysis in digital humanities seems to address its problems with reliability and reproducibility. Those questions became the main argument in the violently received publication by Da (2019), although at that point they had already been addressed many times before, both within the field itself[13] and in discussions between digital and non-digital scholars of particular subjects.[14]

The problem of reliability is often exemplified by the supposed black-boxedness of those digital methods applied in text analysis studies. Various scholars have attempted to provide benchmark studies and explanations of scientific methods which have proven popular in the humanities (e.g., Jockers and Witten 2010; Evert et al. 2009), but there are still methods that remain difficult or almost impossible to explain even by experts. [15] Should this discourage DH scholars from applying computational text analysis methods the complexity or performance of which is not explicable altogether? Not necessarily, I would argue, as the presence of humanists in these debates is crucial if such techniques are to develop with a proper input on matters of language, text, discourse, and expression (cf. Bender 2013 and her other publications). Humanists should not shy away from voicing our concerns, but there needs to be a better effort to explain the methods of DH in the research and publications they generate.

The DH2020 discussion (Schöch et al. 2020) on the subject of replication produced some particularly salient points on the problem of reproducibility. While the complex nature of reproducing humanist research has been discussed in detail—for example, in the case of psychology—the specific challenges of achieving it in the case of text analysis often seem obfuscated. Particularly important in alleviating this issue is the introduction of repeated research typology presented by Christof Schöch (Schöch et al. 2020, 4), which explains in a clear visual way the differences between replication (same research question, same dataset, same method), re-analysis (same question and dataset but with a different method), reproduction (same question and method but different data), and so on.

Schöch argues that the use of such typology:

> can help assess the merits and limitations of a given replication study to assess whether, given the stated objectives of the authors, they have employed a suitable type of replication strategy. And it can support designing a replication study and clarify what data, code and contextual information needs to be obtained or reconstructed in order to perform a specific type of replication.
>
> (Schöch et al. 2020, 4)

The idea that scholars should approach their research conscious of how it replicates previous studies and how it can be further replicated seems like a good way forward for computational text analysis.

In recent years we have also become increasingly aware of another issue related to the development of computational text analysis as a discipline. The increase in studies has led to the suffocation of space at major conferences and in leading journals. Scholars outside computational text analysis think there is little space for other kinds of DH, while scholars inside computational text analysis feel their discipline is becoming too crowded, with a lot of valuable contributions being lost in the noise.

And this has led to observable divisions in the field, the most obvious being the split between papers using "simpler" approaches, and those using "very complex" methods. Quite noticeable, especially in the last decade, is the trend of publications that provide a scrupulous description of all technical details and venture to propose new or significantly improved advanced methods derived from computer science, as one would do submitting a paper to natural language processing or computational linguistics venues.[16] While many of these venues have resisted the digital humanities label, more scholars working on computational text analysis—especially those concerned with the scientific and technical aspects of their methodologies—should consider bringing their work to such communities.

Then we have those papers that barely touch upon the technical aspect, and opt instead to re-use well-proven, established methods, pointing to methodological explanations in external sources, focusing instead on particularly important problems that should be presented to a broader humanities community. Prose less intertwined with technical babble can engage readers from the wider humanities and non-digital disciplines, as evidenced for example by the excellent book *Redlining Culture: A Data History of Racial Inequality and Postwar Fiction* by Richard Jean So (2020). But if we omit methodological jargon in DH papers, do we not rob the readers of an opportunity to learn? Is doing so not akin to saying, "this is our difficult black magic thing, you would not understand it so just take what we are saying on faith"? Do we not perpetuate the belief that digital methods are for the few who hold the privileged knowledge and resources necessary to do computational analysis?

Undoubtedly, the computational text analysis community has had its problems: many of its adepts bemoan a lack of specialized venues for sharing their work, there is disagreement over the most appropriate way to treat the representation of these methods in publications, the community has been accused of perpetuating a divide between the "soft" and "hard" digital humanities, and it has also been accused of being a bit of a "boys' club." This may once have been the case, but in recent years, the community has made strides towards discussing—and where necessary, addressing—these issues.[17]

Still, some problems persist, problems that endanger the future development of computational text analysis. How do we avoid overcrowding DH publications with computer-assisted text analysis studies but at the same time not detach text analysis studies from its important place in digital humanities? What should conferences and journals prioritize in their review expectations: unique and important topics or novel methods? Do we really need to make this choice, and can we not have both? Is the future of digital humanities one in which we have separate communities and journals and conferences that focus either on more computationally intense studies or the more general ones?

It is obvious that the multitude of computer-assisted and digital humanities oriented text analysis studies require more publication venues.[18] We can also observe a change in dynamics of conference publishing with preference towards publishing full papers as conference proceedings rather than traditional long abstracts preferred by ADHO Digital Humanities Conference, like in the case of the new Computational Humanities Research[19] workshop or the new *Journal of Computational Literary Studies* conference.

Of course, as we use increasingly sophisticated methods and fewer of us understand these techniques, the field stands at the risk of creating infallible figures whose studies are rarely doubted, or rather, few feel the courage to express any such doubt. In fact, if we consider the overly emotional and ad-personam responses to Nan Z Da's critiques,[20] perhaps we are already there. As Gavin (2020) writes:

> Whether those admissions are totally discrediting or totally beside the point depends, as such things always do, on one's perspective. Scholars whose work had been impugned tended to react as if the most important thing was whether or not the criticisms leveled at themselves were justified. Several tried to redirect the conversation toward methodological questions regarding the role of statistics in humanistic inquiry, but the sheer drama of mutual accusation swamped such concerns.
>
> (2020, 2)

I subscribe to Gavin's disappointment, so will not cite specific examples of such responses, but I believe they are easy to find and recognize. While not all of Da's arguments were sufficiently grounded and well-formed, the fact that some of the responses turned into a social media spat that seemed more about individuals than methodology made this a disheartening exchange to follow.

The whole Da saga may have a particular impact on early career scholars, discouraging them to openly criticize questionable research performed by leading figures in the field. The reality is that scholars in precarious positions may well fear retaliation, that they may even jeopardize their chances for professional advancement. At a moment when we are all advocating for greater diversity and inclusion, for more openness and multilingualism, we should also ask ourselves: can any of the aforementioned ever really be achieved if we do not first open ourselves to critique—both internal and external? Computational text analysis does have problems with reproducibility, with methodological vagueness, with black-box tools and techniques. If we cannot accept critiques of these things, if we cannot critique them ourselves and as a community, where does that leave us? What will the future of this discipline look like?

LOOKING TO THE FUTURE

The introduction of computers to the field of text analysis has certainly produced new critical possibilities. Text has always been at the heart of the digital humanities, and there is no reason why this should become irrelevant in the future. However, just as linguists suffered as their discipline endured numerous evolutions and fragmentation, the field of text analysis seems to be at a crossroads where it must resolve a great many intellectual, methodological, and individual matters and divisions. Hopefully, the current state of the field can lead to positive change, to new perspectives, and fresh scrutiny of methods applied in this domain.

ACKNOWLEDGMENTS

The research was supported by the Large-Scale Text Analysis and Methodological Foundations of Computational Stylistics project (SONATA-BIS 2017/26/E/HS2/01019) funded by the National Science Centre of Poland.

NOTES

1. Examples of early experiments in stylometry include Mendenhall (1887, 1901); Dittenberg (1881); Campbell (1867); and Lutosławski (1897).
2. Notable tools include *stylo* (Eder et al. 2016) and *JGAAP* (Juola et al. 2009).
3. Though it must be said that summer schools such as the European Summer University in Digital Humanities (https://esu.fdhl.info/) and the Digital Humanities Summer Institute (https://dhsi.org/) offer great opportunities to develop one's skills and projects under expert guidance.
4. See the website of the European Summer University in Digital Humanities, https://esu.fdhl.info/.
5. See the website of Digital Humanities Summer Institute, https://dhsi.org/.
6. See the website of the Digital Humanities Institute in Beirut, dhibeirut.wordpress.com/courses/.
7. See the workshops at the website of the Centre for Digital Humanities, University of Lagos, cedhul.com.ng/.
8. See the website of the Digital Methods in Humanities and Social Sciences workshops in Tartu, digitalmethods.ut.ee/node/22501.
9. See the website of the Moscow-Tartu Digital Humanities School, hum.hse.ru/en/digital/announcements/268278278.html.
10. See also the variety of studies presented at R2L Conference coordinated by David Joseph Wrisley (NYU Abu Dhabi) and Kasra Ghorbaninejad (University of Victoria), co-located with Digital Humanities Summer Institute, https://dhsi.org/dhsi-2021-online-edition/dhsi-2021-online-edition-aligned-conferences-and-events/dhsi-2021-right-to-left/.

11. Note that Stylometric bibliography on Zotero, representing just one of three branches of computational textual analysis (https://www.zotero.org/groups/643516/stylometry_bibliography), curated by Christof Schöch as of late 2021 counts as many as 3,492 items—papers and conference abstracts.
12. See e.g., "Part 6. Forum: Text Analysis at Scale" of 2016 Debates in Digital Humanities, or discussions following Nan Z. Da's (2019) paper published across various media, e.g., *Cultural Analytics* journal.
13. See panel discussion on Replication and Computational Literary Studies at DH2020 conference, Schöch et al. (2020), or activities of Special Interest Group "Digital Literary Stylistics," especially 2019 and 2020 workshops on tool criticism, https://dataverse.nl/api/access/datafile/19409 with slides: https://dls.hypotheses.org/activities/anatomy for 2019, and https://dls.hypotheses.org/1110 for 2020.
14. See many years of discussions by Shakespeareans engaging with digital studies, e.g., Hugh Craig, Brian Vickers, Jonathan Hope.
15. As evidenced by discussions around large-scale language models or the use of neural networks (cf. for example Zarrieß et al. 2021 addressing these problems in the context of natural language generation).
16. For example, ACL or EMNLP conferences.
17. One such example is the Computational Humanities Research initiative, initially criticized as contributing to the divisions, but three years later an example of a community that maintains a relatively lively discussion forum on research and academia issues: discourse.computational-humanities-research.org. It should also be noted that it has just hosted its second online workshop, with numerous female scholars presenting their research and being involved in the program committee both years. The *Journal of Cultural Analytics* also regularly accepts publications addressing the problems of gender or monolingualism.
18. While there are quite a few of language-specific reach, e.g., Umanistica Digitale, and some new open-source initiatives like the *Journal of Computational Literary Studies* (https://jcls.io/) or hybrid *International Journal of Digital Humanities*, the main DH venues such as *Digital Scholarship in the Humanities*, *Digital Humanities Quarterly*, or *Digital Studies* are still popular among adepts of computational text analysis, and given their visibility and the fact that they are among few Scopus, etc., listed journals in the field, their popularity is unlikely to disappear entirely.
19. Computational Humanities Research, https://discourse.computational-humanities-research.org/.
20. "The Computational Case against Computational Literary Studies," the 2019 paper by literary scholar Nan Z. Da criticizing the validity of computational text analysis in applications to literary studies, gained a lot of attention soon after its publication. As Da supported her critique with examples of her attempts at replication of process and results described in the papers she addressed, a number of scholars, including ones whose research was criticized addressed the publication both on social media (especially Twitter) and in publication venues. Following online informal discussions, *Critical Inquiry* opened a forum for debate: https://critinq.wordpress.com/2019/03/31/computational-literary-studies-a-critical-inquiry-online-forum/ and a series of responses and discussions was published in *Cultural Analytics*.

REFERENCES

http://www.digitalhumanities.org/companion/

Argamon, Shlomo. 2008. "Interpreting Burrows's Delta: Geometric and Probabilistic Foundations." *Literary and Linguistic Computing* 23 (2): 131–47. https://doi.org/10.1093/llc/fqn003.

Beausang, Chris. 2020. "A Brief History of the Theory and Practice of Computational Literary Criticism (1963–2020)." *Magazén*, no. 2. https://doi.org/10.30687/mag/2724-3923/2020/02/002.

Bednarek, Monika. 2010. *The Language of Fictional Television: Drama and Identity*. London/New York, NY: Continuum.

Bednarek, Monika. 2018. *Language and Television Series: A Linguistic Approach to TV Dialogue*. Cambridge: Cambridge University Press.

Benatti, Francesca and Justin Tonra. 2015. "English Bards and Unknown Reviewers: a Stylometric Analysis of Thomas Moore and the Christabel Review." *Breac: A Digital Journal of Irish Studies* 4.

Bender, Emily M. 2013. *Linguistic Fundamentals for Natural Language Processing: 100 Essentials from Morphology and Syntax*. San Rafael, CA: Morgan & Claypool Publishers.

Blei, David M. 2012. "Probabilistic Topic Models." *Communications of the ACM* 55 (4): 77. https://doi.org/10.1145/2133806.2133826.

Burrows, John. 1987. *Computation into Criticism: A Study of Jane Austen's Novels and Experiment in Method*. Oxford: Clarendon Press.

Burrows, John. 2002. "'Delta': A Measure of Stylistic Difference and a Guide to Likely Authorship." *Literary and Linguistic Computing* 17 (3): 267–87. https://doi.org/10.1093/llc/17.3.267.

Campbell, Lewis. 1867. *The Sophistes and Politicus of Plato*. Oxford: Oxford Clarendon Press.

Da, Nan Z. 2019. "The Computational Case against Computational Literary Studies." *Critical Inquiry* 45 (3): 601–39. https://doi.org/10.1086/702594.

Daelemans, Walter. 2013. "Explanation in Computational Stylometry." *In Computational Linguistics and Intelligent Text Processing*, edited by Alexander Gelbukh, 7817: 451–62. Lecture Notes in Computer Science. Berlin, Heidelberg: Springer Berlin Heidelberg. https://doi.org/10.1007/978-3-642-37256-8_37.

Dittenberg, Wilhelm. 1881. "Sprachliche Kriterien Für Die Chronologie Der Platonischen Dialoge." *Hermes: Zeitschrift Für Klassische Philologie* 16: 321–45.

Eder, Maciej, Mike Kestemont, and Jan Rybicki. 2016. "Stylometry with R: A Package for Computational Text Analysis." *R Journal* 16 (1): 107–121. https://journal.r-project.org/archive/2016/RJ-2016-007/RJ-2016-007.pdf.

Eichmann-Kalwara, Nickoal, Jeana Jorgensen, and Scott B. Weingart. 2018. "Representation at Digital Humanities Conferences (2000–2015)." In *Bodies of Information: Intersectional Feminism and the Digital Humanities*, edited by Elizabeth Losh and Jacqueline Wernimont, 72–92. Minneapolis, MN: University of Minnesota Press. https://doi.org/10.5749/j.ctv9hj9r9.9.

Eichmann-Kalwara, Nickoal, Jeana Jorgensen, and Scott B. Weingart. 2018. "Representation at Digital Humanities Conferences (2000–2015)." In *Bodies of Information: Intersectional Feminism and the Digital Humanities*, edited by Elizabeth Losh and Jacqueline Wernimont, 72–92. Minneapolis, MN: University of Minnesota Press. https://doi.org/10.5749/j.ctv9hj9r9.9.

Evert, Stefan, Thomas Proisl, Fotis Jannidis, Isabella Reger, Steffen Pielström, Christof Schöch, and Thorsten Vitt. 2017. "Understanding and Explaining Delta Measures for Authorship Attribution." *Digital Scholarship in the Humanities* 32 (supplement 2): ii4–16. https://doi.org/10.1093/llc/fqx023.

Evert, Stefan, Thomas Proisl, Christof Schöch, Fotis Jannidis, Steffen Pielström, and Thorsten Vitt. 2015. "Explaining Delta, or: How Do Distance Measures for Authorship Attribution Work?" In *Corpus Linguistics 2015 Abstract Book*, edited by Federica Formato and Andrew Hardie. Lancaster: UCREL. https://doi.org/10.5281/zenodo.18308.

Evert, Stefan, Thomas Proisl, Thorsten Vitt, Christof Schöch, Fotis Jannidis, and Steffen Pielström. 2015. "Towards a Better Understanding of Burrows's Delta in Literary Authorship Attribution." In Proceedings of the Fourth Workshop on Computational Linguistics for Literature, 79–88. Denver, Colorado, USA: Association for Computational Linguistics. http://www.aclweb.org/anthology/W15-0709.

Gavin, Michael. 2020. "Is There a Text in My Data? (Part 1): On Counting Words." *Journal of Cultural Analytics* 1 (1): 11830. https://doi.org/10.22148/001c.11830.

Gröndahl, Tommi and N. Asokan. 2019. "Text Analysis in Adversarial Settings: Does Deception Leave a Stylistic Trace?" *ArXiv:1902.08939 [Cs]*, February. http://arxiv.org/abs/1902.08939.

Grzybek, Peter. 2014. "The Emergence of Stylometry: Prolegomena to the History of Term and Concept." In Katalin Kroó and Peeter Torop (eds), *Text within Text – Culture within* Culture, 58–75. Budapest, Tartu: L'Harmattan.

Hamilton, William L., Jure Leskovec, and Dan Jurafsky. 2018. "Diachronic Word Embeddings Reveal Statistical Laws of Semantic Change." *ArXiv:1605.09096 [Cs]*, October. http://arxiv.org/abs/1605.09096.

Holmes, David I. 2003. Stylometry and the Civil War: The Case of the Pickett Letters, CHANCE, 16:2, 18–25, doi: 10.1080/09332480.2003.10554844

Holmes, David I. 1998. "The Evolution of Stylometry in Humanities Scholarship." *Literary and Linguistic Computing* 13 (3): 111–17., https://doi.org/10.1093/llc/13.3.111.

Hoover, David L. 2008. "Quantitative Analysis and Literary Studies." In *A Companion to Digital Literary Studies*, edited by Ray Siemens and Susan Schreibman. Oxford: Blackwell. http://digitalhumanities.org:3030/companion/view?docId=blackwell/9781405148641/9781405148641.xml&doc.view=print&chunk.id=ss1-6-9&toc.depth=1&toc.id=0.

Jockers, M. L. and D. M. Witten. 2010. "A Comparative Study of Machine Learning Methods for Authorship Attribution." *Literary and Linguistic Computing* 25 (2): 215–23. https://doi.org/10.1093/llc/fqq001.

Juola, Patrick. (2009). "JGAAP: A System for Comparative Evaluation of Authorship Attribution." Proceedings of the Chicago Colloquium on Digital Humanities and Computer Science 1 (1).

Juola, Patrick, Michael V. Ryan, and Michael Mehok. 2011. "Geographically Localizing Tweets via Stylometric Analysis." American Association for Corpus Linguistics Conference (AACL 2011), Atlanta, Georgia.

Klaussner, Carmen and Carl Vogel. 2015. "Stylochronometry: Timeline Prediction in Stylometric Analysis." In *Research and Development in Intelligent Systems XXXII*, edited by Max Bramer and Miltos Petridis, 91–106. Cham: Springer International Publishing. https://doi.org/10.1007/978-3-319-25032-8_6.

Lutosławski, Witold. 1897. *The Origin and Growth of Plato's Logic*. Hildesheim: Georg Olms.

Mendenhall, Thomas. 1887. "The Characteristic Curves of Composition." *Science* 9 (214S): 237–46. https://doi.org/10.1126/science.ns-9.214S.237.

Mendenhall, Thomas. 1901. "A Mechanical Solution of a Literary Problem." *Popular Science Monthly* 60 (December): 98–105.

Mikolov, Tomas. 2012. "Statistical Language Models Based on Neural Networks." PhD, Brno University of Technology.

Mosteller, Frederick and David L. Wallace. 1963. "Inference in an Authorship Problem." *Journal of the American Statistical Association* 58 (302): 275–309. https://doi.org/10.2307/2283270.

Plecháč, Petr, Klemens Bobenhausen, and Benjamin Hammerich. 2019. "Versification and Authorship Attribution: A Pilot Study on Czech, German, Spanish, and English Poetry." *Studia Metrica et Poetica* 5 (2): 29–54. https://doi.org/10.12697/smp.2018.5.2.02.

Rebora, Simone, Peter Boot, Federico Pianzola, Brigitte Gasser, J. Berenike Herrmann, Maria Kraxenberger, Moniek Kuijpers, et al. 2019. "Digital Humanities and Digital Social Reading." Pre-print. Open Science Framework. https://doi.org/10.31219/osf.io/mf4nj.

Rezaei, Sohrab and Nasim Kashanian. 2017. "A Stylometric Analysis of Iranian Poets." *Theory and Practice in Language Studies* 7 (1): 55. https://doi.org/10.17507/tpls.0701.07.

Rudman, Joseph. 1997. "The State of Authorship Attribution Studies: Some Problems and Solutions." *Computers and the Humanities* 31, no. 4 (1997): 351–65. http://www.jstor.org/stable/30200436.

Schler, Jonathan, Moshe Koppel, Shlomo Argamon, and James Pennebaker. 2006. "Effects of Age and Gender on Blogging." In AAA 1 Spring Symposium on Computational Approaches for Analyzing Weblogs, 6.

Schreibman, Susan, Ray Siemens, and John Unsworth, eds. 2004. *A Companion to Digital Humanities*. Oxford: Blackwell. http://www.digitalhumanities.org/companion/.

Schöch, Christof, Karin van Dalen-Oskam, Maria Antoniak, Fotis Jannidis, and David Mimno. 2020. "Replication and Computational Literary Studies." June 14. https://doi.org/10.5281/ZENODO.3893428.

So, Richard Jean. 2020. *Redlining Culture: A Data History of Racial Inequality and Postwar Fiction*. New York, NY: Columbia University Press.

Soffer, Oren, Zef Segal, Nurit Greidinger, Sinai Rusinek, and Vered Silber-Varod. 2020. "Computational Analysis of Historical Hebrew Newspapers: Proof of Concept." *Zutot* 17 (1): 97–110. https://doi.org/10.1163/18750214-12171087.

Suzuki, Takafumi and Mai Hosoya. 2014. "Computational Stylistic Analysis of Popular Songs of Japanese Female Singer-Songwriters." *Digital Humanities Quarterly* 8 (1).

Terras, Melissa, Julianne Nyhan, and Edward Vanhoutte. 2013. *Defining Digital Humanities*. London and New York, NY: Routledge.

Thompson, Laure and David Mimno. 2018. “Authorless Topic Models: Biasing Models Away from Known Structure.” In *Proceedings of the 27th International Conference on Computational Linguistics*, 3903–14. Santa Fe, New Mexico: Association for Computational Linguistics. https://aclanthology.org/C18-1329.
Zarrieß, Sina, Henrik Voigt, and Simeon Schüz. 2021. “Decoding Methods in Neural Language Generation: A Survey.” *Information* 12 (9). https://pub.uni-bielefeld.de/record/2957091.

CHAPTER SIXTEEN

The Possibilities and Limitations of Natural Language Processing for the Humanities

ALEXANDRA SCHOFIELD (HARVEY MUDD COLLEGE)

In Jonathan Swift's 1726 satire *Gulliver's Travels*, the titular traveler takes an extended tour of the Academy of Projectors in the city of Lagado. According to the narrator, the academy was established to help establish a community of innovation, where professors or "projectors" would be funded to realize the "happy proposals" of projects to improve life in the surrounding nation (Swift 1726). This goal has not come to pass, as witnessed by Gulliver's tour through half-realized and often absurd experiments in the Academy. One project that the protagonist stops to attend more closely is that of the Engine, an enormous machine designed to help create new knowledge and writing through the permutation of words. Though the machine's inventor is identified as a single projector, operating the machine requires forty students to turn cranks and to pore over the resulting permuted words in search of sensible sentence fragments to transcribe. While the transcriptions so far have yet to yield new insight, the projector is too invested in its success to give up hope of a future positive outcome:

> He assured me "that this invention had employed all his thoughts from his youth; that he had emptied the whole vocabulary into his frame, and made the strictest computation of the general proportion there is in books between the numbers of particles, nouns, and verbs, and other parts of speech."
>
> (Swift 1726, III.V)

The core description of the machine itself has drawn praise as providing an early account of a computer (Weiss 1985). This may be a bit of an overstatement, since the previous century had already seen multiple gear-based calculating machine designs in Europe, including those of Blaise Pascal, Thomas de Colmar, Samuel Morland, and Gottfried Leibniz (Jones 2016). However, while mechanization of computation might not have been new, it is hard to find an earlier account of a machine made specifically to manipulate language. The account notably does so without ascribing the Engine its own intelligence: in the account, the professor responsible for the project has attempted to put all of his specialized knowledge of art, science, and language into the machine, and he relies on his students to detect which random emissions might signal something more meaningful. Ultimately, the Engine is only the middle step in a generative process: it can neither create new words nor permute existing ones beyond the specifications of the professor, nor does it exert agency over where the scribes focus their attention in its output.

As a computer scientist by training, I would be out of my depth in making a strong case for or against whether algorithmic approaches belong in the digital humanities, as many with more background in humanistic work have written on the subject; for example, debate pieces from Nan Z. Da and Andrew Piper looking at case studies of these topics (Da 2019; Piper 2019). What I do feel I can speak to, however, is the current practice that defines the labor of text-mining for cultural analysis and, specifically, the middle steps undertaken by a modern *corpus projector* that are often absent in the popular narratives of popular modern work in natural language processing.

I define a *corpus projector* as an expert in a particular text collection interested in using natural language processing technologies to help draw conclusions about that corpus. While the corpus projector's work is not generative in the same sense as the satirical figures from Lagado, both types of projector rely on their own expertise and effort to translate between human and mechanical and back again. In this process, both strive to benefit from the new perspective on text for which the machine is an interface, and both carry the risk of exposing apparent meaning where none may exist.

I want to use this article to shed a bit of light on a crucial aspect of the corpus projector's role: the work it takes to develop the "engine," or the combination of corpus processing and computer programming that ensures that quantities produced about a text are relevant to the intended inquiry. This development work is a central and time-consuming element of work for those interested in developing a corpus, but one that is notably underprivileged as part of the computer science publications from which text analysis techniques are often drawn. My hope in drawing attention to the tasks of corpus projection is to help demystify both why this work takes time and why it is time well-spent in the service of truly exploring the boundaries of what identifies a collection of texts.

To make sense of the work of projectors, I want to start with a broad view of what natural language processing steps are typically demanded to perform a computational analysis of a text corpus. From there, I will distinguish some of the ways that the workflow of a corpus projector becomes complex in ways often unanticipated by someone new to natural language processing, and why I consider the unique decisions a projector might make in response to those complexities to be a strength of the computational approach. Finally, I will provide some brief thoughts on how to more justly characterize the information that comes from the unique engines that corpus projectors design to fit their own inquiry processes.

A CONVENIENT FICTION

What do I mean when I refer to the field of *natural language processing*? Broadly speaking, I use natural language processing (or NLP for short) to refer to any kind of work that uses computers to do *something* ("processing") with human-authored ("natural") language. This is extremely broad by design: much like the digital humanities community, the natural language processing community unites people from a variety of disciplines with shared questions about how to perform programmatic operations on language data, including linguistics, psychology, statistics, library science, and computer science. For the moment, I will focus specifically on the computer science approach to *analytic* tasks in NLP: that is, tasks where the objective is to algorithmically elicit some sort of information about patterns in existing text, such as category labels, scores, predicted attributes, or comparative statistics.

Natural language processing work is usually framed by defining a particular task of interest: for instance, in *sentiment analysis*, the goal is to predict a score of how positive or negative the opinion of an author is of a particular subject based on the author's description. In a computer science venue, the object of interest is how well computers can support these tasks, which necessitates a focus on the algorithmic: the choice or development of a model, the inference strategy for that model, and the numerically driven approach to consistent evaluation. Authors frame their advances using evaluation of performance on text "datasets" that co-exist with structured annotations. In the ideal case, there are existing settings that naturally produce such datasets to support these tasks: while a collection of news articles may be hard to score for sentiment, reviews of restaurants, films, or products are increasingly available online and typically accompanied by a numerical score summarizing sentiment.

A formulaic recipe arises among many of these task-oriented publications: 1) design or select a model or method to address the task; 2) design or select one or more datasets that are compatible with this task (e.g., a collection of restaurant reviews for sentiment analysis); 3) instantiate the model or method in a way that is compatible with the structure and content of these dataset(s); and 4) evaluate how well this method performs with these data. This last step may include both quantitative analyses (e.g., how often was the sentiment score accurately predicted) as well as qualitative analyses (e.g., what type of review content would cause sentiment score to be overestimated by the model).

Researchers tell stories, and this four-step story is a common one not just in computer science's treatment of NLP, but in papers introducing any kind of computational model. It is also a convenient fiction: in this sequence of events, everything works on the first try. In this story, there is very little room to share past generations of rejected models and configurations. Nor is one likely to find much detail about why particular decisions were made in steps 2 and 3: though the practice of providing supplementary material with information to reproduce these steps has become more common, it rarely includes much in the way of rationale. And yet this formula is an effective choice for NLP researchers whose goals are to compare model design and results, as it centers the work of steps 1 and 4 to allow comparison between papers in the same task.

However, the elisions and omissions of this formula prove a formidable obstacle to those from other communities looking to computer science texts for a path forward with a new collection of texts: when there is only room for discussion of the final successful model, key decisions about how to respond to different properties of text or crucial differences in seemingly similar problems are left for those following up on the work to slowly, painfully rediscover. It is already a matter of increasing frustration to NLP researchers themselves, as described in the Proceedings of the Workshop for Insights from Negative Results in NLP, a workshop started in 2020 to respond to this deficiency in the literature: "the development of NLP models often devolves into a product of tinkering and tweaking, rather than science. Furthermore, it increases the time, effort, and carbon emissions spent on developing and tuning models, as the researchers have little opportunity to learn from what has already been tried and failed" (Sedoc et al. 2021).

The task of corpus development is one that takes considerable care and effort. Large-scale projects from corpus linguistics have established well-used collections such as the Brown corpus (Kučera and Francis 1967), which provides samples of annotated written American English in a variety of settings, or the Europarl corpus (Koehn 2005), which has grown from eleven to twenty-one languages of parallel text translated for the European Parliament. The goal of developing these datasets is usually task-oriented, as alluded to in Kim Jensen's analysis contrasting computation corpus linguistics with the corpus work of digital humanities: "The latter tends to operate with unstructured data in the form of 'raw' text whose only structure is often that individual texts are stored in their own documents" (Jensen 2014). Ideally, future researchers will compete to optimize their performance on that task through established comparative metrics (e.g., proportion of documents with correctly predicted sentiments). Events organized around these established task-centered corpora have been a significant driver of innovation in NLP research for decades, from government-funded "bakeoffs" in the 1990s (Anderson, Jurafsky, and McFarland 2012) towards modern events with "shared tasks" such as the annual Workshop for Semantic Evaluation (Herbelot et al. 2020), as well as long-term standardized benchmarks like GLUE (Wang et al. 2019). The broader expectation is that these papers must include at least an abbreviated version of the process above to contextualize whether the task is sufficiently difficult.

In this research environment, there is a strong incentive for researchers to re-use corpora, metrics, and processing steps so they may more easily compare their results with existing work. Code may be recycled and shared in ways that promote collegiality and reduce labor for everyone. However, the centering of a few "tidy" collections has consequences about the broader applicability of the models presented at these venues. Hovy and Spruit describe some of this phenomenon in terms of *topic exposure*, where the relative prominence of certain topics in the research community can lead to naive misconceptions about how well-established models generalize to other data or tasks (Hovy and Spruit 2016). For instance, in *sentiment analysis*, one might suggest that the specific choice of predicting review ratings for English user-generated reviews is overexposed due to the ease of access of these datasets. A consequence of this is the underexposure of other types of sentiment analysis tasks for which data may be scarcer or harder to label. The dominance of English or Western European languages, privileged dialects, or widely known textual venues also can lead to incorrect conclusions about how universal the applications of these models are: The fact that a particular model for sentiment analysis achieves highly accurate results on a standard English dataset of syndicated movie reviews does not guarantee the same model design will be effective for reviews posted in Arabic on an online forum.

The compounding of narrow dataset practice—the emphasis on data re-use, the de-emphasis of describing the how and why of data processing, and the underexposure of "inconvenient" data—already makes the landscape challenging for new digital humanities researchers trying to explore an established task with a new computational model, as there is some learning curve to discovering what processing decisions are appropriate for a pair of task and dataset. This task has been eased by the development of established tutorials and guides directly from researchers in the digital humanities, including textbooks that give step-by-step tutorials on common computational text analysis tasks with R (Jockers 2014; Arnold and Tilton 2015) and Python (Walsh and Dombrowski 2021). Even with these, however, further challenges await a novice corpus projector trying to study a new text collection with a research question not accompanied by a well-worn framing task like classification or ranking prediction.

GRAPPLING WITH THE UNIQUE

In my course on natural language processing, I often ask my students in a brief exercise to see if they can devise an English 4-gram (that is, a sequence of four English words) that could be plausibly used as part of a sentence but is completely devoid of any matching results on their favorite search engine. The task is surprisingly easy: one can construct unusual nouns ("viscounts of cotton candy"), match specific subjects with improbable verbs ("octopus pilots rogue tugboat"), or give one object multiple surprising and disjoint modifiers ("miniature lugubrious pink azaleas"). Even if students are constrained to stick to common words, as Swift's Engine might do, it is still feasible to come up with unseen sequences (e.g., "seven professors compare lawyers"). The exercise has multiple benefits as an instructor: it is quick to do, it tends to produce hilarious imagery, and it illustrates how easy it is to bring something entirely unique into existence with language.

Natural language processing inevitably relies on trying to draw trendlines through the unique, to break down contextualized passages of text into small-enough decontextualized units that repeating patterns among those units can emerge. In particular, when computer scientists do NLP work that centers a new model, the mission is to demonstrate quantitatively how this model improves on other models. In this paradigm of text analysis, authors in a benchmark text collection register as merely above categorical labels for texts, useful for distinguishing source and variability. To focus on a single author or to draw a thesis from the unique contexts of individual texts would run counter to convention: uniqueness signifies an obstacle to generalization, not a positive discovery.

Once the approaches are brought into the realm of humanistic computational text analysis, however, the objective shifts: the goal is no longer the best performance, but to capture something concrete and meaningful about the world and the individuals from which the text arose. Recent proponents of quantitative approaches to text emphasize that they are meant to augment humanist analyses instead of merely reducing them to numbers. Lisa Marie Rhody's critique of Franco Moretti's formalist framing of the concept of "distant reading" mentions the significance of how quantitative approaches to text draw us back as readers: "Inexorably, distant reading will invite us into the spaces left behind in its maps, graphs, and diagrams to explore through close reading and observation" (Rhody 2017). The hope is that this approach adds another layer to the possible arguments to be made about text, as phrased by Andrew Goldstone and Ted Underwood: "… quantitative methods like ours promise not definitive simplicity but new ways of analyzing the complexity of literary and cultural history" (Goldstone and Underwood 2014, 360).

One particular augmentative benefit of computational analysis is the ability to move beyond examples for or against a theory towards a broader sense of how widespread a phenomenon is across more texts than could be discussed in a single article. Andrew Piper, a professor of cultural analytics, summarizes this key benefit of *generalization*: "A more credible practice of generalization requires that we examine both positive and negative examples in order to understand the representativeness (or distinctiveness) of the sample we are observing and the representativeness (or distinctiveness) of the individual observations within those samples" (Piper 2019). The positioning of generalization is not as a simplification of what traditional analysis would do, but as an alternative approach that balances a less nuanced analysis with more coverage of possible texts. Even then, however, Piper and others admit that the approach is caveated by the limitations to completeness of a corpus and the methods of processing:

> The data sets that I use here are thus not to be confused with some stable macrocosm, a larger whole to which they unproblematically gesture. They too are constructions … Future work will want to explore the extent to which these and other observations depend on different kinds of collections. We need to better understand the interactions between methods and archives.
>
> (Piper 2018, 10)

To create a model of text requires trial and error to understand what sort of havoc the unstoppable force of a corpus of texts will wreak when confronted with the immovable object of a computer program. The need for iteration is not unique to text: as Richard Jean So reflects on the application of statistical models to text, he writes that in general "a model allows the researcher to isolate aspects of an interesting phenomenon, and in discovering certain properties of such aspects, the researcher can continue revising the model to identify additional properties" (So 2017, 669). What often is omitted in this description, however, is how messy that iteration process is. The answers to questions such as "what counts as a word?," "what is the author's name?," or "where does the body of the text start?" often require months of trial and error to reach some approximate rule that serves most cases in the context of the corpus.

As an example of the complexity involved, consider the Viral Texts Project's analysis of the reprinting of texts in nineteenth-century newspapers (Smith, Cordell, and Dillon 2013; Cordell and Mullen 2017). An early step in creating the text data corpus for this project was the collection of the text of newspapers and periodicals from the Chronicling America database, which in turn determined the text from scanned newspapers through the use of *optical character recognition,* or OCR, to register the words in the texts. Since OCR engines are typically developed with more modern texts in mind, they are prone to errors in historical text, such as inferring space where none exists, conflation of similar characters such as an A and a 4, or the introduction of rogue punctuation from an ink splatter. Even when working well, OCR engines will intentionally strip spacing information, blurring headlines, bylines, and paragraphs together into one unit. Archives also have some biases as to which papers are present, as the time and technology required to perform this digitization is not equally available to all libraries and archivists.

The Viral Texts authors do a remarkable job of approaching these difficulties one by one, but through interventions that might initially be a shock: since no spacing is available, none is relied on, with passage similarity instead determined through a statistical heuristic based on phrases of text. All words shorter than four characters are omitted, as the likelihood of OCR error is too high: while these engines can correct words based on a model of language probabilities, but these corrections are less likely to work on shorter and more common words. Before the team has even had the opportunity to determine which words one will use to distinguish a poem's editorial introduction, they have already necessarily made choices to filter, edit, and process in order to have a consistent and comparable collection. The results are glowing for Viral Texts, leading to many iterations of new patterns of inquiry and insight. However, identifying processing workflows like those of the Viral Text Project takes a significant amount of time, and puts a heavy price on selecting a new corpus to study: there is little certainty that the time cost to do this preparatory work will pay off with compelling results every time.

Nan Z. Da succinctly critiques computational literary studies based on the limitations of these judgments: "In CLS data work there are decisions made about which words or punctuations to count and decisions made about how to represent those counts. That is all" (Da 2019). Here, I agree: many computational text analyses boil down to choices made by corpus projectors of what counts as a word and how to count words. Corpus projectors iterate through models, text processing pipelines, and data visualizations in search of sturdy signals that can withstand interrogation of their source. The models of text designed to surface these signals can also be useful tools in finding issues, such as incorrectly ingested text or inadvertently included boilerplate; it typically takes a human reader to make sense of where things went wrong and what to do in the next iteration.

However, I do not consider iteration of text processing techniques to be a weakness of using computational methods. Certainly, the instruments that we deploy to render text into a form compatible with a program are not objective: they are calibrated by our judgments of what belongs in the quantities we gather and compare and explore. However, they also represent significant choice about what text is attended to and how. Instead of framing these choices as "pre-processing" steps made before the analysis begins, my past collaborators Laure Thompson and David Mimno use a term I now prefer, "purposeful data modification," to reframe the work of corpus processing as a deliberate act of analysis itself. Deep knowledge and analyses go into the apparently simple decisions of how to capture useful quantities from corpora: the process of determining the vocabulary to consider encompasses iterated close reading of troublesome passages, deep dives into related analyses of similar texts, and long conversations about how to verify that the numerical results that arise correspond to something *real*. It is a process that, to my knowledge, bears much more resemblance to the "classic" work of analysis than anyone may care to admit, leading generally to a rich conversation between the aggregate and the unique. From beginning to end, success in these analyses relies on human judgments to extrapolate unusual results into decisions about what features to retain.

The fact that these choices are also interpretive, that they require a projector to determine suitable operational details of how to frame a question, is not a failure of a humanistic inquiry: it is the inquiry itself proceeding as it should. The choice to dwell on how one word is used while shirking another cannot be a subliminal one; we are forced to explicitly state what "counts" and what does not in the form of unambiguous lines of code. Within this process, it is crucial that we do not minimize the early choices of pre-processing as a chore preceding the work, but instead interrogate it as one of the most fundamental pieces of the work itself, where decisions must be logged and shared.

SUGGESTIONS FOR NEW PROJECTORS

As outlined above, the development process to use NLP for humanistic questions is slow, context-specific, and nonetheless crucial to perform carefully. So how should one conduct this work? I want to stress two values that I believe provide rigor without mischaracterizing the significance of the work.

First, *plan to iterate*. It may take many trials to get a corpus processing workflow right: so, instead of treating each newfound issue as a bug, consider it an opportunity to run an experiment. Take as an example the process of *corpus curation*, or pruning out irrelevant text from a collection. Raw text corpora inevitably include unwanted text, whether it is web formatting code, publisher's notes, or even images that an OCR engine attempted to read as text. When issues in these raw texts arise, the natural response is to try to troubleshoot the immediately visible error: for instance, if one noticed many numbers in what should have been non-numerical prose, one might take an immediate computational approach (such as a regular expression) to remove the offending numbers. However, this visible issue is likely the tip of a deep iceberg of different erroneously included text patterns, such as page numbers accompanying chapter headings and author names. Instead, it can be valuable to take these moments as an opportunity to explore whether an error is accompanied by other less obvious issues of a similar category. For this process, there is no replacement for reading examples of the affected text to determine what is happening, ideally while also introspecting about what signals could help generalize to possible manifestations of that phenomenon. Throughout, one should log these observations and the interventions made in response, both to communicate the full procedure at publication time and to help guide the way on future projects with similar texts. This work need not be tedious. An example group that gives life to the insights from this iteration process is The Data-Sitters Club, a community of computational text analysts who chronicle their text investigations, both successful and unsuccessful, in the style of their corpus of analysis: The Baby-Sitters Club children's book series (Bessette et al. 2019).

There is one caveat in this experimental process: while iteration is helpful, it should be in service of capturing the ideas of your research question, not your intended answer. In standard machine-learning work, people usually quickly learn about the distinction between *training data* used to fit a model and *test data* used to see whether that model is a good match for data not seen when the model was fitted. This isolation is rarely so clean when the content of the corpus is so central to the inquiry: in a small collection, the choice to withhold texts can be significantly more fraught. However, the lack of clean separation between model fit and model evaluation bears a considerable risk: there are enough combinatorial counts to explore and enough unique phenomena in every text collection that if you try enough processing techniques, something will match the trend you seek. Rather than taking the approach of the projector at Lagado, who has scribes filter text to only things they find meaningful, one should take a step back to consider whether there is a broader way to determine what to include shaped by the research question. Taking extra time to establish this makes your later interpretive work and justification easier.

Second, *prioritize simplicity and reproducibility*. Absent the invention of an artificial general intelligence (an outcome I find unlikely in my lifetime), the tools we use to understand text cannot access all of the context and subtext we have as humans interacting with the world. Computational linguists Emily M. Bender and Alexander Koller argue this about the limits of meaning acquisition in programs exposed to "form" without context: "even large language models ... do not learn 'meaning'; they learn some reflection of meaning into the linguistic form which is very useful in applications" (Bender and Koller 2020, 5193). In the context of cultural analytic work, I consider this limitation a blessing in disguise: a computational program that has only ever accessed raw text cannot be a fully formed text critic, and therefore will miss some things about text. Therefore, when engaging with a model of text, the value is not necessarily in having the best-fit model, but instead in having clarity about how the generalizations of the model connect to the particularities of text.

Here, a clear procedure is paramount: if it is not easy to write down what procedure determined the features of a text or what a parameter signifies in the model, it can be impossible to extract what generalizations the models may produce. Counterintuitive as it may be to prefer a model that oversimplifies, these oversimplifications can make it much easier to identify what is misunderstood by the model: where a refined model of language may be a better predictor of word senses, a model unaware that the typical sense of a printed word has shifted over time will help discover poorly fitted words.

Both values share at their core the necessity of documenting the journey to a particular corpus processing procedure: even if the final publication of text analysis does not have room to spare, it is worth setting up a code repository or blog post to detail what procedures were done to select the corpus, vocabulary, and model in order to guide future corpus projectors.

THE HAND OF THE ENCODER

Much of what I have written is about the challenges awaiting someone exploring NLP for digital humanities research, so I want to turn for a final moment to what is joyful and exciting about this work.

In *Is There A Text In This Class?* literary theorist Stanley Fish argues that no order of precedence can be established between identifying context and deciding meaning: "One does not say 'Here I am in a situation; now I can begin to determine what these words mean.' To be in a situation is to see the words, these or any other, as already meaningful" (Fish 1980, 313). As somebody enthusiastic about many topics but expert in few, my work has offered delightful opportunities to play tourist among the many projects of other scholars, to get to see what is already meaningful as a naive observer. Inevitably, these questions yield new information for both the expert corpus projectors and myself: the historical reasons why the projectors thought a particular feature might relate to their question, the concerns the projectors have about what they have realized their analytics are unable to capture, or the surprising facts they have learned about little-known texts in their corpus that have broken all their expected rules.

Even when natural language processing techniques produce unhelpful results, the exercise of generating these models can refresh perspective and beckon new insights. "One can only understand error in a model by analyzing closely the specific texts that induce error," as Richard Jean So describes in his argument for statistical models of text; "close reading here is inseparable from recursively improving one's model" (So 2017). I would go a step further, to say that this process of iteration is valuable even without the intent to improve a model at all. Alison Booth eloquently celebrates the engagement in this imperfect and iterative process in her reflections on her project *Collective Biographies of Women*, a digital investigation leveraging both quantitative analyses and individual development of women's narratives: "It is exciting to collaborate on experimental humanities, recognizing the hand of the encoder, the confirmation of our bias, the beauty of big data and of lurking iotas, the vast spaces reduced to a map, and the notion of a metaphor" (Booth 2017, 625). Even as text inevitably defines the rules that computational processing would define for it, the act of witnessing that defiance helps expose our own assumptions about not only the text but also our process of reading it. The involvement of human hands in building a special-purpose analysis engine for the question at hand should be a feature, not a foible.

Though natural language processing tools still may be simply unsuited for many lines of text inquiry, I hope that those testing the waters of these approaches take some heart from the possible knowledge to be gained simply from the attempt, especially in the form of unresolved questions that may shape a future line of non-computational inquiry. Were the Engine of *Gulliver's Travels* real, I admit I would love to look through the volumes of transcriptions. I doubt I would learn any new science from the words the students transcribed, but I suspect that, from their choices, I might learn something about Lagado and its projectors.

REFERENCES

Anderson, Ashton, Dan Jurafsky, and Daniel A. McFarland. 2012. "Towards a Computational History of the ACL: 1980–2008." *Proceedings of the ACL 2012 Special Workshop on Rediscovering 50 Years of* Discoveries, 13–21. Jeju Island, Korea.

Arnold, Taylor and Lauren Tilton. 2015. *Humanities Data in R*. Cham: Springer International Publishing. https://r4thehumanities.home.blog/.
Bender, Emily M. and Alexander Koller. 2020. "Climbing towards NLU: On Meaning, Form, and Understanding in the Age of Data." *Proceedings of the 58th Annual Meeting of the Association for Computational Linguistics*, 5185–98. Stroudsburg, PA: Association of Computational Linguistics.
Bessette, Lee Skallerup, Katia Bowers, Maria Sachiko Cecire, Quinn Dombrowski, Anouk Lang, and Roopika Risam. 2019. *The Data-Sitters Club*. https://datasittersclub.github.io/site.
Booth, Alison. 2017. "Mid-Range Reading: Not a Manifesto." *PMLA/Publications of the Modern Language Association of America* 132 (3): 620–7.
Cordell, Ryan and Abby Mullen. 2017. "'Fugitive Verses': The Circulation of Poems in Nineteenth-Century American Newspapers." *American Periodicals* 27 (1): 29–52.
Da, Nan Z. 2019. "The Computational Case against Computational Literary Studies." *Critical Inquiry* 45 (3): 601–39.
Fish, Stanley. 1980. *Is There a Text in This Class? The Authority of Interpretive Communities*. Cambridge, MA: Harvard University Press.
Goldstone, Andrew and Ted Underwood. 2014. "The Quiet Transformations of Literary Studies: What Thirteen Thousand Scholars Could Tell Us." *New Literary History* 45 (3): 359–84.
Herbelot, Aurelie, Xiaodan Zhu, Alexis Palmer, Nathan Schneider, Jonathan May, and Ekaterina Shutova. 2020. "Proceedings of the Fourteenth Workshop on Semantic Evaluation." Barcelona, Spain: International Committee for Computational Linguistics.
Hovy, Dirk and Shannon L. Spruit. 2016. "The Social Impact of Natural Language Processing." *Proceedings of the 54th Annual Meeting of the Association for Computational Linguistics*. Berlin.
Jensen, Kim Ebensgaard. 2014. "Linguistics in the Digital Humanities: (Computational) Corpus Linguistics." *MedieKultur: Journal of Media and Communication Research* 30 (57).
Jockers, Matthew. 2014. *Text Analysis with R for Students of Literature*. Cham: Springer International Publishing.
Jones, Matthew L. 2016. *Reckoning with Matter: Calculating Machines, Innovation, and Thinking about Thinking from Pascal to Babbage*. Chicago, IL: University of Chicago Press.
Koehn, Philipp. 2005. "Europarl: A Parallel Corpus for Statistical Machine Translation." *Proceedings of the Tenth Machine Translation Summit (MT Summit X)*. Phuket, Thailand: Asia-Pacific Association for Machine Translation.
Kučera, Henry and W. Nelson Francis. 1967. *Computational Analysis of Present-Day American English*. Providence, RI: Brown University Press.
Piper, Andrew. 2018. *Enumerations: Data and Literary Study*. Chicago, IL: University of Chicago Press.
Piper, Andrew. 2019. "Do We Know What We Are Doing?" *Journal of Cultural Analytics* 5 (1): 11826. doi: 10.22148/001c.11826.
Rhody, Lisa M. 2017. "Beyond Darwinian Distance: Situating Distant Reading in a Feminist Ut Pictura Poesis Tradition." *PMLA/Publications of the Modern Language Association of America* 132 (3): 659–67.
Sedoc, João, Anna Rogers, Anna Rumshisky, and Shabnam Tafreshi. 2021. "Introduction." *Proceedings of the Second Workshop on Insights from Negative Results in NLP*, 3. Online and Punta Cana, Dominican Republic: Association of Computational Linguistics.
Smith, David, Ryan Cordell, and Elizabeth Maddock Dillon. 2013. "Infectious Texts: Modeling Text Reuse in Nineteenth-century Newspapers." *2013 IEEE International Conference on Big* Data, 86–94. IEEE.
So, Richard Jean. 2017. "All Models Are Wrong." *PMLA/Publications of the Modern Language Association of America* 132 (3): 668–73.
Swift, Jonathan. 1726. *Gulliver's Travels*. Project Gutenberg. www.gutenberg.org/ebooks/9272.
Walsh, Melanie, and Quinn Dombrowski. 2021. "Introduction to Cultural Analytics and Python." Vers. 1.1. August. doi: 10.5281/zenodo.5348222.
Wang, Alex, Amanpreet Singh, Julian Michael, Felix Hill, Omer Levy, and Samuel R. Bowman. 2019. "GLUE: A Multi-Task Benchmark and Analysis Platform for Natural Language Understanding." *Proceedings of the International Conference of Learning Representations*. New Orleans.
Weiss, Eric A. 1985. "Anecdotes [Jonathan Swift's Computing Invention]." *Annals of the History of Computing* 7 (2): 164–65. doi: 10.1109/MAHC.1985.10017.

CHAPTER SEVENTEEN

Analyzing Audio/Visual Data in the Digital Humanities

TAYLOR ARNOLD AND LAUREN TILTON
(UNIVERSITY OF RICHMOND, USA)

"With the inauguration of *Computers and the Humanities*, the time has perhaps arrived for a more serious look at the position of the humanistic scholar in the world of data processing," wrote English scholar Louis Milic in the inaugural issue of the journal (Milic 1966). The opening issue's invitation to participate immediately offered a capacious definition of humanistic inquiry. Under the editorship of Joseph Raben at CUNY's Queens College, the opening issue explicitly called for a broad definition that ranged from literature to music to the visual arts as well as "all phases of the social sciences that stress the humane." The centrality of fields such as music and art history were a given. "The music people have been the most imaginative," Milic argued. In the second issue, art historian Kenneth Lindsey laid out the state of "Art, Art History, and the Computer." "Within the past few years, we have witnessed the growth of interest in how sophisticated mechanical instruments can promote both the production of art and a better organisation of the data of art history," he wrote (Lindsay 1966).[1] Audio and visual work as a primary source and therefore as data was central to their configuration of DH, which was known at the time as humanities computing. The new journal sought to harness "the phenomenal growth of the computer during the past decade" and demonstrate what could be possible when researchers across disciplines created, analyzed, and communicated their work at the intersection of their object of study and computers. The first issues radiate with excitement.

Fast forward forty years, and the possibilities of the computer could not be more apparent as signaled by the flourishing of what is now labeled as Digital Humanities (DH). However, the institutional configurations that led to "Humanities Computing" and now "Digital Humanities" consolidated around text and specific fields such as literary studies and linguistics. The logocentrism of the field is best demonstrated by the form and content of the field's publications, which are an important and powerful gauge of credit and prestige because they reflect and shape a field's priorities. For example, the official journal of the European Association of Digital Humanities, *Digital Scholarship in the Humanities*, was titled *Literary and Linguistic Computing* from its founding in 1986 until 2015. *Computing in the Humanities* relaunched as *Language Resources and Evaluation* in 2005 and situated itself squarely in a particular configuration of linguistics (Ide 2005). They reflect the narrowed configuration of the field in disciplinary terms by the early 2000s.

"Text often remains the lingua franca" of DH, argued Tara McPherson (2009). "Digital humanities is text heavy, visualisation light, and simulation poor," stated Erik Champion (2016).

The centrality of text, specifically literary texts, to the term digital humanities led Lev Manovich to offer the term cultural analytics to signal media and contemporary digital culture as the object of study (Manovich 2020, 7). Of course, researchers have been working in these areas for decades. However, as humanities computing became digital humanities and then **D**igital **H**umanities with a capital D and capital H (Johnson 2018), the institutional structures from conferences (Weingart 2017) to journals (Sula and Hill 2019) to hiring have circulated around text and text-heavy fields. Yet, the tide is changing.

We offer an observation, perhaps even provocation: DH is undergoing an A/V turn. The combination of access to data either through digitization or born-digital sources alongside advances in computing from memory to deep learning is resulting in a watershed moment for the analysis of audio and visual data in the field. Decades of work have created streams including scholarship from academic disciplines such as media studies, new and expanded publication avenues such as the *International Journal for Digital Art History* and the renaming of journals like *Literary and Linguistic Computing* (LLC) to *Digital Scholarship in the Humanities* (DSH), and institutional shifts such as the founding of ADHO's AVinDH SIG in 2014. Colleagues are advocating for DH within the structures of another institutional configuration such as specific academic disciplines and advocating within the institutional structures that form the field of Digital Humanities to see and feature A/V work. Our collective efforts to remove dams and enable streams across institutional structures by individuals, groups, and organizations are making space for audio visual data as an object of study as well as a form of scholarship.

The A/V turn brings together two streams of work in DH. The first is what one could call an audio or sound turn. While work on sound is not new in DH, in fact it enjoyed a great deal of prominence in the 1960s and 1970s, the last decade has seen a renewed interest (Sula and Hill 2019). Work includes commitments to sound archives alongside new DH methods such as distant listening. Exciting new publications such as Lingold, Mueller, and Trentien's 2018 edited volume *Digital Sounds Studies* signal the emergence of "a new and interdisciplinary field … at the intersection of sound studies and digital humanities," as Tara McPherson argues (121). While this area of work has not been articulated as a turn, the call to participate and listen is amplifying. As the co-editors write, "we need to bring the insights of sound studies to bear upon the emergent field of digital humanities" (Lingold, Mueller, and Trentien 2018, viii) and vice versa.

The second is the visual turn. In the last few years, there has been a combination of observations and calls claiming that DH is undergoing such a turn. Münster and Terras offer a survey on topics, researchers, and epistemic cultures in what they term the "visual digital humanities" (2020). Wevers and Smits argue that the "visual digital turn" is underway thanks to the affordances of computational advances in computer vision, specifically deep learning and neural networks (2020). For over a decade, media studies scholars such as Lev Manovich and Tara McPherson alongside digital art historians have pioneered the analytical possibilities of visual culture and DH, thereby forging the visual turn. Our own work has been a part of both turns as exemplified by the special issue of *Digital Humanities Quarterly* on A/V data that we co-edited (Arnold, van Gorp, Scagliola, and Tilton 2021). In this chapter, we make explicit what was implicit in our previous work. Bringing both calls together, we argue that an A/V turn is underway.

In the sections that follow, we begin to lay out a topology of A/V work in DH. The next two sections explore developments that have supported A/V data as a primary source. We begin with digitization and annotation and then turn to machine learning and deep learning. The following section focuses on A/V as a form of scholarship. Formats such as mixtapes and video essays offer audio and visual ways to produce and communicate scholarship. We end with a return to our provocation and a brief discussion of the challenges and opportunities moving forward.

DIGITIZATION AND ANNOTATION

For decades, digitization and annotation have been instrumental to the field and opening up A/V data as a primary source of research. From archives to institutions of higher education to community-based nonprofits, groups are developing a steady stream of digital materials. How to organize, discover, and analyze through the affordances of digital technologies is a flourishing area of DH. The development of archives and collections as digital public humanities projects is a reason for and result of digitization. Along with facilitating access, a popular area is the development of annotation tools for adding context and information retrieval (Pustu-Iren, Sittel, Mauer, Bulgakowa, and Ewerth 2020). The development of approaches and tools for custom annotation facilitates close analysis for discipline deep-learning specific inquiry and pedagogy.

A prominent area of work is the digitization of materials. Goals include building digital archives and projects, facilitating access and discovery, and developing datasets for computational analysis. Major sites for digitization are galleries, libraries, archives, and museums, often referred to as GLAM institutions. Their commitment to access and discovery has been foundational to work in A/V in DH. They have pioneered digitization processes and models, researched and promoted the importance of collections, developed metadata frameworks such as schemas and ontologies that DH tools like *Omeka* rely on, and led initiatives to foster open access. For example, the National Film Archive of India has been collecting, researching, and promoting the importance of Indian cinema since 1964. The Akkasah: Center for Photography in Abu Dhabi is working to preserve the visual history of the Arab world while demonstrating how photography has shaped images of the region for over a century. Since 1926, the United States Library of Congress has been collecting sound recordings and remains a pioneer of preservation and digitization methods through the Motion Picture, Broadcasting, and Recorded Sound Division. The final example is the Rijksmuseum, which has set the standard for open access to digital images of museum holdings through their commitment to open access data, with major institutions like the United States Smithsonian Institution following suit.[2] The increased availability of A/V materials and data is spurring research by making A/V data accessible.

As access to digitization equipment and funding has expanded and become less cost prohibitive, digitization and related digital projects are also finding homes in nonprofit organizations and within academia.[3] Given the long history of A/V materials as secondary objects to GLAM institutions, efforts to collect, preserve, and digitize these materials also come from within academia. For example, the Southern Oral History Program at UNC-CH has been digitizing their oral histories of the American South that date back to 1973 to make them accessible. Started in 2015, *PodcastRE* is creating a "searchable, researchable archive of podcasting culture" and is based at the University of Wisconsin-Madison (Hoyt, Bersch, Noh, Hansen, Mertens, and Morris 2021). Among the most ambitious efforts is the *Archives Unleashed Project* housed at the University of Waterloo, which has been making petabytes of historical Internet content accessible for study since 2017. Projects such as these are demonstrating how DH is a key interlocutor in the creation, access, and discovery of audio and visual data and facilitating scholarship while expanding what counts as scholarship and scholarly work in academia.

Efforts to build digital archives and collections around a particular topic now garner academic credibility. One form of these initiatives is digital public humanities projects (Brennan 2016). These projects are built around access, discovery, and interpretation, and designed to be accessible to a particular public. Such work is indebted to areas such as archive studies that have long argued that archives and collections are sites of power that shape memory and thereby which stories are told. DH researchers are coming together, often into teams built around multi-institutional collaborations, to create digital public projects around a shared area of study. Examples include the *Amateur Cinema Database*, *Fashion and Race Database*, and *New Roots/Nueva Raíces: Voices from Carolina del Norte* founded in 2017, 2017, and 2006 respectively as well as the quickly growing Covid projects around the world such as *covidmemory.lu* from C2DH.[4] Importantly, tools developed by DH researchers have responded to and facilitated the development of this area. Among the most prominent is *Omeka*, which is an open-source publishing platform designed for creating archives, scholarly collections, and exhibitions. A repertoire of plugins includes functions for image annotators, mapping, and multimedia.

The final example we draw on is from our own work creating *Photogrammar*, which started in 2010. The project uses interactive maps and graphs to facilitate access and discovery of the 170,000 documentary photographs created during the Great Depression and the Second World War by a unit funded by the United States federal government. The Historic Unit of the Farm Security Administration and then Office of War Information, known as the FSA, offers a view into the era through the lens of the state as well as several of the most acclaimed documentary photographers of the twentieth century. We then use the same features to make scholarly arguments about the collection (Arnold, Leonard, Tilton 2017; Cox and Tilton 2019). Our remixing of the archive through data visualizations is designed to communicate new scholarship about the collection while creating space for users to pursue scholarly inquiry based on their questions.

Another aspect of DH that has relied on archives and digitization is annotation. The approach is driven by a plethora of analytical purposes. Among them are disciplinary-specific analysis, pedagogy, and information retrieval. For example, art historians alongside film and media scholars regularly conduct formal analysis. While art historians harness annotation to study composition and iconography, film and media studies scholars study mise-en-scene alongside features such as shot length and cuts. In the classroom, annotation tools facilitate close readings as participants use the built-in features and categories to learn how to engage in close analysis. Another line of work is using these annotation tools to develop metadata that facilitates access and discovery of A/V material through information retrieval systems such as browse and search.

A major area of scholarship in DH is the development of digital tools to support annotation. A prominent area is film and media annotation tools given the field's focus on close analysis. Building off of linguistic annotation tools like *ELAN*, teams like the *Film Colors Project* have developed tools for the annotation of color in film over the last decade. Their work is part of a large ecosystem of projects such as *Mediate* and the *Semantic Annotation Tool* (SAT) that are developing tools specifically at the intersection of film and media studies and DH for close analysis and media literacy (Burges et al. 2021; Williams and Bell 2021). Many of these are also designed with pedagogical goals.

Another prominent application is sound annotation. Tools like the *Oral History Metadata Synchronizer* (OHMS), which was started in 2008, facilitate the creation of transcripts and metadata from oral interviews. The tool can then be integrated with *Omeka*, which brings us full circle. *Omeka* facilitates the building of a collection, and participants can use the OHMS module to further develop metadata that can then facilitate access and discovery of the collection. The combination brings archive building and annotation together, often through digital public projects such as the Bluegrass Music Hall of Fame and Museum's oral history project and digital exhibit, which is a partnership with the University of Kentucky's Louie B. Nunn Center for Oral History. While we have mostly focused on manual annotation, newer tools such as our *Distant Viewing Toolkit* are now looking to scale up the analysis through machine learning, which we will turn to in the next section.

MACHINE LEARNING/DEEP LEARNING

Thinking of digitized and born-digital items such as music, paintings, and photographs as data has not been an intuitive move in the humanities (Posner 2015). Yet, many of the fields that DH draws on and is in conversation with, such as statistics, use this terminology (Arnold and Tilton 2019). One outcome is researchers involved in DH are increasingly thinking of the primary sources we study as data. The very addition of a "Data Sets" section in *Journal of Cultural Analytics* is a testament to this shift. The journal is also indicative of the prominence of machine learning in DH, which has not been without challenges.

The challenge has been twofold. One is the development of the data and metadata to study. For decades, researchers have been working to digitize and make features they want to study machine-readable, primarily through the construction of metadata or translation of a feature into text. Linguistic scholars have been at the fore, by manually tagging formal elements of spoken and written language through the construction of encoded corpora (Burnard and Aston 1998; Davies 2010). Manual annotation and related tools have been pivotal. The second development is the form of data that computers can process. Advances in storage and processing combined with new computational methods are opening up the scale of analysis as well as what can be analyzed. The advances in audio and image processing combined with the last several decades of digitization have opened new possibilities for large-scale data analysis. One of the most recent additions to the computational repertoire—machine learning and specifically deep learning—are facilitating the A/V turn in DH.

Machine learning involves the use of data to create and improve algorithms. Approaches such as topic modeling, named-entity recognition, and principal component analysis have become commonplace among DH scholars in areas such as computational literary analysis, visual digital humanities, and digital sound studies. A subset of machine learning in artificial intelligence known as deep learning has had a particularly large impact in opening areas of research in A/V DH (Arnold and Tilton 2021). They are expanding the scope and scale of analysis, particularly the kinds of evidence (Piper 2020). For example, art historians studying iconography in art through object detection, media studies scholars studying visual style and narrative structure in TV using face recognition, and historians using deep-learning algorithms to extract machine-readable text from scans of old newspaper pages.

An exciting development is that DH researchers are not just adopting these methods and their embedded theories as is but developing theories and frameworks that draw on and build off machine learning. We are now seeing the development of DH frameworks, methods, and theories that bring together scholarship across fields such as computer science, film studies, media studies, and sound studies under new transdisciplinary spaces, including cultural analytics, distant listening, and distant viewing (Clement 2013a; Bermeitinger et al. 2019). They have led to calls for tools that reflect the interpretative possibilities and commitments of these theories and methods such as *ImagePlot*, *HIPSTAS* (Clement 2013b), and the *Distant Viewing Toolkit* (Arnold and Tilton 2020).

As an example of how researchers are combining method and theory to formulate scholarship that emerges specifically from DH, we turn to our own work. We situate distant viewing as a theory and method for the critical use of computer vision to study a corpus of (moving) images. The *Distant Viewing Toolkit* is built to facilitate two of the four parts of our DV method: annotate and organize. Rather than manual annotation, we draw on machine learning to automatically annotate at scale to pursue our scholarly questions about images, specifically twentieth- and twenty-first-century visual culture from the United States in our case. Along with building custom annotators, we then aggregate the annotations into features such as the categorization of shot length and detection of film cuts. In other words, the toolkit enacts the theory and facilitates the method. Such developments demonstrate how DH scholars are not only borrowing, but adjusting, remixing, and building approaches to machine learning embedded with and facilitating the kind of scholarly inquiry that animates the humanities. This scholarship is also a part of expanding the scope of scholarship in the field, from traditional articles to software, and joins an even larger ecosystem of DH researchers forging new paths of scholarly argumentation.

FORMS OF SCHOLARSHIP

An exciting area of A/V work is how researchers are pushing the forms of scholarship in the field. As scholars ranging from rhetoric to media studies argue, *how* we make our claims shapes the argument. Or, as Marshall McLuhan famously stated, the medium is the message. Or, at least, a part of the message. Exploration of how one can use A/V to argue and communicate insights about A/V is an exciting area in DH, which is forging an expanded notion of what counts as DH.

Work in this area comes in audio, visual, and audio visual formats. As researchers have long lamented, forms such as text articles and books often reduce the aural and visual complexity of A/V to words. There are times when those words approximate the explosion of meaning and feeling that a piece of art or stanza of music can elicit, but they cannot replicate the embodied process of knowing that seeing and hearing garner. A growing area of work now draws on the forms that we employ to communicate our scholarship.

Two areas include audio, specifically podcasting and mixtapes, and moving images, specifically videographic criticism. Along with being a form to reach wider audiences, podcasting is providing a space to weave in the sounds that are often the object of study. The rise of podcasting has created an exciting space to literally voice and listen. Podcasts like *Books Aren't Dead*, which is a part of the Fembot Collective, are using audio technology to produce intersectional feminist DH (Edwards and Hershkowitz 2021). Scholars such as Tyechia Thompson are using DH in the form of pedagogical mixtapes to study and communicate knowledge about Afrofuturism (Thompson and Dash 2021). The second area comprises video essays and videographic criticism. As scholars such as Jason Mittell have argued, videographic criticism is a DH method to study moving images and sound which then uses the same forms to convey scholarly arguments (Mittell 2019, 2021). DH on A/V is pushing boundaries by expanding our forms of scholarship and *how* we communicate scholarship in new and more expansive directions.

The final section that we will highlight is visualization. We return to Eric Champion's point from 2016 that DH is "visualisation light and simulation poor." An exciting development is that this is changing as forms of scholarship such as interactive visualizations become more prominent. The change is the result of a combination of factors that include new frameworks like React.js that have lowered the barrier to building interactive graphics, accessing A/V data, and increased credit in the academy that shapes which forms and kinds of scholarship are prioritized. Publications such as *Reviews in DH*, more capacious ideas of peer review such as grants, and awards like the American Historical Associations' Rosenzweig Prize are expanding the kinds of scholarship that DH counts. Interactive data visualization projects such as American Panorama's *Mapping Inequality* and *Kinometrics*, interactive 3D visualizations such as *Virtual Angkor Wat* and *Victoria's Lost Pavilion*, and tools such as *Image Plot* are just several examples of projects that are pushing the boundaries of visualization as a way of knowing and form of scholarship in DH.

CONCLUSION

The sections above are just a few of the parts of the topology of A/V in DH. It is not exhaustive and reflects our angle of view, which comes with its own limits due to our positionality and subjectivity. However, we see this work as a part of recognizing and naming an exciting development in the field. The work outlined above alongside those that have been featured and organized in other spaces such as *Digital Humanities Quarterly* support our observation that DH is undergoing an A/V turn. It is an exciting time.

We end by returning to our framing of the A/V turn as a provocation. In spaces such as conferences, we have argued that DH has been focused on text and left little space for other ways of knowing, such as images, sound, and time-based media. The call has been met with discomfort and even hostility. Often, it is interpreted as a threat to text-based scholarship and specifically text analysis, which has enjoyed copious space in formal DH structures such as conferences and journals. We love this area of DH, and actively engage with it. However, we are calling for a bit more space for other kinds of knowledge and ways of knowing in these same DH structures. Often, there is pressure not to provoke but to be "nice," which has created an affective and discursive culture where telling others to be nicer is a way to diminish or silence other calls within the field as we have seen with efforts such as #transformDH and #pocoDH. We think the tension that our provocation may produce can be productive and generative for it calls into question the status quo and offers other possible directions for the field.

So, we end with calling not just for more space, but for an A/V turn in DH. The consolidation of the field around text by the early 2000s led scholars like McPherson and Manovich to call for fields such as film studies and new frames such as cultural analytics to disrupt the logocentrism of DH. This groundwork combined with technological, institutional, and financial developments has facilitated a quickly growing area of scholarship around A/V in DH. The development and analysis of A/V data, and therefore its use as a primary source, as well as the creation of A/V work as a form and medium of scholarly interpretation and argumentation are two ways that DH is seeing an A/V turn. There is more work to do to imagine the areas of inquiry we could address and questions we could ask. What version of DH could we build if we moved beyond the field's logocentrism?

What kinds of scholarship could we pursue? What kinds of collaborations and partnerships could we imagine? What if we then thought across text, image, and sound? What version of the field could we create? We won't know unless we make more spaces to hear from other parts of DH as well as expand what counts. Let's explore this turn together.

NOTES

1. For a longer history of the relationship between Art History and DH in the twentieth century, see Greenhalgh (2004).
2. In 2016, the Riksmuseum announced that 250,000 images of artworks would be released for free.
3. For example, see projects such as the Early African Film Database, the Amateur Movie Database, and Cinema, and the Santa Barbara Corpus of Spoken American English.
4. To visit the projects, visit http://www.amateurcinema.org, https://fashionandrace.org, and https://newroots.lib.unc.edu (links are active as of December 2021).

REFERENCES

Arnold, Taylor and Lauren Tilton. 2019. "New Data: The Role of Statistics in DH." In *Debates in the Digital Humanities 2019*, edited by Matthew Gold and Lauren Klein. Minneapolis, MN: University of Minnesota Press.

Arnold, Taylor and Lauren Tilton. 2020. "Distant Viewing Toolkit: A Python Package for the Analysis of Visual Culture." *Journal of Open Source Software* 5 (45): 1800.

Arnold, Taylor and Lauren Tilton. 2021. "Depth in Deep Learning: Knowledgeable, Layered, Impenetrable." In *Deep Mediations: Thinking Space in Cinema and Digital Cultures*, edited by Karen Redrobe and Jeff Schrebel, 309–28. Minneapolis, MN: University of Minnesota Press.

Arnold, Taylor, Peter Leonard, and Lauren Tilton. 2017. "Knowledge Creation through Recommender Systems." *Digital Scholarship in the Humanities* 32 (3): 151–7.

Arnold, Taylor, Jasmijn van Gorp, Stefania Scagliola, and Lauren Tilton. 2021. "Introduction: Special Issue on AudioVisual Data in DH." *Digital Humanities Quarterly* 15 (1).

Bermeitinger, Bernhard, Sebastian Gassner, Siegfried Handschuh, Gernot Howanitz, Erik Radisch, and Malte Rehbein. 2019. "Deep Watching: Towards New Methods of Analyzing Visual Media in Cultural Studies." White Paper.

Brennan, Sheila. 2016. "Public, First." In *Debates in the Digital Humanities 2016*, edited by Matthew Gold and Lauren Klein. Minneapolis, MN: University of Minnesota Press.

Burges, Joel, Solvegia Armoskaite, Tiamat Fox, Darren Mueller, Joshua Romphf, Emily Sherwood, and Madeline Ullrich. 2021. "Audiovisualities out of Annotation: Three Case Studies in Teaching Digital Annotation with Mediate." *Digital Humanities Quarterly* 15 (1).

Burnard, Lou and Guy Aston. 1998. *The BNC Handbook: Exploring the British National Corpus*. Edinburgh: Edinburgh University Press.

Champion, Erik. 2016. "Digital Humanities is Text Heavy, Visualization Light, and Simulation Poor." *Digital Scholarship in the Humanities*, 32: i25–i32.

Clement, Tanya. 2013a. "Introducing High Performance Sound Technologies for Access and Scholarship." *The International Association of Sound and Audiovisual Archives Journal* 41: 21–8.

Clement, Tanya. 2013b. "Distant Listening or Playing Visualisations Pleasantly with the Eyes and Ears." *Digital Studies/Le champ numérique* 3 (2).

Cox, Jordana and Lauren Tilton. 2019. "The Digital Public Humanities: Giving New Arguments and New Ways to Argue." *Review of Communication* 19 (2): 127–46.

Davies, Mark. 2010. "The Corpus of Contemporary American English as the First Reliable Monitor Corpus of English." *Literary and Linguistic Computing* 25 (4): 447–65.

Edwards, Emily and Robin Hershkowitz. 2021. "Books Aren't Dead: Resurrecting Audio Technology and Feminist Digital Humanities Approaches to Publication and Authorship." *Digital Humanities Quarterly*, 15 (1).

Greenhalgh, Michael. 2004. "Art History." In *Companion to Digital Humanities*, edited by Susan Schreibman, Ray Siemens, and John Unsworth, 31–45. Oxford: Blackwell.

Hoyt, Eric, J. J. Bersch, Susan Noh, Samuel Hansen, Jacob Mertens, and Jeremy Wade Morris. 2021. "PodcastRE Analytics: Using RSS to Study the Cultures and Norms of Podcasting." *Digital Humanities Quarterly* 15 (1).

Ide, Nancy and Nicoletta Calzolari. 2005. "Introduction to the Special Inaugural Issue." *Language Resources and Evaluation* 39 (1): 1–7.

Johnson, Jessica Marie. 2018. "4DH + 1 Black Code / Black Femme Forms of Knowledge and Practice." *American Quarterly* 70 (3): 665–70.

Lindsay, Kenneth C. 1966. "Art, Art History, and the Computer." *Computers and the Humanities* 1 (2): 27–30.

Lingold, Mary Caton, Darren Mueller, and Whitney Trentien. 2018. *Digital Sounds Studies*. Durham, NC: Duke University Press.

Manovich, Lev. 2020. *Cultural Analytics*. Cambridge, MA: MIT Press.

McPherson, Tara. 2009. "Introduction: Media Studies and the Digital Humanities." *Cinema Journal* 48 (2) (Winter): 119–23.

Milic, Louis T. 1966. "The Next Step." *Computers and the Humanities* 1 (1): 3–6.

Mittell, Jason. 2019. "Videographic Criticism as a Digital Humanities Method." In *Debates in the Digital Humanities*, edited by Matthew Gold and Lauren Klein. 224–42. Minneapolis, MN: University of Minnesota Press.

Mittell, Jason. 2021. "Deformin' in the Rain: How (and Why) to Break a Classic Film." *Digital Humanities Quarterly* 15 (1).

Münster, Sander and Melissa Terras. 2020. "The Visual Side of Digital Humanities: A survey on Topics, Researchers, and Epistemic Cultures." *Digital Scholarship in the Humanities* 35 (2): 366–89.

Piper, Andrew. 2020. *Can We Be Wrong? The Problem of Textual Evidence in a Time of Data*. Cambridge: Cambridge University Press.

Posner, Miriam. 2015. "Humanities Data: A Necessary Contradiction." Self-published blog post. http://miriamposner.com/blog/humanities-data-a-necessary-contradiction/.

Pustu-Iren, Kader, Julian Sittel, Roman Mauer, Oksana Bulgakowa, and Ralph Ewerth. 2020. "Automated Visual Content Analysis for Film Studies: Current Status and Challenges." *Digital Humanities Quarterly* 14 (4).

Sula, Chris Alen and Heather V. Hill. 2019. "The Early History of Digital Humanities: An Analysis of Computers and the Humanities (1966–2004) and Literary and Linguistic Computing (1986–2004)." *Digital Scholarship in the Humanities* 34 (1): i190–i206.

Thompson, Tyechia L. and Dashiel Carrera. 2021. "Afrofuturist Intellectual Mixtapes: A Classroom Case Study." *Digital Humanities Quarterly* 15 (1).

Weingart, Scott and Nickoal Eichmann-Kalwara. 2017. "What's under the Big Tent? A Study of ADHO Conference Abstracts." *Digital Studies/Le champ numérique* 7 (1): 1–17.

Wevers, Melvin and Thomas Smits. 2020. "The Visual Digital Turn: Using Neural Networks to Study Historical Images." *Digital Scholarship in the Humanities* 35 (1): 194–207.

Williams, Mark and John Bell. 2021. "The Media Ecology Project: Collaborative DH Synergies to Produce New Research in Visual Culture History." *Digital Humanities Quarterly* 15 (1).

CHAPTER EIGHTEEN

Social Media, Research, and the Digital Humanities

NAOMI WELLS (SCHOOL OF ADVANCED STUDY, UNIVERSITY OF LONDON)

Social media is an increasing if still relatively marginal focus of research in the Digital Humanities (DH), as reflected in recent conferences in the field. Nevertheless, particularly given the ways media and communication studies have taken the lead in establishing the theoretical and methodological foundations for social media research, critical questions remain about what digital humanists can bring to the study of social media. With those humanities researchers who do engage with social media often influenced by qualitative approaches such as digital ethnography, social media research further adds new complexities to discussions about what we understand by the digital in relation to the humanities, and more specifically the extent to which the field is or should be defined by the use of digital, or more specifically computational, methods.

This chapter addresses these often implicit, and at times explicit, questions that risk positioning social media researchers in the humanities on the periphery of the field. While not always productive to overly define, or at worse police, disciplinary boundaries, it is nevertheless important to acknowledge the uncertain position of social media research in DH and to attempt to uncover some of the dynamics that underlie this reality. This chapter is addressed both at the broader humanities community and those already located within DH structures, in recognition of the fact that despite DH now becoming a more established field, there remains a continued need to further expand it in ways that broaden the perspectives on and approaches to digital research in the humanities. The ever increasing uses of social media—as well as its archival instability that may mean significant content is no longer available to study in the near future—highlight the urgent task of rapidly growing the field of humanities researchers from both within and beyond DH who are willing and able to apply their expertise to these contemporary cultural materials.

TAKING SOCIAL MEDIA SERIOUSLY AS CULTURAL TEXTS

In terms of the wider humanities community and its traditional objects of study, one of the primary obstacles to engagement with social media materials continues to be a hesitancy concerning the legitimacy of social media content as cultural texts deserving of serious consideration and analysis. This may appear to be an overly pessimistic view of the field, and it is interesting to note how in relation to video games, Coltrain and Ramsay present a more optimistic viewpoint: "in the end, it was not hard for an academy well adjusted to the long shadow of cultural studies to conform itself to these new objects of inquiry" (2019, 36). While I would question their more positive framing of the academy, I agree that it is cultural studies we should (re)turn to when articulating the necessity for humanities researchers to engage with these new forms of digital text that play a central role in any attempt to define or understand contemporary culture.

Indeed, when attempting to explain why we should take social media seriously as cultural texts, it can feel like rehearsing the same arguments that arose when cultural studies established itself as a discipline in the 1960s and 1970s.[1] The fact that similar discussions are taking place highlights how the recent rapid expansion in the uses of digital media and technologies, and the processes of social and cultural change this has triggered, may reflect a similar turning point as that identified by Hall in post-war Britain, when new forms of mass culture associated also with technological change provoked what he described as a "crisis in the humanities" (1990).

More specifically, contemporary digital culture once again calls on us to reflect on our understandings and definitions of culture. The participatory nature of social media and the lower barriers to "public artistic expression and civic engagement" (Giaccardi 2012, 3) have provoked an increased blurring of more professional or intentional forms of cultural production and more anthropological "everyday" forms of cultural practice and texts. Significantly, at least in my own disciplinary area of Modern Languages, many of those who have engaged with social media, including myself, often originate from a primarily linguistics background. The probable explanation is that more "everyday" cultural and linguistic practices are a more established object of study for linguists. Nevertheless, while linguists certainly bring an important perspective, researchers with a background of studying more formal cultural production associated with literature and the arts have much to contribute to the study of social media.

Social media platforms have become an important space in which professional, semi-professional, and more grassroots forms of cultural production are published, circulated, and remediated. The rich intertextuality of social media content (Varis and Blommaert 2018), as well as the ways social media users evaluate and make use of the distinct aesthetic and narrative affordances of different platforms, further point to the potential contributions of humanities researchers in offering valuable insight into the cultural practices and texts associated with social media. This is not, however, to attempt to collapse different cultural forms and practices, or to make glib equivalences between a novel crafted over several months or years, and short tweets rapidly composed and sent out over the course of a day. Rather, the aim is to better acknowledge and understand the undeniably important cultural "work" that occurs in the everyday, even if it is undeniably distinct in both form and content, as well as to recognize that literary and other arts must be understood in relation to everyday practices (Puri 2016, 40). Equally, recognizing the cultural value of social media content should not be mistaken for an idealistic lens that erases the social and material conditions in which social media texts are created or, as with the earlier equally contentious study of popular culture (Hall [1981] 2020), that ignores the extent to which social media content may conform to or perpetuate imperialist and capitalist cultural forms and ideologies.

At this point, it is important to reflect on what the relationship of social media research in DH might be with the more established body of research in (new) media and communication studies. In reality, the boundaries may well be necessarily blurred, with cultural studies often underpinning work in that field and media studies operating similarly at the intersection of the humanities and social sciences. Any attempt to articulate or distinguish between a humanities, media studies or social sciences approach will inevitably be partial and cannot entirely avoid simplistic reductions of complex heterogeneous fields. Nevertheless, while allowing for, and indeed advocating for, porous disciplinary boundaries rather than fixed or rigid definitions, it is I believe still productive to identify broad trends and the different emphases that researchers' distinct disciplinary training and analytical frameworks bring to similar objects of study.

As Moreshead and Salter highlight in this volume (Chapter 9), media studies pays particular attention to infrastructural changes, which in social media research is reflected in an emphasis on platforms and their features or "affordances," with a particularly important body of critical platform studies research that addresses the capitalist logics that underpin them (see for example Fuchs 2015). While there is also a body of media studies research focusing more on the content produced through these platforms, a stronger humanities emphasis on social media "texts" has the potential to significantly expand our understanding of the different ways groups and individuals appropriate these platforms for distinct representational purposes. By text here, I refer not specifically to the textual, but rather to a more expansive understanding of cultural text that encompasses audio, visual, and multimodal texts. In his own reflections on the relationship between DH and media studies, Sayers argues that engaging with media studies encourages the humanities to go "beyond text" (2018, 4), although he appears to be referring more specifically to the textual, rather than the broader notion of cultural text. Nevertheless, in the rush to emphasize the novelty and distinctiveness of social media—as the term "new media studies" makes most explicit—what may have at times been obscured is the extent of continuity with earlier cultural texts, including the "purely" textual,[2] and more importantly the continued relevance of existing theoretical and analytical frameworks in the humanities.

This is not to deny the value of (critical) platform studies, and in particular the strong emphasis in media studies on understanding the specific affordances of different social media platforms (Bucher and Helmond 2018), which undeniably play an important role in facilitating or constraining the production of different forms and types of social media text. As with the study of electronic literature in DH, the multimodal affordances of social media platforms challenge humanities researchers who previously worked on predominantly textual material to develop new interpretive frameworks that encompass the "rich combinations of semiotic modes" in social media texts (Jones, Chik, and Hafner 2015, 7). Equally, as Moreshead and Salter emphasize, it is vital for DH researchers to acknowledge and confront the role of the algorithmic in constructing digital spaces and constraining who can participate in them. A humanities emphasis on texts should consequently not mean extracting social media content from its contexts of production and reception, or ignoring the specificities of these new cultural materials. Instead, a humanities lens that draws on both this existing body of platform-focused research and its own history of theoretical and analytical approaches to the study of cultural texts has the potential to significantly expand our understanding of the different uses groups and individuals make of these platforms.

More specifically, humanities scholars may bring new perspectives to social media content that challenge the emphasis on social media as primarily communicative tools. Again, social media undeniably performs extremely important communicative roles, but the dominance of a communication studies framing risks obscuring the importance of social media platforms as spaces of cultural representation. Discussions of social media affordances, for example, often refer primarily to "communicative" or "social" affordances, rather than the often implicit but less explored representational affordances of different platforms. Communication and representation are inevitably blurred concepts on social media and in everyday life, but humanities researchers who bring a stronger emphasis on representational practices in literature, art, or other cultural forms are likely to bring a different lens to these materials that foregrounds, for example, aesthetic and narrative choices. Again, such research should be understood as complementary rather than as competing with earlier communication studies research. The intent here is not to critique media and communication studies approaches in themselves, but rather to highlight that humanities researchers have an under-explored potential to bring different emphases and analytical lenses to their counterparts in other fields. More importantly, rather than an attempt to substitute media and communication studies perspectives, cross-disciplinary forums and collaborations between researchers working across these fields can together provide a more holistic understanding of the multiplicity of cultural, social, political, and economic functions of social media platforms and texts.

It is important at this juncture to acknowledge work that exemplifies this more humanities-based approach to researching social media. In my own disciplinary area of Modern Languages, for example, researchers such as Ogden (2021) and Sadler (2018) exemplify the ways language- and culture-focused social media research can simultaneously address both traditional humanities themes of narrativity and visual representation, while equally considering questions of power and the specific conditions of production and reception of social media content.

Furthermore, there is a growing emphasis on understanding social media content as a form of born-digital cultural heritage, which can help to foreground connections with key areas of humanities research such as cultural memory studies. In particular, the focus on social media as a form of heritage provokes questions concerning the longer-term preservation of social media content that memory institutions have begun to grapple with, in collaboration with the community of Internet history and web archive researchers that sit at least partially within DH structures.[3] Nevertheless, further expanding the contributions of humanities researchers to exploring the cultural significance of social media materials, particularly in relation to questions of heritage and memory, is vital for supporting the work of those attempting to archive social media materials. In particular, a stronger body of humanities research on social media could play a vital role both in terms of strengthening the case for the preservation of these materials, and in terms of supporting and advising archivists on the extremely challenging but unavoidable task of selecting materials of particular significance for preservation from the vast range of social media content potentially available.[4] In turn, the work of (web) archivists can help to further social media research, with archival preservation another route to ensuring wider acceptance of the cultural significance of social media materials, particularly for the humanities community given its close associations with forms of archival research. Nevertheless, given the challenges associated with archiving social media at scale (Thomson 2016),[5] it is vital that such archiving initiatives are developed alongside a more expansive body of humanities research on these materials in their "live" form.

METHODOLOGICAL REFLEXIVITY IN DH

To turn to the position of social media research within DH structures, one of the key questions it raises is in relation to how humanities researchers who study social media may engage in a more limited way with the types of computational methods that at least for some researchers are considered integral to the field's disciplinary identity. This is not to say that it is not possible for humanists to apply computational methods to social media content in productive and valuable ways. However, what I want to suggest is that the use of computational methods should not be considered a prerequisite for DH research. Here then I would suggest that while Sayers's explanation that "if media studies is *about* media and technologies, then digital humanities works *with* them" (2018, 1) undeniably reflects earlier trends in both fields, DH should be sufficiently expansive to more explicitly incorporate both those who apply digital methods to texts that previously existed in non-digital forms and those who apply more traditional humanities approaches to born-digital materials.[6]

This is not to argue that methodologies are not an important aspect of disciplinary identity, and one of the valuable contributions that DH has brought to the humanities community is to prompt greater reflection on humanities methodologies more broadly. In particular, the need to justify the applications of computational methods to humanities research, while inevitably at times frustrating, has led to a rich body of work that reflects more explicitly on the value and implications of different methodological approaches in the humanities. What I consequently want to propose is that, rather than DH being defined by the application of a specific and clearly defined set of methodologies, we should instead emphasize a heightened methodological reflexivity as a defining strength of the discipline in ways that still allow for an openness of approaches to digital research.

To apply this explicitly to social media materials, while I have earlier emphasized continuities with earlier cultural forms, this is not to deny the specificity of these materials and the need to think critically about the effects of transferring and applying specific methods and approaches to them. In particular, and returning to the question of the doubted cultural value of born-digital social media content, this distinguishes social media texts from more traditional literary or historical objects of study. When applying computational methods to literary or historical texts, the cultural value or significance of these texts is often taken for granted. This is partly due to the prestige associated with certain cultural forms in the case of literature or with the accrual of value over time for more "everyday" historical documents that is often reinforced by archival preservation. Computational work on these materials also often built on a pre-existing body of more traditional qualitative humanities research that further helped to demonstrate the cultural value of these materials.

What I want to emphasize here is that, before rushing to transfer the same DH computational methods over to social media texts, it is important to reflect on the potential effects this may have for materials whose cultural value continues to be questioned and where computational research is not building on an existing body of qualitative humanities research. There are of course also practical obstacles to easily applying computational approaches to social media content due to the limits placed by many platforms on the forms of scraping necessary for large-scale computational research. Nevertheless, given the scale of social media data and the emphasis on metrics underpinning many of these platforms (Grosser 2014), large-scale computational approaches may in fact appear to be the more obvious or even conventional approach to studying these materials. Again, this distinguishes social media content from earlier literary or historical objects of study, where the groundbreaking work of digital humanists was to apply new computational methods in ways that challenged, or at least proposed an alternative to, the more conventional ways of studying such texts. In contrast, the application of computational methods to social media research in the humanities carries with it the risk of conforming to, and potentially even reinforcing, the corporate logics that underpin these platforms. There are of course still critical ways of applying computational methods to social media content, but it may be the application of seemingly "traditional" smaller-scale humanities approaches such as close reading that goes against the grain of the seemingly overwhelming mass of data and metrics thrown out by these platforms. To put it another way, just as referring to literary or historical texts as "data" may have challenged many in the humanities community to reflect more explicitly on their objects of study, might referring to social media "content" and "users" as "texts" and "authors" have a similar effect in challenging wider understandings of the uses and value of these materials?

Again, here we need to acknowledge an existing body of qualitative social media research associated in particular with forms of digital ethnography that aim to understand the significance of digital media and technologies in the daily lives of individuals and groups, with many studies focusing specifically on social media.[7] Such methods have been deployed primarily by anthropologists and media studies and linguistics scholars who again may at least partially identify with the humanities but who are rarely embedded within DH structures. Digital ethnographic approaches have nevertheless revealed how the close analysis of social media contexts and content can provide rich insight into contemporary cultural, linguistic, and social practices. Again, humanities researchers need not merely replicate such research but while ethnography may not be as embedded as a strictly humanities approach, there are close affinities between ethnographic approaches and their emphasis on "thick description," and more established humanities approaches such as close reading. As digital ethnography research has further revealed, the ways it relies on developing close relationships with social media users can provide ethical routes to researching more private forms of social media content, particularly on increasingly important platforms like WhatsApp (Tagg et al. 2017), than larger-scale and for practical reasons often necessarily more distanced computational approaches to social media research. For example, the more public nature of Twitter and the related ways it makes its data more freely available at scale is a primary reason why it dominates social media research, even if significant demographics are underrepresented on the platform. Again, smaller-scale humanities approaches that similarly rely on developing close relationships of trust with research participants, such as oral history methods for example, may have a similar potential to broaden our understanding of a much wider range of social media platforms than those that more readily support computational approaches to research.

A more expansive view of DH as more than "just" a method further has the potential to bring in a wider range of scholars from across the humanities, many of whom may lack the expertise or the opportunities to develop the high-level technical skills necessary for much computational research. This is not to deny the value of developing such skills, but there remain undeniable obstacles, particularly for researchers located in academic structures without either the training or technological infrastructure and support to make such research possible. While there are other important initiatives in DH to address such inequalities,[8] an additional route to expanding the field is to better support and incorporate into DH structures smaller-scale qualitative approaches to the study of digital materials such as social media content that might be more easily applied in such contexts. This is particularly important in terms of dramatically expanding research from beyond the Global North on social media, given the ways social media platforms are appropriated into different cultural contexts for distinct communicative and representational purposes.[9]

A broader definition that moves DH further beyond the practical application of computational methods is also an opportunity to foreground the ways the DH research community as a whole contributes to wider understandings and in particular theorizations of the digital. This echoes longer-term calls for the DH community to more actively contribute to discussions and debates on the role of digital media and technologies in relation to "the full register of society, economics, politics, or culture" (Liu 2012, 491). Equally, while there may appear to be an obvious division between those applying computational approaches to pre-digital texts, and those studying born-digital materials through predominantly qualitative approaches, there are many areas of productive overlap between the two that justify their inclusion within the same broad field of DH. "Born-digital" research materials are in practice hard to easily define or isolate, particularly since social media users are themselves often active participants in the remediation of pre-digital cultural materials. Equally, even where social media researchers may not themselves engage with computational methods associated with machine learning and natural language processing tools, participation in discussions on the potential uses and misuses of such tools is of undeniable value for "non-computational" digital researchers. In particular, social media researchers should ideally incorporate into their qualitative analysis an at least partial understanding of the computational tools, and associated algorithmic culture, underpinning these platforms.[10] Even for social media researchers who apply primarily qualitative approaches to analyzing their research materials, computational tools can play a valuable role in facilitating the collection and identification of relevant research materials, both when deployed by the researcher themselves or where qualitative Internet research may rely on the earlier computational labor of others, such as web archivists. These are just a few examples of existing areas of overlap, but they point to the necessity of an expansive definition of DH research that facilitates and strengthens such connections, and that ensures the field recognizes and values the multiplicity of ways humanities researchers engage with the digital in terms of both how and what they study.

CONCLUSION

This essay has aimed to articulate a potential route towards embedding social media research in DH in ways that can also contribute to further expanding the field to incorporate a wider range of perspectives and approaches from across the humanities. It is an inevitably partial account and undoubtedly reflects my own positionality as a UK-based humanities researcher with training and a background in primarily qualitative approaches to contemporary cultural and linguistic practices in both their digital and non-digital forms. The intention was not, however, to suggest there is anything "better" about a humanities approach than a media studies approach, or a qualitative rather than quantitative approach. Instead, this should be seen as an invitation to reflect on how we as humanities scholars respond to the undeniable importance of social media in relation to contemporary cultural forms and practices, building productively and collaboratively on a more established body of existing research in other disciplines. Equally, the emphasis on qualitative approaches is aimed not at diminishing the rich body of computational research in the humanities produced by existing DH researchers, but rather at advocating for a more expansive definition of digital research in the humanities that furthers our field's potential to make a vital and wide-ranging contribution to understanding the many roles that digital media and technologies play in our lives.

NOTES

1. Although cultural studies has been taken up in different ways in different national contexts, and is itself influenced by theorists from other contexts, the discussion here refers primarily to the UK context in which the field first arose.
2. Although arguably humanities research has never been purely textual, with for example a rich tradition of studies focusing on the materiality of historical and literary documents, and the archives in which they are stored.

3. See for example the work of the Web Archive studies network WARCnet: https://cc.au.dk/en/warcnet/.
4. The archiving of social media materials, particularly at scale, also presents major ethical challenges (Jules, Summers, and Mitchell 2018). Again, there is an opportunity here for the field of DH to contribute more actively to the ethics of social media and Internet research in ways that both complement the existing work of groups such as the Association of Internet Researchers and better support those working within DH structures undertaking research in this area.
5. I would like to thank Beatrice Cannelli (School of Advanced Study) who through her doctoral research on archiving social media has furthered my own awareness and understanding of these challenges.
6. It is important to note here that there are groups and initiatives that do adopt this more expansive view of DH, with for example the US-based Association for Computers and the Humanities explicitly incorporating into their 2021 conference Call for Papers "Humanistic research on digital objects and cultures" as an area of interest alongside more computational approaches (December 29, 2020).
7. Other labels such as Internet, virtual or online ethnography have also been used (see for example Hine 2000), with some differences in approaches, although they tend to have in common persistent forms of observation and immersion in specific online contexts, and a focus on the richly contextualized analysis of online texts and (inter)actions.
8. See for example Risam (2018), and the work of the GO::DH Minimal Computing working group to address computing infrastructural constraints, https://go-dh.github.io/mincomp/.
9. See for example Dovchin (2015). I would also like to thank Nayana Dhavan, a doctoral researcher at King's College London, for highlighting this point at a recent seminar on social media research and DH.
10. To offer an example from my own research focusing on multilingualism online, while my primary interest is in how individuals use and move across languages in their social media content, the increasing uses of machine translation are an unavoidable feature of social media platforms that inevitably will have some effect on individual choices to use or mix languages, or to engage in content originally posted in languages other than those they know or use.

REFERENCES

Bucher, Taina and Anne Helmond. 2018. "The Affordances of Social Media Platforms." In *The SAGE Handbook of Social Media*, edited by Jean Burgess, Alice Marwick, and Thomas Poell, 233–53. London: SAGE Publications Ltd. https://doi.org/10.4135/9781473984066.n14.

Coltrain, James and Stephen Ramsay. 2019. "Can Video Games Be Humanities Scholarship?" In *Debates in the Digital Humanities 2019*, edited by Matthew K. Gold and Lauren F. Klein, 36–45. Minneapolis, MN: University of Minnesota Press. https://doi.org/10.5749/j.ctvg251hk.

Dovchin, Sender. 2015. "Language, Multiple Authenticities and Social Media: The Online Language Practices of University Students in Mongolia." *Journal of Sociolinguistics* 19 (4): 437–59. https://doi.org/10.1111/josl.12134.

Fuchs, Christian. 2015. *Culture and Economy in the Age of Social Media*. New York, NY: Routledge.

Giaccardi, Elisa. 2012. "Introduction: Reframing Heritage in a Participatory Culture." In *Heritage and Social Media: Understanding Heritage in a Participatory Culture*, edited by Elisa Giaccardi, 1–10. New York, NY: Routledge.

Grosser, Benjamin. 2014. "What Do Metrics Want? How Quantification Prescribes Social Interaction on Facebook." *Computational Culture*, no. 4. http://computationalculture.net/what-do-metrics-want/.

Hall, Stuart. 1990. "The Emergence of Cultural Studies and the Crisis of the Humanities." *October*, 53: 11. https://doi.org/10.2307/778912.

Hall, Stuart. 2020. "Notes on Deconstructing 'the Popular' [1981]." In *Essential Essays*, vol. 1, edited by David Morley, 347–61. Durham, NC: Duke University Press.

Hine, Christine. 2000. *Virtual Ethnography*. London and Thousand Oaks, CA: SAGE.

Jones, Rodney H., Alice Chik, and Christoph A. Hafner. 2015. "Introduction: Discourse Analysis and Digital Practices." In *Discourse and Digital Practices: Doing Discourse Analysis in the Digital Age*, edited by Rodney H. Jones, Alice Chik, and Christoph A. Hafner, 1–17. London and New York, NY: Routledge, Taylor & Francis Group.

Jules, Bergis, Ed Summers, and Vernon Mitchell. 2018. "Ethical Considerations for Archiving Social Media Content Generated by Contemporary Social Movements: Challenges, Opportunities, and Recommendations." Documenting The Now White Paper. https://www.docnow.io/docs/docnow-whitepaper-2018.pdf.

Liu, Alan. 2012. "Where Is Cultural Criticism in the Digital Humanities?" In *Debates in the Digital Humanities*, edited by Matthew K. Gold, 490–510. Minneapolis, MN: University of Minnesota Press. https://doi.org/10.5749/minnesota/9780816677948.003.0049.

Ogden, Rebecca. 2021. "Instagram Photography of Havana: Nostalgia, Digital Imperialism and the Tourist Gaze." *Bulletin of Hispanic Studies* 98 (1): 87–107. https://doi.org/10.3828/bhs.2021.6.

Puri, Shalini. 2016. "Finding the Field: Notes on Caribbean Cultural Criticism, Area Studies, and the Forms of Engagement." In *Theorizing Fieldwork in the Humanities*, edited by Debra A. Castillo and Shalini Puri, 29–50. New York, NY: Palgrave Macmillan.

Risam, Roopika. 2018. *New Digital Worlds: Postcolonial Digital Humanities in Theory, Praxis, and Pedagogy*. Evanston, IL: Northwestern University Press.

Sadler, Neil. 2018. "Narrative and Interpretation on Twitter: Reading Tweets by Telling Stories." *New Media & Society* 20 (9): 3266–82. https://doi.org/10.1177/1461444817745018.

Sayers, Jentery. 2018. "Introduction: Studying Media through New Media." In *The Routledge Companion to Media Studies and Digital Humanities*, edited by Jentery Sayers, 1–6. New York, NY: Routledge, Taylor & Francis Group.

Tagg, Caroline, Agnieszka Lyons, Rachel Hu, and Frances Rock. 2017. "The Ethics of Digital Ethnography in a Team Project." *Applied Linguistics Review* 8 (2–3). https://doi.org/10.1515/applirev-2016-1040.

The Association for Computers and the Humanities. 2020. "Call for Proposals: Association for Computers and the Humanities 2021." *The Association for Computers and the Humanities* (blog), December 29. https://ach.org/blog/2020/12/29/call-for-proposals-association-for-computers-and-the-humanities-2021/.

Thomson, Sara Day. 2016. "Preserving Social Media." Digital Preservation Coalition. https://doi.org/10.7207/twr16-01.

Varis, Piia and Jan Blommaert. 2018. "Conviviality and Collectives on Social Media: Virality, Memes, and New Social Structures." *Multilingual Margins: A Journal of Multilingualism from the Periphery* 2 (1): 31. https://doi.org/10.14426/mm.v2i1.55.

CHAPTER NINETEEN

Spatializing the Humanities

STUART DUNN (KING'S COLLEGE LONDON)

The "spatial humanities" is a convenient umbrella term for a range of approaches to the human record, broadly defined, which situate place and space as a primary subject of study. Following this "spatial turn," place is (or rather can be) more than just an ancillary consideration in research whose main concern is the understanding of texts, historical events, objects, and so on (Bodenhamer et al. 2010). However most applications of the spatial humanities (or at least applications which identify themselves as such) concern relatively "traditional" humanities fields and research questions, as far as they are generally understood in academia. There have been relatively few attempts to link the theoretical and critical frameworks to the geography of the contemporary digital world, the Internet and the World Wide Web. Most treatments of the latter, predominantly, are informed by the social sciences and new media studies. At first sight this seems surprising, for the human understanding and construction of place have always been framed by the development and use of technology for visualizing, analyzing, and interpreting geographical information. Whether it is the depiction of the celestial spheres of Babylon incised on clay tablets (Brotton 2012, 2), the development of mapping on paper, the use of trigonometry to triangulate relative positions, or the precise determination of locations on the earth's surface by satellites, the organization of geographic information beyond the immediate arc of the individual's observation and experience has always provoked particular questions for the study of the human record, and has always been shaped by the media and technologies of the day. The Geographical Web, or "Digiplace" (Graham 2013), is simply the most recent manifestation of this.

"Geographic information" is in itself not a straightforward thing to define, but broadly, it may be taken to mean any piece of information that contains one or more references to location, whether directly, by referring to one or more places, or indirectly, referring to place relatively ("north of ..." "a long way from ..." etc.) (Hill 2007). The prevalence and importance of spatial information in making sense of data—of turning it into information—has resulted in the oft-quoted, and possibly apocryphal statement, that 80 percent of all data is georeferenced.[1] Whether this statistic is true or not, new research questions have emerged in the last two decades concerning how space, place, and location are both conceived and perceived in the contemporary world as well as in historical and cultural settings. However, there has been relatively little effort to explore possible connections between these two views of geocomputation in the human record. It is therefore worth expending some effort teasing out ways in which the use of spatial technology to explore the *past* human record may be informed by those used to mediate the *present* one. That is what this chapter seeks to achieve.

THE ROLE OF TECHNOLOGY AND "READING PLACE"

As noted, the human ability to navigate, explore, and understand place is enabled by technology, which is also a key driver of spatial humanities research. Many "geotechnologies" designed to analyze, visualize, and compute location are applicable in multiple domains. Indeed, in many cases in the spatial humanities, most such technologies have their origins *outside* the communities of practice using them there. Early adoption of geotechnology in humanities scholarship is often framed in terms of the use of analytical software such as Geographical Information Systems (GIS) to explore humanities research questions, or to enable the emergence of new questions altogether. In certain subfields, especially archaeology, this early adoption attracted criticism for encouraging over-reduction, and regressive, overly positive approaches to questions of qualitative humanistic interpretation (Conolly and Lake 2006). Human views of place are fuzzy, diverse, and subjective, and the humanities have their own theoretical mechanisms for dealing with them. Any attempt to impose the certainty of present-day mapping on to past perceptions of place bypasses those mechanisms, and risks "false positives." Reducing (for example) the range of a prehistoric community's hunter-gathering grounds to a fixed polygon, using a frame of geographic reference which that community had no understanding of whatsoever, is theoretically problematic, as that community would, of course, have had no such conception of such a polygon constraining their activities. What use would it be for present-day interpretation of these?

More recently, "informal" approaches to georeferencing, which take account of subjective human and interpretive views of place have been theorized. These "offer opportunities for extending concepts of space, new tools of representation and breaks down some of the barriers within processes of reasoning" (Lock and Pouncett 2017, 6), thus deliberately deconstructing the certainties of computational geodata. Such "processes of reasoning" are core to the epistemologies and ways of working with which humanists, with or without digital technology, are long familiar. The key challenge in applying the spatial humanities to the contemporary world therefore is how to co-opt these processes in such a way as to frame computational techniques for processing geodata derived from the human record. The use of the term "reading" to describe the process of consuming geographical data from a map (whether digital or not) illustrates the connection. "Reading the map" is a familiar phrase, referring to the process of establishing a route from A to B using a map. This is a quantitative navigational exercise, drawing on cartographic symbols, transitions, and connections. However, "reading" is also a concept taken for granted in the humanities, covering also deep interrogation of primary and secondary textual sources, and of knowledge creation. The spatial humanities require a similarly interrogative application of the concept of reading on the construction of geographic *information* from geographic *data*.

In parallel with this, the phrase "reading the map" calls attention to the challenges presented by geodata in contemporary culture and society, given that digital mapping platforms underpin large sections of day-to-day interaction with the digital world. Due to the automation of map interfaces on smartphones, and the ability to geolocate oneself immediately in real time using GPS, map reading is a skill rarely called upon in this world. Despite this, or perhaps because of it, our lives, environments, homes, workplaces, cities, and worldviews are increasingly mediated by spatial technology, most obviously via platforms such as Google Maps (Fink 2011, 3) or open-source mapping platforms such as OpenStreetMap (Mooney 2015, 319). As a result, it is now widely accepted in contemporary theory on information infrastructures and science and technology studies that, given its pervasiveness in modern life, geo-information should be treated as a special category (Noone 2020, 169–70). There is some parallel, at least in chronological terms, between the emergence of the "spatial humanities" in the academy and this "digitization of place" in contemporary culture. As with criticism of geographical analytics in the humanities, some scholars have critiqued the pervasive adoption of quantitative views of place in the everyday.

PLACE AND THE HUMAN RECORD: TECHNOLOGICAL ORIGINS

Interaction at large scales between contemporary society and geodata is technologically driven, by platforms such as Google Maps and Open Street Map. But the filtering and mediation of geographical knowledge by technology is by no means a purely modern phenomenon. The technical history of cartography itself tells us that the process of spatializing the human record, whether in the past or the present, is intimately connected with the development of geospatial technologies and media. Abraham Ortelius states in the prologue to his *Theatrum Orbis Terrarum*:

> Seeing that as I thinke, there is no man, gentle Reader, but knoweth what, and how great profit the knowledge of Histories doth bring to those which are serious students therein ... there is almost no man be it that he have made neuver so little an entrance in to the same ... for the understanding of them aright, the knowledge of Geography, which, in that respect is therefore of some – and not without just cause called **The eye of History**.
>
> (Ortelius 1606, n.f.)

This claim that a knowledge of geography as the "Eye of History" is a distinctive and important perspective (see also Avery 1990, 270–1 for further discussion of this passage) represents a departure from assumptions widely accepted since the Renaissance: that, to understand the Ancient World, which was represented in the European mind at the time by Greece, Rome, and the Holy Land, one must understand the physical and political geographies of the contemporary landscape in which the narratives of history occurred. It was not enough to simply read the texts which described those events. Hitherto, text had been the main medium for understanding the Classics and Classical history: to understand the ancient Roman world, one would have needed to read Polybius, Tacitus, or Livy perhaps; for Greece, Thucydides or Herodotus. And for the Holy Land of course, the testimony of the Bible was the main source of historical, as well as theological, evidence. In the context of the printing revolutions which drove this text-centric approach to the past, Abraham Ortelius is, in this prologue, offering a radical perspective on what it meant to create and share historical knowledge in the present, knowledge which took physical form in the *Theatrum* (which was itself the outcome of several new technological processes in printing, engraving, coloring, etc.).

Ortelius was not, himself, a great innovator of cartographic or scientific method; in fact, the *Theatrum* is itself merely a representative summary of the finest works of *other* sixteenth-century cartographers, most notably Gerhadus Mercator (of whom more below). What was new about it was the sheer volume of formal geographical knowledge assembled in one place, and the projection of that knowledge into visual (mapped) form. By editing a canon of contemporary and historical cartographical knowledge together in the *Theatrum* and including maps of the Holy Land and the Roman Empire and other historical landscapes, Ortelius presented these places as formal, *visual* historical constructs, defined by their geography as well as by the events familiar from the textual canon. "The understanding of them aright" therefore both depended upon knowledge of these landscapes and enabled and broadcast this knowledge to contemporary audiences. This new perspective on historical geography as a field worthy of study in its own right must, in turn, be seen in the context of the contemporaneous European Age of Exploration, when commerce, trade, and proto-colonial projects brought Dutch, and other European societies, into contact with land and cultures that had not previously experienced one another.

As in the present day, advances in geotechnology provoked further issues; and the opening up of historical method beyond the perspective of the written word was not the only problem which Ortelius describes in his prologue. He also identified two further, more practical problems. The first was financial:

> There are many that are much delighted with Geography or Chorography, and especially with Mappes or Tables containing the Plotts and Descriptions of Countreys, such as there are many now adayes extant and everywhere to be Sold, But because they have either not that, that should buy them, or if they have so much as they are worth, yet they neglect them, neither do they anyway satisfy them.
>
> (Ortelius 1606, n.f.)

Ortelius was concerned that the cost of maps and nautical charts would be beyond many of those who would stand to benefit from the information they contained. His Atlas, he hoped, would put these rich sources of geographical information within the economic reach of his customers, the moneyed middle classes and the new, wealthy mercantile elite (many of whom will have made their money, directly or not, as a result of sea voyages during the Age of Exploration, which in turn drove the boom in Flemish cartography of this time).

Secondly, the physical size of maps presented difficulties for his customers:

> Others there are who when they have that which will buy them would very willingly lay out the money, were it not by reason of the narrownesse of the roomes and places, broad and large Mappes cannot be so open'd or spread so that everything in them may be easily and well be seen and discern'd.
>
> (Ortelius 1606, n.f.)

Even his customers who did have the necessary wherewithal to purchase "Mappes or Tables" were constrained by the physical space available in their houses. There was simply not enough room available in the living spaces of middle-class early Modern Europe for proper use to be made of them. We may draw a comparison between Ortelius's statements on the economics of cartography to the so-called "digital divide" of the present day, where access to geotechnologies by the relatively wealthy acts to separate them from those without the wherewithal for such access.

SPATIALIZATION AS MULTITASKING

This concern with affordability and accessibility has resonances with current discourses in the spatial humanities and the role of public participation in creating geodata. As noted above, the increased use of GIS applications and related technologies in relation to archaeology, history, literature, and textual studies and languages, bought a raft of new modes of thinking about space and place in the humanities, and new methodological insights, while at the same time inviting criticisms of reduction and oversimplification. The causes of this "spatial turn" (Sui and DeLyser 2012, 112), as a set of connected strands which bring both further intellectual insight and technological limitation, may be partly traced to widened access, both scholarly and non-scholarly, to geotechnologies and geodata—just as Ortelius's adoption of then-new technologies of printing, copperplate engraving, and book production did in 1606.

The term "spatial turn" has been in use since at least 2010 (Bodenhamer et al. 2010, viii). In large part, it was driven by increased availability of relatively inexpensive geospatial analysis tools, often accessible to non-coders, which enable forms of collaboration between computer scientists and humanists that had not previously been possible, or at least very difficult to realize. These are technologies which, echoing the example of Ortelius, enable historians, literary scholars, archaeologists, etc., to engage with place as an independent critical concept. This was the primary motivation for applying techniques and technologies that had previously been developed for other tasks, which are *not* related to geography or geographical information, and appropriating them for analysis of the human record. For example, text parsing and analysis software which had been developed for processing and extracting information from digitized text was employed to extract placename information, a process which became known as *geoparsing* (Scharl 2007, 6). Another example is the use of semantic annotation tools, developed to annotate textual corpora, to manually read, identify, describe, and mark up place in text. This approach has in turn fueled a substantial field of research in the creation, use, and application of online gazetteers (Mostern and Johnson 2008). This came together with ideas of information management and retrieval developed by web engineers for managing information on the so-called "Semantic Web," or linked open data (LOD) which enables online information to be linked as semantic units rather than as webpages. Most notably, the Pelagios project employs these principles of semantic annotation to link together gazetteers of historical and cultural geo-information, and to enable humanities scholars to use them to explore and interrogate their own content (Palladino 2016, 67). These developments are all results of the barriers being lowered, not just to geotechnologies in terms of their cost and technical complexity but also between different application areas.

BEYOND THE GEOGRAPHY OF TEXT(?)

As with many early examples of the Digital Humanities, most early manifestations of the spatial humanities were, and are, concerned with the processing, manipulation, and mapping of text. The medium of the gazetteer—formal lists of places, structured references to single places—allows parallel references to places in text to be treated as points of data, rather than semantic objects. However, it relies conceptually on the manifestation of place in written form, the placename.

Text is a form of content which lends itself well to automated processing techniques, which includes the identification of place through the creation of gazetteers, annotation, and the digital tools and methods upon which, historically, much of the Digital Humanities is based. As Matt Kirschenbaum has written:

> First, after numeric input, text has been by far the most tractable data type for computers to manipulate. Unlike images, audio, video, and so on, there is a long tradition of text-based data processing that was within the capabilities of even some of the earliest computer systems

and that has for decades fed research in fields like stylistics, linguistics, and author attribution studies, all heavily associated with English departments.

(Kirschenbaum 2012, 8)

Text therefore exerts a powerful and logical attraction for methods associated with converting the human record into digital form. The digitization of *text* does not entail the kind of interpretive ambiguities for the digitization of *place* that are evident in the archaeological example above, or indeed Ortelius's "Eye of History." Unlike abstract events or interpretations of the past, there are clear processes behind converting text into binary, machine-readable form. This dates to the foundational projects of Humanities Computing/Digital Humanities, most notably Roberto Busa's *Index Thomisticus*, the punch-card-based concordance containing a full lemmatized collection of every word in the works of St Thomas Aquinas. This work, enabled in the middle of the twentieth century by emergent digital processing technologies, was inextricably linked to the storage, manipulation, and retrieval of text-based information. The developments described above, however, show that place is a more abstract concept than text, which has undergone a much more rapid and much less structured interaction with "the digital," especially since the mid-2000s. I return to this topic below. However, in the context of the argument made above about the qualitative complexity of spatial knowledge creation, digitizing text is no more a process of reading it than digitizing geodata is a process of reading a map.

This much becomes clear when one looks at geographic information beyond text in the spatial humanities. Despite the predominance of text in many early adoptions in Digital Humanities, spatial humanities has always extended far beyond text. Creative practitioners, for example, have moved to employ critical understandings of space and place as means of creating performative and artistic practices *in situ*. A major element of this is the use of such practices to express change and dynamic interaction between humans, human consciousness, and the landscape, emphasizing connections between past place and present place. In his located work drawing on his "home" landscapes of Lincolnshire, *In Comes I*, the theater artist Mike Pearson comments that "certain elements of past lifeworlds persist through time, fully formed in the present one and the next, a continuity that is nevertheless perpetually under threat" (2007, 216). Another example of embodied practice connecting past landscapes with present ones are community-facing archaeology activities, such as the Cambridge Community Heritage Project, which enabled local volunteers to explore local archaeology in a highly structured way through pre-planned test pits, and on-site access to expertise to help them interpret the finds they made. Creative, practical, and *in situ* processes such as this, which garner embodied, experiential understanding of human space and place are sometimes referred to as "deep mapping," or "vertical pathways of discovery," as Carenza Lewis puts it (Lewis 2015, 414), which both physically and conceptually explore *under the surface* of the landscape.

Mapping the nuances and subjectivities of the human record beyond text is sometimes referred to as "deep mapping." Bodenhamer (2007) says of deep mapping:

[i]n its methods it conflates oral testimony, anthology, memoir, biography, images, natural history and everything you might ever want to say about a place, resulting in an eclectic work akin to 18th and early 19th century gazetteers and travel accounts. Its best form results in a subtle and multilayered view of a small area of the earth.

(Bodenhamer 2007, 105)

This encapsulates many of the complex, discursive, multi-layered descriptions of the idea of reading that many humanists apply to texts, manuscripts, and archives, and situates them firmly in the context of human-geographic information. Indeed it was a writer, William Least Heat Moon, who first popularized the term in his *PrairyErth: a Deep Map*, a traveler's account of the region around Strong City, Kansas. This has in turn been linked by some to the idea of the "Spatial Narrative," or what the artist Ian Biggs has called the "essaying of place."[2]

Rather than being a single method, therefore, or a definable approach, we can see deep mapping as the expression of a fundamental human need to understand one's immediate place in a literal, geographical, sensory, physical, and psychological sense. Text is merely one way in which this sense of place can be conveyed. Deep mapping can be best understood as the range of tools and methods that exist—some digital and some not—which enable this. It is, in other words, a paradigm of method driven partly, but certainly not wholly, by geospatial technology, and in a set of applications of these technologies to humanities materials. Usually, the outputs of these are published as conventional scholarly literature.

This highlights the key epistemological break which spatial humanities makes with the longer tradition of Digital Humanities. As noted above, historically much work in the Digital Humanities has been concerned with text. Key initiatives, such as the Text Encoding Initiative (TEI), founded in 1987, were synonymous with large areas of Digital Humanities, then known by other terms such as "Humanities Computing." This is because text is, itself, structured in a granular manner which lends itself to the grammars and stylistics of digitization. The Digital Humanities provides the critical frameworks, apparati, and tools to make sense of that text and, through the use of technological ideas such as the Semantic Web, geoparsing, digital gazetteers, etc., the *spatial humanities* provides equivalent frameworks for processing text into geodata and geovisualizations. The key question is how can we draw on those frameworks to understand geodata created in the present day?

DIGITAL PLACE IN THE PAST AND PRESENT

At this point we must circle back to the centrality, and long history, of geotechnology in humans' relationship with place. The very idea of deep mapping itself exposes the complexity of studying technology-driven views of place in the humanities. As with many early examples of spatial (and Digital) humanities, the term has its roots in the production and reception of place in text; and the representation, consumption, and use of place in present-day digital culture and society is similarly multi-layered. The wider public outside the academy has started to engage with place and space through the medium of the digital since the mid-2000s, a process which reflects the development of the interactive World Wide Web and so-called "Web 2.0." The launch of Google Maps in 2005 ushered into the popular consciousness a new way of viewing and experiencing place, in which the user was the "placed" at the center of the map, thus enabled to navigate their environment in real time, rather than having to identify, construct, and operationalize a "pre-read" course by understanding and relating the features depicted on a paper map. Just as geotechnology opened up new, problematic, and challenging vistas on place in the humanities, so mass direct interaction with place, mediated by the digital, has opened up new vistas of ethnographic, science and technology studies and social sciences research, exploring the impact of geospatial technologies on human behavior. This includes the capacity to navigate and human engagement with digital place, as well as human ability to visualize and conceptualize it.

Debate about *digital place* itself has shifted in its critical thinking about space and place in the last ten to twenty years across the humanities, literature, and creative practice. As well as being driven by new tools for geographical analysis, and by wider public update of mapping technologies as argued above, it is also enabled by the way digital technology shapes our views of space and place itself; and the way in which we break place down into intelligible units for analysis, understanding, navigation, and interpretation. The activity (and the verb) "to map" implies a process of ontologizing place, breaking it down into discrete and logical components which human understanding can easily embrace. As Sébastien Caquard puts it, "[A]n ontology reduces the world into entities and categories of entities that can be clearly defined in order to mine them, retrieve them, and map them at will" (2011, 4). Maps, whether digital or not, are thus structured to make space and place retrievable and manageable, to "ontologize" if for human consumption, based on extrinsic and abstract spatial definitions and indicators, which need have nothing to do with the human user's previous experience and knowledge. These non-viewpoint dependent modes of mapping, or *allocentric* mapping, which depends on cues and reference points to navigate, can be distinguished from *egocentric* mapping, which depends on the map-user's location being known independently (e.g., through GPS). Navigation using platforms such as Google Maps gives the impression of egocentric wayfinding, however extrinsic, allocentric data is being streamed to the user in real time.

The interpretive complexity of allocentric mapping is closer to a humanist's idea of close reading or textual analysis than many forms of egocentric digital mapping. It requires layers of representation, processing, and interpretation. This led Trevor Harris, in his discussion of spatial narratives, to reject the term "shallow map" as an antonym for "deep map," preferring instead "thin map." The term "shallow map," he writes, "intentionally or otherwise, implies a meaning of superficiality and inconsequentiality … there is overwhelming evidence to disprove these … descriptions and any consideration of the impact of GIS on contemporary society validates the value and contribution of these maps across a range of societal endeavours" (2015, 30).

The invisible malleability of egocentric digital mapping, and the absence of humanistic theoretical frameworks and safeguards, renders it vulnerable to manipulation and appropriation by those with large resources invested in the platforms, such as multinational corporations. As Agnieszka Leszczynski (2014) has pointed out, the appropriation of user-generated mapping and real time human interactions with digital maps at scale by large technology companies further their commercial and cultural interests. They do this, she argues, precisely by masking the ambiguity and complexity of digital place, delivering it seemingly straightforwardly and seamlessly to the user's smartphone.

COMPRESSING SPACE AND TIME

A key distinction between place mapped and read in the digital world and place mapped and read in physical media is that "cyberspace," "human space" egocentric mapping versus allocentric mapping, is agnostic in scale. Scale in the digital world is not fixed, it adapts and fits according to the user's deployment of the zoom button. This ability to shrink the world to fit any perspective was foreshadowed by discussions of the impact of network technologies before they became ubiquitous in Western societies. In 1991, the pioneering feminist geographer Doreen Massey described the process of "time space compression," a conceptual shrinking of place caused by accelerating communication and travel in the last quarter of the twentieth century. Echoing the perspectives on deep mapping given by Lewis and Bodenhamer (above), Massey frames the idea of space time compression happening to places which are, themselves, fluid, and which thus are resistant to objective description or ontologizing: “One of the great one-liners in marxist [*sic*] exchanges has for long been ‘ah, but capital is not a thing, it’s a process’. Perhaps this should be said also about places; that places are processes, too ... places do not have to have boundaries in the sense of divisions which frame simple enclosures” (Massey 1991, 29).

This is a key point: ever since the idea of “digital place” was first theorized, a governing principle of it has been that place *is* framed by “simple enclosures”—the point, the line, the polygon, processable by binary systems and readable by machines. The key difference of this from “ground truth,” where the geometries of geographical features follow what is on the ground, is that these points, lines, and polygons are entirely abstract. Depending on the scale, whole cities or regions can be represented by points or pixels. This has caused controversy in the past. The rollout of Microsoft’s Windows 95 package included a map of world time zones, in which part of the Kashmir region was represented as lying inside Pakistan, leading to diplomatic objections from the Indian government.[3] The advent of digital mapping, which has characterized so much of the emergence of digital culture and society since Massey described the idea of time space compression in 1991, *requires* definition, enclosure, inclusion, and exclusion even where—perhaps especially where—they did not previously exist. The numerical abstraction of geography is thus a major challenge facing present-day digital and spatial humanities. It represents, in the terms described by Caquard above, a particularly aggressive form of the ontologizing of place.

Numerical abstraction is not a product of the age of digital mapping itself. Indeed the famous Mercator projection system, developed at the same time that Ortelius was stitching together his Atlas for the purposes of creating new forms of geographical knowledge in “the Age of Exploration,” is a case in point. The system, based on the principle of distorting the globe’s surface cylindrically, so that it can be spread out flat, is a mathematical one: it preserves the true angular relationship between points (and thus land masses), but distorts land masses themselves, shrinking them as they move away from the Equator; so that Greenland looks to be about as large as Australia, when in fact Australia is at least four times the size of Greenland. Therefore, the fact that place online is *numerical* does not mean that it is *objective*. Perceptions of objectivity are often given by digital maps, whether purposely or not. The same is even more true of maps drawn by human hand, on paper, or some other surface.

CONCLUSION

In the last twenty-five years or so, the Digital Humanities more broadly has shifted from a focus on the use of technology to explore research questions *about* the human record, and the study of technology *as a part of* the human record. A core challenge for the spatial humanities is how to disentangle these two perspectives. Ortelius was dealing both with contemporary demand for mapping products, stemming from new social, cultural, and political awareness of distant cultures, *and* rethinking what such awareness meant for “the Eye of History.” How might co-created views of digital place in the wider world help us inform historical, cultural, and performative views of place in the humanities?

Answering this question convincingly means moving beyond asking what computational approaches might achieve for study of the historical, literary, and creative human records, and how they are relevant to the digital world of today. The primary intellectual assets of the spatial humanities are the materials of the human record in all its forms, and the geographical, or georeferenced, information that they contain. It is text, and the ways in which text encodes and conveys spatial information have received much of the attention of this discourse, reinforcing the dominant place of text in the broader field of the Digital Humanities. Other fields of the humanities that are *not* concerned with text raise interpretive challenges for the digital treatment of place which are harder to address. A "core" area of literature has emerged in the spatial humanities which addresses these challenges. This chapter has attempted to connect this with more recent areas of literature, which deal with digital place in real time in the present day. Digital place mediates digital culture and society by location-enabled devices, which enable the immediate quantification of human interaction with place, via platforms such as Google Maps. The theory-rich literature of spatial humanities has much to contribute to this, with its understandings of openness, collaboration, the critical siting of technology in place discourse, and its critique of the positivist reductivity of the point, the line, and the polygon. In such a world, Ortelius's Eye of History is also the Eye of the Contemporary.

NOTES

1. See Caitlin Dempsey, "Where is the Phrase '80% of Data is Geographic' From?", GIS Lounge, October 28, 2012, https://www.gislounge.com/80-percent-data-is-geographic.
2. See https://web.archive.org/web/20210227215800/https://www.iainbiggs.co.uk/text-deep-mapping-as-an-essaying-of-place/.
3. See "Pixel this – and computer giant loses millions," *The Sydney Morning Herald*, August 20, 2004, https://www.smh.com.au/world/pixel-this-and-computer-giant-loses-millions-20040820-gdjl2l.html.

REFERENCES

Avery, Bruce. 1990. "Mapping the Irish Other: Spenser's a View of the Present State of Ireland." *English Literary History* 57: 263–79.

Bodenhamer, David J. 2007. "Creating a Landscape of Memory: The Potential of Humanities GIS." *International Journal of Humanities and Arts Computing* 1 (2): 97–110.

Bodenhamer, David J., John Corrigan, and Trevor M. Harris. 2010. *The Spatial Humanities: GIS and the Future of Humanities Scholarship*. Bloomington and Indianapolis: Indiana University Press.

Brotton, Jeremy. 2012. *A History of the World in Twelve Maps*. London: Penguin Books.

Caquard, Sébastien. 2011. "Cartography I: Mapping Narrative Cartography." *Progress in Human Geography* 37 (1): 135–44. https://doi.org/10.1177/0309132511423796.

Conolly, James and Mark Lake. 2006. *Geographical Information Systems in Archaeology*. Cambridge: Cambridge University Press.

Fink, Christoph. 2011. "Mapping Together: On Collaborative Implicit Cartographies, their Discourses and Space Construction." *Journal for Theoretical Cartography* 4: 1–14.

Graham, Mark. 2013. "The Virtual Dimension." In *Global City Challenges*, 117–39. London: Palgrave Macmillan.

Harris, Trevor M. 2015. "Deep Geography, Deep Mapping: Spatial Storytelling and a Sense of Place." *Deep Maps and Spatial Narratives* 28–53. Bloomington and Indianapolis: Indiana University Press.

Hill, Linda L. 2007. *Georeferencing: The Geographic Associations of Information*. Cambridge, MA: MIT Press.
Kirschenbaum, Matthew. 2012. "What is Digital Humanities and What's It Doing in English Departments?" In *Debates in the Digital Humanities*, edited by Matthew K. Gold, 3–11. Minneapolis, MN: University of Minnesota Press.
Leszczynski, Agnieszka. 2014. "On the Neo in Neogeography." *Annals of the Association of American Geographers* 104 (1): 60–79. https://doi.org/10.1080/00045608.2013.846159.
Lewis, Carenza. 2015. "Archaeological Excavation and Deep Mapping in Historic Rural Communities." *Humanities* 4 (3): 393–417. https://doi.org/10.3390/h4030393.
Lock, Gary and John Pouncett. 2017. "Spatial Thinking in Archaeology: Is GIS the Answer?" *Journal of Archaeological Science*. https://doi.org/10.1016/j.jas.2017.06.002.
Massey, Doreen. 1991. "A Global Sense of Place." *Marxism Today* 35: 24–9. https://www.unz.com/print/MarxismToday-1991jun-00024
Mooney, Peter. 2015. "An Outlook for OpenStreetMap." *OpenStreetMap in GIScience*, 319–24. Cham: Springer. https://doi.org/10.1007/978-3-319-14280-7_16.
Mostern, Ruth and Ian Johnson. 2008. "From Named Place to Naming Event: Creating Gazetteers for History." *International Journal of Geographical Information Science* 22 (10): 1091–108. https://doi.org/10.1080/13658810701851438.
Noone, Rebecca. 2020. "Navigating the Thresholds of Information Spaces: Drawing and Performance in Action." In *Visual Research Methods: An Introduction for Library and Information Studie*s, edited by S. Bedi and J. Webb, 169–88. London: Facet.
Ortelius, Abraham. 1606. *The Theatre Of The Whole World Set Forth By that Excellent Geographer Abraham Ortelius*. John Norton, Printer to the King's Majesty In Hebrew, Greeke and Lataine.
Palladino, Chiara. 2016. "New Approaches to Ancient Spatial Models: Digital Humanities and Classical Geography." *Bulletin of the Institute of Classical Studies* 59 (2): 56–70. https://doi.org/10.1111/j.2041-5370.2016.12038.x.
Pearson, Mike. 2007. *In Comes I: Performance, Memory and Landscape*. Exeter: University of Exeter Press.
Scharl, Arno. 2007. "Towards the Geospatial Web: Media Platforms for Managing Geotagged Knowledge Repositories." *The Geospatial Web – How Geo-Browsers, Social Software and Web 2.0 are Shaping the Network Society* 18 (5): 3–14. https://www.geospatialweb.com/sites/geo/files/sample-chapter.pdf.
Sui, Daniel and Dydia DeLyser. 2012. "Crossing the Qualitative-Quantitative Chasm I: Hybrid Geographies, the Spatial Turn, and Volunteered Geographic Information (VGI)." *Progress in Human Geography* 36 (1): 111–24. https://doi.org/10.1177/0309132510392164.

CHAPTER TWENTY

Visualizing Humanities Data

SHAWN DAY (UNIVERSITY COLLEGE CORK)

Visualization of data is undertaken for a variety of reasons, uses, and purposes in the humanities. Ultimately this forms part of a process of knowledge construction through exploration and discovery. The act of visualizing data as information is both an individual inward pursuit as well as an external performance. Engagement with the viewer/participant and audience raises questions, provokes discussion, and can stimulate activism. Traditionally non-humanistic disciplines have tended to often focus on using data visualization specifically for analysis and definitive substantiation. Until recently, few data visualization tools have been created specifically to fulfill the humanities' unique needs, which has led to adoption and adaptation, often involving conscious or unconscious compromise towards heuristic ends. As a result, these otherwise-engineered tools and methods pose challenges to visualizing humanities data. This chapter explores these challenges and issues to encourage reflection and possibly inspire effective remedy.

Humanistic scholarly practice has a natural aversion to drawing boundaries and putting things into conveniently defined boxes. Non-humanistic disciplines regularly define and construct models as frames of study, which has informed the tools built to visualize data in these spheres. This process implicitly leads to this acceptance of clearly delineated objects of study that follow definable rules. Any sense of inductive reasoning in the humanities pauses before seeking to expound a generalized and abstracted principle. As a result, humanities scholars have sought and devised new and innovative means and coupled them with parallel processes that engage collaboration, crowdsourcing, and empower the public humanities for social change. Discussion of visualizing humanities data is premised on the assumption that this process is unique in many ways from that carried out in other spheres and disciplines. However, this assertion demands critical reflection and consideration of practice.

Visualization of humanities data is an increasingly valuable pursuit, and a "playful iterative approach to quantitative tools," as suggested by Van Zundert, "can provoke new questions and explorations" (Van Zundert 2016). However, the less than critical adoption of the tools, techniques, and methods devised to visualize humanities data (whether developed for specific purpose or adapted from outside) and the potentially compelling but misleading conclusions that might be drawn based on the authority of visual presentation of data pose challenges to scholarly reception (D'Ignazio and Bhargava 2018). This tension demands exploration to raise both awareness and conscious and deliberate activities to mitigate the dangers.

This chapter considers the visualization of humanities data across five axes of tension to expand on the body of scholarship providing critical appreciation:

1. defining versus provoking
2. engaging versus convincing

3. performing versus processing
4. ambiguity versus precision
5. innovating versus demonstrating/adhering.

In *Dear Data*, Giorgia Lupi and Stephanie Posavec (2016) describe a year in which they actively sought to represent their everyday lives through data and creative visualization of this data. The correspondents had only met one another in person twice previously. So, they undertook an experiment to discover how they could get to know one another by exchanging these creative visualizations of each other's lives.

The data visualization process is recounted as playful engagement wherein the correspondents create and share unique and bespoke visualizations of their everyday lives. This particularly engaging book sparks two critical observations. First, human activity (personal or collective) can and is being increasingly metricized. Second, this data (and the seriously playful part) can be represented in various visual ways that bring the data back to life and allow for the rhythms and cycles of ordinary existence to be engagingly explored and shared for deeper understanding.

Lupi and Posavec highlight how humanities scholars today engage in play with data that often results in forms of visualization that create new forms of *representation* and similarly connect this with *performance*.

Early reflections on the particularities of humanities engagement with data, such as Drucker's seminal reflection on the constructivist origins of captured "data," have led to a critical realization of how humanities and social science scholars must make explicit the connection to the sources and the representational nature of data (Drucker 2011). These questions around data as subjective and "given" (Galloway 2011) in contrast with information as "something which is taken" have critical implications for what we represent visually, let alone how we choose to do so. Lavin has recently defended the term data as being "far more rich etymologically" than Drucker credited it but reaffirmed nonetheless the critical engagement with data that marks a unique humanities approach (Lavin 2021).

Similarly, data can be repeatedly re-represented through consequent processes and undergo a reductionist transformation (much as imagery does through the xerographic process). However, it can also be distorted and enhanced through collation and conjunction with related data—simplification and legibility, making the underlying data less transparent to the viewer (Drucker 2011).

So, what makes the visualization of *humanities* data unique?

One lately emerging observation is the unique value that humanities scholars can bring to the scientific process. Where data becomes sacrosanct and its existence as *capta* is overlooked, the perceived authority of numbers and data—once captured—is rarely reflected upon. Instead, humanities scholars by practice are encouraged to bring a natural skepticism or curiosity to the origins as a corrective, to step back and consider sources and acquisition to identify misuse through bad data practice.

Rising to Drucker's call for "humanists to think differently about the graphical expressions in use in digital environments" (2010), the subsequent decade witnessed an appreciation of information visualization as a tool not merely for new ways to conceptualize data through graphical means but also in the social affordances of the process of visualization. Visualization is the shiny, distracting toy that entices but distracts.

In 2010 Joanna Drucker challenged the discipline to "a critical frame for understanding visualisation as a primary mode of knowledge production" as she presciently pointed out, at the time, "critical understanding of visual knowledge production remains oddly underdeveloped." So, where do we find ourselves today? How have we met this challenge in the humanities? Jennifer Edmond posed a challenging counterpoint to the ordering of knowledge, asserting that much like messy desks, our modes of reading in the humanities are where the serendipity happens (2018). We need to maintain multiple layers of engagement to make the mental leaps that result in synthesis and knowledge assembly. So how do we find the use of visualization in the humanities has evolved? How has self-reflection informed our engagement with the multivariate forms of data that we work with in the humanities today? So, what does the hackneyed phrase "appealing directly to visual cortex" really mean? Bypassing language centers and avoiding overt translation and processing, plugging directly in cerebral appreciation suggests a rawness, but simultaneously a blunt but genuine reality—unprocessed, unrefined, and yet with an appealing authenticity.

DEFINING VERSUS PROVOKING

The allegation that data visualization provides a means for STEM practitioners to pack conclusions into tidy boxes contrasts with a distinctly different way in which similar processes are employed in the humanities to inspire new questions.

As a reflection on humanities practice, Edmond explores the practice of reading in academic practice and argues that distraction and the "noise" of the deluge of data is an intrinsic part of how we make connections and are inspired in knowledge creation (2018). She further demonstrates how this practice has changed and is changing by the impact of dealing with increasing quantities of data in our everyday life, whether on our digital media or in the way in which traditional media has shifted to data-driven (and visually presented) narrative.

Likewise, this ability to playfully engage with the data through imaginative visualization has called upon the artistic and creative manner in scientific practice that is not limited to the humanities scholar; William Allen points to this playful engagement as a form of brokerage practiced throughout society (2018). He highlights the nature of the relationship that develops through engagement between the creator, the disseminator, and the recipient and demonstrates how this is a reciprocal process of essentially play. Anecdotally, the author recalls a surreal night on the town in Cambridge, MA, where he met a researcher at MIT. Although a practicing fine artist, he was also undertaking DNA sequencing in a lab at MIT, where his imagination was seen as crucial to the scientific process. However, where this creativity or a unique appreciation of the human condition may be seen as unique to humanities scholarship (it is not), embracing what is perceived as a scientifically rigorous means of engaging with data is also integral and thoughtful collaboration.

We need to play in creation but also play in engagement. Edward Tufte affirmed the role of visualization of data to "create engagement" in his seminal *The Visual Display of Quantitative Information* (Tufte 2001). The use of visualization to analyze and present data in purely scientific pursuit is used to substantiate and reinforce findings. The humanities, however, adopt a seeming counterpose. Visualization of humanities data (as illustrated by Lupi and Posavec) invites questioning and discovery, provoking engagement and seeking to inspire questions instead of substantiating findings. This is indeed a far different approach and intent.

Likewise, there remain aspects of humanities data and how it is conceptualized for visualization. Humanities data is often inherently messier than that in scientific pursuits due to the human and the condition being considered. Representation of ambiguity, for example, is of increasing focus in many spatial humanities circles. These have led to conscientious approaches to develop and explore new techniques to visualize particular to the humanities addressing gray areas and recognize the need not to imply greater precision or certainty to the data through the way it is portrayed.

ENGAGING VERSUS CONVINCING

However, visualizing data in the humanities is not merely about discovery and exploration. Still, it perhaps uniquely has an implicit responsibility to stimulate engagement and interaction with the broader community (Fitzpatrick 2019) to activate phenomenological reflection on data. Thus, humanities information visualization, although a form of representation for realizing, identifying, or extracting knowledge, is equally employed for stimulating critical thought to enable action, participation, etc.

Although Helen Kennedy et al. (2016) assert that visualization primarily attempts to make data represented objective, this is contrary to the underlying appreciation in the humanities that data is unambiguously subjective, and that visualization seeks to affirm that fact and provoke subsequent reflection. Objectivity remains a STEM determination, an attempt to convince that discussion has been exhaustively pursued and concluded beyond a reasonable doubt. For the humanities scholar, the visualization is meant to suggest, connect, provoke, and propose in the hope of instigating further discussion, living up to the adage that *when the conversation stops, the visualization fails.*

The ability to carry out effective data visualization is integral to engaging with different layers of abstraction. Simply presenting all the available data often hinders the process and overwhelms the viewer (Galloway 2011). Abstracting allows for a broad visual perspective of a large dataset (a God's eye view) but utilizing emerging technological affordances allows for data visualization that permits examination in detail where desired and allows for critical engagement with the underlying data (Tufte 2006). Emerging forms of visualization such as multi-layered tree-maps (Bederson et al. 2002), sunbursts (Feitelson 2004), or packed circles (Wang et al. 2006) are particularly engineered to afford this engaged mode of visual interaction.

This trend towards the use of hierarchical data visualization as a visual interface to collections has been growingly apparent in the realm of cultural heritage where collection artifacts (grounded in well-constructed metadata) have engendered new forms of visual browsing techniques. For example, collection browsing platforms, such as the 100 Archive[1] or Vikus Viewer,[2] offer "generous interfaces" and allow for leveraging metadata schema for enhanced exploration and discovery.

The affordance of visualizing humanities data densely, as a means of cultural observation, has inspired the questions: What if our visual representations showed all the available datapoints? Could this allow us a more informed perception of previously invisible social phenomena? And if so, what new forms of visual representation would this demand? (Manovich 2011). All of which suggest the pronounced power of data visualization to allow individuals to bring their own experiences and contexts to richer, deeper, and larger unlocked datasets which were previously unfathomable in tabular or textual forms.

These affordances further amplify and empower the call by Kathleen Fitzpatrick for the active involvement of scholars and, in particular, those in the humanities to ensure its survival and that of humankind through the reinvention of the university's role (2019). The ability to communicate research and share datasets more effectively through visualization can amplify access to scholarly output by wider society. It reaffirms the role of the humanities scholar (and not humanities alone) outside of the environs of academia.

However, this, in turn, demands a digital criticality that allows the public sphere to be drawn into a renewed engagement eased through the responsible and informed use of data visualization. Visualization of humanities data provides the power and the opportunity for engagement, and it harnesses the traditional humanities performative practice.

PERFORMING VERSUS PROCESSING

In the humanities, novel or innovative approaches are expected practice and innovation itself is often a demanded part of the process. Arguably, there are fewer convenient frameworks to be applied—or in fact, demanded in the humanities—but a unique approach is readily acknowledged and accepted. Although there are defined methods and a general epistemic approach, the beauty of humanities practice, especially concerning the visualization of data, is in the performance and is often in the way we journey (Ellenberg 2021). Thus, the process of visualizing data and the aspects of the process undertaken are not just critical to its outcome but lend an experience throughout the performance of visualizing humanities data (Kräutli and Davis 2016).

There has been a perceptible increase in scholarship on both how performance has a role in what we would typically see as a purely scientific pursuit and necessary critical reflection on the nature of the process and how the original data were compiled. The intrinsic link between the representations and the represented is a vital aspect of visualization and one which Lev Manovich raises by characterizing tool and method development as a conversation. His work challenges a rote "process" of visualization of data. It suggests the process as a way of thinking and of "talking out" the process of deliberating and making manifest our thought processes, "a challenge to traditional forms of cultural analytics" (Manovich 2009). The development of tools and methods to visualize humanities data demonstrates how algorithms and interfaces are crafted through code—the means of approaching the problem is a form of the language of representation and analysis. Manovich's Software Studies lab reimagines and reinvents the visual interface to access and analyze massive amounts of human behavioral data. These interactive visualizations permit analysis and questions such as "What happens when you can see an entire body of artistic output within a single frame?" or "Can we explore cross-cultural behaviors across time and space through social media?" (SelfieExploratory[3]). Big questions that are left open-ended and to the viewer to answer through their own exploratory experience.

In a manner, this recognition that the journey (process) has much to contribute to the outcomes of an exercise of visualization is not unique or new but connects to the practice of the performative in the humanities and an openness and willingness in many humanities disciplines to the open-ended and the serendipitous as a recognized form of scholarship. Access to, and appreciation of, foundational artistic aspects highlight the importance of the process, whether in notational representation as a form of visualization of data, or in finding means to develop visual interfaces to navigate the immensity of social media data to explore human behavior.

AMBIGUITY VERSUS PRECISION

Perceived end use determines the varying levels and degrees of precision required in visualization (as well as the underlying data itself). The term "perceived" is particularly noted here as the publisher certainly cannot imagine all potential ways a visualization may be interpreted. The need for a means to represent ambiguity or fuzziness in visualization to more honestly convey uncertain or ambiguous data has long been realized. However, the responsibility incumbent on the creator, author, or publisher of the visualization to represent this uncertainty has only been more recently demanded. Geospatial visualization has struggled with this challenge (Plewe 2002; Grossner and Meeks 2014). This need for "fuzziness" in geographic information systems used to visualize both spatial and temporal data is challenged by digital representation. The binary and fixed nature of static forms of Cartesian display is inadequate to represent disputed boundaries, imperfect memory, or contested interpretation. Over time we have devised means to attempt to represent ambiguity, displaying contradictory claims, using transparency as a proxy for precision, or even utilizing new visualization models, such as augmented reality, allowing the user to experience diverse "realities." However, in many of these cases, the digital nature of the data itself lends itself to a determined precision, precision that may not, in fact, represent the true nature of humanity or human phenomena. Likewise, how precision—or ambiguity—is represented has massive implications for interpreting and using the represented (or mis represented) data (Grossner 2020). One of the most significant challenges confronting most consumers of visualized data is realizing that it is not definitive, that it is based on fuzzy data and needs to be recognized as such when attempting to draw conclusions. The increasing prevalence of what is referred to as "deterministic construal error" must be addressed but requires a digital competency that remains unrealized mainly (Joslyn and Savelli 2021) not just within academia but throughout society.

Digital methods are increasingly embracing ambiguity as being genuinely representative and thus a crucial aspect of being, but this is not an aspect of traditional or analog visualization practice. Although some forms of visualizations such as the whisker diagram—used to represent fluctuating stock prices over a particular period—have allowed for a degree of explicit brackets of representation, even this demands a border/fixed frame of reference.

The digital challenges humanities scholars, but in pursuit of bringing honesty and transparency to emerging powerful tools, it demands unique and innovative means to allow for varied precisions and representations that avoid implying structure or definition where none exists.

INNOVATING VERSUS DEMONSTRATING/ADHERING

There is a perceptible impetus in the humanities subtly demanding that the creation of new knowledge requires innovation in process, where non-humanistic practice demands greater adherence to specific methods. Although ultimately stemming from data supporting replicability as part of the scientific process, in the humanities, replication has not typically been seen as an essential part of practice, and conclusions are most often open-ended. The performative aspect mentioned previously often demands innovation in process or at least in application—counter to traditional non-humanistic—and a habit amongst those visualizing humanities data.

All the challenges we have identified lead us to consider where innovation exists in the humanities today and how other spheres are innovating to deal with emerging challenges. The scientific method proposes a rigid methodology that can be employed to systematically explore phenomena through data to reach a replicable and determined solution to a hypothesis. The challenge in the humanities is that there is a less rigid application of the methodology and an embracing of exploration and discovery—often without a predetermined hypothesis. Unlike STEM practice, there is not a demand for a conclusive and definitive outcome. A conclusive definition demands adherence and more rigid adherence to demonstrated practice than to innovative engagement (D'Ignazio and Bhargava 2018). Ambiguity is embraced, represented, and made transparent—if well visualized.

When the process becomes performance and developing new forms of visualization becomes an end in itself, the culture of reflection marries with the culture of innovation.

CONCLUDING

Data from the outset has been and remains contentious, not just in the humanities. Before even considering visualizing for whatever intent, the nature of the object under observation and analysis has raised questions about bias, significance, completeness, form, etc. Issues raised by Drucker and others have sparked critical discussion about the underlying layer of the visualization process, particularly in the humanities. The result of this discussion which we do not engage with in depth in this chapter, is that we often (usually?) start with a flawed object, and then we visualize.

In visualizing data, we then make decisions around methods, aesthetics, intent that may distort further but certainly cannot repair what was not already there. If the data is flawed in any manifold ways mentioned earlier, we only move away from mimetic truth. The challenge then becomes whether we can stimulate discussion around the represented data, to illustrate or at least make transparent its faults and thereby use it to bring greater understanding if not just recognition that we need better data.

Visualization of data in any discipline serves various purposes, and the humanities are not distinct in this. However, this chapter has sought to illustrate that there are unique uses and processes in the humanities that distinguish practice and demand ongoing reflection, especially considering the ever-growing authority and pervasive power of visualization in modern society.

In the simplest structural form of practice, information visualization can be bifurcated into analysis and presentation. Within these spheres, it serves additional and varied functions. But ultimately, after the analysis and the results, how information is visualized is based on a scholarly propensity to share. We all feel this necessity to translate and share knowledge to give it a broader and longer life and find an impactful purpose. This simply reaffirms the mutually beneficial collaboration between diverse approaches to visualizing data in the humanities that invoke performance, invite engagement and provoke rather than palliate.

NOTES

1. See The 100 Archive, https://www.100archive.com.
2. See The Vikus Viewer – Urban Complexity Lab – University of Applied Sciences Potsdam, vikusviewer.fh-potsdam.de.
3. Software Studies Laboratory, http://selfiecity.net/selfiexploratory/.

REFERENCES

Allen, William L. 2018. "Visual Brokerage: Communicating Data and Research through Visualisation." *Public Understanding of Science* 27 (8): 906–22. https://doi.org/10/gf7wd4.

Bederson, Benjamin B., Ben Shneiderman, and Martin Wattenberg. 2002. "Ordered and Quantum Treemaps: Making Effective Use of 2D Space to Display Hierarchies." *ACM Transactions on Graphics* 21 (4): 833–54.

Brüggemann, Viktoria, Mark-Jan Bludau and Marian Dörk. 2020. "The Fold: Rethinking Interactivity in Data Visualization." *Digital Humanities Quarterly* 14 (3).

D'Ignazio, Catherine and Rahul Bhargava. 2018. "Creative Data Literacy: A Constructionist Approach to Teaching Information Visualization." *Digital Humanities Quarterly* 12 (4).

Drucker, Johanna. 2010. Graphesis: Visual Knowledge Production and Representation. http://scholar.google.com/scholar_url?url=https://journals.tdl.org/paj/index.php/paj/article/download/4/50/0&hl=en&sa=X&ei=Ny5JYNH7McbYmQHZl554&scisig=AAGBfm22wpsUJ0DUArb5ogFJBUQAXS-OJQ&nossl=1&oi=scholarr.

Drucker, Johanna. 2011. "Humanities Approaches to Graphical Display." *Digital Humanities Quarterly* 5 (1).

Edmond, Jennifer. 2018. "How Scholars Read Now: When the Signal Is the Noise." *Digital Humanities Quarterly* 12 (1).

Ellenberg, Jordan. 2021. *Shape: The Hidden Geometry of Absolutely Everything*. London: Penguin.

Feitelson, Dror G. 2004. "Comparing Partitions with Spie Charts." *Occasional Paper, School of Computer Science and Engineering, The Hebrew School of Jerusalem*. https://www.cs.huji.ac.il/~feit/papers/Spie03TR.pdf.

Fitzpatrick, Kathleen. 2019. *Generous Thinking: A Radical Approach to Saving the University*. Baltimore, MD: Johns Hopkins University Press.

Galloway, Alexander. 2011. "Are Some Things Unrepresentable?" *Theory, Culture & Society* 28 (7–8): 85–102. https://doi.org/10/fxjbmv.

Grossner, Karl and Elijah Meeks. 2014. "Topotime: Representing Historical Temporality." Conference Paper DH, Lausanne. July. https://www.researchgate.net/publication/272827034_Topotime_Representing_historical_temporality.

Grossner, Karl. 2020. "Towards Geospatial Humanities: Reflections from Two Panels." *International Journal of Humanities and Arts Computing* 14 (1–2): 6–26.

Joslyn, Susan and Sonia Savelli. 2021. "Visualising Uncertainty for Non-Expert End Users: The Challenge of the Deterministic Construal Error." *Computer Science* 27 (January). https://doi.org/10.3389/fcomp.2020.590232.

Kennedy, Helen, Rosemary Lucy Hill, Giorgia Aiello, and William Allen. 2016. "The Work that Visualisation Conventions do." *Information, Communication & Society* 19 (6): 715–35. https://doi.org/10.1080/1369118X.2016.1153126.

Kräutli, Florian and Stephen Boyd Davis. 2016. "Digital Humanities Research through Design." *DH Early Career Conference 2016 – Mapping the Scope & Reach of the Digital Humanities*. King's College London, May 20.

Lavin, Matthew. 2021. "Why Digital Humanists Should Emphasise Situated Data over Capta." *Digital Humanities Quarterly* 15 (2).

Lupi, Giorgia and Stephanie Posavec. 2016. *Dear Data: A Friendship in 52 Weeks of Postcards*. Hudson, NY: Princeton Architectural Press.

Manovich, Lev. 2009. "Cultural Analytics: Visualizing Cultural Patterns in the Era of 'More Media'." *DOMUS*. http://manovich.net/content/04-projects/063-cultural-analytics-visualizing-cultural-patterns/60_article_2009.pdf.

Manovich, Lev. 2011. "What is Visualisation?" *Visual Studies* 26 (1): 36–49. https://doi.org/10/dg3k7t.

Manovich, Lev. 2015. "Data Science and Digital Art History." *International Journal for Digital Art History* 1. https://journals.ub.uni-heidelberg.de/index.php/dah/article/view/21631.

McCosker, Anthony and Rowan Wilken. 2014. "Rethinking 'Big Data' as Visual Knowledge: The Sublime and the Diagrammatic in Data Visualisation." *Visual Studies* 29 (2): 155–64. https://doi.org/10/gh8w2z.

Plewe, Brandon. 2002. "The Nature of Uncertainty in Historical Geographic Information." *Transactions in GIS* 6: 431–56. doi: 10.1111/1467-9671.00121.

Software Studies. 2014. The Selfieexploratory. http://selfiecity.net/selfiexploratory/.

Tufte, Edward R. 2001. *The Visual Display of Quantitative Information*, 2nd edn. Cheshire, CT: Graphics Press.

Tufte, Edward R. 2006. *Beautiful Evidence*. Cheshire, CT: Graphics Press.

van Zundert, Joris. 2016. "Screwmeneutics and Hermenumericals: The Computationality of Hermeneutics." In *A New Companion to Digital Humanities*, edited by S. Schreibman, R. Siemens, and J. Unsworth, 331–47. New York, NY: John Wiley and Sons.

Wang, Weixen, Hui Wang, Dai Guozhong, and Hongan Wang. 2006. "Visualisation of Large Hierarchical Data by Circle Packing." *Conference Proceedings of the 2006 Conference on Human Factors in Computing Systems*, CHI 2006, Montréal, Québec, Canada, April 22–27. doi: 10.1145/1124772.1124851.

Whitelaw, Michael. 2015. "Generous Interfaces for Digital Cultural Collections." *Digital Humanities Quarterly* 9(1).

PART THREE

Public Digital Humanities

CHAPTER TWENTY-ONE

Open Access in the Humanities Disciplines

MARTIN PAUL EVE (BIRKBECK, UNIVERSITY OF LONDON)

It is easy to overlook the magnitude of change involved in the shift to the digital publication of scholarship. In moving all costs to first copy and reducing those of all subsequent versions to almost zero, digital publication, when coupled with academic systems of remuneration, carries to some startling logical conclusions. Namely: that if we can find a way to pay for all the labor of publishing the first copy, we could give anybody access to read the published version, without having to charge them. This is called "open access" (OA) and it refers to conditions under which peer-reviewed scholarship is made free to read and free to re-use with attribution (for the seminal work on this subject, see Suber 2012).

Open access was formalized in approximately 2002, with the triple signing of the Budapest Open Access Initiative, the Bethesda Statement on Open Access Publishing, and the Berlin Declaration on Open Access to Knowledge in the Sciences and Humanities (Chan et al. 2002; "Berlin Declaration on Open Access to Knowledge in the Sciences and Humanities" 2003; Suber et al. 2003). Each of these declarations, in its own way, notes that there are educational benefits in allowing people to read work without having to pay. These range from ensuring that the poorest in the world can have access to the scientific literature to ensuring that interested third parties can read about their own histories without facing unaffordable subscription fees. The last of these declarations, in particular, specifies that these benefits can be found across all disciplines: in the sciences *and the humanities*.

There are several pieces of terminology around open access that are worth spelling out up front. OA comes in different "color" flavors. Green open access refers to conditions under which an author publishes in a journal and then deposits their author's accepted manuscript (or later version if permitted) in an institutional or subject repository. This is the version of open access that the UK's Research Excellence Framework uses. It does so since the predecessor to Research England, HEFCE, found that 96 percent of journal articles submitted in REF2014 *could* have been made openly accessible under this route (Poynder 2015). That is, many publishers have liberal policies that will allow academics to deposit a version of their paper for open dissemination. However, the green route often does not provide access to the version of record (the final PDF of a paper, for instance). There is, then, a further strand of open access, called gold open access, in which the publisher makes material openly available at source. This could mean that, for instance, a user lands

---

Much of this chapter is an update of Eve 2014a.

on a journal article's page and, instead of having to login, can download the article immediately without providing credentials or any fee. Gold open access implies a different business model for a publisher. After all, if one cannot sell access to scholarship, the implication is that there must be another route for funding its publication.

OA also comes in various shades of licensing. "Gratis" work refers to those pieces that are free to read, but that come with no additional permissions for the reader. One cannot legally share these with others, nor quote more extended passages than is permitted by fair dealings or fair use legislations. On the other hand, "libre" OA refers to conditions under which an open license—such as the Creative Commons Attribution License—is applied to works. These licenses grant a series of additional permissions to downstream users to permit re-use. Most of the Creative Commons licenses that are used for open access (except for the CC0 license) require attribution, but also specify that a re-user may not imply any endorsement from the original author. As I will go on to note, this is a highly controversial area in the humanities disciplines.

Which brings me, after this definitional throat clearing, to the subject of this chapter. It might seem, from the above, that OA is an ideal and obvious solution to the global inequality of access to scholarship in the humanities disciplines. After all, in my country, the United Kingdom (often said to be a forerunner in OA policy terms, but one that actually lags behind South America by some distance), most of my colleagues marched in protest against the introduction and then dramatic rise of student fees for access to university. What then could be controversial about allowing anybody to read scholarship for free? Yet, in its practical implementation, open access—particularly in the humanities disciplines—remains a controversial and difficult subject. Open access is by no means universally accepted and it finds critics from both the left and right of the political spectrum. In the remainder of this chapter I will attempt to set out and unpick some of these controversies.

HUMANITIES ECONOMICS 101

In 2010 the British politician, Liam Byrne, Chief Secretary to the Treasury at the time, left a note for his Liberal Democrat successor, David Laws. The note read simply: "I'm afraid there is no money" (Byrne 2015). The same could be said for funding in the humanities disciplines, which Peter Suber has called a "dry climate" (Suber 2014). This has implications for open access.

Indeed, the first thing to note is the difficulty of the economics of humanities scholarship and publishing. Academics are paid to produce research and then are free to publish this wheresoever they choose. Those with secure research and teaching contracts at universities—admittedly a rarer and rarer breed—are nonetheless accorded an academic *freedom from the market*. That is, they are not required to sell their research work, en masse, in order to earn a living. Instead, the university pays them a salary, and they can give away the work, if they so choose.

Systems of accreditation and structured norms of academia, though, require that academics publish their work in peer-reviewed academic journals or with reputable book publishers. These venues, usually independent third-party publishers, oversee the review process (though do not themselves conduct the reviews), sometimes provide copyediting, proofreading, typesetting, a digital and/or print platform, digital preservation, permanent identifiers, ingest into indexing and discovery systems, and a whole host of other activities and services. Traditionally, these venues have sold access either as a subscription, in the case of journals, or for purchase, in the case of books. The typical customer for these is the academic library acting under instruction from faculty.

And make no mistake: these venues can be very profitable at the expense of academic libraries. For many years expenditure on serials (journals) has reached hyperinflationary proportions. As the Association for Research Libraries puts it, "Spending for ongoing resources, which includes print and electronic journals, continues to skyrocket, showing a 521% increase since 1986" (Association of Research Libraries 2017). In the same period, the consumer price index rose by only 118 percent. The most profitable of academic publishers—Elsevier, Wiley, Taylor & Francis—also make staggeringly large margins on their journal sales that dwarf the returns seen by Big Pharma and oil companies (Larivière, Haustein, and Mongeon 2015). Yet the term "academic publishers" harbors many entities, of different types. They range from the giants of the industry, who make almost obscene profits, down to small, independent university presses, who are subsidized internally by their host institutions.

In any case, one of the erroneous criticisms leveled at the open access movement is that it does not respect publishing labor. It is wrongly claimed that open access advocates wish to put all publishers out of business and do not respect the labor of publishing. David Golumbia, for instance, writes that "[o]ne searches OA literature in vain for discussions of the labor issues" (Golumbia 2016). While some OA advocates may indeed disregard this issue and wish for the destruction of academic publishers, Golumbia has clearly not searched the literature as thoroughly as he claims, as articles by Paul Boshears (2013), Emily Drabinski and Korey Jackson (2015), Christopher Kelty (2014), and myself (2014a, 62–7; 2014c, 2016) have all directly addressed this issue.

Nonetheless, this debate aside, it is clear that publishers who provide a professional service require a revenue stream if they are not to be operated as voluntary organizations. While green OA does not require a new business model for journal articles, if a publisher wishes to go "gold," it requires a new way of generating revenue. The most common—and notorious—model for gold open access is the "article processing charge" (APC). In this model, instead of asking readers to pay, the author, their institution, or their funder pays an upfront fee to the publisher. This fee is not any kind of way of bypassing peer review as a form of vanity publishing but is rather there to cover the labor and business expenses of publishing the article. This appears to make sense. For imagine a world in which academic libraries no longer subscribed to journals. In this world that budget could be used to pay for the *outputs* of academics to be openly available. After all, it seems that the same amount of money would be in the system, it would just be used to fund OA.

What this model overlooks, though, is the basic distributional economics of the forms. In a subscription system, the cost is spread between many actors. Each library pays only a fraction of the revenue that a publisher needs to receive. In the APC model a single purchaser is made to bear the entire price of publication. This can work in the natural sciences where large grants can easily bundle \$2,000–\$3,000 on a budget line. In the humanities, this project funding is far harder to come by and most humanists would be laughed out of their Dean's office for suggesting that the institution front this fee. Indeed, most humanities work is subsidized not by external grant funding, but by cross-subsidy from tuition or ongoing unhypothecated research-funding streams.

The APC model of open access has been branded as iniquitous and, indeed, even colonial (Mboa Nkoudou 2020). (Though it is not clear that it is *worse* than a subscription model on this front.) It is feared that such a business model would restrict all but the most elite, wealthy/funded scholars from publishing in high-prestige humanities titles, while also making the high level of selectivity in such titles (with a large amount of labor going into rejecting work) unviable. This is often couched as an attack on academic freedom, although those making this charge usually hypocritically neglect to mention that there are many curbs on publishing in such titles even in the subscription environment (for more on this, see for instance Shrieber 2009; Holbrook 2015; Johnston 2017). The same figures rarely criticize those elite humanities titles—such as *boundary2*, for instance—that are invitation only. Nonetheless, APCs cause concern.

Yet if APCs are bad in the eyes of many humanists, then their equivalent in the book field—the Book Processing Charge (BPC)—is many times worse. Indeed, it is also the high-cost media forms in which the humanities disciplines circulate their work (monographs) that have led to challenges for OA uptake in these spaces.

OPEN PAGES AND NEW LEAVES

The importance of monographs to humanities dissemination—and, therefore, the importance of ensuring that in a future "open" ecosystem these forms are not left behind—has not been lost on funders, even while recognizing the difficulties. Consider, for instance, the requirements imposed by the group of funders known as cOAlition S. While this group's headline policy—the quasi-Bond villainesque-sounding "Plan S"—was uncompromising on journals, it acknowledged that while its "principles shall apply to all types of scholarly publications" it is also "understood that the timeline to achieve Open Access for monographs and book chapters will be longer and requires a separate and due process" (cOAlition S 2018). At least part of the problem that cOAlition S recognized is the fact that "books" are not just one thing: "It is expected that how and whether Plan S applies to different forms of monograph or books, such as trade books, will be considered as part of the future guidance," with said guidance projected by the end of 2021 (cOAlition S 2020).

To understand the challenges of OA for books, one needs, first, to ask: how much does it cost to publish an academic book? This simple question yields a range of answers. A Mellon Ithaka study from 2016 found that, at US university presses, the figure was, at the lowest, $15,140, while the most expensive title cost $129,909 to produce (Maron et al. 2016). These costs are debatable. Open Book Publishers, a younger, born-OA press, works on a different model and estimates its costs to be "$1k for distribution" and "$6.5k for 'first copy' title setup costs" (Gatti 2015). This is achieved, in part, by lowering the cost of "acquisitions." In the US university press scene, acquisitions and developmental editing are seen as core activities of these entities. Yet these are expensive activities and some have questioned their value.

In any case, the high cost of book production has meant that concentrating economic models for open access, such as book processing charges, yield utterly unaffordable prices. For instance, Palgrave Macmillan, as part of SpringerNature, charges a fee of 13,000 euros per book (SpringerNature n.d.). Upon hearing this, many scholars' jaws hit the floor. However, Palgrave charges approximately 93 euros for at least some of its hardback volumes (such as for my own, Eve 2014b, as of September 2020). This means that, in such a case, the BPC would be the equivalent of selling 140 copies, part of a longer trend of smaller unit sales of academic monographs that is still unfolding (for more on this, see Crossick 2015). Nevertheless, with departmental book purchasing budgets often smaller than the cost of a single BPC, the economic distribution of this model is prohibitive (for more on the landscape of OA business models for books, see Pinter 2018). One estimate for the cost of requiring all monographs to be open access in the UK's Research Excellence Framework put the figure at £96m (Eve et al. 2017).

That said, other models are available. The breadth of business models that can support open monograph publication has been sampled in a range of recent reports (Crow 2009; Kwan 2010;

London Economics 2015; Ferwerda, Pinter, and Stern 2017; Speicher et al. 2018; Penier, Eve, and Grady 2020). One of the most striking findings in this area is that the desire for print still remains strong and many younger OA publishers—such as Open Book Publishers, punctum books, and Open Humanities Press—find that they are able to sell enough print copies to support their operation, even though a copy was available to download for free. In other words, open access does not replace print and the affordances of that medium (and reader desires for paper and ink) mean that selling print could remain a viable revenue source.

There are also more interesting revenue models for open access that do not concentrate the cost and that do not load this onto the author. These are called consortial models. In these models many academic libraries pool their resources into a central membership fund so that a publisher can make the work openly accessible, without charging authors or readers directly (Eve 2014c). Examples of this model include the Open Library of Humanities (of which I am a CEO)—which publishes twenty-eight journals using this funding model—and Knowledge Unlatched, which has released over two thousand openly accessible monographs through a similar system (Look and Pinter 2010; Pinter and Kenneally 2013). Such so-called "diamond" models for OA—in which there are fees neither to authors nor readers—eliminate many of the fears that authors have of the APC and BPC models, though they come with other challenges (such as, who decides which titles, at a book publisher, for instance, are made OA?). These models nonetheless equitably spread the cost burden among a large number of actors and achieve open access without ever excluding an author on the inability to pay. However, such models have not found universal favor as they are sometimes less accountable or transparent, and funders, for instance, often wish to pay *for their authors*, rather than to support a publishing infrastructure in general.

OA AND ITS DISCONTENTS

There are other objections to open access in the humanities disciplines that are more fundamental and not related to the economics of publishing. One of the strongest of these is the backlash against open licensing (Holmwood 2013, 2018; Mandler 2013, 2014). There are many reasons for this distrust that mostly stem from the permission to create derivative works. These anxieties range from historians (for instance) fearing that others will corrupt their words to worries that for-profit educational providers will simply appropriate freely disseminated work and use it to undercut the traditional research university. I will briefly cover these objections and the responses to them.

As noted, these fears come from the fact that, although the Creative Commons licenses almost always demand that it be clear that the work has been modified if a third party is re-using it, they do not demand that re-users signal how that work has been altered. The concerns here from historians, in particular, center around political re-use of their material outside the academy by extreme political groups. They are concerned that their words will be altered and attributed to them, by, for example, neo-Nazi groups, with only a footnote specifying that the work has been changed, resulting in reputational damage and historical distortion.

Prominent libel suits, such as *David Irving v Penguin Books and Deborah Lipstadt* over Holocaust denial, indicate that there are consequential and important uses for public history that can result in problems that require recourse to legal remedy. The situation that is here posited is not totally unrealistic. It is also fueled by the fact that we have insufficient publicly available legal advice on the extent to which defamation and libel suits remain viable with respect to work under the attribution clause of the CC BY 4.0 license. The license, for instance, requires that creators waive their moral rights in order for the rights granted by the license to be exercised. The license allows modification and requires attribution. It therefore makes sense that a modification must be attributed unless waived, albeit with modification noted (but there is no requirement to notify the author of modified attribution). It is possible, then, that reputational damage/defamation could ensue from such attribution but that an author would have waived the moral right to pursue such a claim (the right "to object to any distortion, modification of, or other derogatory action in relation to the said work, which would be prejudicial to the author's honor or reputation" specified in the Berne Convention). It is also possible, though, that defamation rests separately from these matters of moral rights within copyright in many jurisdictions. The question might hinge, though, in a court case on whether the attribution was wrongful if it indicated that the text had been modified ("'I hate open access' – Martin Paul Eve, wording modified from original"). If this were the case, though, the CC BY license might allow the attribution of anything to anyone, which seems unlikely to be held up in court.

Further to these objections about licensing, which are mired in legal technicalities, there is also a group that believes that open access is a solution without a problem. Namely, some people believe that everyone who needs access to research already has that access and that allowing access to the general public will result in misinterpretation of these outputs (Osborne 2013, 2015). This seems a difficult argument to sustain, given that many people graduate from humanities courses every year and could get access to research articles, if they were willing to pay. At present, we see low levels of continued engagement with those who have left university and it seems likely that at least one reason for this is that the material that they were able to access while they were at university is now prohibitively expensive. It remains my belief that if we could provide greater access to our research outputs—without charging readers—we would be contributing to the general education of the world in ways that would be beneficial to everyone.

Finally, there are concerns—mostly among those who are fresh to the debate—about the quality of open access publications. Will such material be peer-reviewed? How will we know if these openly published scholarly works are of the same quality as their subscription and purchase-based rivals? It is worth noting, up front, that there are problems with peer review in itself (see Eve et al. 2021). But the fundamental point remains that whether material has been subject to academic peer review or not is *totally unrelated* to whether that material is then sold or given away for free.

DIGITAL HUMANITIES AND ITS FORMS

I am mindful, in the closing moments of this chapter, that I have been speaking mainly about conventional humanistic output forms—the journal article, the academic monograph—in a volume that is concerned with digital humanities practices. Often, DH produces outputs that are unconventional: software, datasets, websites. Sometimes, these outputs have been OA by default, without there ever having been a discussion about it. Yet, I feel it is important that those in the DH community acknowledge and understand the often-vicious and tricky debates about open access and its economics in the more traditional humanistic space. For debates about openness are engendered and made possible by the shift to digital publishing.

Indeed, this debate only grows more intense. I will close with an anecdote that seems emblematic of this growing divide. In 2016 I attended two events, two days apart. One was a panel event in front of a group of conventional historians. The other was a digital history panel. I was speaking, on both occasions, about open publication and the talk was broadly the same. At the first panel, the response was outrage. I was asked why I wanted to destroy all that was held sacred by this community. I was seen as an outside radical with dangerous ideas that would bring down the walls of a long-established publication culture. At the second—the digital history panel—I received the opposite reaction. "Why," I was asked, "are you so conservative in your views?" Why, it was posited, should we not aim bigger in our aspirations for the global accessibility of knowledge?

Such an anecdote is, at the end of the day, just one instance. It is also, though, a marker of the strong sentiments that the debate around OA can stir.

ACKNOWLEDGMENTS

Martin Paul Eve's work on his chapter was supported by a Philip Leverhulme Prize from the Leverhulme Trust, grant PLP-2019-023.

REFERENCES

Association of Research Libraries. 2017. "ARL Statistics 2014–2015 and Updated Trends Graphs Published." *Association of Research Libraries* (blog). https://www.arl.org/news/arl-statistics-2014-2015-and-updated-trends-graphs-published/.

"Berlin Declaration on Open Access to Knowledge in the Sciences and Humanities." 2003. October 22. http://oa.mpg.de/lang/en-uk/berlin-prozess/berliner-erklarung.

Boshears, Paul F. 2013. "Open Access Publishing as a Para-Academic Proposition: OA as Labour Relation." *TripleC: Communication, Capitalism & Critique. Open Access Journal for a Global Sustainable Information Society* 11 (2): 614–19.

Byrne, Liam. 2015. "'I'm Afraid There Is No Money.' The Letter I Will Regret for Ever." *The Guardian*. May 9. http://www.theguardian.com/commentisfree/2015/may/09/liam-byrne-apology-letter-there-is-no-money-labour-general-election.

Chan, Leslie, Darius Cuplinskas, Michael Eisen, Fred Friend, Yana Genova, Jean-Claude Guédon, Melissa Hagemann et al. 2002. "Budapest Open Access Initiative." February 14.

cOAlition S. 2018. "Plan S." Plan S and COAlition S. https://www.coalition-s.org/.

cOAlition S. 2020. "OA Monographs." Plan S and COAlition S. https://www.coalition-s.org/faq-theme/oa-monographs/.

Crossick, Geoffrey. 2015. "Monographs and Open Access: A Report for the Higher Education Funding Council for England." Higher Education Funding Council for England. https://dera.ioe.ac.uk/21921/1/2014_monographs.pdf.

Crow, Raym. 2009. "Income Models for Open Access: An Overview of Current Practice." SPARC. https://sparcopen.org/wp-content/uploads/2016/01/incomemodels_v1.pdf.

Drabinski, Emily and Korey Jackson. 2015. "Session: Open Access, Labor, and Knowledge Production." *Critlib Unconference*, March. http://pdxscholar.library.pdx.edu/critlib/2015/Conference/11.

Eve, Martin Paul. 2014a. *Open Access and the Humanities: Contexts, Controversies and the Future*. Cambridge: Cambridge University Press. https://doi.org/10.1017/CBO9781316161012.

Eve, Martin Paul. 2014b. *Pynchon and Philosophy: Wittgenstein, Foucault and Adorno*. London: Palgrave Macmillan.

Eve, Martin Paul. 2014c. "All that Glisters: Investigating Collective Funding Mechanisms for Gold Open Access in Humanities Disciplines." *Journal of Librarianship and Scholarly Communication* 2 (3). http://dx.doi.org/10.7710/2162-3309.1131.

Eve, Martin Paul. 2016. "An Old Tradition and a New Technology: Notes on Why Open Access Remains Hard." Martin Paul Eve. February 9. http://eprints.bbk.ac.uk/17024/.

Eve, Martin Paul, Kitty Inglis, David Prosser, Lara Speicher, and Graham Stone. 2017. "Cost Estimates of an Open Access Mandate for Monographs in the UK's Third Research Excellence Framework." *Insights: The UKSG Journal* 30 (3). https://doi.org/10.1629/uksg.392.

Eve, Martin Paul, Cameron Neylon, Daniel O'Donnell, Samuel Moore, Robert Gadie, Victoria Odeniyi, and Shahina Parvin. 2021. *Reading Peer Review:* PLOS ONE *and Institutional Change in Academia*. Cambridge: Cambridge University Press.

Ferwerda, Eelco, Frances Pinter, and Niels Stern. 2017. "A Landscape Study on Open Access and Monographs: Policies, Funding and Publishing in Eight European Countries." Zenodo. https://doi.org/10.5281/zenodo.815932.

Gatti, Rupert. 2015. "Introducing Data to the Open Access Debate: OBP's Business Model (Part Three)." Open Book Publishers (blog). October 15. https://doi.org/10.11647/OBP.0173.0016.

Golumbia, David. 2016. "Marxism and Open Access in the Humanities: Turning Academic Labor against Itself." *Workplace: A Journal for Academic Labor* 28. http://ices.library.ubc.ca/index.php/workplace/article/view/186213.

Holbrook, J. Britt. 2015. "We Scholars: How Libraries Could Help Us with Scholarly Publishing, if Only We'd Let Them." Georgia Institute of Technology. https://smartech.gatech.edu/handle/1853/53207.

Holmwood, John. 2013. "Markets versus Dialogue: The Debate over Open Access Ignores Competing Philosophies of Openness." *Impact of Social Sciences* (blog). October 21. http://blogs.lse.ac.uk/impactofsocialsciences/2013/10/21/markets-versus-dialogue/.

Holmwood, John. 2018. "Commercial Enclosure: Whatever Happened to Open Access?" *Radical Philosophy* 181: 2–5.

Johnston, David James. 2017. "Open Access Policies and Academic Freedom: Understanding and Addressing Conflicts." *Journal of Librarianship and Scholarly Communication* 5 (General Issue). https://doi.org/10.7710/2162-3309.2104.

Kelty, Christopher. 2014. "Beyond Copyright and Technology: What Open Access Can Tell Us about Precarity, Authority, Innovation, and Automation in the University Today." *Cultural Anthropology* 29 (2): 203–15. https://doi.org/10.14506/ca29.2.02.

Kwan, Andrea. 2010. "Open Access and Canadian University Presses: A White Paper." http://blogs.ubc.ca/universitypublishing/files/2010/03/ACUP-White-Paper-Open-Access_Kwan.pdf.

Larivière, Vincent, Stefanie Haustein, and Philippe Mongeon. 2015. "The Oligopoly of Academic Publishers in the Digital Era." *PLOS ONE* 10 (6): e0127502. https://doi.org/10.1371/journal.pone.0127502.

London Economics. 2015. "Economic Analysis of Business Models for Open Access Monographs." London Economics. January. http://londoneconomics.co.uk/blog/publication/economic-analysis-business-models-open-access-monographs/.

Look, Hugh, and Frances Pinter. 2010. "Open Access and Humanities and Social Science Monograph Publishing." *New Review of Academic Librarianship* 16 (supplement 1): 90–7. https://doi.org/10.1080/13614533.2010.512244.

Mandler, Peter. 2013. "Open Access for the Humanities: Not for Funders, Scientists or Publishers." *Journal of Victorian Culture* 18 (4): 551–7. https://doi.org/10.1080/13555502.2013.865981.

Mandler, Peter. 2014. "Open Access: A Perspective from the Humanities." *Insights: The UKSG Journal* 27 (2): 166–70. https://doi.org/10.1629/2048-7754.89.

Maron, Nancy, Christine Mulhern, Daniel Rossman, and Kimberly Schmelzinger. 2016. "The Costs of Publishing Monographs: Toward a Transparent Methodology." New York, NY: Ithaka S+R. https://doi.org/10.18665/sr.276785.

Mboa Nkoudou and Thomas Hervé. 2020. "Epistemic Alienation in African Scholarly Communications: Open Access as a *Pharmakon*." In *Reassembling Scholarly Communications: Histories, Infrastructures, and Global Politics of Open Access*, edited by Martin Paul Eve and Jonathan Gray, 25–40. Cambridge, MA: The MIT Press.

Osborne, Robin. 2013. "Why Open Access Makes No Sense." In *Debating Open Access*, edited by Nigel Vincent and Chris Wickham, 96–105. London: British Academy.

Osborne, Robin. 2015. "Open Access Publishing, Academic Research and Scholarly Communication." *Online Information Review* 39 (5): 637–48. https://doi.org/10.1108/OIR-03-2015-0083.

Penier, Izabella, Martin Paul Eve, and Tom Grady. 2020. "COPIM – Revenue Models for Open Access Monographs 2020." *Community-Led Open Publication Infrastructures for Monographs*. https://doi.org/10.5281/zenodo.4011836.

Pinter, Frances. 2018. "Why Book Processing Charges (BPCs) Vary So Much." *Journal of Electronic Publishing* 21 (1). https://doi.org/10.3998/3336451.0021.101.

Pinter, Frances and Christopher Kenneally. 2013. "Publishing Pioneer Seeks Knowledge Unlatched." http://beyondthebookcast.com/transcripts/publishing-pioneer-seeks-knowledge-unlatched/.

Poynder, Richard. 2015. "Open Access and the Research Excellence Framework: Strange Bedfellows Yoked Together by HEFCE." February 19. http://www.richardpoynder.co.uk/REF_and_OA.pdf.

Shrieber, Stuart. 2009. "Open-Access Policies and Academic Freedom." *The Occasional Pamphlet* (blog), May 28. https://blogs.harvard.edu/pamphlet/2009/05/28/open-access-policies-and-academic-freedom/.

Speicher, Lara, Lorenzo Armando, Margo Bargheer, Martin Paul Eve, Sven Fund, Leão Delfim, Max Mosterd, Frances Pinter, and Irakleitos Souyioultzoglou. 2018. "OPERAS Open Access Business Models White Paper." *Zenodo*, July. https://doi.org/10.5281/zenodo.1323708.

SpringerNature. n.d. "Open Access Books Pricing." https://www.springernature.com/gp/open-research/journals-books/books/pricing.

Suber, Peter. 2012. *Open Access*. Essential Knowledge Series. Cambridge, MA: MIT Press. http://bit.ly/oa-book.

Suber, Peter. 2014. "Preface." In *Open Access and the Humanities: Contexts, Controversies and the Future*, edited by Martin Paul Eve, ix–xi. Cambridge: Cambridge University Press. http://dx.doi.org/10.1017/CBO9781316161012.

Suber, Peter, Patrick O. Brown, Diane Cabell, Aravinda Chakravarti, Barbara Cohen, Tony Delamothe, Michael Eisen et al. 2003. "Bethesda Statement on Open Access Publishing." http://dash.harvard.edu/handle/1/4725199.

CHAPTER TWENTY-TWO

Old Books, New Books, and Digital Publishing

ELENA PIERAZZO (UNIVERSITÉ DE TOURS) AND PETER STOKES (ÉCOLE PRATIQUE DES HAUTES ÉTUDES – UNIVERSITÉ PARIS SCIENCES ET LETTRES)

Digital publishing is a big topic that means different things to different people, and the Digital Humanities has an important theoretical and critical contribution to offer. An important part of this involves reflecting on how the digitization of old books and documents is done and what is changing in the way we read and use these digitized books and editions; and, on the other hand, on the way we digitize scholarly practices and methods. In a previous contribution (Blanke et al. 2014) we concentrated our attention on the aspects of modelization, standardization, and data infrastructure. While some of these aspects are still very central to our discourse (and in fact we will return to them shortly), other aspects have emerged as also being fundamental, such as the sharing and publication of research data as scholarship as well as or instead of polished websites, articles, or monographs, and all the questions that this entails. Digital publications are models, insofar as they are (more or less) conscious selections and representations of certain elements. The digitized object may be presented on a screen as text (a sequence of characters) or as images of the physical object, but it may also comprise data communicated through application programming interfaces (APIs), in mashups and metadata. It follows that the digital object both embeds and is the result of scholarly practice, and therefore that academic output now extends to computer software (code) and even platforms for presentation, archiving and "informal" publications such as blogs and social networks. We therefore argue for a continuity between the digitization of a medieval manuscript, the sharing of the metadata, and the publication of gray literature, namely the so-called "mesotext" (Boot 2009). However, this position raises further questions about academic practices, such as how to publish, and the status of these activities and their outputs. Practices and methods that come from other disciplines are being used in new contexts; the speed of scholarship is also changing, not to mention the context in which it is practiced. Nevertheless, in some circles the publication of an article in an online-only journal is still frowned upon, let alone the recognition of data or code as valid (Digital) Humanities scholarship deserving of academic recognition. The use of these new practices, including new publication formats, should not compromise the quality of discussion, but these transformations are profound and they *can* lead to such problems. The question for us is therefore how to understand and use these changes in a positive and productive way, in which scholarly rigor and the advancement of knowledge still remain at the center of our work. The Digital Humanities have been defined as "project-based scholarship" (Budrick et al. 2012), a definition that indicates a change in the heuristics of the research itself, as well as the implication that such research can be sustained only for the duration of a (funded) project. We still lack the distance to judge this evolution in full, but we can and must continue to survey its impact and the relevance of these changes. A full analysis is of course impossible in a short contribution like this, but in the discussion that follows we will attempt to raise some of the key points and provide some indication of where the issues lie.

DIGITAL PUBLISHING OF OBJECTS FROM THE PAST

Modeling objects, texts, documents, and books

The digitization and publishing of historical books may seem simple: one only needs an appropriate camera and setup, one takes better-than-life photographs of the pages, and presto, the work is done. However, the reality is of course very much more complex than this. Among other things, digitizing requires first establishing what we want to digitize (the book? the text?), then deciding for whom we are digitizing it (scholars? of which disciplines?), how this digitization will be published and accessed (on an existing platform? from a home computer or mobile device?), and for how long this digitization will be needed (for a quick one-off reading? the foreseeable future?). All these questions and others demand answers, particularly when the object is fragile or restricted and may not be available for digitization again.

In the past two decades, solid guidelines and recommendations have been developed for the creation of digital surrogates able to serve different purposes and to last for as long as one can reasonably expect. Metadata standards and protocols now allow for much wider and easier sharing of digitized images than was imaginable only a few years ago: stable and open standards help to ensure longevity and interoperability of images, while protocols for sharing and archiving are helping findability and interoperability. Despite this progress, obstacles still remain, and technical improvements cannot address the need for scholarly reflection around the digitization of historical objects. In particular, the large-scale and easy availability of digital surrogates has opened the door to theoretical discussions around the nature of these representations with respect to the original, as well as about the different uses and cognitive experiences that they provide. For instance, recent publications have brought attention to the need of critically reflecting on the modeling of the objects that we digitize (Pierazzo 2015; Eggert 2019; Flanders and Jannidis 2019). Following McCarty, modeling in this sense can be defined as "the heuristic process of constructing and manipulating models," where a "model" is "take[n] to be either a representation of something for purposes of study, or a design for realizing something new" (2005, 24). McCarty has added that models "are by nature simplified and therefore fictional or idealized representations" of real objects (24). The process of modeling is also an epistemological process: when modeling an object, through this activity we come to know this object, and this activity. Modeling is therefore interpretative, and it is required both to better know the object we model, and to build something new (the object we publish, for instance). By asking which are the main features of books, texts, and works that we need to preserve for their digital publication we investigate the object and we anticipate what the user will do with the digitized resource. In other words, our interpretation of the physical object is reflected in the digital object we publish and will influence how the resulting publication can be used.

In the domain of texts and books, the scholarly community has produced a large number of models, the two best-known being perhaps the TEI and FRBR.[1] The former is focused primarily on the text (hence *Text* Encoding Initiative) and the latter on the relationship between published books as objects in a collection and the content of these books, namely the works produced by authors. As is the case for all models, the two mentioned here each have strengths and shortcomings, and their widespread use should not be confused with their fitness for the purpose for which they have been used. Nevertheless, their usefulness has been clearly demonstrated in practice. A model, even when incomplete and partial, is also a cognitive tool and the iterative nature of the modeling activity is a powerful heuristic device, as demonstrated by the rich literature surrounding these (and other) models (McCarty 2005; Flanders and Jannidis 2019).

Conceptual models should ideally be independent of the technologies used to implement them, but in practice this is rarely the case. The critical apparatus, which can be considered a model for representing textual transmission, has been heavily shaped by the constraints of the printed book: the scarcity of space and the dimension of the page have led to the invention of a system of conventions that are economic and synthetic but hard to grasp for the uninitiated. The now familiar, if not standard, format for the publication of the digital scholarly editions sees the edited text on one side of the screen and the facsimile image on the other. This may seem "natural" but is a result of technological factors, namely the rectangular dimension and "landscape" orientation of the standard computer screen, and this dependency is demonstrated by how poorly it works on mobile devices, for instance. Another example is the method of photographing old books which usually relies on cradles able to flatten pages even for the tightest bindings. This leads to a digitization which proceeds page by page, rather than by two-page opening which is in fact how the original object is seen and used. It is important here to note, then, how many of the features of print and digital publishing which we consider "natural" or which represent scholarly pillars are indeed the results of compromises due to technological constraints.

Multifold Editions

While digital editions are slowly making their way into the academic community, print editions are still the most likely outlet for textual scholarship. This is partly because publishing houses provide not only professional support in the production of the edition but also because a printed edition is stable and more easily accepted as a scholarly output. These considerations have led several digital editorial projects to provide multifold embodiments of their work, that is, publishing the "same" work in different formats such as print and perhaps several different digital forms. This tendency has now become almost a given for any form of new book publication, with publishing houses offering the same "book" as printed objects, print-on-screen (normally PDF) and ePub; digital scholarly editions tend to add the source files as well, often in the form of TEI XML as mentioned above. In addition to these more or less "official" publications, one often finds pre-prints deposited in various open archives, or earlier versions of editions that have been later updated. In fact, one of the most striking features of digital publications is that they can be easily and seamlessly updated at any moment, and while some editions provide access to previous versions for the sake of transparency and citability, most of them do not, leaving this task to tools like the Way Back Machine offered by the Internet Archive.[2] This proliferation of formats does not apply only to digital editions of old books, but is found in all sorts of digital publication. In fact it is complicated and magnified by the fact that all these embodiments can potentially be multiplied over and over, since in practice people can often easily download and republish the content (legally or not), but also thanks to the fact that search engines often provide snippets of content as results of users' queries, decontextualizing and recontextualizing the content each time, a point we will come back to shortly.

Editions as Data, Metadata, and APIs

As suggested here, a digital edition is more than just a representation of a textual object from the past that is shaped by the scholarship of its editor. While printed editions are most easily treated one at a time, and at close inspection, the digital environment allows (and almost invites) a wider set of usages. In a seminal contribution of 2005, Franco Moretti has introduced the concept of "distant reading," namely the possibility of mining large quantities of texts in order to study phenomena that escape close investigation. Beside being texts to be read, digital editions can also be used as data to be investigated and queried, potentially alongside other data from elsewhere. This activity goes well beyond "counting words," a simplistic label that has sometimes been used to define these computational approaches, extending instead to disciplines such as stylometry and authorship attribution. The call to publish the "raw" data that underlies digital editions (for instance the XML source) alongside the finalized edited versions is more current than ever, in a moment where the sharing and publishing of the data and metadata of one's resource are becoming increasingly normal and even required. Even scholars that were afraid to let people look into their encoded texts a few years ago are now publishing their pre-prints and preliminary ideas in institutional or private repositories (see infra) or on blogs, a practice that shows how much our culture has changed in a very short span of time. The availability of source files (XML or otherwise), as well as that of metadata (Dublin Core, METS or other formats) implies also that expectations about the deliverables of editions have changed as well.[3] The publication of FAIR principles and their recommendations concerning metadata (Wilkinson et al. 2016) is changing radically the face of digital publishing of cultural heritage texts, where the availability of standardized data and metadata is in many ways more important than the publication of the edited text within a self-contained, dedicated website.[4]

This principle of "editions as data" can be taken even further through application programming interfaces (APIs). The concept here is to publish the content in a format that is adapted not for people, but rather for other software and computers. It is then up to others to produce the necessary tools for interacting with the content, be that an interface for (human) reading, or any number of other possibilities such as distant reading, stylometry, mashups, and so on. Such an approach is very flexible and potentially very powerful, but it is also a very different and sometimes unsettling form of publication. Principles that seem essential to the very foundation of scholarly publishing are quickly undermined, citation being just one example of this. This question is helped in part by initiatives such as the Canonical Text Service (CTS), which defines a minimal protocol for requesting and therefore referring to specific parts of texts (including sections, chapters, words, and so on).[5] In this way, the text can be published not (or not only) as a complete entity, but (also) as fragments, which can then be republished in new interfaces, combined with other texts from elsewhere, or presented not as text at all but as graphs, charts, or other analyses. The principle of APIs is of course not limited to CTS but is very widespread in the informatics community and is becoming increasingly so also for the Digital Humanities. In particular, another important standard for publication is the International Image Interoperability Framework, or IIIF.[6] As the name suggests, IIIF is primarily focused on the image rather than the text, and so is often (but by no means only) used as a means of publishing images of books and documents. Repositories can publish their content via the IIIF standard, and this content can then be collected automatically by software and treated in some way, for instance shown to the user. A number of different viewers already exist for IIIF content, meaning that one can very easily work with images of books in different environments with different software.

Publishing Datasets and Code

An increasingly important area of change in publishing for the Digital Humanities is the publication not only of final articles and monographs but also the raw data, code, images, and other research outputs that make up the whole. This is important for several reasons: on the one hand, the ideal at least of research is that work should be as transparent and reproducible as possible, so that others can understand exactly what was done, verify the methods and data, and so on. Indeed, this question of the scientific method applied to at least some parts of the humanities has been discussed for well over a century, with palaeography and philology being two examples among many (Derolez 2003, 1–10; Canart 2006; Stokes 2015). Although many aspects of humanities research cannot be quantified and are not meaningfully reproducible, nevertheless it is an ideal that is often sought where possible, and this implies that conclusions based on digital analyses should be supported by publishing not only the article or monograph but also all of the code, data, and the precise steps and parameters used to produce the results. This is already a requirement in at least some branches of the "hard" sciences and is being discussed more and more in computer science as well as in the Digital Humanities.[7]

The line between code and article is also blurred in the principle of "literate programming," which was first proposed by Donald Knuth (1984) but which did not find widespread use until the last decade or so. The principle here is to reverse the normal paradigm in programming: rather than centering the code and adding occasional comments to the users to explain the content, the principle of literate programming is instead to focus on the explanatory text and then embed the code into this. This means that the human thought process takes priority over the machine code structure, and Knuth argued that this in turn would help support intelligibility and transparency in coding. Probably the most common form of this at present, at least for Digital Humanities and data science, is Jupyter Notebooks which allows one to include both formatted text and lines of code, mixing the two in a way that allows one to produce fully functional software embedded in the context of a written discussion.[8] This hybrid form of publication is still not used for software of any complexity but is extremely widespread for discussions particularly of data analysis as well as other uses in Digital Humanities and elsewhere, so much so that a search of GitHub at the time of writing returns close to nine million examples of Jupyter Notebooks.[9]

Despite this very widespread use of notebooks, blogs, and other less formal publication formats, scholars can still be reluctant to publish their raw research data, for various reasons including that they may be concerned that others will use the data to publish before the original authors, or even that the data is "too messy" for publication. Indeed, many questions arise here, one of which is what should be published, and indeed what even should be considered as "research data," particularly for the humanities. Few researchers in the humanities would consider publishing their private notes taken during a visit to the library, for instance, and in general this seems reasonable as publication is traditionally reserved for the final analysis rather than working notes. However, as we have seen, the role of publishing and indeed the notion of finality in research is changing, such that the question of what should be published becomes less and less clear.

SCHOLARLY INFRASTRUCTURE FOR DIGITAL PUBLICATIONS

The changing nature of scholarly publication discussed above raises a number of practical and theoretical questions about the nature and format of publication, and the impact of this on scholarship and vice versa. Furthermore, publication is at the core of academic practice and evaluation, and so these changes necessarily have an impact on scholarship and how it is carried out. Digital publishing of scholarly outputs, whether digital editions of books of the past or new contributions, is still struggling to find recognition in the appropriate academic venues, particularly when it comes to early career scholars. Digital publications can be tainted with being self- or, worse, vanity publications, since in many cases they do not go through the filter and quality checks that characterize print. To publish, it is enough to have an Internet connection and an account in some publishing service or have some server space. From here, it is easy to generalize, thinking that all digital publications are just the result of a self-evaluation, or of our inner circles. Even project websites hosted by universities do not escape this logic, in the sense that while securing some funding requires going through sometimes extremely strict processes, the publication of the website is in a sense left to the care of the project teams and their internal editorial workflow with little or no external verification.

The challenges of establishing scholarly infrastructures for digital scholarly output go well beyond the creation of an equivalent to peer review, since academic trust and quality require a series of facilities like the possibility of citing texts which do not disappear in the meantime, of verifying who did what and what one can do with the outputs that are offered online, all points that will be discussed here.

Existing Infrastructures

One of the many challenges in digital publication is the degree to which standard frameworks can be used for publication. In some ways this "prêt-à-porter" model would be ideal, since it would enable scholars and indeed publishers to publish their material with a relatively modest investment in terms of funding, programming expertise, and so on.[10] On the other hand, practice has shown that this is very difficult to achieve, since editions and indeed digital publications in general are very different, with different needs, requirements, and data models as discussed above, and this means that most editions take a "haute couture" approach of custom-made software. This has implications for cost, as discussed above, and sustainability, since as a general principle that the more a publication is based on ad hoc data formats and software, the more difficult it is to sustain in the longer term. It is also becoming increasingly clear that the existing body of highly specialized and idiosyncratic publications of the last twenty years is no longer sustainable, and publishers (including academic centers) are increasingly being forced to make difficult decisions around closing down publications even when in some cases they are still important and in regular use (Smithies et al. 2019). Once again, the problem is somewhat easier to manage if one considers only the data, but even here significant difficulties arise, for instance around who should bear the responsibility of maintaining this content.

Many attempts have been made to balance this tension between "prêt-à-porter" and "haute couture," and in practice one can now go a relatively long way towards standard frameworks, with sophisticated systems that do most of the work with standard templates and can then be customized for the final details (examples of which include the TEI Publisher, the Edition Visualisation Technology, and TEI plugins for Omeka and Omeka S, among others) (Pierazzo 2019). Another approach is to focus on the "publication as data." This approach is more limited but is very much more manageable in practice. Here, rather than attempting to preserve the full interface and "experience" of the publication, the decision is instead to focus on preserving the data and perhaps the accompanying code, but not in a way that users can interact with directly. This is becoming increasingly easy to achieve, given the number of infrastructures for scientific data which are being used as archival and publication formats for the humanities. In a European context, the best-known example is probably Zenodo, which is supported by CERN, the OpenAire initiative and the European Commission's research program. Described as a "catch-all repository for EC funded research,"[11] Zenodo is specifically aimed to encourage "Open Science" by providing near-unlimited storage for researchers regardless of their institution, country, or source of funding. Researchers are explicitly encouraged to submit "data, software and other artifacts in support of publications," but also other content such as "the materials associated with the conferences, projects or the institutions themselves, all of which are necessary to understand the scholarly process." Upon submission, a dataset is automatically given a stable identifier (a DOI), and updates to the dataset are automatically versioned and given new identifiers in order to ensure long-term citability and stability. Although "Science" is here used in the European sense of scholarly research of any type (including the humanities), the reality is nevertheless that Zenodo had its origins in the "hard" sciences, as revealed by its basis in CERN, but use by other researchers is nevertheless rapidly increasing.

A different approach is provided by GitHub which is a private company that has been used for some time now to publish code and, increasingly, other forms of data and even text-based publications.[12] The Git model explicitly encourages the principle of "commit [or publish] early, commit often," an approach that stands strongly in contrast to the traditional approach of publishing only a "final" version after a long period of careful reflection, writing, and checking. Indeed, an interesting example here is the parallel operation of GitHub (and GitLab), Zenodo, and the Software Heritage Archive. Summarizing somewhat crudely, the model is that one more or less continually publishes to GitHub or an instance of GitLab, with an implicit expectation but no guarantee that this will remain available for the foreseeable future. At specific moments, when the authors decide that the content has reached a sufficiently stable point, it can be automatically harvested by Zenodo where it receives a DOI and is published in an archived form for the long term. At the same time, all the small updates that are made to GitHub are also automatically harvested by the Software Heritage Archive, where again they are assigned long-term identifiers which can be resolved to URLs in much the same way as DOIs. This then results in three separate copies of the material, hosted on three distinct infrastructures, two of which have permanent identifiers and are designed along FAIR and archival principles, and the third intended more for day-to-day use. This seems to bring the advantage of three different forms of archiving and publication, for three different purposes, as well as the improved likelihood of at least one of these remaining available. However, it in turn poses further questions such as which version, if any, is "the" definitive one.

Questions arise here also around the use of public or private services for publication and archiving. On the one hand, one may wish to avoid private companies due to concerns around the control and ownership of data, the guarantee that data will remain freely accessible for research, and the perception that public money means at least theoretical accountability to the public and more specifically to voters, as opposed to the very large technology companies which increasingly give the impression of being accountable to no one. On the other hand, many custom-built local setups lead to fragmentation of data, the difficulty of finding content, and the risk of having to duplicate content to ensure visibility or even compliance with different and sometimes conflicting requirements of funding bodies, home institutions, and so on. The question of sustainability is more complex: even very large multinational corporations can still go bankrupt or (perhaps more likely) decide that a given service is no longer profitable and so terminate it at short notice. However, much the same can be observed in public institutions, as funding priorities and indeed governments change over time, such as the Arts and Humanities Data Service in the United Kingdom which began in 1996 but stopped receiving deposits in 2008 and was decommissioned entirely in 2017.[13]

Peer Review

Peer review is sometimes considered one of the pillars of scholarship: the fact that the arguments and findings of a scholar or a group of scholars must be validated by their own peers has been in place since the seventeenth century and constitutes arguably the single most important method for building trust and knowledge. However, digital publications have challenged this assumption in many ways, arguing that digital scholarship does not need it, or that it should be substituted by commentaries, or by online social practices, or proposing new, open, and more transparent forms of validation and so on. Borgman (2007, 84) has broadened the question by discussing how legitimization is built in digital form and maintains that while the traditional forms of validation ("authority, quality control, certification, registration in the scholarly priority and trustworthiness") remain substantially the same, "new technologies are enabling these functions to be accomplished in new ways." For example, in February 2015 Matthew Jockers (2015a) published a contribution on his blog maintaining that his toolset (called the "Syuzhet package") was able to analyze plots of novels and thanks to this he had been able to detect the existence of six or seven novel plot shapes; this claim was picked up by several online journals and made a bit of a sensation. However, weeks after this publication, one scholar, Annie Swafford, started to point out several issues with the package; several other scholars followed her lead and by the beginning of April, Jockers declared the "requiem" of this part of the package (2015b). As pointed out by Jockers himself, it is worthwhile noting how these flaws were discovered within just a few weeks, after more than daily online contributions on blog posts and tweets, while offline such discussion would have taken "years to unfold." Digital scholarship can indeed be fast, and scholarly validation can be reached within a few weeks and a few hundred tweets, a fact that might prompt one to suggest that formal peer review is finished.

The alleged obsolescence of peer review is based on considerations that it was invented in times of scarcity, namely when the limited physical space in scholarly journals forced one to choose what to print. It can then be argued that, once those constraints are removed, there is no reason why contributions should be rejected, since scholarly contributions will go through a natural selection, with good contributions being cited and used, while bad contributions will be substantially ignored. Ford (2008) reflects on the mechanism of building scholarly authoritativeness and concludes that this is achieved through a form of accountability that is established on a disciplinary basis, namely that not all disciplines build their trustworthiness in the same ways. This process is indeed multifold, in the humanities as in any other sector, but for the most part it relies on the way we "allow" some content to be present in the scholarly discourse. The publications of gray literature in various types of disciplinary, national, or even private repositories are challenging this assumption (for which see above), since what we "allow" to be published here is indeed "whatever their authors think fit." However, while many scholars are equipped to evaluate the quality of publications, official or not, this may not apply to students or less experienced researchers, and least of all by those with no expertise in the field, with the risk that unfinished or "half-cooked" material may be cited and trusted beyond its worth. These considerations have and still are fueling a general distrust of online publications, with the results that some people may see their career expectations being jeopardized by the fact of having published online.

The scholarly community has reacted to the issues in different ways. Several professional organizations such as the MLA (2012) have published series of guidelines for assessors and career panelists on how to evaluate online resources. In France, the National Scientific Research Council (CNRS) has sponsored several tools to improve the quality of online publications, favoring the reaction of digital and hybrid publishing houses, a hub of open access online journals and a national pre-print repository now used for the assessment of careers and research centers, forcing the latter then to self-regulate.[14] Some other professional organizations have developed forms of accreditation such as badges that can be added to websites, or lists of approved or otherwise certified sites, examples here including the Medieval Electronic Scholarly Alliance (MESA) and NINES projects, or the Medieval Academy of America's database of digital resources, which includes publication of the standards for evaluation and criteria for inclusion.[15] The situation is changing rapidly, then, but a long road still remains.

Citation, Credit, and Intellectual Property

One important question that arises here is how to properly manage credit and attribution, particularly when so many different people are contributing to publications in many different ways and at different levels. One may well argue that this is not a new problem: after all, we are increasingly recognizing that the print model favors the "headline" author but ignores entirely the contribution of all the other people who have also influenced and sometimes even directly written the text, such as the editors, copyeditors, typesetters, publishers, research assistants, librarians and archivists, artists or draftspeople, and more. Nevertheless, the fact remains that better systems for credit and attribution are necessary, particularly since an increasing number of early-career positions are focused on producing this source of content for digital projects, and it is precisely these people for whom proper attribution and visibility is the most important. It is therefore imperative that everyone but especially those in senior positions do all that is possible to ensure that credit is properly given, both within projects and when external content is being harvested and re-used in any way. Doing so is not as straightforward as one might hope, however: for instance, proper crediting and attribution also requires proper citability, since the one is impossible without the other. As we have seen above, citing texts is complicated when the text itself is fluid, and although systems such as CTS are significant contributions here, many difficulties remain including not only that these systems need to be implemented in practice. Similarly complex are questions of intellectual property, copyright, and licensing. On the one hand, it may be very unclear who owns the intellectual property, and this can be a significant barrier to the forms of publication discussed here. The principle is becoming increasingly widespread that cultural heritage and even information should be free to all, and one can certainly dispute the practices of some publishing houses which have fees and revenues that seem disproportionate to their services. Indeed, the increasing role of open access publishing has been implicit throughout this chapter but not expressly addressed. However, publishers continue to play a very important role and are likely to do so for the foreseeable future, since authors normally have neither the desire nor the aptitude to fulfill these tasks, and their labor must of course be paid for like all others. How to manage this in a world of constant re-use, re-publication, mashups, and so on is by no means clear.

Citation, credit, and intellectual property are all complicated by these new forms of publication. If a transcript is prepared by one person in one project, then elaborated with different layers of markup potentially by different people in different projects, then integrated into a collection by someone else and finally processed and analyzed by someone else, then properly crediting all these people requires significant care and effort and assumes that this history of use and re-use is properly documented and available, something that is by no means always the case. In principle, one can imagine some form of tracking, whereby authorship in its different senses is labeled in a fine-grained way and re-use and re-publication of data can thereby be followed, quantified, and rewarded, whether that reward is in terms of payment, career progression, citation metrics, or so on. The technology here would be relatively straightforward, but the details are complex and potentially very problematic, so a great deal of thought and care would be required.

CONCLUSIONS

The discussion here touches on just some of the important issues that have arisen due to the changing nature of publication. Many others remain which are no less pressing or difficult and which can only be hinted at here. Overall, then, the changing nature of publication has in many ways moved the center of authority and control, since publishers and authors can no longer predict the forms in which their material will be accessed. This situation has led Vitali-Rosati to the elaboration of the theory of "editorialisation" (2018), in which he claims that authors and editors in a digital space have ceased to be central to the dissemination of knowledge and have been replaced by APIs and content management systems. While this conclusion may feel a bit extreme, it is certainly true that nobody can predict the way any given content, including a digital scholarly edition, will be accessed and consumed, and if this could be true to some extent also for print publications, the digital format has taken this affordance very much further. The many lives and embodiments of a text prove its relevance and impact, for instance, but the risk of some or all of one's publication being taken out of context and bent to new, unexpected meanings is indeed something that can make a scholar very uncomfortable. It is clear that we need to reflect on this and indeed all these points, and potentially to take action if we want to have a role in the publication of our content, and in the forms of dissemination and use of the work of the editor and the edited texts.

NOTES

1. The Functional Requirement for Bibliographic Records (FRBR) is a model developed by the International Federation of Library Associations and Institutions (IFLA) which is intended to describe and facilitate the retrieval of bibliographic objects. Currently, the last version available from the IFLA website is dated 2008: see http://www.ifla.org/VII/s13/frbr/. For the TEI, see further http://www.tei-c.org.

2. See https://web.archive.org/.
3. The set of metadata developed by the Dublin Core Metadata Initiative is perhaps the most used format of metadata of the web: see https://dublincore.org/. The Metadata Encoding & Transmission Standard (METS) is one of many metadata standards developed, distributed and maintained by the Library of Congress: see https://www.loc.gov/standards/mets/.
4. The "FAIR" principles state that data should be Findable, Accessible, Interoperable and Reusable: see, for example, https://www.go-fair.org/fair-principles/.
5. The formal specification for the CTS URN is available at https://cite-architecture.github.io/ctsurn_spec/. Similar to the CTS is the Distributed Text Service, DTS: the details are not relevant to this discussion, but for more see https://distributed-text-services.github.io/specifications/.
6. See http://iiif.io/.
7. One very broad example is the OpenAIRE initiative (https://www.openaire.eu), and for a very specific one see OCR-d (https://ocr-d.de).
8. See further Jupyter, https://jupyter.org.
9. This figure is obtained by searching GitHub for files with extension:ipynb (https://github.com/search?&q=extension%3Aipynb, but note that this search requires a GitHub account). A search for repositories containing "ipynb" where the language is "Jupyter Notebook" gives approximately 2,600 results (https://github.com/search?l=Jupyter±Notebook&q=ipynb&type=Repositories); this search is open to those without accounts but gives only repositories (collections of files), not individual notebooks.
10. The first part of this section, including the terms "prêt-à-porter" and "haute couture," draws heavily on Pierazzo (2019).
11. This and the other citations on Zenodo that follow are from https://about.zenodo.org.
12. Summarizing somewhat crudely, git is the name of software which can be used to manage distributed copies of files; GitHub is a commercial service and website which can be used to host and publish content managed via git. Also of interest is GitLab, another commercial service which is very similar to GitHub but can be used either as a service or installed on an institutional or other server. See further https://git-scm.com, https://github.com and https://about.gitlab.com.
13. The former site is http://www.ahds.ac.uk, which is currently occupied by a placeholder page. For the original plan of the AHDS see Greenstein and Trant (1996), and for general process of decommissioning, applied to a significant number of sites, see Smithies et al. (2019).
14. Open Edition is a publicly funded digital publishing infrastructure for scholarship; it publishes both Freemium and Open Access books as well as a large number of peer-reviewed journals (https://www.openedition.org). HAL-Archives Ouvertes is a nationwide, pluridisciplinary open archive for all sorts of scholarly outputs (https://hal.archives-ouvertes.fr/).
15. See https://mesa-medieval.org, https://nines.org and http://mdr-maa.org, respectively.

REFERENCES

Blanke, Tobias, E. Pierazzo, and P. A. Stokes. 2014. "Digital Publishing Seen from the Digital Humanities." *Logos* 25: 16–27. https://doi.org/10.1163/1878-4712-11112041.

Boot, Peter. 2009, *Mesotext: Digitised Emblems, Modelled Annotations and Humanities Scholarship*. Amsterdam: Pallas Publications.

Borgman, Christine L. 2007. *Scholarship in the Digital* Age: *Information, Infrastructure and the Internet*. Cambridge, MA: MIT Press.

Budrick, Anna, Johanna Drucker, Peter Lunenfeld, Todd Presner and Jeffrey Schnapp. 2012. *Digital Humanities*. Cambridge, MA: MIT Press.

Canart, Paul. 2006. "La Paléographie est-elle un art ou une science?" *Scriptorium* 60 (2): 159–85.

Derolez, Albert. 2003. *The Palaeography of Gothic Manuscript Books from the Twelfth to the Early Sixteenth Century*. Cambridge Studies in Palaeography and Codicology, 9. Cambridge: Cambridge University Press.

Eggert, Paul. 2019. *The Work and the Reader in Literary Studies: Scholarly Editing and Book History*. Cambridge: Cambridge University Press.

Flanders, Julia and Fotis Jannidis, eds. 2019. *The Shape of Data in Digital Humanities: Modeling Texts and Text-Based Resources*. London: Routledge.

Ford, Michael. 2008. "Disciplinary Authority and Accountability in Scientific Practice and Learning." *Science Education* 92 (3): 404–23. https://doi.org/10.1002/sce.20263.

Greenstein, Daniel and Jennifer Trant. 1996. "Arts and Humanities Data Service." *Computers & Texts* 13. http://users.ox.ac.uk/~ctitext2/publish/comtxt/ct13/ahds.html.

Jockers, Matthew L. 2015a. "Revealing Sentiment and Plot Arcs with the Syuzhet Package," "Some Thoughts on Annie's Thoughts ... about Syuzhet," "Is that your Syuzhet Ringing?" https://www.matthewjockers.net//?s=Syuzhet.

Jockers, Matthew L. 2015b. "Requiem for a Low Pass Filter." https://www.matthewjockers.net/2015/04/06/epilogue/.

Knuth, D. E. 1984. "Literate Programming." *The Computer Journal* 27 (2): 97–111. https://doi.org/10.1093/comjnl/27.2.97.

McCarty, Willard. 2005. *Humanities Computing*. Basingstoke: Palgrave Macmillan.

MLA (The Modern Language Association). 2012. "Guidelines for Evaluating Work in Digital Humanities and Digital Media." https://www.mla.org/About-Us/Governance/Committees/Committee-Listings/Professional-Issues/Committee-on-Information-Technology/Guidelines-for-Evaluating-Work-in-Digital-Humanities-and-Digital-Media.

Moretti, Franco. 2005. *Graphs, Maps, Trees: Abstract Models for a Literary History*. London: Verso. http://www.loc.gov/catdir/toc/ecip0514/2005017437.html.

Pierazzo, Elena. 2015. *Digital Scholarly Editing. Theories, Models and Methods*. London: Routledge.

Pierazzo, Elena. 2019. "What Future for Digital Scholarly Editions? From Haute Couture to Prêt-à-Porter." *International Journal of Digital Humanities* 1: 209–20. https://doi.org/10.1007/s42803-019-00019-3.

Smithies, James, Carina Westling, Anna-Maria Sichani, Pam Mellen, and Arianna Ciula. 2019. "Managing 100 Digital Humanities Projects: Digital Scholarship and Archiving in King's Digital Lab." *Digital Humanities Quarterly* 13 (1). http://www.digitalhumanities.org/dhq/vol/13/1/000411/000411.html.

Stokes, Peter A. 2015. "Digital Approaches to Palaeography and Book History: Some Challenges, Present and Future." *Frontiers in Digital Humanities* 2 (5). https://doi.org/10.3389/fdigh.2015.00005.

Swafford, Annie. 2015. "Problems with the Syuzhet Package," "Continuing the Syuzhet Discussion," and "Why Syuzhet Doesn't Work and How We Know." *Anglophile in America: Annie Swafford's Blog*. https://annieswafford.wordpress.com/category/syuzhet/.

Vitali-Rosati, Marcello. 2018. *On Editorialization: Structuring Space and Authority in the Digital Age*. Amsterdam: Institute of Network Cultures. https://papyrus.bib.umontreal.ca/xmlui/handle/1866/19868.

Wilkinson, Mark D., Michel Dumontier, IJsbrand Jan Aalbersberg, Gabrielle Appleton, Myles Axton, Arie Baak, Niklas Blomberg et al. 2016. "The FAIR Guiding Principles for Scientific Data Management and Stewardship." *Scientific Data* 3 (1): 160018. https://doi.org/10.1038/sdata.2016.18.

CHAPTER TWENTY-THREE

Digital Humanities and the Academic Books of the Future

JANE WINTERS (SCHOOL OF ADVANCED STUDY, UNIVERSITY OF LONDON)

On December 15, 1999, the Institute of Historical Research (IHR) at the University of London ran a one-day conference which asked bluntly "Is the monograph dead?" (Institute of Historical Research 1999). Unfortunately—this was the 1990s after all—the conference program was never published online, and delegates were instead encouraged to contact the IHR's conference secretary for a printed program and registration form. The event was, however, covered by *Times Higher Education*, under the equally gloomy rubric "Rumours of a death that may be true." This short report, which makes it clear that the apparent crisis was long in the making, ends with a comment from Michael Prestwich, who suggests that "One day academics may be able to download a monograph from their computers ... Far from killing it off, this could mean a new lease of life for the monograph." This is in sharp contrast to the opening sentence of the piece, which floats the idea that the end of the scholarly monograph might be "hastened by the internet," as well as by the pressures on academics to publish more frequently in order to meet the requirements of national assessment exercises (Swain 1999).

More than twenty years later, this ambivalence about the impact of digital technologies on "the book" remains. The most striking chapter in an edited collection arising from the Academic Book of the Future project (funded by the Arts and Humanities Research Council in the UK from 2014 to 2016) is a playful and provocative imagining of a future where "wearable books" are the norm. It is framed as a dystopia, in which "Linked Ideas" dominate: "Linked Ideas meant that the old distinction between articles, monographs and co-authored books disappeared. Text was text. It was just a question of the length of an academic debate around an idea; the value of what was being said rather than how long it took for you to say it." The chapter concludes by envisioning a world, time-stamped to 2038, where the discipline of print humanities is beginning to capture headlines (and funding) and the printed book has a newly "disruptive" role to play in scholarly communication (Pidd 2016). The deliberate inversion of contemporary debates about the impact of the digital, and of digital humanities, is a useful exercise in teasing out what we do and do not value in books, although the chapter does retain the shorthand of "book equals printed object" for broad parodic purposes. In this scenario, it is the materiality of the printed book which dominates and not the content of the book or the ideas that it communicates.

The volume in which Pidd's chapter appears is an example of one kind of experimentation with the form of the book that has become increasingly familiar. *The Academic Book of the Future* is a short-form monograph published in the Palgrave Pivot series (Lyons and Rayner 2016). There have always been books that are shorter than might usually be expected, but within many humanities disciplines, the norm has for some time been around 80,000–100,000 words. Many commissioning and academic editors will have had the experience of asking authors to cut perhaps as many as 30,000 words in order to achieve this limit. There is a marked tendency to write "long," especially from first-time authors who are often adding new material to their already long PhD research. The short-form monograph, or "minigraph," of which Palgrave was an early adopter in 2012, was a breath of fresh air. Here was something a bit different, which promised speedier publication and flexibility (features similarly emphasized by the Cambridge Elements series launched in 2019 (Page 2019)). Except that a short book which is published either in print or as a PDF only differs from a traditional monograph by virtue of its length. An author may have to wait less time to see their book in print, but the processes of commissioning, review, production, and publication remain the same.

So why even mention short-form monographs? There are two reasons for starting this chapter by talking about minigraphs: first, while they do not represent the kind of innovation that many in the digital humanities would like to see, they are indicative of a willingness among both authors and publishers to rethink what they do; and second, the length of written outputs is a key question for humanities researchers, and one which begins at the dissertation. This is not the place to discuss the form of the humanities PhD—that is a whole different project—but the constraints of the book format, and the disadvantages faced by researchers engaging with digital methods, are present from the very beginning of a digital humanities research career. Let us take the example of a PhD student who is working with a combination of web and social media archives to study the impact of the 2020 Covid-19 pandemic on the cultural heritage sector in the UK. That student will very likely have to learn a number of new digital methods, become expert in the use of one or more digital tools (and perhaps even develop their own), collect, clean, and analyze their data, and visualize large and complex datasets. The data, the tools, and the data representations are a significant part of the intellectual contribution of the thesis, yet they are more often than not relegated to appendices. They are effectively footnotes to the 80,000-word thesis that accompanies them. This is both a huge amount of work—developing and documenting a digital project while also writing a book-length thesis—and an undervaluing of the outputs of digital research. These outputs simply do not fit within the common (and regulatory) understanding of a PhD thesis. Instead, the student is forced to shrink their interactive visualization of social media hashtags so that it can be printed in two dimensions on an A4 page. And this will just be the first time that they have to compromise, to deliberately make their data harder, if not impossible, to decipher. What if we reimagined the humanities PhD to allow a student using digital methods to write a shorter description of their research and findings, accompanied by a database, visualization, or other digital output? This is far from uncharted territory, and there are examples that can be drawn from other practice-based disciplines.[1]

This is one way of accommodating the requirements of digital humanities researchers, but it perpetuates the division between a book-like piece of writing, on the one hand, and digital outputs, on the other. Even if both are valued, and this is not a given, we are doing authors and readers a disservice by insisting on their separation at the point of publication. The mere act of digital publication does not solve the problem, even if it opens up opportunities that the printed book cannot support. As Mole 2016 argues, the challenge is not just to develop new forms of digital publishing but also "to ensure that the most valuable qualities of the academic book as printed codex migrate to the new media environment without being devalued." Publishers and researchers have for some time been taking halting steps along the pathway to different modes of publishing and different kinds of academic books. A willingness to experiment is admirable and necessary, but to date this has been relatively, and perhaps necessarily, constrained. It is quite common to see digital material appended to a book which is still delivered online as a PDF (see, for example, Nicholas Piercey's history of early Dutch football, published by UCL Press, which is accompanied by an Excel spreadsheet containing data about players and their clubs). The next step on from this is the embedding of multimedia content. Open Book Publishers, for example, has a prominent statement on its website that it is "happy to embed images, audio and video material in any publication," linking to two titles which have embedded audio resources (Open Book Publishers n.d.). More complex material, however, is dealt with through the provision of QR codes (print) and hyperlinks (digital) which take the reader to the author's website or to external sources (see, for example, Erik Ringmar's *History of International Relations*).

These are useful digital enhancements, but they are in no sense indicative of a definitive move away from the constraints of print. The print-on-demand book still holds sway. Given the challenges of developing sustainable business models for open access books, this should come as no surprise. There are significant costs associated with publishing books, and of moving to a fully open access landscape (Eve et al. 2017). At the same time, there is a demonstrable audience (of libraries, readers and authors) for printed copies of open access material which helps to offset some of those costs. We are likely to get to future forms of the academic book through evolution rather than revolution. But there are some striking examples of small-scale innovation, either at individual university presses, or through the development of experimental publishing platforms which may be used by publishers but also by authors. In some cases, these quite deliberately blur the boundary between books and digital projects. Stanford University Press, for example, aims "to advance a publishing process that helps authors develop their concept (in both content and form) and reach their market effectively in order to confer the same level of academic credibility on digital projects as print books receive" (Stanford University Press n.d.). The explicit linkage of content and form marks a step change from the addition of digital data to a written text that is presented in a relatively standard way, and encourages authors to think about presentation, form, and structure as they are writing. This is a very different mode of working, for both authors and publishers—one which is both costly and time-consuming. The results, however, offer a glimpse into a different future for the book.

Stanford University Press has published just seven of these digital projects since 2016, but they showcase how we might think about the forms in which we publish digital outputs. They also encourage us to consider the relationship between projects of this kind—which will resonate with many people working in the digital humanities—and our understandings of book-type objects. The words book and monograph appear nowhere on these pages, but the publications do have ISBNs and are garlanded with what look very much like book reviews; some have a Table of Contents and an Index. All seven publications are different in how they have approached the digital, but one example should be sufficient to indicate the ambition of the initiative, and the possibilities of innovative digital publishing. Matthew Delmont's *Black Quotidian*, published in 2019, includes introductory "chapters" which frame the more experimental content, but at its heart is "a trove of short posts on individual newspaper stories [which] brings the rich archive of African-American newspapers to life, giving readers access to a variety of media objects, including videos, photographs, and music." True to the publisher's ambitions, the form of the publication makes an important intellectual statement: there are 365 "daily" posts presented as a blog which, in the author's words, "offers a critique of Black History Month as a limiting initiative" (Delmont 2019).

The relationship between form and content that is on display here suggests very early and close engagement between author and publisher. This is not the normal way in which humanities monographs are written and published, and it is extremely hard to scale. Most authors cannot expect to approach their publisher with such bespoke requirements at the proposal stage, but there are other platforms and services which offer something of a middle way. These might provide greater flexibility for publishers, or seek primarily to empower authors. Among its many features, the Manifold publishing platform provides options for publishers to convert EPUB files, or even Google Docs, and transform them into enhanced digital editions. All that is required is an author with a "willingness to think beyond the normal confines of the traditional print strictures" (Manifold n.d.). For the team behind Scalar, the focus is directly on authors: it is "an open source authoring and publishing platform that's designed to make it easy for authors to write long-form, born-digital scholarship online." The embracing of the born-digital is an indication of a very different approach to long-form writing, which does not envisage a printed book as even a subsidiary output. This is readily apparent from the ambition "to take advantage of the unique capabilities of digital writing, including nested, recursive, and non-linear formats" (The Alliance for Networking Visual Culture n.d.). There is no doubt that a degree of technical expertise is needed if authors are to make full use of the affordances of these publishing platforms, but cloud services which do not require researchers to install software locally are significantly lowering the barriers to entry. It is perhaps more of a challenge for many writers fundamentally to rewire the way that they write, to have to reconsider form, and to think differently about audience.

Both Manifold and Scalar emphasize collaboration, which is often viewed as a hallmark of the digital humanities, but within the humanities generally remains more a feature of project work and the writing of journal articles than book production. Collaborative writing and publishing, and a rethinking of what constitutes authorship, are likely to play an important role in the books of the future. Co-authorship, of course, is nothing new, and the twenty-first century has seen the development of the book sprint as an aid to rapid and collaborative authorship. This handbook itself is a form of collaborative writing. While the chapters may have been written independently, the editor encouraged all authors not just to review selected contributions but to read the book at an early draft stage in order to draw out connections and crossovers. This kind of collaboration is built into the edited collection, which remains a mainstay of humanities publishing, even if its death has been predicted over the years with even more certainty than the monograph's (Webster 2020). But digital books have the potential significantly to expand the circles of collaboration, and qualitatively to change the nature of the interactions and engagements that occur there. An early example of this expansion took the edited collection as its starting point, but opened up the conversation to a much wider group than simply the authors. It also exposed the process of editing a book in ways that can remain uncomfortable for researchers who are not used to making public their drafts and corrections, let alone their rejections. *Writing History in the Digital Age* (2012), edited by Jack Dougherty and Kristen Nawrotzki, began life as an open call for essay ideas in the summer of 2011, which elicited more than 60 contributions and 261 comments. This list was publicly whittled down to twenty-eight essays which went out for open peer review, of which twenty made their way into the final book. There is still a version of the book online which presents each of these stages, and publishes peer review comments alongside the draft text. In a striking example of the limitations of print, all of the additional material was stripped away from the printed version of the book which appeared in 2013 (Dougherty and Nawrotzki 2013). The requirement to format a book for printing has erased the traces of complex collaboration which are clearly visible in the version that appears online.

There are other examples of this kind of experimentation with crowdsourcing and collaboration in book publishing, for example the ambitious *Hacking the Academy* (Cohen and Scheinfeldt 2010), but as is so often the case, innovation has spread more rapidly in journal than book publishing. Individual books, as described above, can embody a unique digital vision, but it is much easier (and crucially more cost effective) to implement change across a journal platform that publishes hundreds of titles. Open peer review in various forms is beginning to become more common in journal publishing. This might be something as simple as naming the peer reviewers whose input has helped to shape a particular article, a model adopted by *Frontiers in Digital Humanities*; or, as in *Wellcome Open Research*, the reviewers' verdicts might be indicated on what is effectively a scorecard, complete with the occasional red cross. For books, the input of those who are known to the author will largely only be evident from the acknowledgments, with perhaps a nod to the "anonymous reviewers" whose feedback proved invaluable. The growing emphasis in the academy on both interdisciplinary and participatory research has widened the scope of possible collaboration. Alongside the increasing (and welcome) emphasis on ethical citation practices, it is to be hoped that this will be reflected in book publishing.

There are other examples to be drawn from journal publishing too. The Executable Research Article (ERA) offers an intriguing possible way forward for books. In August 2020, eLife announced that it would be launching ERAs, "allowing authors to post a computationally reproducible version of their published paper" in the eLife open access journal. This moves the publishing options decisively away from the appended downloadable spreadsheet towards a form of publication which allows authors to present their data and code as the products of scholarship alongside their written analysis. And, as is pointed out in the press release, this is not just something of value for authors: "Readers can ... inspect, modify and re-execute the code directly in their browser, enabling them to better understand how figures have been generated, change a plot from one format to another, alter the data range of a specific analysis, and more" (eLife 2020). For digital humanities researchers who are already using Jupyter Notebooks, for example, an Executable Book would be a welcome next step and an effective harnessing of the potential of the digital. It would certainly be a more useful means of scholarly communication than writing a book which references datasets published on the author's own website and/or hosted in an institutional repository.

The monograph, as Geoffrey Crossick argues, "is an extended work that exists as an integral whole in which argument and evidence weave together in a long and structured presentation" (2016). But as he acknowledges, this does not mean that readers will work through an academic book from cover to cover, however much the author might want them to. Depending on their particular needs at the time, they might begin with the index, read the introduction and conclusion first, read a single chapter, or dip in to multiple chapters. They might return to a book many times, drawing different things from it on the second or third visit. Books are not easily segmented, or at least not without damage, but they do allow for multiple points of entry. This is something that the printed book supports very well, but digital monographs have managed much less successfully, precisely because rather than in spite of the ongoing influence of print formats. Once again, some of the most interesting explorations of different modes of interaction with digital content are happening in the field of journal publishing. The *Journal of Digital History* (*JDH*), a collaboration between the Centre for Digital and Contemporary History at the University of Luxembourg and the publisher

De Gruyter, is experimenting with transmedia storytelling to allow readers to choose their path through an article according to their interests. Each article has three layers—a narration layer, a hermeneutic layer, and a data layer—and the reader's engagement with some or all of these will be unique to them. It is a complex exercise—each layer is peer-reviewed separately and authors have to work very closely with the journal editors—but this form of publication admirably reflects both digital humanities methods and reader behavior. And it is perhaps uniquely suited to representing the myriad ways in which people read books.

The multi-layered, networked book is one vision of the future. It unites form and content; allows readers to choose their modes of interaction; connects text, data, sound, image, and film; and includes executable code and interactive visualizations—all without sacrificing those qualities of the printed book which are so valued in the humanities. But what if the digital academic book of the future is actually something much more stripped back, even minimal? Reducing the size and complexity of digital products delivers clear gains in terms of accessibility, and indeed of environmental impact. This was the argument made by the team behind the Digital Archimedes Palimpsest (n.d.): "It is important to make clear what this digital product is not. It does not come with a GUI (Graphic User Interface), and this means that it is unwieldy to use. Our thinking behind this is that GUI's [*sic*] come and go, and our mission at this stage in the project is to provide a permanent document." The Melville Electronic Library has begun to follow suit, producing minimal, and it is hoped more sustainable, versions of its editions, for example Melville's *Battle Pieces* (Melville Electronic Library n.d.). These scholarly editions, which are an important type of book in the humanities ecosystem, are not often beautiful but they are functional, flexible, and more susceptible to long-term preservation than some of the highly customized publications described previously. Sometimes the decision to adopt a simpler approach is born out of necessity. Johanna Drucker, for example, describes how she and her colleagues working on "The History of the Book Online" project spent considerable time learning to use the content management system Drupal in order to fit within local library infrastructure. It was only sometime later that they realized ongoing software upgrades "were going to make [the] entire project obsolete." Instead, all of the content was migrated "into a simple HTML/CSS format in which links, index, and navigation are hand-coded." The key lesson taken from this experience, and one which those responsible for designing future academic books would do well to bear in mind, is "the principle that you should never wed intellectual content to a platform structure as getting it out is a time-consuming and tedious process" (Drucker 2021).

As the team behind the Digital Archimedes Palimpsest point out, simplicity does not necessarily equate to ease of use, but readers will not find themselves struggling to download content if they have a poor Internet connection or on an outdated computer. They may also find themselves able to access content long after more eye-catching books have become inaccessible. Questions of sustainability and long-term preservation loom large when considering digital expressions of academic research, although arguably still not as large as they should. Digital content disappears, of course, but it is also susceptible to significant and damaging change. In a large-scale study of scientific articles, Jones et al. (2016) found, for example, that "for over 75% of references the content has drifted away from what it was when referenced … [raising] significant concerns regarding the long-term integrity of the web-based scholarly record." And this is just textual content online. How much more vulnerable will be multimedia publications, which not only contain locally hosted data but link to or embed audio and video content located elsewhere on the web? The idea of the book as a networked digital object holds huge promise, but how can we ensure that any network remains intact, especially when it might consist of hundreds of separately hosted elements? A minimal computing approach is one response to this problem, but highly sophisticated digital books are already being published and the future looks likely to be one which includes even greater diversity and complexity.

Sustainability is a perennial concern for researchers in the digital humanities, and one which is closely linked to modes of publication (see, for example, Edmond and Morselli 2020). If at least some of the academic books of the future eschew the minimal, how can we ensure that they remain available to readers? Web archives are one solution, and the designers of digital books can take steps to make their products as easy as possible for web crawlers to harvest in full (see, for example, Library of Congress n.d.). Most of the books published by Stanford University Press under the umbrella of Stanford Digital Projects include a section titled "Web Archive." At the time of writing, however, each of these sections consists solely of a short paragraph noting that "This page will be updated with archival information within twelve months of the release." This is true for works published as early as 2018, like *Black Quotidian*. Multiple captures of the latter are, in fact, accessible using the Internet Archive's Wayback Machine, but on further investigation, these turn out only to be partial copies. Key functionality, and content, has not been successfully preserved. This is just one example, but it is an example where thought has clearly been given to long-term sustainability and preservation and still an archival copy is not readily accessible.

These are some of the problems facing books that are published on the web, but academic books are made available on other digital platforms and in other digital forms. Ebooks of the type typically purchased by academic libraries do not pose a particular preservation challenge, nor indeed offer anything particularly innovative (quite the reverse in many instances). But there are emerging digital book formats which offer intriguing new functionality and as a consequence further complicate the preservation landscape. It is for this reason that the British Library has established an "Emerging Formats" project, with the aim of developing its capacity "to collect publications designed for mobile devices that respond to reader interaction or are structured as databases." The notion of reading a full-length academic monograph, even an interactive one, on a mobile phone may not seem very attractive, but it is much easier to imagine reading a shorter-form book using an app. For the moment, innovation in this area is largely the province of fiction and popular non-fiction—"genres of writing, from horror and science fiction to documentary and memoir"—but where the crossover history title leads, a more squarely academic book may well follow (British Library 2020). The intersection between the digital and public humanities has significant potential to generate format innovation of this kind. The "Ambient Literature" project, for example, aimed "to investigate the locational and technological future of the book," focusing on "the study of emergent forms of literature that make use of novel technologies and social practices in order to create robust and evocative experiences for readers." Among the outputs of the project was *The Cartographer's Confession* (2017), by James Attlee, which combined fiction and non-fiction and included "audio, prose, illustrations, an original collection of 1940's [*sic*] London photographs, 3D soundscapes, and a bespoke musical score" (Ambient Literature n.d.). It is currently downloadable from the Apple App Store, but has already disappeared from Google Play. The examples of *Black Quotidian* and *The Cartographer's Confession*, both recent experimental titles which have demonstrably resisted archiving, highlight the inherent ephemerality of the new academic books and serve as an indicator of future problems with the coherence of the scholarly record. Digital preservation is not an insurmountable problem, but its consideration is crucial. If we are not careful, the academic books of the future will be characterized by absence, by broken links, missing data and images, and 404 error messages.

This question of digital absence connects to another important consideration for readers and authors of academic books: citation. Hyland (1999) asserts that "One of the most important realizations of the research writer's concern for audience is that of reporting, or reference to prior research ... Citation is central to the social context of persuasion." His analysis of a corpus of eighty research articles in different disciplines found that researchers in the humanities and social sciences included more citations than their peers in the sciences and engineering (Hyland 1999, 341–2). Citation is important for academic writing. Humanities citation practices will have to be accommodated within future academic books, but they will also have to adapt so that they do not constrain what is possible when the digital book becomes the version of record (a convention which is now well established in journal publishing). In disciplines such as history, for example, the footnote is still a dominant form of referencing. This makes sense when the reading unit is a printed page, but is not a viable option for scrolling or non-linear content. Hyperlinking mitigates the annoyance for some readers of flicking backwards and forwards from text to endnote, and the full details of inline citations may be revealed by hovering over a link to bring up a text box. These are interesting evolutions of current behaviors, which are beginning to make use of the structure of the web in particular, but perhaps revolution awaits us in the future? The dominance of the page is also evident within references, where pagination guides a reader to a specific quotation from a larger book or article. But many of the works I have cited in this chapter do not have pages, or they have a pre-print pagination which will be altered when the digital version of record appears in print. When this volume reaches the copyediting stage, convention dictates that I will be asked to supply "missing" page numbers. As forms of academic publishing, including the book, begin to move away from the organizing principle of the page, authors, readers, and publishers will have to rethink how they acknowledge and/or locate the writing and research that has influenced them. Persistent identifiers that are both human- and machine-readable are one suggestion (see, for example, Blaney 2012), but for the moment we are in a transitional period, where new forms are squeezed uncomfortably into old structures.

Old and new publishing structures and modes of publication can and will co-exist. As with the academic monograph, the death of the printed book has long been predicted. In 2014, Mandal noted that "As early as 1994 the death knell of print was being sounded in Sven Birkert's *The Gutenberg Elegies* ... And yet, the printed book remains a ubiquitous presence in our supermarkets, our libraries and our bookshelves, healthily outliving its predicted demise." There is room for both print and digital, whether they complement each other or fundamentally diverge. We should also not assume that there is only one way forward for the digital academic book of the future, that there is a binary choice to be made between the minimal and the expansive, for example. One of the advantages of the digital book over its print predecessor is precisely that there are many different paths that can be followed, and the same content can be presented in multiple different ways for different audiences.

Then, too, there is the idea of the "living book" (Deegan 2017, 85), which evolves over time and might eventually come to include types of content that were not conceived of when the first material was commissioned. A notable example is UCL Press's BOOC (Books as Open Online Content) initiative, the modular presentation of which allows for the addition of new material and new forms without disruption to the overall structure of the publication (University College

London n.d.). This is publishing as continuum, which is simultaneously challenging (what if my book is never finished?) and inspiring (just think of what I might be able to do in five years' time!). The liminal space between first idea and final publication is interestingly occupied by the "Forerunners" series published by the University of Minnesota Press. These are short, "thought-in-process" books, which draw on "scholarly work initiated in notable blogs, social media, conference plenaries, journal articles, and the synergy of academic exchange" (University of Minnesota Press n.d.). Here the digital book is not the end of a research journey, but an agile step on the way.

These are just some examples of possible book futures, but whatever else the academic book becomes, let us hope that it will be more accessible. It may support non-consumptive research rather than the download of large datasets; it may be translated in real time based on individual language preferences; it may be available as text to speech for those with visual impairment—and it will be open to anyone who wants to read it. That is a future to look forward to.

NOTE

1. At the time of writing, the Royal College of Art in London, for example, gives PhD students two options for a PhD: it can "take the form of a thesis (60,000–80,000 words) or by project (a body of work and thesis 25,000–40,000 words)," https://www.rca.ac.uk/research-innovation/research-degrees/.

REFERENCES

The Alliance for Networking Visual Culture. n.d. "About Scalar." https://scalar.me/anvc/scalar/.

Ambient Literature. n.d. "About the Project." https://research.ambientlit.com.

Attlee, James. 2017. *The Cartographer's Confession*. https://apps.apple.com/gb/app/the-cartographers-confession/id1263461799.

Blaney, Jonathan. 2012. "The Problem of Citation in the Digital Humanities." In *Proceedings of the Digital Humanities Congress 2012*, edited by Clare Mills, Michael Pidd and Esther Ward. Sheffield: University of Sheffield.

British Library. 2020. "Emerging Formats." https://www.bl.uk/projects/emerging-formats.

Cohen, Daniel J. and Tom Scheinfeldt. 2010. *Hacking the Academy: New Approaches to Digital Scholarship and Teaching from Digital Humanities*. Ann Arbor, MI: University of Michigan Press.

Crossick, Geoffrey. 2016. "Monographs and Open Access." *Insights* 29 (1): 14–19. http://doi.org/10.1629/uksg.280.

Deegan, Marilyn. 2017. *Academic Book of the Future Project Report*. London: University College London.

Delmont, Matthew F. 2019. *Black Quotidian: Everyday History in African-American Newspapers*. Stanford: Stanford University Press. http://blackquotidian.org/.

The Digital Archimedes Palimpsest. n.d. http://archimedespalimpsest.org/digital/.

Dougherty, Jack, and Kirsten Nawrotzki. 2012. *Writing History in the Digital Age*. https://writinghistory.trincoll.edu/.

Dougherty, Jack, and Kirsten Nawrotzki. 2013. *Writing History in the Digital Age*. Ann Arbor, MI: University of Michigan Press.

Drucker, Johanna. 2021. "Sustainability and Complexity: Knowledge and Authority in the Digital Humanities." *Digital Scholarship in the Humanities*, fqab025. https://academic.oup.com/dsh/article/36/Supplement_2/ii86/6205948.

Edmond, Jennifer, and Francesca Morselli. 2020. "Sustainability of Digital Humanities Projects as a Publication and Documentation Challenge." *Journal of Documentation* 76 (5): 1019–31. https://doi.org/10.1108/JD-12-2019-0232.

eLife. 2020. "eLife Launches Executable Research Articles for Publishing Computationally Reproducible Results." https://elifesciences.org/for-the-press/eb096af1/elife-launches-executable-research-articles-for-publishing-computationally-reproducible-results.

Eve, Martin Paul, Kitty Inglis, David Prosser, Lara Speicher, and Graham Stone. 2017. "Cost Estimates of an Open Access Mandate for Monographs in the UK's Third Research Excellence Framework." *Insights* 30 (3): 89–102. http://doi.org/10.1629/uksg.392.

Hyland, Ken. 1999. "Academic Attribution: Citation and the Construction of Disciplinary Knowledge." *Applied Linguistics* 20 (3): 341–67. https://doi.org/10.1093/applin/20.3.341.

Institute of Historical Research. 1999. "Is the Monograph Dead?" https://web.archive.org/web/20000311210234/http://ihr.sas.ac.uk/ihr/mono.html.

Jones, Shawn M., Herbert van de Sompel, Harihar Shankar, Martin Klein, Richard Tobin, and Claire Grover. 2016. "Scholarly Context Adrift: Three out of Four URI References Lead to Changed Content." *PLOS One*, December 2. https://doi.org/10.1371/journal.pone.0167475.

Journal of Digital History. https://journalofdigitalhistory.org/en/about.

Library of Congress. n.d. "Creating Preservable Websites." https://www.loc.gov/programs/web-archiving/for-site-owners/creating-preservable-websites/.

Lyons, Rebecca E. and Samantha J. Rayner, ed. 2016. *The Academic Book of the Future*. Basingstoke: Palgrave Macmillan.

Mandal, Anthony. 2014. "Hachette v Amazon, the Death of Print and the Future of the Book." *The Conversation*, June 23. https://theconversation.com/hachette-v-amazon-the-death-of-print-and-the-future-of-the-book-22278.

Manifold. n.d. https://manifoldapp.org/.

Melville Electronic Library. n.d. *Battle Pieces*. https://melville.electroniclibrary.org/editions/battle-pieces/01--prefatory-note.

Open Book Publishers. n.d. "Images and Audio/Video Material." https://www.openbookpublishers.com/section/85/1.

Page, Benedicte. 2019. "CUP Launches Big Programme of Speedy, Mid-Length Publishing, Cambridge Elements." *The Bookseller*, January 15. https://www.thebookseller.com/news/cup-launches-big-programme-speedy-mid-length-publishing-cambridge-elements-933356.

Pidd, Michael. 2016. "Wearable Books." In *The Academic Book of the Future*, edited by Rebecca E. Lyons and Samantha J. Rayner, 18–23. Basingstoke: Palgrave Macmillan.

Piercey, Nicholas. 2016. *Four Histories about Early Dutch Football 1910–1920*. London: UCL Press.

Ringmar, Erik. 2019. *History of International Relations: A Non-European Perspective*. Cambridge: Open Book Publishers.

Stanford University Press. n.d. "Publishing Digital Scholarship." https://www.sup.org/digital/.

Swain, Harriet. 1999. "Rumours of a Death that May be True." *Times Higher Education*, December 10. https://www.timeshighereducation.com/features/rumours-of-a-death-that-may-be-true/149251.article (accessed April 13, 2021).

University College London. n.d. *BOOC (Books as Open Online Content)*. London: UCL Press. https://ucldigitalpress.co.uk/BOOC.

University of Minnesota Press. n.d. *Forerunners: Ideas First*. https://manifold.umn.edu/projects/project-collection/forerunners-ideas-first.

Webster, Peter. 2020. *The Edited Collection: Past, Present and Futures*. Cambridge: Cambridge University Press.

CHAPTER TWENTY-FOUR

Digital Humanities and Digitized Cultural Heritage

MELISSA TERRAS (UNIVERSITY OF EDINBURGH)

The Gallery, Library, Archive, and Museum (GLAM) sector have been involved in a concerted—if not evenly distributed—effort to mass-digitize collections for over thirty years. Over this period, digitization processes and best practices have been established, and there have been various phases and mechanisms of delivery which now see a vast, if patchy, digital cultural heritage landscape. Throughout, the Digital Humanities (DH) community has engaged with digitized primary historical sources, using the *products* of the digitization *process* for research, teaching, and engagement. The DH community has built corpora, infrastructure, tools, and experimental interventions upon digitized cultural material, while also computationally mining and examining content. This chapter will analyze and synthesize the different activities those in DH undertake with digitized cultural heritage, stressing that those wishing to utilize it in research and teaching must engage with digitization processes as well as the product to best understand the data's epistemic foundations. Developing a holistic understanding of the digitization environment that provides the data DH activities are built upon can allow DH to contribute to it in return, influencing and impacting future digitization activities. This chapter sketches out an agenda for the Digital Humanities in considering digitized content, while also reflecting on how best to create a feedback loop to undertake activities and produce research that will be of interest to digitization providers. By working together with collections and their owners, DH activities can help construct a roadmap towards building inclusive digital heritage datasets that are useful, re-usable, and point to the benefits of user engagement with our digitized past, while also more fully understanding and influencing the processes which create digital resources, and the wider digital scholarship landscape.

DIGITIZATION IN THE GLAM SECTOR

Since the 1970s, GLAM have been undertaking digitization, using the affordances of digital networked infrastructure to improve the management of, increase engagement with, and provide access to heritage collections (Hughes 2004; Parry 2013; Terras 2011). Mass-digitization began in earnest in the 1990s (Lee 2001, 160). An increase in digital activity across society as a whole has meant that users now expect heritage to also be available in digital form (Falk and Dierking 2016, 122), with a related increase in digital interactive and immersive technologies in heritage contexts (Pittock 2018, 5). The technical aspects of routine, service, or mass-digitization are now mostly a solved problem, with best practice guidelines available for others to consult.[1] There are expert communities of practice across the heritage sector regarding data acquisition and recording.[2] The importance of digitization to the heritage landscape was highlighted by the Covid-19 epidemic, when many institutions worldwide had to restrict physical access: in response, 86 percent of museums increased their online presence and/or the amount of content they were placing online (ArtFund 2020), with emerging opportunities regarding the reframing of digitized content as an essential part of cultural memory (Kahn 2020).

What is the relationship of the DH to this digitization landscape? How best can researchers respond to it, not only to use the *products* of heritage digitization, but to engage in the *process*, to encourage inclusion and engagement? Most obviously, DH techniques such as text and network analysis are dependent on machine-processable data, usually generated from the digitization of primary historical sources.[3] These data sources may be small scale, and possibly creatable by a lone researcher undertaking digitization themselves for a particular project. Both the copyright exceptions that allow digital copies of material to be created for nonprofit data-mining research purposes, and the advancing technological infrastructures mean that, even with a modern smartphone and a few choice apps, high-quality data regarding historical collections can now be efficiently created for later processing and analysis (Rutner and Schonfeld 2012). However, more often than not, researchers use the products of digitization carried out by a GLAM institution, or a digitization service provider working with material from a GLAM collection.

The choice of which materials are selected for digitization may be driven by the needs of a particular research project, with allied external funding. Nowadays, in a changing resource landscape, it is becoming more common for GLAM institutions to be expected to provide their own internal budgets for digitization. Researchers may build questions around data from primary sources that were pre-emptively digitized as they were believed by GLAM curators and their institutions to be critical material, in the hope that if you build it they will come (Warwick et al. 2008). Institutions may also engage with their research communities to understand how best to triage and schedule digitization of their collections (Wingo, Heppler, and Schadewald 2020). Alternatively, DH researchers can sometimes access content that was mass-digitized because there was a suitable business case for monetizing it, for example, local newspapers that were primarily digitized by publishers for paid-for access by the online genealogy community (Hauswedell et al. 2020).

The products of digitization have changed over the past thirty years, and with that, so has the way DH activities interact with digitized content. There initially was naivety and techno-futurist excitement surrounding digitization (which saw a lot of material subject to "scan-and-dump," where large collections were placed online although only accessible one item at a time, and there was a hope that people would find this transformative (Warwick et al. 2008)). Those in the Digital Humanities have always been interested in re-using available content, and ascertaining how these platforms were being used by others, to best frame how future resources may be deployed. The mid- to late 2000s saw the rise of Museum 2.0, with a growth in participatory culture and co-creation, aligned with providing content via a growing range of social media platforms (including Facebook, Twitter, Tumblr, Instagram, and more recently SnapChat, and TikTok) as well as data and information sharing platforms (such as Flickr Commons and Wikipedia). Research demonstrated that this type of activity both reached and broadened GLAM audiences (Finnis 2011; Hughes 2012). Over the past decade, digital cultural heritage aggregators, such as Europeana,[4] Google Arts and Culture,[5] the Internet Archive,[6] and the Digital Public Library of America[7] have become the central point of call for a user wishing to browse cultural or heritage content, and attention has been paid to the politics and biases that are embedded within these infrastructures and mass-digitization programs (Thylstrup 2019; Kizhner et al. 2020). There has also been interest in how to extract further information from digitized heritage: mechanisms for users to engage with digitized resources in order to complete more complex tasks such as identification, cataloging, and transcription via crowdsourced volunteering, are now in the mature phase of delivery and study by those in DH (Ridge 2014; Terras 2016). An increasing range of approaches to produce digital scholarly editions that build upon digitized images of texts are now used (Bleier et al. 2018). Over the past decade, the OpenGlam[8] movement has made these types of activities easier, providing frameworks, pilot studies, and use-cases for institutions aiming to provide "ethical open access to cultural heritage" (OpenGlam n.d.) by releasing resources onto the open web with a range of licenses (usually via the Creative Commons[9] framework). Moving beyond the relationship of a user encountering one heritage item in digital form at a time, to encountering and synthesizing many, the "Collections as Data" movement, where large-scale data deposits of complete collections are made available under open licensing to encourage users to take, manipulate, analyze, and reframe digital collections (Padilla 2018), is of much interest to many in the DH. As technologies that can produce machine-processable data have reached maturity, such as handwritten text recognition (Muehlberger et al. 2019), and voice to text speech recognition (Picheny et al. 2019), the volume of data derived from GLAM material available for scholars is rapidly expanding, and requires understanding in best practice for both GLAM institutions in creating, hosting, delivering, and stewarding content, and DH approaches in utilizing it at this scale.

However, it should always be remembered that although digitization is the starting point for these DH activities only a fraction of content has been digitized. While 82 percent of GLAM institutions across Europe have a digital collection or are engaged in digitization (Nauta et al. 2017, 5) after thirty years of large-scale investment in digital, on average 22 percent of heritage collections across Europe have been digitized, and only 58 percent of collections have even been cataloged in a collections database (Nauta et al. 2017, 6). There is still much to do, then, in both digitizing, understanding how best to deliver and use digitized content in research and teaching, and extracting further value from digitized content (Terras et al. 2021). The DH community is well placed to both utilize the *products* of such activities, and advise and become integrated into the *processes*, questioning the selection choices that are made along the way, and highlighting the impacts that these have on diversity and inclusion, while also being able to use digital tools to reflect on the make-up, history, and constitution of collections which steward our cultural heritage.

ENGAGING WITH THE DIGITIZATION PROCESS

GLAM digitization processes and resources do not exist solely to provide data for further analysis by those within DH. Established reasons given for mass-digitization of heritage content include: increasing access to wider communities; supporting preservation; collections development; raising the profile of collections and institutions; while also supporting research, education, and engagement (Hughes 2004, 8–17). There is nowadays much focus on engagement and its related metrics in digital GLAM strategies, which is partly driven by funders and the quantified nature of feedback regarding online estates. For example, in the UK, the Department for Digital, Culture, Media and Sport launched a "Culture is Digital" policy paper in 2018 that frames the best use of technology in GLAM as driving "audience engagement, boosting the digital capability of cultural organizations and unleashing the creative potential of technology." The effect is that many organizations in the sector are, by necessity, focusing rather shallowly on reporting on clicks and likes, "using digital technology to engage audiences" (DCMS 2018). This is at the expense of thinking more holistically about digital strategies that can benefit both the organization and their communities in various ways (Finnis 2020).

Digitization is just part of a suite of digital activities transforming the GLAM sector (including "Strategy and Governance; Cultural Programme; Places and Spaces; Marketing & Communications; HR; IT; and Finance & Operations," see Digital Compass (2020) for a toolkit designed to assess and expand an organization's use of digital). Amongst this changing landscape, it has been noted that it is not clear how GLAM organizations ever intended their audiences to re-use mass-digitized heritage content (Harrison et al. 2017). Additionally, "institutions do not have the resources of major technology providers for research and development, and are not as easily able to recover from costly technological failures in design, delivery, or public relations" (Terras et al. 2021). As a result, making the most of digitized cultural heritage data for DH research purposes (or an approach now more commonly known as "Digital Scholarship" within the library and heritage sector (Greenhall 2019)) requires sensitivity from scholars who often have to navigate very different infrastructures and processes, institutional focus, and available resources. Equipping a GLAM institution "for digital scholarship, navigating a world of Big Data, and understanding the implications of this shift, requires organisation-wide re-evaluation" (Ames and Lewis 2020), at a time of much austerity for the sector. There is therefore an opportunity for DH to contribute to this space, working alongside the creators of digitized content, in establishing how best to allow researchers to gain access to content, and undertake the type of advanced digital methods that are espoused by DH.

DIGITIZATION AND THE DIGITAL HUMANITIES AGENDA

DH researchers have much to contribute in understanding, quantifying, critiquing, elucidating, and contributing to both the process and products of digitization, closing the gap between institutions and researchers in this data-led space. GLAM institutions themselves are grappling with the challenges that the volume of digitized content they create now presents, given that "new services, processes and tools also need to be established to enable these emerging forms of research, and new modes of working need to be established to take into account an increasing need for transparency around the creation and presentation of digital collections" (Ames and Lewis 2020). How best should the DH respond to this? In their analysis of "Big Data Stakeholders," Zwitter (2014) identifies three categories: collectors, utilizers, and generators, noting that there are "power relationships" between the three, and that the role of the collector is essential in determining what is collected and made available to others. Therefore, an intervention is needed from DH, to engage as and with collectors, utilizers, and generators, bridging and reframing the relationship of GLAM as provider and scholar as consumer of digitized content. Digital humanists should be responding to this call to arms by not only passively using GLAM content, but in producing research and other digital outputs that engage in discussions with the GLAM sector regarding digitization processes, deployment, engagement, and adoption.

What, then, should be the DH agenda, in the digitization space? Points of meaningful tension and possible contribution exist in: keeping abreast of technological developments and possible deployment in the GLAM sector; understanding institutional, political, and financial drivers of digitization, and the legal frameworks which bind and contain activities, to understand how these affect availability and usefulness of its products; making best use of the mass of digitized content at scale now emerging from the sector; and understanding how social media, and other external digital platforms, can best be used to promote engagement and inclusion.

Although the technical processes of routine digitization are now understood,[10] there are always new developments which require reception studies to understand how they may be best used within the GLAM sector (as well as interdisciplinary projects to be undertaken with the computational and engineering sciences in developing advanced capture and processing techniques). Currently, there is much interest in applying machine-learning approaches to digitized content (Jaillant 2021), capture techniques such as hyper-spectral imaging (Howell and Snijders 2020), and the implications regarding the recent affordability of 3D scanning techniques such as LiDAR (Collins 2018). Additionally, the Foundation Projects of the UK's "Towards a National Collection"[11] are investigating current best practices, presenting roadmaps for required research in: cross-walking catalogs and other digitized content sources;[12] best practice in sharing digitized images via IIIF,[13] implementing a system of Persistent Identifiers that can provide long-lasting references to digital heritage objects;[14] utilizing computer vision and deep-learning methods to identify patterns and linkages across collections using image processing;[15] making best use of geospatial data to connect diverse collections;[16] preserving and sharing born-digital and hybrid objects;[17] understanding the best way to engage with volunteers in citizen research projects;[18] and developing ethical methodologies and technical approaches to address outmoded and offensive racist, sexist, colonial, and imperial language in digital heritage information landscapes.[19] Those in the DH community are well placed to work with a broad range of collaborators in these many areas to establish how developing technologies will impact, and could be efficiently deployed, in a GLAM context, influencing DH research further down the line regarding how we teach and research with these techniques and their data.

However, as well as the technical aspects to creation of digitized content, DH could do much to elucidate the context regarding *which* digital resources are created within the GLAM sector. Choices made about digitization are a matter related to social justice: if we can only analyze historical and archival information that has been digitized, it is worth paying closer attention to the politics of selection, and how that relates to issues of diversity and inclusion (Thylstrup 2019; Zaagsma 2019). There are various tensions at play when decisions are made about resource allocation: cultural heritage organizations are inherently risk-averse, with often conflicting requirements from funders, a board of trustees, senior management team, curatorial staff, conservators, and digitization professionals. The sector, which is moving into post Covid-19 operationalizing, is still reeling after years of austerity, and resources for digitization and related digital activities are scarce. There are ethical dimensions regarding digitization funding: "the availability of funding, public or private, plays a key role in enabling digitization projects in the first place ... Funding is ... influenced by ... memory politics and the way in which a given country's or group's past, or aspects thereof, resonate in public discourses and debates" (Zaagsma 2019). For example, what, pray tell, could possibly be the current political drivers for government funding for a new digital "National Collection: Opening UK Heritage to the World" in recently post-Brexit Britain? (AHRC 2021).

How will the choices made in this investment impact what we can access, research, and study, for years to come? Furthermore, as standard digitization processes continue to scale, and given cultural heritage institutions have limited resources, the relationship with technology providers and major publishers, and the constraints and opportunities that arise with regard to commercial digitization of heritage, is worthy of further thought and consideration (Thylstrup 2019; Hauswedell et al. 2020). All of these aspects are worthy of appraisal and investigation by DH scholars when using the products of digitization processes, which will be of use to a broad constituency, from museum studies, to library science, to critical heritage studies, to DH itself. Only in understanding the choices made in the creation of digitized content, can it be fully appraised and utilized.

Attention is also needed regarding the legal frameworks which bound and contain digitization activities, and therefore shape the online GLAM environment. Copyright is a major barrier to digitization, particularly around restrictions placed on the digitization of modern in-copyright materials (Kariyawasam and Adesanya 2019) and "orphan works," the vast majority of cultural heritage holdings where the copyright owner of an item cannot be identified, which has ramifications for choices made around digitization (Korn 2009). Other legal frameworks such as Intellectual Property Rights, and General Data Protection Regulations, likewise affect digitization selection. Understanding how these legal frameworks (and the business models that grow from them, see Terras et al. (2021)) intersect with risk-averse institutions and the digitization process will help understand—and improve—the digitized cultural heritage environment.

Understanding GLAM collections and material at scale is an emerging research area. There is much potential for DH in the appraisal of scale, focus, and topic of previously digitized content, as well as elucidating biases contained within them. In this current phase of digitization, we are realizing that the choices previously made over the past four decades have ethical and often colonial aspects, and that there is a duty of care that comes with making cultural and heritage collections available to a wider audience (Odumosu 2020). These biases play out at scale, with unforeseen consequences for audiences and researchers (Kizhner et al. 2020). There is much more that can be done with digitized cultural heritage to foreground and support minority voices within collections, in order to "mainstream equality," for example around aspects of gender, race, or LGBTQ+ issues (National Galleries Scotland 2019). The DH community is well placed to analyze and synthesize "Collections as Data" (Padilla 2018), understanding the substance, and biases of cultural heritage collections, as well as searching and analyzing them. Catalog metadata itself is a crucial but overlooked information resource that is ripe for mining, visualization, and further analysis to understand collection structure, coverage, themes, and lacunae (Lavoie 2018). Additionally, the recursive process of applied data science means that new questions that *could* be asked regarding institutional collections are sure to emerge, and understanding and making these explicit would be a contribution from DH research, itself.

Digitized cultural heritage often intersects with social media (in GLAM policies as well as activities). This is a fascinating space, as the first point of contact with heritage with a wide audience is now often over these distributed channels. The specific institutional context, resources, opportunities, struggles, and approach to risk management, is all set out for public display (Koszary 2018), and resulting analysis. For example, the Digital Footprints and Search Pathways[20] project is researching the effect of the Covid-19 pandemic on the audiences of online collections, to help GLAM institutions develop engagement plans and policies for crisis situations. There is much ongoing work to be done in DH classes and projects around the intersection of digitized heritage and ever-changing online platforms.

IMPLEMENTING THE DIGITAL HUMANITIES DIGITIZATION AGENDA

How practically should the DH community undertake activities that center digitization, to engage with, feedback to, and improve it? Firstly, it is necessary to ask of any dataset derived from historic primary resources how it came into being—how, where, and why was it digitized? The flip-side of this is to ask, how, where and what was *not* digitized that may be aligned to the research question, or would increase accessibility and inclusion if it were to be made available in machine-processable form. This should be part of the discussion and presentation of any DH project that depends and builds upon digitized material, so that the reader can fully understand the limitations and possibilities framing the undertaking.

Secondly, DH teaching programs should also be integrating the study of digitization processes into the curriculum, gaining both practical technical experience, and understanding application and theory.[21] This is another skill set that will allow our graduates to go on to find useful and gainful employment, where their critical and technical skills can contribute to the "digitalisation of society" (Bowen and Giannini 2014), but from a different intellectual tradition than the computational sciences (which in itself is a matter of diversity and inclusion within technology industries (Lachman 2018)). The processes and products of digitization of cultural heritage are mirrored in current efforts to digitize medical, banking, educational, legal, and judicial records (which can have social consequences: see for example Favish (2019) on the effect of using OCR to release the backlog of Deeds Registry records in post-apartheid South Africa). There is much a DH lens can introduce into learning about digitization processes and how they overlap with wider societal concerns.

Thirdly, given the scale of DH relevant digitized content is rapidly expanding (particularly with transcripts automatically generated from manuscript material via HTR, or from audio material via speech recognition, and from the Collections as Data movement), skill sets are needed to be able to best use this data. Developing best practice approaches to historical text and data mining at scale should be an aspiration for DH (including learning and applying prior work from related fields such as statistics, Natural Language Processing, and High Performance Computing, or working in conjunction with individuals from those domains who have the skill sets to methodically tackle data at scale). As ever, it is difficult to recommend one particular tool, methodology, or method that represents "Digital Humanities"—the selection will be entirely dependent on the research question being asked and the dataset it is being applied to. However, there is much innovation happening in this space, for example Tim Sherratt's *GLAM Workbench*[22] tools, which provide a set of Jupyter Notebooks for exploration and re-use of data from cultural heritage institutions. For example, in working with data from Trove[23] (which aggregates collections from Australian GLAM providers) there are notebooks for visualizing: searches in Trove's digitized newspapers; the total number of newspaper articles in Trove by year and state; the geographical spread of content mapped by place of publication in Trove; and to analyze rates of OCR (Optical Character Recognition), amongst others.[24] Another set of crucial resources for learning this type of processing is the Programming Historian[25] which hosts rigorously peer-reviewed tutorials for digital research skills, which are openly licensed for re-use. The Data Carpentries[26] lesson programs, which teach foundational coding and data science skills, are also a resource to be drawn down on. In learning what is possible via these approaches, scholars can respond best to the opportunities presented by the growing body of machine-processable GLAM data. DH researchers should also keep a close eye on the developing Collections as Data community and their datasets; for example, the Data Foundry[27] at the National Library of Scotland has made available large-scale, openly licensed digitized collections with transparent documentation for others to build upon, which now underpin the Library Carpentry Text and Data Mining[28] lesson. There are many cultural heritage institutions moving in this direction.

Fourthly, in developing skills more commonly associated with information rather than computer scientists, such as qualitative interview, survey, and observation skills (Gorman and Clayton 2005), as well as user experience (UX) design skills (Kuniavsky 2010), DH as a field will be able to ask broader questions about the creation, usability, use, and impact of digitized resources, particularly within their chosen research area (such as analyzing social media activity). In doing these, DH scholars will be able to gauge the "epistemic affordances" of digitized content: the abilities, possibilities, and limitations of this environment when used in knowledge creation (Markauskaite and Goodyear 2017). Qualitative methods will be useful here, including action research (Schön 1983), grounded theory (Strauss and Corbin 1997), and content analysis (Krippendorff 2018) to synthesize different reports or analyze public-facing information. These are useful options to add to and expand the DH curriculum, as well as being crucial research skills where navigating complex information about decision-making environments with intersecting power structures, in systematic and rigorous ways.

Fifthly, it is also probable that these different skill sets need different individuals working together as part of a project team to undertake this type of research (interdisciplinary collaborative work being a perceived norm in many DH projects (Klein 2015)). It is worth stressing that information and museum professionals including librarians, curators, conservators, social media managers, and digitization practitioners have much expertise that can both benefit DH projects and allow professional development and opportunities for DH scholars: this is never one-way skills transfer. Individuals within DH have a lot to learn from our professional collaborators.

Finally, to ensure the impact of activities in this space, this crucial research should be published in venues that interact with a wider audience rather than the core DH publication and conference venues. (I have always wondered why, if DH is so interdisciplinary, so many of its practitioners mainly present their work in DH silos, rather than publishing in overlapping venues, where librarians, curators, digitization experts, heritage professionals, information and computational scientists, and wider industry contacts will more easily come across it.) Such venues are not hard to find, and many have overlaps with the Digital Humanities conference,[29] such as iConference,[30] or Museums and the Web.[31] Likewise, museum, critical heritage, library and information science, and digital sociology journals are all apposite venues for DH researchers to publish in, that will allow others grappling with building the digital scholarship landscape to become more aware of the types and range of extant and future DH activities.

CONCLUSION

This chapter has sketched a forward-facing agenda for the Digital Humanities community regarding how it perceives, considers, uses, analyzes, and reports on activities with digitized cultural heritage content. In doing so, it has made concrete recommendations regarding how digitized content can continue to underpin while also be considered part of core Digital Humanities research and teaching activities. There is much useful activity still to be done at the nexus of DH and GLAM digitization practices. In doing so, the DH community will be able to inform and work with GLAM institutions in best responding to, delivering, and guiding the future of digitized content. In navigating together the "new challenges around data management, storage, rights, formats, skills and access" that these mass-digitized resources present (Ames and Lewis 2020), the DH community can contribute to the wider, emerging digital scholarship landscape.

NOTES

1. http://www.digitizationguidelines.gov.
2. See for example the Association for Historical and Fine Art Photography (https://ahfap.org.uk) which has an active mailing list, or the ImageMuse discussion group for cultural heritage imaging (http://www.imagemuse.org).
3. An alternative to this is utilizing born-digital content (Kirschenbaum 2013): we concentrate on digitized heritage content within this chapter to allow focus.
4. Europeana, https://www.europeana.eu/en.
5. Google Arts and Culture, https://artsandculture.google.com.
6. Internet Archive, https://archive.org.
7. Digital Public Library of America, https://dp.la.
8. OpenGlam Network, https://openglam.org.
9. Creative Commons, https://creativecommons.org.
10. Federal Agencies Digital Guidelines Initiative, http://www.digitizationguidelines.gov.
11. Towards A National Collection, https://www.nationalcollection.org.uk/projects.
12. Heritage Connector, Science Museum, UK, https://www.sciencemuseumgroup.org.uk/project/heritage-connector/.
13. Practical applications of IIIF as a building block towards a digital National Collection, National Gallery, UK, https://tanc-ahrc.github.io/IIIF-TNC/ and https://iiif.io.
14. Towards a National Collection – HeritagePIDs, British Library, UK, https://tanc-ahrc.github.io/HeritagePIDs/.
15. Towards a National Collection – Deep Discoveries, National Archives, UK, https://tanc-ahrc.github.io/DeepDiscoveries/.
16. Locating a National Collection, British Library, UK, https://tanc-ahrc.github.io/LocatingTANC/.
17. Preserving and sharing born-digital and hybrid objects from and across the National Collection, Victoria & Albert Museum, UK, https://gtr.ukri.org/projects?ref=AH%2FT01122X%2F1.
18. Engaging Crowds: Citizen research and heritage data at scale, National Archives, UK, https://tanc-ahrc.github.io/EngagingCrowds/.
19. Provisional Semantics, Addressing the challenges of representing multiple perspectives within an evolving digitised national collection, Tate, UK, https://www.tate.org.uk/about-us/projects/provisional-semantics.
20. Digital Footprints and Search Pathways, Centre for Research in Digital Education, University of Edinburgh, https://www.de.ed.ac.uk/project/digital-footprints-and-search-pathways.
21. Where this isn't done already: some courses, such as the MA/MSc in Digital Humanities in the Information Studies department at UCL (established by the author in 2015, forgive me) have a practice-led "Introduction to Digitisation" module, that teaches the principles and practices of the digitization life cycle in a hands-on environment (UCL 2021).
22. GLAM Workbench, https://glam-workbench.github.io/about/.
23. Trove, National Library of Australia, https://trove.nla.gov.au.
24. GLAM Workbench, Trove Newspapers, https://glam-workbench.github.io/trove-newspapers/.
25. Programming Historian, https://programminghistorian.org.
26. Data Carpentry, https://datacarpentry.org.
27. Data Foundry, National Library of Scotland, https://data.nls.uk.

28. Library Carpentry: Text and Data Mining, http://librarycarpentry.org/lc-tdm/.
29. Digital Humanities Annual Conference, Alliance of Digital Humanities Organisations, https://adho.org/conference.
30. iSchools Annual Conference, iSchools, https://ischools.org/iConference.
31. Museums and the Web Annual Conference, https://www.museweb.net/conferences/.

REFERENCES

Ames, Sarah and Stuart Lewis. 2020. "Disrupting the library: Digital scholarship and Big Data at the National Library of Scotland." *Big Data & Society* 7 (2): 2053951720970576.

ArtFund. 2020. "Covid-19 Impact: Museum Sector Research Findings. Summary Report." https://www.artfund.org/assets/downloads/art-fund-covid19-research-report-final.pdf.

Arts and Humanities Research Council. 2021. "Towards a National Collection: Opening UK Heritage to the World." https://ahrc.ukri.org/research/fundedthemesandprogrammes/tanc-opening-uk-heritage-to-the-world/.

Bleier, Roman, Martina Bürgermeister, Helmut W. Klug, Frederike Neuber, and Gerlinde Schneider, eds. 2018. *Digital Scholarly Editions as Interfaces*, vol. 12. BoD–Books on Demand.

Bowen, Jonathan P. and Tuia Giannini. 2014. "Digitalism: The New Realism?" *Electronic Visualisation and the Arts (EVA 2014)*: 324–31.

Collins, Lori D. 2018. "Terrestrial Lidar." *The Encyclopedia of Archaeological Sciences*. Wiley Online Library. https://doi.org/10.1002/9781119188230.saseas0575.

Department for Digital, Culture, Media and Sport. 2018. "Culture is Digital." https://assets.publishing.service.gov.uk/government/uploads/system/uploads/attachment_data/file/687519/TT_v4.pdf.

Digital Culture Compass. 2020. "Tracker: Quick Guide." https://digitalculturecompass.org.uk/using-the-tracker/.

Falk, John H. and Lynn D. Dierking. 2016. *The Museum Experience Revisited*. New York, NY: Routledge.

Favish, Ashleigh. 2019. "Data Capture Automation in the South African Deeds Registry using Optical Character Recognition (OCR)." Master's thesis, Faculty of Commerce, African Institute of Financial Markets and Risk Management. http://hdl.handle.net/11427/31389.

Finnis, Jane. 2011. "Let's Get Real 1: How to Evaluate Success Online." Culture24, Brighton, UK. https://www.keepandshare.com/doc/3148918/culture24-howtoevaluateonlinesuccess-2-pdf-september-19-2011-11-15-am-2-5-meg?da=y&dnad=y.

Finnis, Jane. 2020. "The Digital Transformation Agenda and GLAMs. A Quick Scan Report for Europeana." Culture24, July. https://pro.europeana.eu/files/Europeana_Professional/Publications/Digital%20transformation%20reports/The%20digital%20transformation%20agenda%20and%20GLAMs%20-%20Culture24%20findings%20and%20outcomes.pdf.

Gorman, Gary Eugene and Peter Robert Clayton. 2005. *Qualitative Research for the Information Professional: A Practical Handbook*. London: Facet Publishing.

Greenhall, Matt. 2019. "Digital Scholarship and the Role of the Research Library. The Results of the RLUK Digital Scholarship Survey." Research Libraries UK. https://www.rluk.ac.uk/wp-content/uploads/2019/07/RLUK-Digital-Scholarship-report-July-2019.pdf.

Harrison, Rodney, Hana Morel, Maja Maricevic, and Sefryn Penrose. 2017. "Heritage and Data: Challenges and Opportunities for the Heritage Sector." Report of the Heritage Data Research Workshop, British Library, London, June 23. https://heritage-research.org/app/uploads/2017/11/Heritage-Data-Challenges-Opportunities-Report.pdf.

Hauswedell, Tessa, Julianne Nyhan, Melodee H. Beals, Melissa Terras, and Emily Bell. 2020. "Of Global Reach yet of Situated Contexts: An Examination of the Implicit and Explicit Selection Criteria that Shape Digital Archives of Historical Newspapers." *Archival Science*: 139–65. https://doi.org/10.1007/s10502-020-09332-1.

Howell, David and Ludo Snijders. 2020. "Spectral Imaging." In *Conservation Research in Libraries*, 172–96. Current Topics in Library and Information Practice. De Gruyter Saur

Hughes, Lorna M. 2004. *Digitizing Collections: Strategic Issues for the Information Manager*.London: Facet Publishing.

Hughes, Lorna M., ed. 2012. *Evaluating and Measuring the Value, Use and Impact of Digital Collections*. London: Facet Publishing.

Jaillant, Lisa, ed. 2021. *Archives, Access and Artificial Intelligence: Working with Born-Digital and Digitised Archival Collections*. Berlin: Transcript.

Kahn, Rebecca. 2020. "Locked Down Not Locked Out – Assessing the Digital Response of Museums to COVID-19." *LSE Impact of Social Sciences Blog*, May 8. https://blogs.lse.ac.uk/impactofsocialsciences/2020/05/08/locked-down-not-locked-out-assessing-the-digital-response-of-museums-to-covid-19/.

Kariyawasam, Kanchana and Olumayowa O. Adesanya. 2019. "Impact of Copyright Law on Mass Digitisation." *New Zealand Business Law Quarterly* 25 (1): 46–68.

Kirschenbaum, Matthew. 2013. "The.txtual Condition: Digital Humanities, Born-Digital Archives, and the Future Literary." *Digital Humanities Quarterly* 7 (1). http://www.digitalhumanities.org/dhq/vol/7/1/000151/000151.html.

Kizhner, Inna, Melissa Terras, Maxim Rumyantsev, Valentina Khokhlova, Elisaveta Demeshkova, Ivan Rudov, and Julia Afanasieva. 2020. "Digital Cultural Colonialism: Measuring Bias in Aggregated Digitized Content Held in Google Arts and Culture." *Digital Scholarship in the Humanities* 36 (3): 607–40.

Klein, Julie T. 2015. *Interdisciplining Digital Humanities: Boundary Work in an Emerging Field*. Minneapolis, MN: University of Michigan Press.

Korn, Naomi. 2009. "In from the Cold: An Assessment of the Scope of Orphan Works and its Impact on the Delivery of Services to the Public." JISC Content, April. https://naomikorn.com/wp-content/uploads/2020/09/SCA_CollTrust_Orphan_Works_v1-final.pdf.

Koszary, Adam. 2018. "Seven Broad Statements that May or May Not Help Your Museum Do a Bit Better at Social Media." November 12. https://medium.com/@adamkoszary/seven-broad-statements-that-may-or-may-not-help-your-museum-do-a-bit-better-at-social-media-6173c3b3af0c.

Krippendorff, Klaus. 2018. *Content Analysis: An Introduction to Its Methodology*, 4th edn. London: SAGE Publications.

Kuniavsky, Mike. 2010. *Smart Things: Ubiquitous Computing User Experience Design*. Burlington, MA: Elsevier.

Lachman, Richard. 2018. "STEAM Not STEM: Why Scientists Need Arts Training." The Conversation, January 17. https://theconversation.com/steam-not-stem-why-scientists-need-arts-training-89788.

Lavoie, Brian. 2018. "How Information about Library Collections Represents a Treasure Trove for Research in the Humanities and Social Sciences." *LSE Impact of Social Sciences Blog*, September 19. https://blogs.lse.ac.uk/impactofsocialsciences/2018/09/19/how-information-about-library-collections-represents-a-treasure-trove-for-research-in-the-humanities-and-social-sciences/.

Lee, Stuart D. 2001. *Digital Imaging: A Practical Handbook*. London: Facet Publishing.

Markauskaite, Lina, and Peter Goodyear. 2017. "Epistemic Thinking." In *Epistemic Fluency and Professional Education*, 167–94. Dordrecht, Springer.

Muehlberger, Guenter, Louise Seaward, Melissa Terras, et al. 2019. "Transforming Scholarship in the Archives through Handwritten Text Recognition: Transkribus as a Case Study." *Journal of Documentation* 75: (5). https://www.emerald.com/insight/content/doi/10.1108/JD-07-2018-0114/full/html.

Nauta Gerhard Jan, Wietske van den Heuvel, and Stephanie Teunisse. 2017. D4.4 Report on ENUMERATE Core Survey 4. Europeana DSI 2- Access to Digital Resources of European Heritage, Europeana. https://pro.europeana.eu/files/Europeana_Professional/Projects/Project_list/ENUMERATE/deliverables/DSI-2_Deliverable%20D4.4_Europeana_Report%20on%20ENUMERATE%20Core%20Survey%204.pdf.

National Galleries Scotland. 2019. "Mainstreaming Equality at the National Galleries of Scotland." April. Edinburgh: National Galleries of Scotland. https://www.nationalgalleries.org/sites/default/files/features/pdfs/Mainstreaming_Equality_Report_April_2019.pdf.

Odumosu, Temi. 2020. "The Crying Child: On Colonial Archives, Digitization, and Ethics of Care in the Cultural Commons." *Current Anthropology* 61 (S22): S289–S302.

OpenGlam. n.d. "A Global Network on Sharing Cultural Heritage." https://openglam.org.

Padilla, Thomas G. 2018. "Collections as Data: Implications for Enclosure." *College and Research Libraries News* 79 (6): 296.

Parry, Ross, ed. 2013. *Museums in a Digital Age*. New York, NY: Routledge.

Picheny, Michael, Zóltan Tüske, Brian Kingsbury, Kartik Audhkhasi, Xiaodong Cui, and George Saon. 2019. "Challenging the Boundaries of Speech Recognition: The MALACH Corpus." *arXiv preprint* arXiv: 1908.03455.

Pittock, Murray. 2018. "The Scottish Heritage Partnership Immersive Experiences: Policy Report." University of Glasgow. http://eprints.gla.ac.uk/198201/1/198201.pdf.

Ridge, Mia, ed. 2014. *Crowdsourcing our Cultural Heritage*. Farnham: Ashgate Publishing.

Rutner, Jennifer and Roger C. Schonfeld. 2012. "Supporting the Changing Research Practices of Historians." Ithaka S&R. https://sr.ithaka.org/wp-content/uploads/2015/08/supporting-the-changing-research-practices-of-historians.pdf.

Schön, Donald A. 1983. *The Reflective Practitioner: How Professionals Think in Action*. New York, NY: Basic Books.

Strauss, Anselm L. and Juliet M. Corbin. 1997. *Grounded Theory in Practice*. London: SAGE Publications.

Terras, Melissa. 2011. "The Rise of Digitization." In R. Rikowski (ed.), *Digitisation Perspectives*, 3–20. Rotterdam, Sense Publishers.

Terras, Melissa. 2016. "Crowdsourcing in the Digital Humanities." In *A New Companion to Digital Humanities*, edited by S. Schreibman, R. Siemens, and J. Unsworth, 420–39. Hoboken, NJ: Wiley-Blackwell. http://eu.wiley.com/WileyCDA/WileyTitle/productCd-1118680596.html.

Terras, Melissa, Stephen Coleman, Steven Drost, Chris Elsden, Ingi Helgason, Susan Lechelt, Nicola Osborne, Inge Panneels, Briana Pegado, Burkhard Schafer, Michael Smyth, Pip Thornton, and Chris Speed. 2021. "The Value of Mass-Digitised Cultural Heritage content in Creative Contexts." *Big Data and Society*. doi:10.1177/20539517211006165.

Thylstrup, Nanna Bonde. 2019. *The Politics of Mass Digitization*. Cambridge, MA: MIT Press.

University College London. 2021. "INST0044: Introduction to Digitisation." Information Studies. https://www.ucl.ac.uk/information-studies/inst0044-introduction-digitisation.

Warwick, Claire, Melissa Terras, Paul Huntington, and Nikoleta Pappa. 2008. "If You Build It Will They Come? The LAIRAH Study: Quantifying the Use of Online Resources in the Arts and Humanities through Statistical Analysis of User Log Data." *Literary and Linguistic Computing* 23 (1): 85–102.

Wingo, Rebecca S., Jason Heppler, and Paul Schadewald. 2020. *Digital Community Engagement: Partnering Communities with the Academy*. Cincinnati, OH: University of Cincinnati Press.

Zaagsma, Gerben. 2019. "Digital History and the Politics of Digitization." Digital Humanities, Utrecht University, July 8–12. https://dataverse.nl/dataset.xhtml?persistentId=doi:10.34894/USARQ6.

Zwitter, Andrej. 2014. "Big Data Ethics." *Big Data & Society* 1 (2). doi: 2053951714559253.

CHAPTER TWENTY-FIVE

Sharing as CARE and FAIR in the Digital Humanities

PATRICK EGAN AND ÓRLA MURPHY (UNIVERSITY COLLEGE CORK)

Sharing co-exists with the scholarly primitives as a foundational aspect of digital humanities and cultural heritage. Best practice digitization, representation, dissemination, and preservation of any cultural object necessitate working in interdisciplinary teams as co-equal generators of knowledge work. Spectral engineers, spatial engineers, color scientists, sound engineers, art historians, conservators, information scientists, and subject matter experts must all contribute to how objects are realized on the public's many devices. Contemporary research cultures must move beyond museological confines, encompassing intangible heritages and the communities where they originate and reside as integral to a shared inclusive future (CARE 2021). A profoundly intersectional cultural heritage for digital humanities is one that challenges the canonical acceptance of previous paradigms, hierarchies, elites, and centers. Such a digital humanities will combine CARE principles (Collective Benefit, Authority to Control, Responsibility, Ethics) with FAIR principles (Findable, Accessible, Interoperable, Reusable), repudiating "power differentials and historical contexts" and "encouraging open and other data movements to consider both people and purpose in their advocacy and pursuit" (CARE 2021).

Central to this research agenda is the person, and scholars seeking to share cultural heritage must continue to place an increasing amount of focus on connecting with communities where artifacts originate. With music collections, for example, renewed efforts are being made to avoid the ethical tensions about ownership of collected material, but also to improve upon some problematic interactions that arose between collectors and communities in the past. Scholars such as Ní Chonghaile (2021, 11) describe ways to become more balanced in sharing—to disrupt dynamics of power—while others recognize the imperative for professionals to rely upon relationships with the communities that already have procedures for dealing with their cultural materials which are transferred from cultural institutions (Reed 2019, 627). Such shifts, coupled with the slow but steady move towards open publishing, are bringing about truly "community-led" approaches to knowledge sharing (Imbler 2021). As a result, a more engaged attitude to sharing can allow scholars to infuse their research with the needs and expectations of the communities who their materials represent.

Community-led approaches to research are in many ways generous, but there is tension between the ideals of generous thinking (see Fitzpatrick 2021) and the often not-so-generous demands of the academy for researchers to pursue as solo and siloed works of scholarship. Nowhere is such tension more apparent than in how scholars are expected to share—in print, in monographs, with prestige publishers, rather than openly. The issues that are found in sharing print have also been seen in tangible and intangible cultural heritage work. This occurred initially because of cost and latterly due to infrastructural and sustainability concerns; concerns that are evident with many other digital humanities projects.

Mark Sample argues that the opportunity presented by the digital is in how it transforms how we share: "We are no longer bound by the physical demands of printed books and paper journals, no longer constrained by production costs and distribution friction, no longer hampered by a top-down and unsustainable business model" (2013, 256). Sample's vision for DH is one in which knowledge building and sharing are one and the same. Others agree: "... we should no longer be content to make our work public achingly slowly along ingrained routes, authors and readers alike delayed by innumerable gateways limiting knowledge production and sharing" (Gold 2011).

A variety of digital humanities projects have reconfigured some of our old ways of sharing—and thus seeing—cultural heritage, publishing full-color plates of Blake,[1] taking 3D scans of manuscript pages (Endres 2020), and working with galleries, libraries, archives, and museums to make our understanding of a great many artworks more accessible (Kenderdine 2015, 22–42). Cultural heritage institutions were amongst the first to share photographic archives, ephemera, and sound. However, the primacy of print modes delayed the onset of the integration of cultural heritage with the digital humanities more broadly. Where scholars wrote about songs, they did not hear the music from the printed page, nor had they a sense of a sculpture's volume when shown a single flat image intended to illustrate a written point. Cultural heritage encounters in digital humanities move beyond the page, allowing interactive experiences that are far less limited.

Cultural heritage operates in a scholarly space that is at the interstices of disciplinary cultures that often do not share the same language but must acquire a lingua franca to work together. FAIR provides such a framework, advocating that research models prioritize findable, accessible, interoperable, and re-usable practices (FORCE11 2016). Standards-based approaches like the International Image Interoperability Format (IIIF) exemplify these ideals, exhorting us to "break down silos with open APIs" (IIIF 2021). In Europe, open access has become integral to funding success, not just in terms of the publication of traditional papers, articles, and books, but in terms of the research data and the digital objects at the core of the work.

Sample has called for a similar ethos to be adopted in the digital humanities, where instead of just writing, classrooms would be filled with students who are both "making" *and* "sharing" with each other and the world outside the academy. From the perspective of the university lecturer, Sample saw this as a "critical intervention into the enclosed experience most students have in higher education" (2011). In this type of DH, opening research to communities outside of the academy becomes less of a perceived nod to "outreach" activity for the common good, but instead initiates a more engaged commitment to truly sharing research, building openness to research publication from the outset.[2] Unfortunately, there is still a long way to go before scholarship in the digital humanities becomes truly open and engaged (see Davis and Dombrowski 2011).

Sharing culture needs to also be sustainable. How many digital projects can guarantee sustainability beyond five years? In terms of time and effort, what is the return on this investment for the scholar in engaging with the digital potentialities for their research beyond a traditional focus and methodology? When the work is complete, where will it reside, and how long will it be accessible? If Sketchfab closes, where will the 3D models live? The perceived impermanence of digital objects becomes a self-fulfilling prophecy when a haphazard approach or funding opportunity prevails over a solid foundation with workflows based on FAIR and CARE principles, accessibility, and sound research data management. Simply put, there is a greater chance that an object will survive if created in an open form, amongst a shared community of researcher practitioners, ideally with a stable URI or DOI.

This is also at the heart of the persistent "how to evaluate a digital humanist for tenure" debate. When is a digital humanities website not a digital humanities website? When it fails to build FAIR content. A short-term website for news or for a short-term project is not the same as a serious digital humanities endeavor with sharing and sustainability as core values—one is marketing, the other knowledge creation.

Many scholars have sought to coalesce projects in the digital humanities and share practice and insight. Discussing the now-retired Project Bamboo, a cyberinfrastructure initiative for the arts and humanities, Quinn Dombrowski explains:

> Bamboo brought together scholars, librarians, and technologists at a crucial moment for the emergence of digital humanities. The conversations that ensued may not have been what the Bamboo program staff expected, but they led to relationships, ideas, and plans that have blossomed in the years that followed (e.g. DiRT and TAPAS project), even as Bamboo itself struggled to find a path forward.
>
> (2014)

Writing in 2014, Dombrowski was hopeful for the future, for the "growth of large-scale collaborative efforts," in initiatives and infrastructures that would address the "shared technical needs" of the "community" (2014). Sadly, many of the projects that attempted to realize such hopes have not survived, and are now subject to the jibe of being ephemeral, not a monograph.

It was this vision of the web as a shared knowledge environment that prompted Tim Berners-Lee's "Information Management: A Proposal" (1989), born of a desire to communicate expert knowledge. However, directories or platforms that achieve this scholarly communication for cultural heritage—for symphonies, for paintings, for sculpture, for installations, for large-scale cultural heritage buildings and landscapes—are difficult to build, fund, and maintain.

There have been many false infrastructural dawns. In the US, from Project Bamboo to DiRT, Dombrowski and others have grappled with "the difficult decisions that fall-out from the 'directory paradox', where the DH community's praise of directories is wildly incommensurate with the interest or resources available for sustaining them" (see Papaki 2021). Dombrowski's "Cowboys and Consortia" talk for DARIAH beyond Europe explores that dichotomy of the solo maverick builder and the sustainability capacity of consortia without a long-term home (2019). Theirs is a prescient reflection that is a macrocosm of the dichotomies at play between the digital project within the infrastructure, and the scholar within the traditional discipline. How may a cowboy's work of building an infrastructure survive to be sustained within a long-term consortium solution? What work might a persistent, integrated research infrastructure do for the scholarly community and society? Sustainability and visibility are critical to success in the longer term. "If you can't see it, you can't be it" is as useful a saying for digital humanities students as for aspiring sports stars—deprecated sites with dead and decaying links are not much use to the DH future.

But hope remains. Integrated examples of European Research Infrastructure Consortia (ERICs) include the Digital Research Infrastructure for the Arts and Humanities in Europe, DARIAH-EU, which proffers a holistic suite of solutions.[3] These include APIs, discovery portals, training, national cultural heritage aggregators, standardization tools, data repositories, analysis, and enrichment tools, amongst a long list. DARIAH-EU is building itself into a uniquely transdisciplinary international and sustainable infrastructure for digital humanities activities. This comprehensive vision includes the European Research Infrastructure for Heritage Science (HS) that similarly engages with the open agenda and promotes sharing within the field of cultural heritage. It builds on decades of work in projects like ARIADNE, CARARE, and 3D Icons that created a network of scholarship in archaeology, conservation, curation, architecture, preservation, and art history. Intrinsic to many of these pan-European projects were open concepts around linked open data ecosystems with the aim of building linked projects capable (albeit metaphorically), via semantic technologies, of organic growth instead of the static databases of previous scholarly instances positing a viable open future. Put simply, in Europe if your work is not going to be made openly shared and sustainable, it will not be funded. The cultural heritage community as a whole has more than embraced this paradigm shift. Exemplary institutions include the Rijksmuseum with its totally open data policy.[4]

These vicissitudes of survival and opportunity are replicated globally and locally as traditional computer centers in universities, and funding bodies are trapped in old project models that militate against innovative forms of open sharing of cultural heritage resources in more sustainable ways. Collaboration to leverage change is possible (see Chang 2021). Multi-layered scholarship of many dimensions exists as our communities have gathered together over decades, many with shared principles of excellent scholarship to communicate expert knowledge rendered in ever newly emerging digital representations. Privileging and using principles such as FAIR and CARE with our knowledges in a multiplicity of forms enriches our shared humanity, our songs and our stories, our haptic responses to 3D sculpture, our immersive experiences that enable meaning-making and understanding. Those of us in a position to further such enrichment must advocate for research and education practices that encourage innovation, CARE-ing and thinking beyond the page, and make evident the possibilities for excellent scholarship and excellent science inherent in new representations of knowledge work.

NOTES

1. See http://blakearchive.org/.
2. For example, see the Domain of One's Own model (Wired Insider 2012).
3. See dariahopen.hypotheses.org.
4. See rijksmuseum.nl/en/research/conduct-research/data/policy.

REFERENCES

Berners-Lee, Tim. 1989. "Information Management: A Proposal." CERN. https://cds.cern.ch/record/1405411/files/ARCH-WWW-4-010.pdf.

CARE. 2021. "CARE Principles of Indigenous Data Governance — Global Indigenous Data Alliance." https://www.gida-global.org/care/.

Chang, Tao (2021) "Arts and humanities infrastructure enabling knowledge with impact." UK Research and Innovation Blog. https://beta.ukri.org/blog/arts-and-humanities-infrastructure-enabling-knowledge-with-impact/.

Chonghaile, Deirdre Ní. 2021. *Collecting Music in the Aran Islands: A Century of History and Practice.* Madison, WI: University of Wisconsin Press.

Davis, Rebecca Frost and Quinn Dombrowski. 2011. "Divided and Conquered: How Multivarious Isolation Is Suppressing Digital Humanities Scholarship." National Institute for Technology in Liberal Education.

Dombrowski, Quinn. 2014. "What Ever Happened to Project Bamboo?" *Literary and Linguistic Computing* 29 (3): 326–39. https://doi.org/10.1093/llc/fqu026.
Dombrowski, Quinn. 2019. *Cowboys and Consortia: Thoughts on DH Infrastructure*. YouTube. Stanford University. https://www.youtube.com/watch?v=0vkHutf_UxY.
Endres, Bill. 2020. *Digitizing Medieval Manuscripts: The St Chad Gospels, Materiality, Recoveries, and Representation in 2D & 3D*. Kalamazoo, MI: ARC Humanities Press.
Fitzpatrick, Kathleen. 2021. *Generous Thinking: A Radical Approach to Saving the University*. Baltimore, MD: Johns Hopkins University Press.
FORCE11. 2016. "FAIR Principles. FORCE11 | The future of research communications and e-scholarship." https://force11.org/info/the-fair-data-principles/.
Gold, Matthew K. 2011. "Introduction: The Digital Humanities Moment." Debates in the Digital Humanities. https://dhdebates.gc.cuny.edu/projects/debates-in-the-digital-humanities/.
IIIF. 2021. "IIIF -International Image Operability Framework." Break down silos with open APIs. https://iiif.io/.
Imbler, Sabrina. 2021. "Training the Next Generation of Indigenous Data Scientists." *New York Times*, June 29. https://www.nytimes.com/2021/06/29/science/indigenous-data-microbiome-science.html.
Kenderdine, Sarah. 2015. "Embodiment, Entanglement, and Immersion in Digital Cultural Heritage." In *A New Companion to Digital Humanities*, edited by Susan Schreibman, Ray Siemens, and John Unsworth, 22–41. Wiley Online Library. https://onlinelibrary.wiley.com/doi/abs/10.1002/9781118680605.ch2
Papaki, Eliza. 2021. "DARIAH » Who Needs Directories: A Forum on Sustaining Discovery Portals Large and Small." https://www.dariah.eu/2020/07/30/who-needs-directories-a-forum-on-sustaining-discovery-portals-large-and-small/.
Reed, Trevor. 2019. "Reclaiming Ownership of the Indigenous Voice: The Hopi Music Repatriation Project." In *Oxford Handbook of Musical Repatriation*. Oxford: Oxford University Press.
Sample, Mark. 2011. "Oct 18: DH in the Classroom." https://cunydhi.commons.gc.cuny.edu/2011/10/13/oct-18-dh-in-the-classroom-shannon-mattern-mark-sample/.
Sample, Mark. 2013. "The Digital Humanities Is Not about Building, It's about Sharing." In *Defining Digital Humanities: A Reader*, edited by Melissa Terras, Julianne Nyhan, and Edward Vanhoutte, 255–57. Farnham: Ashgate.
Wired Insider. 2012. "A Domain of One's Own." Wired. https://www.wired.com/insights/2012/07/a-domain-of-ones-own/.

CHAPTER TWENTY-SIX

Digital Archives as Socially and Civically Just Public Resources

KENT GERBER (BETHEL UNIVERSITY)

How can the digital humanities community ensure that its digital archives are public resources that live up to the best potential of digital humanities without repeating or perpetuating power imbalances, silences, or injustice? A framework for anti-racist action, the "ARC of racial justice," developed by historian Jemar Tisby in his study of the complicity of the Christian church in perpetuating racism in the United States, is one way that this goal can be accomplished. The ARC is an acronym for three kinds of interrelated and interdependent kinds of actions one can take to fight racism and work for change: Awareness (building knowledge), Relationships (building connections in community), and Commitment (systemic change and a way of life) (2019, 194–7). This chapter applies Tisby's ARC framework to the context of publicly available digital archives and how they can become more socially and civically just by making sure the "silences in the archive" are identified, the variety of stories are told, and injustices are addressed (Thomas, Fowler, and Johnson 2017). The interaction between digital humanities scholars, community members, and cultural heritage professionals, such as librarians, archivists, and museum curators, is an important dynamic for digital archives that serve as public resources. When digital humanities projects result in digital archives they are often the result of collaboration and conversations between digital humanities scholars and cultural heritage professionals because of shared core values of these professions and their institutions to provide wide and equitable access, as well as to "advance knowledge, foster innovation, and serve the public" (Spiro 2012; Vandegrift and Varner 2013; Gerber 2017). The public-facing function of these collections and projects is also deepened by the engagement with the public humanities and public history communities (Brennan 2016, "Public History Roots").

AWARENESS

One central concept that drives librarians involved in these public-facing collections is the mission "to improve society through facilitating knowledge creation in their communities" (Lankes 2011, 15). In order to improve society through public resources and community engagement that is more just, inclusive, and mutually beneficial, it is important to be informed by the A, Awareness, of Tisby's ARC. To begin, we must be self-aware of how our personal identities and social positions within our institutions shape how we act and are perceived by our colleagues and community partners, especially with historically marginalized or exploited communities (Earhart 2018). With this in mind, I am a librarian who specializes in creating and managing the digital library, or digital archive, of the cultural heritage and scholarship of a mid-sized private liberal arts university located in the United States. The personal and social identities that influence my perspective and context include being white, straight, Christian, middle-class, cis-male, and located in the Global North, all placing me within a social context with privileges, blind spots, or vulnerabilities. These professional, institutional, social, and personal contexts are important to name and acknowledge in order to authentically discuss the role that perspective and power dynamics play in this topic. If digital archives are going to be a public good that more fully address inequality and are mutually beneficial to the communities they serve, those who create them must be aware of their biases and context.

There is a rich body of intersectional, critical work surveyed by Roopika Risam that engages how the digital humanities interact with the theoretical and practical factors of "race, gender, disability, class, sexuality, or a combination thereof" and colonialism (2015, para. 4; 2018b) that can be drawn upon to notice the silences in the archive. There is also a rich body of work in the *Debates in the Digital Humanities* series[1] in which the first two chapters of the 2019 edition discuss "Gender and Cultural Analytics" by Lisa Mandell and "Critical Black Digital Humanities" by Safiya Umoja Noble. Additionally, the volume *Bodies of Information* is a collection of chapters centered on Intersectional Feminism in the Digital Humanities (Wernimont and Losh 2018). Unfortunately, too many in the digital humanities and cultural heritage community have overlooked these bodies of work. One notable example regarding digital archives was triggered by a digital humanities conference keynote by a white male that ignored these works resulting in the affective labor of expressions of anger, sadness, and frustration on social media and the collaborative development of "angry bibliographies" like the "Justice and Digital Archive Bibliography"[2] by Jaqueline Wernimont to demonstrate that the resources do exist (Risam 2018a). As a white male, it is important for me (and colleagues like me) to engage with these resources and learn from them in order to avoid these past mistakes and to also ensure that my theory and practice is informed by these perspectives in order to recognize "one's own experience in relationship to complex positionality is crucial to understanding how we, as digital humanities scholars, might work in ethical, nonexploitive ways, attending to what might be missteps due to lack of consideration" (Earhart 2018). These critical treatments can lead to better ways of doing digital humanities. In *Digital Community Engagement*, digital humanities and public history scholars Wingo, Heppler, and Schadewald believe "that digital humanities has the capacity to positively shape the study of the arts, culture, and social sciences. We believe it can do so while promoting inclusion, justice, and recovery with beneficial impact for communities" (2020, "Introduction").

Just like the term Digital Humanities has a varied and elusive definition,[3] the definition of digital archive varies based on the people and organizations that are responsible for its creation and its purpose. For the purposes of this chapter, "digital archives" will represent the variety of digital humanities projects that result in collections involving various interactions between cultural heritage institutions, researchers investigating a particular theme or "thematic research collections" (Palmer 2004) and collaborations with community groups or an archive as "an ecosystem of individuals, communities, and institutions that care for and use these materials" (Hubbard 2020, "Communities, Individuals, and Institutions: Building Archives Through Relationships with Care"). This definition will include a spectrum bounded by the narrowest definition on one end—a digital version of the holdings of one particular cultural heritage institution, research group, or governmental body—and including a variety of digital collections that are collaborations between the above-mentioned groups and have shared meaning and purpose between them beyond a single collector. The limit of the broader definition of digital archive stops short of being "random collections of objects and documents that bring pleasure to the collector but have little or no impact on the larger order of things" (Eichhorn 2008). While I make some effort to broaden my perspective beyond examples based in the United States, the majority of the archives covered in this chapter are US-focused.

Digital archives as public resources can include some digital archives that are developed for a specific academic audience, like the *Networked Infrastructure for Nineteenth Century Electronic Scholarship* or *NINES*[4] that features nineteenth-century British and American literature, or for a specific purpose like addressing the injustice of slavery with the *Georgetown Slavery Archive*,[5] but are still considered public resources because they are still open for anyone to access on the web without charge. I am excluding projects or archives that are commercially generated or have barriers to access such as subscription fees or require a user to sign in for access. Others are intentionally designed to be public resources through collaboration with the community like the *Remembering Rondo History Harvest*[6] or for a general audience centered on a geographic region like the *Digital Library of the Caribbean*.[7] Because digital archives in this definition all have some connection with institutions like universities and cultural heritage organizations who are collaborating with communities it is important to point out that these same institutions can often perpetuate or exacerbate exploitation, oppression, or marginalization in these projects instead of improving these conditions (Earhart 2018). Also, how can we make sure that we are aware of the "frozen social orders" in the archives themselves and that the innovations that we foster are "sociological as well as technological" (Nell Smith 2014, 404)?

The future of digital archives is also shaped by the items and the tools that are used to curate them. Some content management software tools have been developed with the digital humanities values of access, openness, and humanities-focused inquiry and narratives in mind, like Omeka, and Scalar,[8] and have made creating projects and digital archives much easier (Leon 2017, 47). In order to avoid past mistakes of colonialism and exploitation, creators of the archive must be more aware of how an item or piece of data is embedded in communities, and ultimately connected to human beings and shape the archive around this condition (Nell Smith 2014; Earhart 2018; Risam 2018b). For example, some of these existing content management systems did not meet the needs of the Warumungu community, an indigenous group in Australia, when they were working on creating a digital archive with a team from Washington State University. The Warumungu community needed "cultural protocol driven metadata fields, differential user access based on cultural and social relationships, and functionality to include layered narratives at the item level" which was also a conclusion reached by communicating with US Native communities by the Smithsonian Institution National Museum of the American Indian (Christen, Merrill, and Wynne 2017, section 1). Mukurtu[9] was the result of this design relationship with the community and led to these important system requirements that were built in to work under the framework of these community needs and protocols. This relationship and development process is a good model to consider as we choose tools or develop new ones, even if they challenge some of the values of providing the widest access possible due to important community values such as indigenous groups or activists who are concerned about public exposure. We need to continue to explore partnerships like this to bring the influence of the humanities and particular knowledge of our communities to shape our practice and our tools.

Items in digital space depend on the technology of file formats and it is important to be aware of the variety of formats that an item in a digital archive can take. The Library of Congress' Recommended Formats Statement for Preservation[10] is one important tool to keep track of those changes and applying these changes into practices and tools that ensure that these archives will last as public resources for traditional materials like text and images while also addressing newer forms like 3D objects, datasets, and video games and how these new forms might meet community needs. We will need to continue to learn how to manage and represent these kinds of digital objects in our practice as the way that information is shared and represented continues to change. As we push boundaries with technology we also need to be aware of how accessible these items and archives are for people with disabilities and how to implement universal design principles in our projects, especially with tools that have been developed to help with this like Scripto and Anthologize (Williams 2012).

We each have our circles of influence and we can build awareness on several levels as discussed above. One can engage with the literature and conversations with colleagues to learn more deeply about the variety of perspectives and identities in the digital humanities and shape our theory and practices accordingly. It is important to acknowledge the predominance of whiteness in digital humanities (McPherson 2012) and cultural heritage fields (Hathcock 2015; Schlesselman-Tarango 2017; Leung and López-McKnight 2021) and take critiques like the ones represented by #ArchivesSoWhite seriously so that we can improve our practice, hiring, and training, "From the collections our repositories acquire to the outreach we conduct, exhibits we mount, and classes we teach, a fundamental shift in how archivists conceptualize their mandate is coming. In addition, we need to re-evaluate how we train, hire, support, and retain diverse staff who truly represent the materials for which they care" (Oswald et al. 2016). Archivists must be aware of their own biases and the power dynamics involved as Calahan explains:

> The archivist's role in deciding what is kept as part of the historical record for society is more crucial with the accrual of digital records, and it is important to be aware of the implication of making acquisition and appraisal decisions in a profession that is predominantly white, in which decision makers are in positions of political, social, and economic power.
>
> (2019, 5)

The community around the Archives for Black Lives in Philadelphia[11] is a project inspired by Jarrett Drake,[12] former archivist at Princeton and PhD candidate in anthropology, and is a good model of how awareness can lead to action in archival practice.

Beyond our own awareness of our personal and professional biases and the context of our tools and materials we also need to continue to raise the public's awareness of the existence of these archives and how they can find them. A large barrier between the public and archive is due to the scattered and opaque nature of the institutions that create and host archives. Once users actually find a digital archive there is an additional issue of how much they actually understand how to use and navigate the website, and if necessary, the traditional finding aid. One solution is to more deeply involve the users in the creation of an archive's web presence and address the four main concerns: "archival terminology, hierarchical structure of descriptions, searching tools, and content visualisation" (Feliciati 2018, 131). Further muddying the waters is the relationship between multiple libraries, archives, and museums and what a casual searcher may find on the web. Digital projects can also be scattered across the web and also receive lower priority in search results than commercial or more "popular" sources of information. Collaborative networked efforts like the Digital Public Library of America (DPLA)[13] and its regional hubs or Europeana[14] help ensure a wider spread of the items in an archive and allow for a larger scope of coverage and themes than a single institution or regional center could hold. The DPLA and the University of Minnesota also developed a tool to pull together disparate sources about African American history and search for them in the *Umbra Search*[15] platform. Networked collaborations like this with attention to metadata, content, and interface design can help to broaden the scope and access to items, emphasize certain themes within larger collections, or provide some context to an item in a digital archive or found on the web.

Whether an archive is digital or not it is a necessary skill to understand not only the holdings of an archive but also its "silences" including "the absence of records from the public view, the absence of certain details in records that are available, or the absence of records altogether" often as a result of privileging written records, informality (not creating records), conflict and oppression, selection policies, privileging the powerful and rich over the ordinary and marginalized, secrecy, and intentional destruction of records (Fowler 2017; Gilliland 2017, xv). If digital humanists can recognize these silences in digital archives they can address them by filling gaps or creating new archives altogether.

One model for addressing these silences with digital archives is to collect items that are ephemeral and would otherwise be lost to the public record especially in spaces of conflict and crisis like the protests after George Floyd was killed by police in Minneapolis. The *George Floyd & Anti-Racist Street Art Archive*[16] was created by professors and students who are part of the Urban Art Mapping Research Project at the University of St. Thomas in Saint Paul, Minnesota and collects images of street art from around the world responding to the call for justice and equality. Much of the items in the archive are already physically gone from their original locations or erased and this archive preserves the energy, art, and expressions of the immediate aftermath for others to view. A few years earlier in 2015 *A People's Archive of Police Violence in Cleveland*[17] (PAPVC) was created in response to the killing of twelve-year-old Tamir Rice by Cleveland police and to document the community's experiences and fill the silence that was in the police, government, and local news narrative. The *Mapping Police Violence Archive*[18] is another example of this same theme that was created to track and visualize incidents of police violence across the United States from 2013 to the present. We can continue to learn from the content of these archives and also from the process in which they were made to create new archives that continue to address more silences and create counter-narratives.

We can raise public awareness of issues of representation, justice, and technology in our role as educators by intentionally engaging our students and guiding them through archives with these issues in mind. As McPherson pointed out, we need "graduate and undergraduate education that hone[s] both critical and digital literacies" (2012, "Moving Beyond Our Boxes") and Miriam Posner's generous sharing of her tutorials and curriculum materials[19] are a model of this combination of literacies. Risam, Snow and Edwards have created an undergraduate digital humanities program at Salem State University that is marked by "a strong commitment to social justice through attention to the ethics of library and faculty collaboration, student labor, and public scholarship that seeks to tell stories that are underrepresented in local history" and to serve as a good model for smaller, teaching-focused institutions in contrast to the more prevalent models in larger research institutions (2017, 342). They do this through two "interwoven initiatives—Digital Salem, a university-wide umbrella digital humanities project to house digital scholarship by faculty and students on the history, culture, and literature of Salem, Massachusetts, and the Digital Scholars Program, an undergraduate research program that introduced students to digital humanities using the university's archival holdings" (Risam, Snow, and Edwards 2017, 341). In the Bethel University Digital Humanities program, which is also a collaboration between a librarian, archivist, and digital humanities scholar, students are introduced to the physical archive by the archivist and then, as the librarian, I teach them the concepts and process of assigning metadata and introduce them to the technology and process of digitizing materials for a digital archive (Gerber, Goldberg, and Magnuson 2019). I also discuss who gets to create a narrative of the archive and how certain narratives are constructed. As a predominantly white campus, meaning the student body is 50 percent white or more, it is important to expose these students to groups different from them so they are assigned to explore archives that emphasize the stories of BIPOC people like the DPLA and Minnesota Digital Library's *History of Survivance: Upper Midwest 19th-Century Native American Narratives* exhibit[20] featuring the Dakota and Anishinaabe (Ojibwe) people on whose homelands Minnesota is located, help transcribe Rosa Parks' papers in the Library of Congress,[21] explore a counter-narrative timeline of our institution's history of discrimination and racism,[22] or explore gender and the overlooked history of the creation of the web by reading and encoding a chapter in TEI XML about computer scientist and hypertext researcher, Dame Wendy Hall,[23] in addition to Tim Berners-Lee who is usually credited with creating the web without further context (Evans 2018, 153–74). Engaging our students with these concepts, communities, and ways of working will hopefully produce citizens who have both critical and digital literacies and are more able to see silences as they use or create digital archives.

We also need to study and learn from digital archives that were created as a response to critical public needs in the US for information about the spread of the coronavirus or on social media. Organizations like Johns Hopkins University[24] and the *New York Times*[25] stepped in to create the publicly available archives that the federal government entities like the Center for Disease Control and Prevention (CDC) were not willing or able to do. Projects like the *COVID Racial Data Tracker*[26] by *The COVID Tracking Project at The Atlantic* called out the need for racial data to be able to understand the impact on different sections of the community. These projects used technical tools like GitHub for the data repository and a variety of visualization tools and dashboards to quickly and clearly communicate the archive's contents and critical tools to understand the social and political, and ultimately human, impact of the Covid-19 pandemic. Rieger (2020) surveyed the institutional and individual efforts of researchers and archivists to document the pandemic for the near future as well as future generations in the early stages of the pandemic, and the digital humanities community and digital archives community will need to continue to analyze and compare these efforts to seek meaning from this rich and challenging time in the world's history. While collecting items from social media is out of scope for this chapter it is important to know about the distributed digital archive model exemplified by the collaborative project between archivists and activists, *Documenting the Now*,[27] which explains on its homepage that it is "a tool and a community developed around supporting the ethical collection, use, and preservation of social media content."

RELATIONSHIPS

The next phase in Tisby's ARC framework is Relationships. To enact change for more socially and civically just digital archives, awareness is not enough. One must take what they now know and enact it within their community and within relationships. Hubbard proposes that we must understand archives as part of an "ecosystem" and integrate the "individuals, communities, and institutions" more deeply into LIS professional development and education training. Hubbard continues that we should center that view of an ecosystem "when we think about archival custody and stewardship we move away from the binary construction of institutional or community ownership and control" (2020).

Efforts within the archival community to broaden the participation in selecting and describing the items and collections are a start to a more collaborative integration of individuals, communities, and institutions. For instance, Fowler explains how collecting policies "favoured the acquisition of records that reflected the perspective of governments or rich and powerful organizations, families and individuals. Only from the 1970s has there been a genuine desire at most archival institutions to reflect wider aspects of the society outside the reading room door, by interacting with groups which do not traditionally use the archive" (Fowler 2017, 34). Fowler goes on to describe efforts to further include users of archives in the description process for National Archive in the UK that seek to move beyond limited feedback to catalogers to include oral history or witness statements to add to records about people who are included in the archives and to make the archives more user-friendly (2017, 57).

One step further in the relationship with communities is for institutions to share or even release ownership of materials in digital archives back to their original owners in a process called "digital repatriation." The largest example of this by the United States federal government is when the Library of Congress gave digitized recordings of wax cylinders to the Passamaquoddy tribe located in what is now the state of Maine and adjusted any remaining access to the wishes of the tribe (Kim 2019). New projects could go even further and start from a place of shared, reciprocal responsibility or even with the community taking the lead while institutions follow. Wingo, Heppler, and Schadewald in their edited volume, *Digital Community Engagement*, have published an equitable model resource for any organization seeking to partner with their community and also for any community seeking to partner with a university. In chapter 2 of *Digital Community Engagement* Hubbard explains in more detail how the *People's Archive of Police Violence* in Cleveland began with archivists offering their services to the community after the killing of Tamir Rice and taking the lead from community organization's needs and agreeing to do whatever the community asked without a preconceived agenda (2020). Chapter 3 of this volume explains how a clear, equal partnership from the beginning between Macalester College and the community organization, Rondo Ave. Inc in Saint Paul, MN, led to a fruitful relationship that created the *Remembering Rondo History Harvest* program and digital archive that was truly a public resource shaped by and useful to the community (Anderson and Wingo 2020). Community leader and project partner Marvin Anderson explained that he was looking for three main things in a college partner: "a. depth of understanding about Rondo's unique history, b. level of advance preparation, and c. clarity of course objectives" (2020). These kinds of partnerships should be more of the model for existing and future digital archives that are meant to be a public resource of any kind.

COMMITMENT

The last phase of Tisby's ARC framework is Commitment. To show commitment to these efforts for more socially and civically just digital archives, changes must be informed by awareness, made in relationship, but must ultimately be implemented in systematic ways that impact policy and have longevity. Engaging with the kind of critical literature and conversations, of which a small sample is included in this chapter and volume, should be a regular part of professional and academic training for digital humanities scholars and cultural heritage professionals. If we are to improve society through our digital archives as public resources we should emphasize the full breadth of communities and experiences while also acknowledging the disproportional presence of whiteness in order to recognize the silences that white supremacy and colonialism have produced. We should make it a life habit and regular practice to listen and learn from the voices speaking into this gap and take the opportunity to grow. This effort does not come without challenge or without cost as most of the resources cited here are in response to backlash or resistance to wider representation, inclusion, or social and civic change.

We can also embed some of these ideas and practices into our tools and follow the example of the Mukurtu project. One example is how the software built in shared authority integrates institutional records alongside tribal metadata for the same digital item and each has independent authority to manage their own version of the records exemplified in the *Plateau People's Web portal*[28] (Christen, Merrill, and Wynne 2017, section 2). The software also incorporates the Traditional Knowledge Labels created by the *Local Contexts project*[29] that were developed in intimate collaboration with indigenous communities that "reflect ongoing relationships and authority including proper use, guidelines for action, or responsible stewardship and re-use." The challenge is to continue to develop and share the ways that we can continue to build or modify digital tools that are compassionate, just, and humanities-influenced resources for the public.

Commitment also implies longevity and projects and digital archives that engage the public long-term can lead to change in public policy. The *Mapping Prejudice Project*[30] initially raised awareness of the systematic practice of adding racist language to housing deeds that excluded people of color from buying the home in the future by creating a digital archive of those deeds and layering them on top of the map of Minneapolis. This arrangement of materials visualized how the practice of redlining in partnership with these racial covenants shaped where people of color could live in the city. In order to process all of the deeds the project invited the public to help identify and tag the language of the racial covenants in the deed documents. Although the language in these deeds was ruled unenforceable in 1948 and illegal in 1968 there was no way to remove the covenants from the deed unless the person who added it was contacted and agreed to remove it. After years of this work, the Minnesota state legislature passed a law in 2019 to enable current property owners to legally discharge the language which then led to the *Just Deeds Project*,[31] which provides free legal services to permanently discharge the racial covenants. While these are small steps in housing equity it serves as a public education resource as well as working to repair an unjust practice. This kind of digital archive is one that does improve society on the awareness level as well as the more systematic commitment to change public policy and legal structures.

There is a remaining challenge of the tension between having space for community groups and members to lead and own their materials and projects including their digital items and spaces while ensuring the preservation and persistence of a digital archive. In the case of the *Remembering Rondo History Harvest* the community was given ownership of the web space but that web domain, rememberrondo.org, at the writing of this chapter is no longer active and is only available through a snapshot from the Internet Archive Wayback Machine.[32] The Omeka-based archive is still available in the Macalester technical infrastructure and this kind of technical and management issue remains a challenge for smaller community organizations that do not want to or do not have the resources to maintain a public web presence for the long-term. This is an opportunity to serve our communities by committing our resources for sustainability but without pushing our own agenda over the needs of the community. An earlier example of this situation is the *September 11 Digital Archive*,[33] which was initially a project of the Roy Rosenzweig Center for History and New Media built with Omeka to document people's responses to the terrorist attacks on the World Trade Center buildings in New York City and the Pentagon in Washington DC in 2001. The archive was acquired by the Library of Congress in 2003 and now is held there for long-term access and preservation. The many digital archives considered in this chapter and in this volume will also have decisions to make regarding sustainability to consider how to manage the archive and how to ensure long-term access for the public. Often when a project does not involve a cultural heritage professional or institution there is an increased risk that the metadata strategy overlooks standards and multidisciplinary vocabularies and that the items in the archive, or the whole archive, are not part of a long-term preservation strategy. This will have to be navigated in a way that respects the needs and concerns of the communities that created them or are a part of the archive ecosystem. Colleges, universities, and cultural heritage institutions will need to consider how they will systematically commit resources to support community and public initiatives like these in their staffing, training, and budget through the whole ARC framework of awareness, relationships, and commitment.

CONCLUSION

Moving through Tisby's ARC framework for racial justice of Awareness, Relationships, and Commitment can help the ecosystem of individuals, communities, and institutions adjust existing digital archives and create new ones that address silences and are more socially and civically just. Practitioners and scholars, particularly ones like myself who are situated in socially dominant identities (white, cis-male), can engage with the rich, intersectional literature and projects to grow our awareness and inform our actions as well as listen to and center colleagues who speak from a variety of traditionally marginalized, non-dominant social identities. Through a deeper awareness of ourselves, our professions and institutions, and our communities we can seek out and deepen relationships that include multiple narratives and create more inclusive and mutually beneficial digital archives using some of these models of community engagement and education. We can establish an awareness-informed and relationship-informed Commitment by seeking ways to make sure that we encourage and challenge our institutions to develop systems and policies that ensure the sustainability of these digital humanities projects and practices in the form of digital archives.

NOTES

1. https://dhdebates.gc.cuny.edu/.
2. https://jwernimont.com/justice-and-digital-archives-a-working-bibliography/.
3. What is Digital Humanities? Made by Jason Heppler and contains 817 definitions submitted by digital humanities scholars and practitioners, https://whatisdigitalhumanities.com/.

4. https://nines.org/about/what-is-nines/.
5. https://slaveryarchive.georgetown.edu/.
6. https://omeka.macalester.edu/rondo/.
7. https://www.dloc.com.
8. The Alliance for Networking Visual Culture, https://scalar.me/anvc/scalar/.
9. Mukurtu Content Management System, https://mukurtu.org/about/.
10. https://www.loc.gov/preservation/resources/rfs/TOC.html.
11. Archives For Black Lives: Archivists responding to Black Lives Matter, https://archivesforblacklives.wordpress.com/.
12. Jarrett Drake's Harvard PhD Candidate page, https://scholar.harvard.edu/drake; Jarrett M. Drake's Writings on Medium, https://medium.com/@jmddrake.
13. dp.la.
14. https://www.europeana.eu/en.
15. https://www.umbrasearch.org/.
16. https://georgefloydstreetart.omeka.net/about.
17. https://www.archivingpoliceviolence.org/.
18. https://mappingpoliceviolence.org/.
19. Miriam Posner's Blog, Tutorials and Other Curricular Material, http://miriamposner.com/blog/tutorials-ive-written/.
20. https://dp.la/exhibitions/history-of-survivance.
21. Rosa Parks: In Her Own Words Crowdsourced Transcription Project, https://crowd.loc.gov/campaigns/rosa-parks-in-her-own-words/.
22. Looking Back to Move Forward Timeline: Selected Clarion Articles about Discrimination, Inequality, Race, and Social Justice at Bethel University 1959–1993, https://cdm16120.contentdm.oclc.org/digital/collection/p15186coll6/custom/looking-back-timeline1.
23. Wendy Hall was also a digital humanities pioneer with the archivist at the University of Southampton when they collaborated to create an interlinked multimedia digital archive of the Mountbatten collection in 1989.
24. Covid-19 Dashboard by the Center for Systems Science and Engineering (CSSE) at Johns Hopkins University (JHU), https://coronavirus.jhu.edu/map.html.
25. *New York Times* GitHub Ongoing Repository of Data on Coronavirus Cases and Deaths in the U.S., https://github.com/nytimes/covid-19-data.
26. https://covidtracking.com/race.
27. https://www.docnow.io/.
28. Example of a multiple perspective record of the Chemawa School Bakery, *c.* 1909 in the Plateau Peoples' Web Portal, https://plateauportal.libraries.wsu.edu/digital-heritage/chemawa-school-bakery-circa-1909.
29. https://localcontexts.org/.
30. https://mappingprejudice.umn.edu/index.html.
31. https://justdeeds.org/.
32. https://web.archive.org/web/20180903043142/; http://omeka.rememberingrondo.org/.
33. https://911digitalarchive.org/.

REFERENCES

Anderson, Marvin Roger and Rebecca S. Wingo. 2020. "Harvesting History, Remembering Rondo." In *Digital Community Engagement: Partnering Communities with the Academy*, edited by Rebecca S. Wingo, Jason A. Heppler and Paul Schadewald, Ch. 3. Cincinnati, OH: University of Cincinnati Press.

Brennan, Sheila A. 2016. *Public, First. Debates in the Digital Humanities 2016*, edited by Matthew K. Gold and Lauren F. Klein. Minneapolis, MN: University of Minnesota Press. https://dhdebates.gc.cuny.edu/read/untitled/section/11b9805a-a8e0-42e3-9a1c-fad46e4b78e5#ch32.

Calahan, Lisa. 2019. "Acquisitions, Appraisal, and Arrangement." In *The Digital Archives Handbook: A Guide to Creation, Management, and Preservation*, edited by Aaron D. Purcell, 3–18. Lanham, MD: Rowman & Littlefield.

Christen, Kimberly, Alex Merrill, and Michael Wynne. 2017. "A Community of Relations: Mukurtu Hubs and Spokes." *D-Lib Magazine* 23 (5/6) (May/June). https://doi.org/10.1045/may2017-christen.

Earhart, Amy E. 2018. "Can we Trust the University? Digital Humanities Collaborations with Historically Exploited Cultural Communities." In *Bodies of Information: Intersectional Feminism and Digital Humanities*, edited by Elizabeth M. Losh and Jaqueline Wernimont, chap. 20. Minneapolis, MN: University of Minnesota Press.

Eichhorn, Kate. 2008. "Archival Genres: Gathering Texts and Reading Spaces." *Invisible Culture: An Electronic Journal for Visual Studies* (12). http://hdl.handle.net/1802/5788.

Evans, Claire Lisa. 2018. *Broad Band: The Untold Story of the Women Who made the Internet*. New York, NY: Portfolio/Penguin.

Feliciati, Pierluigi. 2018. "Access to Digital Archives: Studying Users' Expectations and Behaviours." In *Digital Archives: Management, access and use*, edited by Milena Dobreva, 121–36. London: Facet Publishing.

Fowler, Simon. 2017. "Enforced Silences." In *The Silence of the Archive*, edited by David Thomas, Simon Fowler, and Valerie Johnson, 1–39. Chicago, IL: ALA Neal-Schuman.

Gerber, Kent. 2017. "Conversation as a Model to Build the Relationship among Libraries, Digital Humanities, and Campus Leadership." *College & Undergraduate Libraries* 24 (2–4): 418–33. DOI: 10.1080/10691316.2017.1328296.

Gerber, Kent, Charlie Goldberg, and Diana Magnuson. 2019. "Creating Dynamic Undergraduate Learning Laboratories through Collaboration between Archives, Libraries, and Digital Humanities." *The Journal of Interactive Technology and Pedagogy* (15). https://jitp.commons.gc.cuny.edu/creating-dynamic-undergraduate-learning-laboratories-through-collaboration-between-archives-libraries-and-digital-humanities/.

Gilliland, Anne J. 2017. "Foreword." In *The Silence of the Archive*, edited by David Thomas, Simon Fowler and Valerie Johnson, xv–xvii. Chicago, IL: ALA Neal-Schuman.

Hathcock, April. 2015. "White Librarianship in Blackface: Diversity Initiatives in LIS." *In the Library with the Lead Pipe*. http://www.inthelibrarywiththeleadpipe.org/2015/lis-diversity/.

Hubbard, Melissa A. 2020. "Archival Resistance to Structural Racism: A People's Archive of Police Violence in Cleveland." In *Digital Community Engagement: Partnering Communities with the Academy*, edited by Rebecca S. Wingo, Jason A. Heppler and Paul Schadewald, chap. 2. Cincinnati, OH: University of Cincinnati Press.

Kim, E. Tammy. 2019. "The Passamaquoddy Reclaim their Culture through Digital Repatriation." *The New Yorker*, January 30.

Lankes, R. David. 2011. *The Atlas of New Librarianship*. Cambridge, MA; [Chicago]: MIT Press; Association of College & Research Libraries.

Leon, Sharon. 2017. "Complexity and Collaboration: Doing Public History in Digital Environments." In *The Oxford Handbook of Public History*, 1st edn, edited by Paula Hamilton and James B. Gardner, 44–66. Oxford: Oxford University Press.

Leung, Sofia Y. and Jorge R. López-McKnight. 2021. "Introduction: This is Only the Beginning." In *Knowledge Justice: Disrupting Library and Information Studies through Critical Race Theory*, edited by Sofia Y. Leung and Jorge R. López-McKnight, 1–41. Cambridge, MA: MIT Press.

McPherson, Tara. 2012. "Why are the Digital Humanities so White? Or Thinking the Histories of Race and Computation." In *Debates in the Digital Humanities*, edited by Matthew K. Gold, chap. 9. Minneapolis, MN: University of Minnesota Press. https://dhdebates.gc.cuny.edu/read/untitled-88c11800-9446-469b-a3be-3fdb36bfbd1e/section/20df8acd-9ab9-4f35-8a5d-e91aa5f4a0ea.

Nell Smith, Martha. 2014. "Frozen Social Relations and Time for a Thaw: Visibility, Exclusions, and Consideration for Postcolonial Digital Archives." *Journal of Victorian Culture* 19 (3): 403–10. doi: 10.1080/13555502.2014.947189.

Noble, Safiya Umoja. 2019. "Toward a Critical Black Digital Humanities." In *Debates in the Digital Humanities*, edited by Matthew K. Gold and Lauren F. Klein, chap. 2. Minneapolis, MN: University of Minnesota Press.

Oswald, Heather, Christine Anne George, Stephanie Bennett, Maureen Harlow, Linda Reynolds, and Kristen Weischedel. 2016. *#ArchivesSoWhite Intro & Bibliography*. Issues & Advocacy. Society of American Archivists. https://issuesandadvocacy.wordpress.com/2016/04/18/archivessowhite-intro-bibliography/.

Palmer, Carole L. 2004. "Thematic Research Collections." In *A Companion to Digital Humanities*, edited by Ray Siemens, Susan Schreibman, and John Unsworth, 348–65. Hoboken, NJ: Blackwell Publishing.

Rieger, Oya Y. 2020. *Documenting the COVID-19 Pandemic: Archiving the Present for Future Research*. Ithaka S+R. https://sr.ithaka.org/blog/documenting-the-covid-19-pandemic/.

Risam, Roopika. 2015. "Beyond the Margins: Intersectionality and the Digital Humanities." *Digital Humanities Quarterly* 9 (2). www.digitalhumanities.org/dhq/vol/9/2/000208/000208.html.

Risam, Roopika. 2018a. "Diversity Work and Digital Carework in Higher Education." *First Monday* 23 (3). DOI: 10.5210/fm.v23i3.8241.

Risam, Roopika. 2018b. *New Digital Worlds: Postcolonial Digital Humanities in Theory, Praxis, and Pedagogy*. Chicago, IL: Northwestern University Press.

Risam, Roopika, Justin Snow, and Susan Edwards. 2017. "Building an Ethical Digital Humanities Community: Librarian, Faculty, and Student Collaboration." *College & Undergraduate Libraries* 24 (2–4): 337–49. DOI: 10.1080/10691316.2017.1337530.

Schlesselman-Tarango, Gina. 2017. *Topographies of Whiteness: Mapping Whiteness in Library and Information Science*. Sacramento, CA: Library Juice Press.

Spiro, Lisa. 2012. "This is Why We Fight: Defining the Values of the Digital Humanities." In *Debates in the Digital Humanities*, edited by Matthew K. Gold, chap. 3. Minneapolis, MN: University of Minnesota Press.

Thomas, David, Simon Fowler, and Valerie Johnson. 2017. *The Silence of the Archive: Principles and Practice in Records Management and Archives*. Chicago, IL: Neal-Schuman, an imprint of the American Library Association.

Tisby, Jemar. 2019. *The Color of Compromise: The Truth about the American Church's Complicity in Racism*. Grand Rapids, MI: Zondervan.

Vandegrift, Micah and Stewart Varner. 2013. "Evolving in Common: Creating Mutually Supportive Relationships between Libraries and the Digital Humanities." *Journal of Library Administration* 53 (1): 67–78.

Wernimont, Jacqueline and Elizabeth Losh, eds. 2018. *Bodies of Information: Intersectional Feminism and Digital Humanities*. Minneapolis, MN: University of Minnesota Press.

Williams, George H. 2012. "Disability, Universal Design, and the Digital Humanities." In *Debates in the Digital Humanities*, edited by Matthew K. Gold, chap. 12. Minneapolis, MN: University of Minnesota Press.

Wingo, Rebecca S., Jason A. Heppler, and Paul Schadewald, eds. 2020. *Digital Community Engagement: Partnering Communities with the Academy*. Cincinnati, OH: University of Cincinnati Press. https://ucincinnatipress.manifoldapp.org/read/digital-community-engagement/section/26d4ed07-2e29-420a-bebd-9b6204763802.

PART FOUR

Institutional Contexts

CHAPTER TWENTY-SEVEN

Tool Criticism through Playful Digital Humanities Pedagogy

MAX KEMMAN (DIALOGIC)

As source material for humanities research is increasingly digitally available, students too increasingly consult and analyze source material using digital methods. Some of those methodologies may currently be associated with digital humanities, such as topic modeling or other distant reading approaches (e.g., Brauer and Fridlund 2013; Jockers 2013). Yet other methods may instead be understood as digital scholarship in the humanities, closer to traditional practices of research but through digital interfaces such as spreadsheets or qualitative data analysis tools (Antonijević 2015). What many of these methods have in common is that they are conducted through graphical user interfaces that consist of overviews and buttons. Since tools play such a prominent role in digital humanities methodologies, a main concern for digital humanities pedagogy should therefore be to teach students how to critically engage with tools, how to assess what effect each button has on analysis and which buttons subsequently ought to be pushed in what order.

Hands-on experience with tools is subsequently regularly a learning objective of digital humanities courses. Instructions are a common approach, where students are shown how to move through a graphical interface, demonstrating how an analysis may be conducted. For an example outside of the digital humanities, when I was an undergraduate I learned how to conduct statistical analyses through SPSS Statistics software by means of such instructions. After a semester of statistical analyses, I knew how to load data that was fit for purpose, how to create the charts that are supposedly relevant, how to give the command for the student's t-test and some other tests, and finally where to find the *p*-value and write this all down. Yet I had very little understanding of why I was pushing one button but not another or why I was writing down one value but not another.

OPENING THE BLACK BOXES OF TOOLS

Confronted with such problems, where one acquires results without fully understanding how those results came about, some authors in digital humanities have argued against using graphical interfaces for digital analyses. Buttons are said to hide the underlying algorithms, providing a black box to the student who cannot discern how one button might do something entirely different from another button. Perhaps even more damning, Joris van Zundert writes that buttons hide the transformative aspects of digital methods (van Zundert 2016). Complex and innovative methodologies end up behind buttons marked by just one or two words that are aligned with what the user is already expected to know. As such, van Zundert fears that such graphical interfaces leave the impression that little has changed compared to earlier analog methodologies.

Yet another criticism of graphical interfaces is that they force specific predetermined practices and perspectives. Johanna Drucker compares tools to Trojan horses, importing assumptions and epistemological biases from the sciences and business into the humanities (Drucker 2011). Mel Stanfill writes that graphical interfaces actively communicate how an analysis should be conducted and which forms of use are right or wrong (Stanfill 2014). This may be especially true in the case of instructions that explicate how a tool ought to be used and how the correct results may be achieved. Instructions tell students that the correct way of using a tool is to follow a specific path of buttons and settings to produce a specific desired result.

Following such criticisms, some authors have argued that students should instead learn to write and read code if they are to understand what buttons do and how digital analyses are constructed (e.g., Kirschenbaum 2009; Montfort 2015; Öhman 2019). But is it really necessary to understand the software's underlying code? Bernhard Rieder and Theo Röhle claim that understanding the underlying code does little to understand the implemented methodologies. As per my earlier example of SPSS, they note that understanding SPSS does not require a student to learn Java (the programming language in which SPSS Statistics is written), but to develop an understanding of statistics and probability theory (Rieder and Röhle 2017). Ben Schmidt distinguishes between "algorithms," the specified steps for a computer to produce output, and "transformations," the way data is reconfigured on the screen by means of the algorithm (Schmidt 2016). He writes that "a transformation expresses a coherent goal that can be understood independently of the algorithm that produces it." That is, even without reading or understanding the code executed when clicking a button, it is the transformation of the data that is performed after clicking the button that can and should be understood. For example, Schmidt explains that it is not necessary to understand sorting as an algorithm or as code. What is instead important is that students see and understand the difference between lists that are sorted by alphabetical, chronological, or order of frequency. Quinn Dombrowski suggests in her chapter in this volume that digital humanities research should be thought of as workflows from data selection until interpretation of results, rather than (just) coding. While code may be relevant or important for some steps in a workflow, it is much more important to understand the workflow and decide how code fits into that. Furthermore, Karin van Es and her co-authors note that isolating the code from other interactive elements in the software and the datasets that are being analyzed may not reveal how transformations are achieved at all (van Es, Schäfer, and Wieringa 2021). They instead argue that tools are relational and that results are produced through the interaction between tools and users.

Marijn Koolen and his co-authors add that "tools are intimately related to the data" (Koolen, van Gorp, and van Ossenbruggen 2019, 12). They write this in response to some scholars that hope to circumvent black-box tools by engaging with the "raw data." However, as long as the material under consideration is stored in a digital format, one will always need some tool to view the data. The data can never be accessed and consulted without a tool mediating between the stored bits and the screen. Elijah Meeks has consequently argued that even spreadsheets, commonly thought of as raw data, are actually information visualizations, ordering, and representing data in human-readable form (Meeks 2014). This might tempt some to distinguish tools by rawness from less to more ordering of data, as a means of moving away from tools that introduce too many interpretations. Yet one might ask whether students would gain a better understanding of the data if they were to open csv files in a text editor or command-line interface, rather than spreadsheet software. I would not expect this to be the case, except to demonstrate the utility of a graphical spreadsheet interface.

DEVELOPING TOOL CRITICISM

To posit that students can either learn to code or suffer dependence on black-box interfaces is therefore an oversimplification. Instead, I argue that it is central to a digital humanities pedagogy that students learn to explore and interrogate tools and interfaces. Doing so, students learn to critically use available tools without falling into overreliance on tools as co-producers of knowledge (van Dijck 2010). Yet equally importantly, they learn not to fear tools as complex environments where students lack confidence in their own performance and fear breaking everything if they hit the wrong button (Rosen, Sears, and Weil 1993). In contrast, Maura Ives has noted that when a tool breaks down, this is a sign that students have successfully explored its limitations (Ives 2014). Rather than a moment to be feared, errors are the moments that students have successfully explored the boundaries of a tool.

Such a critical exploration and interrogation of tools and interfaces constitutes what has been called *(digital) tool criticism*. Karin van Es and her co-authors define tool criticism as "the critical inquiry into digital tools and how they are used for various purposes within the research process" (van Es, Schäfer, and Wieringa 2021, 52). Marijn Koolen and his co-authors similarly define it as

> the reflection on the role of digital tools in the research methodology and the evaluation of the suitability of a given digital tool for a specific research goal. The aim is to understand the impact of any limitation of the tool on the specific goal, not to improve a tool's performance. That is, ensuring as a scholar to be aware of the impact of a tool on research design, methods, interpretations and outcomes.
>
> (Koolen, van Gorp, and van Ossenbruggen 2019, 14–15)

Tool criticism moves beyond the matter of tools as practical and efficient, to consider them as interventions in the research process that enable or confine specific research practices or questions. Tool criticism raises awareness of tools as intellectual products where someone has worked out how methods can be implemented and how different options provide flexibility. As such, tool criticism explicates the relation between students (and scholars) and tools as co-producing results and necessitates reflection on this relationship.

How can one develop a pedagogy of tool criticism? Van Es and her co-authors warn that tool criticism is essentially not a checklist of actions to perform or questions to ask that can be followed. Tool criticism is instead a general mentality towards tools and one's research process to critically reflect on what one is doing and why. One can subsequently not only produce results but explain why those results were produced and not others. According to Koolen and his co-authors the process of tool criticism and understanding the transformations of software develops by means of *experimentation*. They write that "[o]ne way to critically evaluate a tool for a given purpose is to experiment with different ways of applying the tool" (2019, 15). Rather than providing a structured tutorial or a checklist of questions to ask, a pedagogy of tool criticism requires students to face a tool and experiment with buttons and settings in a different order to see what happens.

EXPERIMENTING WITH TOOLS

To develop a mentality of tool criticism students should first develop a mentality of experimenting with tools. This aligns with general practices of digital humanities, where experimentation has been argued to be one of the central values (Spiro 2012). Jentery Sayers embedded experimentation as a central value in what he called a *tinker-centric pedagogy* (Sayers 2011). Students collaborate to test and tinker with tools and digital media and document what changes from one experiment to another. Through this documentation, students can subsequently reflect on the impact of one choice as compared to another. Students can see how one option or button led to a result as compared to other results from alternative buttons and options. Writing down how different results were produced, through which buttons and settings, students simultaneously learn how to document their research process, developing a reflective and replicable methodology.

Following this practice of reflective tinkering, Andreas Fickers and Tim van der Heijden write of *thinkering* (coined earlier by Huhtamo 2010) to combine tinkering with technology and critical reflection as a hermeneutic process and thinking (Fickers and van der Heijden 2020). On the website of their research institute, this combination of reflection and tinkering is presented as the tenet of digital history and digital humanities:

> "Thinkering" – composed of the verbs "tinkering" and "thinking" – describes the action of playful experimentation with technological and digital tools for the interpretation and presentation of history. Finding new historical insights and innovative forms of narrative are at the centre of the C²DH's activities.[1]

Thinkering in their view is not just an activity while using a specific tool, but the very process of innovating historical and humanistic narratives by experimentation with digital tools. The question is thereby not only what opportunities or limitations a tool may have, or what effect a specific button has, but a reflection on how the produced results impact scholarly insights and narratives and what new perspective is added to the scholarly debate.

Sayers, Fickers, and van der Heijden agree that the process of experimentation should essentially be *play*. As play, Sayers notes that tinkering implies a lack of expertise, where one is not yet entirely sure what is the proper way of doing things. In developing a playful pedagogy, Mark Sample notes that the process of doing things is emphasized instead of the result (Sample 2020). A playful pedagogy is not centrally about producing the right results by means of carefully clicking the right buttons and selecting the correct options, but is about the process of how a result may be achieved. Mistakes consequently become recognized as more informative than success, where each mistake is a lesson learned and each failed path enables a new path to be taken. Play thereby lowers the stakes, where an error is not a failure, inviting risk-taking.

Furthermore, Sample argues play moves away from pursuing a single objective. Playing with tools thereby becomes a means of discovering, rather than a practical and efficient way of reaching a predetermined goal. Students are immediately confronted with the fact that digital humanities methodologies are not necessarily more efficient or easier than analog methodologies, but instead offer different views and interpretations. There may not be a single correct answer. Since different options and buttons produce various results, experiments produce a multitude of perspectives rather than a single authoritative one. Stefan Werning adds that a playful pedagogy confronts students with different ways that data can be analyzed, visualized, and used. Rather than following standardized methods of data analysis, playing with data diversifies data practices, developing not just a critical mindset to the tool but to the underlying data as well (Werning 2020). Play thereby mitigates the problems of graphical interfaces identified by Drucker and Stanfill that tools force specific practices and perspectives. By playing and experimenting with buttons and settings and reflecting on how these experiments lead to different views, students are encouraged to explore their own actions within the tool, rather than follow a predetermined path of how the tool ought to be used.

Finally, Sample notes that research practices become recognized as ambiguous rather than certain, where ambiguity and uncertainty make the process more interesting, rather than problematic. That students become comfortable with uncertainty in digital environments is a necessity when digital environments are increasingly messy. Tools are constantly changed with small or big updates that move around buttons, reorder settings, or introduce new forms of interaction with the environment. Explicating this messiness teaches students less to trust tools as refined and finished products, instead recognizing them as a plethora of possibilities that move around.

Comfort with uncertainty is, moreover, a necessity when digital environments become increasingly complex as they become increasingly powerful. For example, when opening the text analysis suite Voyant Tools,[2] a user is confronted with five windows, each with different buttons, settings, and views on the data. This visual confrontation with the plethora of possibilities is in my opinion an oft-overlooked *advantage* that tools with graphical interfaces have over coding. While technically the possibilities of analysis may be larger with coding, a student is not confronted with these possibilities when asked to type commands in an empty window. For example, the popular network analysis software Gephi lists the different layout algorithms below one another, presenting them as visibly distinct alternatives. This prompted Bernhard Rieder to run multiple layout algorithms on the same underlying data to demonstrate how different algorithms produce different interpretations (Rieder 2010). When writing code, such a list of layout algorithms is not necessarily shown or may be embedded in the documentation. Students may not even think of trying different algorithms if they are not first confronted with their existence.

Recognizing ambiguity and uncertainty is not to expect students to exhaust all possible permutations of settings and buttons and results. Instead, playful experimenting with buttons constitutes what Stephen Ramsay has called a "hermeneutics of screwing around" (Ramsay 2014). Ramsay's context is that a scholar may be confronted with a million books, where a scholar cannot be expected to have read them all. Instead, scholars develop an understanding of the path that is taken as a situational, relational, and playful means of learning. In digital tools, students may not necessarily be confronted with a million buttons, but here too the goal is for students to understand their path through the interface and how other paths may be possible. Students thereby learn to become comfortable with overwhelming possibilities, as they learn to take and reflect on decision points and experiments in the context of alternative routes.

In summary, I posit that a pedagogy of tool criticism consists of confronting students with the messiness of tools, letting them play around with buttons and settings, and to learn through experimentation how buttons and settings in different constellations and orders produce various results. By encouraging students to reflect on the differences between results they learn that each result, each "data transformation," is dependent on the choices made by a researcher when using a tool. Instead of being entirely dependent on tools to bring about research results, students learn how to discover the limits of tools and to what extent the researcher is in control of the outcomes of analysis.

NO REFLECTION WITHOUT INTERACTION

The way students are traditionally taught to use tools has insufficiently included tool criticism as a skill. Students are simply taught which tool is the right one and how to apply it, taking them one button to the next setting until the right result is produced. Students may therefore be confused when they are simply tasked to play around with a tool. In my classes, when students played with tools such as Voyant Tools, many students would stick to the default views and play around with settings and buttons in those. Some students would try to play with other views. The feeling of being overwhelmed was not uncommon. In one of my classes, one student finally stood up and stated that while I spoke of playing, they were not having much fun.

This underscores that experimentation and critical reflection are necessary skills that students have not sufficiently developed thus far. As a field characterized by the import of tools, concepts, and methods from multiple different fields (Klein 2014), becoming comfortable with overwhelming (digital) possibilities, experimenting and reflecting on what happens should be a central concern in our pedagogies. Digital humanities engagement with tools should extend beyond mere instrumentalism (Liu 2012), towards a mindset of critically reflecting on how software affects research practices. Yet there can be no reflection without interaction and no tool criticism without deeply engaging and experimenting with those tools.

ACKNOWLEDGMENTS

Thanks to Karin van Es and the anonymous reviewers for their feedback on an earlier version of this chapter.

NOTES

1. Thinkering. https://www.c2dh.uni.lu/thinkering.
2. Voyant Tools is a web-based suite of tools to analyze text corpora, see https://voyant-tools.org/.

REFERENCES

Antonijević, Smiljana. 2015. *Amongst Digital Humanists: An Ethnographic Study of Digital Knowledge Production*. Pre-Print. Basingstoke and New York, NY: Palgrave Macmillan.

Brauer, René and Mats Fridlund. 2013. “Historicizing Topic Models. A Distant Reading of Topic Modeling Texts within Historical Studies.” In *Cultural Research in the Context of “Digital Humanities”: Proceedings of International Conference 3–5 October 2013, St Petersburg*, edited by L. V. Nikiforova and N.V. Nikiforova, 152–63.

van Dijck, José. 2010. “Search Engines and the Production of Academic Knowledge.” *International Journal of Cultural Studies* 13 (6): 574–92. https://doi.org/10.1177/1367877910376582.

Drucker, Johanna. 2011. “Humanities Approaches to Graphical Display.” *Digital Humanities Quarterly* 5 (1): 1–21.

van Es, Karin, Mirko Tobias Schäfer, and Maranke Wieringa. 2021. “Tool Criticism and the Computational Turn: A ‘Methodological Moment’ in Media and Communication Studies.” In *M&K Medien & Kommunikationswissenschaft* 69 (1): 46–64. https://doi.org/10.5771/1615-634X-2021-1-46.

Fickers, Andreas and Tim van der Heijden. 2020. “Inside the Trading Zone: Thinkering in a Digital History Lab.” *Digital Humanities Quarterly* 14 (3).

Huhtamo, Erkki. 2010. "Thinkering with Media: On The Art of Paul DeMarinis." In *Paul DeMarinis: Buried in Noise*, edited by Ingrid Beirer, Sabine Himmelsbach, and Carsten Seiffairth, 33–46. Heidelberg: Kehrer Verlag.

Ives, Maura. 2014. '"Digital Humanities Pedagogy: Hitting the Wall and Bouncing Back." *CEA Critic* 76 (2): 221–4. https://doi.org/10.1353/cea.2014.0016.

Jockers, Matthew L. 2013. *Macroanalysis: Digital Methods and Literary History*. Champaign, IL: University of Illinois Press.

Kirschenbaum, Matthew. 2009. "Hello Worlds: Why Humanities Students Should Learn to Program." *Chronicle Review*, January 23. https://www.chronicle.com/article/hello-worlds/.

Klein, Julie Thompson. 2014. *Interdisciplining Digital Humanities: Boundary Work in an Emerging Field*. Online. University of Michigan Press. https://doi.org/10.3998/dh.12869322.0001.001.

Koolen, Marijn, Jasmijn van Gorp, and Jacco van Ossenbruggen. 2019. "Toward a Model for Digital Tool Criticism: Reflection as Integrative Practice." *Digital Scholarship in the Humanities* 34 (2): 368–85. https://doi.org/10.1093/llc/fqy048.

Liu, Alan. 2012. "Where Is Cultural Criticism in the Digital Humanities?" In *Debates in the Digital Humanities*, edited by Matthew K. Gold, 490–509. Minneapolis, MN: University of Minnesota Press.

Meeks, Elijah. 2014. "Spreadsheets Are Information Visualization." Stanford | Digital Humanities. April 28. https://digitalhumanities.stanford.edu/spreadsheets-are-information-visualization.

Montfort, Nick. 2015. "Exploratory Programming in Digital Humanities Pedagogy and Research." In *A New Companion to Digital Humanities*, edited by Susan Schreibman, Ray Siemens, and John Unsworth, 98–109. Chichester: John Wiley & Sons. https://doi.org/10.1002/9781118680605.ch7.

Öhman, Emily Sofi. 2019. "Teaching Computational Methods to Humanities Students." In *Digital Humanities in the Nordic Countries Proceedings of the Digital Humanities in the Nordic Countries 4th Conference*, CEUR-WS.org.

Ramsay, Stephen. 2014. "The Hermeneutics of Screwing Around; or What You Do with a Million Books." In *Pastplay: Teaching and Learning History with Technology*, edited by Kevin Kee, 111–20. Minneapolis, MN: University of Michigan Press.

Rieder, Bernhard. 2010. "One Network and Four Algorithms." *The Politics of Systems* (blog). October 6. http://thepoliticsofsystems.net/2010/10/one-network-and-four-algorithms/.

Rieder, Bernhard and Theo Röhle. 2017. "Digital Methods: From Challenges to Bildung." In *The Datafied Society*, edited by Mirko Tobias Schäfer, and Karin van Es, 109–24. Amsterdam: Amsterdam University Press. https://doi.org/10.1515/9789048531011-010.

Rosen, Larry D., Deborah C. Sears, and Michelle M. Weil. 1993. "Treating Technophobia: A Longitudinal Evaluation of the Computerphobia Reduction Program." *Computers in Human Behavior* 9 (1): 27–50. https://doi.org/10.1016/0747-5632(93)90019-O.

Sample, Mark. 2020. "Play." In *Digital Pedagogy in the Humanities*. Modern Language Association. https://digitalpedagogy.hcommons.org/keyword/play/.

Sayers, Jentery. 2011. "Tinker-Centric Pedagogy in Literature and Language Classrooms." In *Collaborative Approaches to the Digital in English Studies*, edited by Laura McGrath, 279–300. Logan, UT: Utah State University Press: Computers and Composition Digital Press.

Schmidt, Benjamin M. 2016. "Do Digital Humanists Need to Understand Algorithms?" In *Debates in the Digital Humanities*. Online. University of Minnesota Press. http://dhdebates.gc.cuny.edu/debates/text/99.

Spiro, Lisa. 2012. "'This Is Why We Fight': Defining the Values of the Digital Humanities." In *Debates in Digital Humanities*, edited by Matthew K. Gold. Online. University of Minnesota Press.

Stanfill, Mel. 2014. "The Interface as Discourse: The Production of Norms through Web Design." *New Media & Society* 17 (7): 1059–74. https://doi.org/10.1177/1461444814520873.

Werning, Stefan. 2020. "Making Data Playable: A Game Co-Creation Method to Promote Creative Data Literacy." *Journal of Media Literacy Education* 12 (3): 88–101. https://doi.org/10.23860/JMLE-2020-12-3-8.

van Zundert, Joris. 2016. "The Case of the Bold Button: Social Shaping of Technology and the Digital Scholarly Edition." *Digital Scholarship in the Humanities* 31 (4): 898–910. https://doi.org/10.1093/llc/fqw012.

CHAPTER TWENTY-EIGHT

The Invisible Labor of DH Pedagogy

BRIAN CROXALL (BRIGHAM YOUNG UNIVERSITY) AND DIANE K. JAKACKI (BUCKNELL UNIVERSITY)

There are as many different digital humanities pedagogies as there are people who teach digital humanities (DH). Indeed, we have the data to prove it: during summer 2019, we conducted an international survey of DH pedagogues, asking them about their teaching practices (Croxall and Jakacki 2019, 2020). At the same time, we found that there are broad similarities in how the subject is taught. For the sake of our argument, we'll name three. First, it's not uncommon for someone teaching a DH class to call on a friend or colleague to be a guest in their classroom. Second, given the highly technical nature of work within DH, we have also observed that many classes go beyond the one-off guest and are taught in a long-term, cooperative manner. Third, DH pedagogues frequently publish and share their teaching materials: syllabi, assignments, and even student ratings. Reflecting upon these three patterns of teaching, each of which we engage in ourselves, we find it remarkable that these norms are so uncommon outside DH. What's more, we have observed that digital humanities practitioners—including us—do not always employ these teaching practices when they teach a different subject. While they clearly have the experience and skills to implement them in any course, there's something particular about DH that seems to lead to individuals taking on new, often invisible, labor.

Oftentimes the term *labor* connotes something not-pleasant: one goes into labor, is punished by years at hard labor; other times it is productive and ultimately rewarding but still toilsome: the fruit of one's labors, a labor of love. Perhaps a blend of these two aspects can be seen in the labors of Hercules—unquestionably heroic but still requiring the de-mucking of the Augean stables. In academic contexts, labor is complex, and that is undeniably so when it comes to pedagogy. But one characteristic that is simple to understand is that the work of teaching goes largely unnoticed, that it is, in fact, invisible. One reason for this is that in all but rare circumstances, academics do not publish about their pedagogy. Another reason, surely connected to the first, is that while institutions require teaching, they do not care to count it for too much. The work that instructors perform to make their teaching happen, then, becomes invisible. If this is true of pedagogy in general in the academy, it seems all the more unusual that teaching DH so often takes on additional, invisible, and uncompensated labor, piling another hecatomb or so into the stables, just for fun. Clever classical references aside, how do the above-mentioned patterns within DH teaching add more work to its teaching? What types of invisible pedagogical labor seem especially endemic to DH? And, more importantly, what is it about the field of DH and its culture that has led people to take on these further burdens?

GUESTS AND HOSTS IN THE CLASSROOM, OR *XENIA* AMONG THE DIGITAL HUMANISTS

Perhaps there is no better model of invisible pedagogical labor within DH than the invited guests who so frequently join our classrooms. The functions that these guests fulfill vary. Sometimes they act as an expert on a particular methodology, filling the role of the non-existent textbook. Sometimes they play visiting author, students having read one of their essays or blog posts. These guests may arrive from our own campuses or they may join us remotely via one or another video platform, a practice that was common well before the onset of Covid-19. Often the guests are friends, but at other times they are people we only know through online and/or scholarly interactions. Regardless of why or from where they are joining a class, tremendous effort goes into these in-class appearances for both guests and hosts. But this effort is, for all intents and purposes, impossible to see.

Before we examine how this labor of hospitality—what the ancient Greeks called *xenia*—is rendered invisible, it's worth asking whether the reciprocal relation between guest and host is specific to DH pedagogy. Our survey revealed that a majority of the 340 participants (54 percent) had been asked to be a guest instructor in a for-credit DH class during the two-year period between July 2017 and July 2019 and had been a guest in, on average, more than five courses (Croxall and Jakacki 2020). What's more, a significant portion of respondents (43 percent) indicated that they had invited guests to their for-credit DH classes during this same time period, welcoming an average of 3.58 guests to their classrooms (Croxall and Jakacki 2020). In short, it's extremely common for those who teach DH to invite others who teach DH to join them in their classrooms.

When we have been guests in a colleague's DH class, we have found that it requires significant labor, but it is labor that we are willing to take on as members of a famously collaborative field. The most obvious of the guest's work is the preparation and delivery of whatever content they plan to share with the class. But there are additional forms of labor that guests undertake when agreeing to join a class. Making time in their schedule for the presentation and its preparation is often non-trivial. Depending on their employment status, they might have to get clearance from a supervisor to participate or to cancel other obligations. And they may have to find time outside of their day-to-day work schedule to prepare and/or participate. If a guest is local and will join the class in-person, they may spend time reconnoitering the teaching space and testing any technology they plan to use. If the guest is remote, they will have to prepare the space from which they will broadcast and ensure that they both have and understand the software that the host will use to bring their telepresence into the classroom.

But it's not just the guest speakers who perform additional labor; the host must also exert themself. First, they must determine where a guest might enhance their class's learning outcome, becoming attuned to their own inadequacies and coming up with a plan (the guest) to compensate or augment. Second, they have to extend an invitation to a guest. And whether the guest is a total stranger or a close friend, time and effort are needed to craft an invite that is rhetorically sound and specific. Third, there will need to be a discussion or two about the structure of the interaction, all of which, again, takes time. The most critical part of these discussions involves the host helping their guest understand the context of the class. What are its aims and trajectory? What will the students have learned before the guest arrives and what are they to learn later? And how can the guest help forward those aims? The work to provide this context to guests ensures that the students learn as much as possible from the visit and that it fits the broader scope of the course.

Fourth, a wise instructor will need to prepare their class for the guest's arrival. While it will be customary to introduce the guest when they appear in the classroom, it's equally important to let the students know ahead of time that a guest will be coming and what the general shape of that guest's visit will be.

Perhaps the most difficult labor that the host must perform happens during the visit itself. Despite all the planning that they have engaged in with their students and their guest, the host must remain carefully attuned to both parties. If, for example, the guest is teaching a skill or methodology, the host needs to be ready to spot students who need assistance. If the guest is discussing their work, the host needs to be ready to ask questions if the students are shy. The host must simultaneously be ready to intervene and steer the conversation so that it best meets the class's needs. Effectively managing the guest, who is, after all, doing the host a favor, is an important task.

At this point, a reader might justifiably observe that little of the above references DH pedagogy. This is absolutely true: bringing in a guest speaker to class always requires this sort of labor from both the guest and the host. But since the DH classroom is a site in which guest speakers appear with potentially high frequency, *xenia*, or the invisible, mutual care of hosts and guests tends to betide DH pedagogues.

MULTIPLE INSTRUCTORS IN THE DH CLASSROOM

If DH pedagogy often relies on guests from inside the community, we have also observed that it depends even more heavily on support from outside the community but within our own institutions. One complication in this labor is the way in which these individuals occupy different roles, have different prestige, and are rewarded differently. In these contexts, invisible labor takes on another context: evaluating that combined labor is a difficult nut to crack.

This kind of multi-functionary teaching should be distinguished from a straightforward workshop or tutorial session, in which a specialist attends a class session to help set up a WordPress instance or introduce Voyant. In these situations, the instructor of record is not necessarily a participant in such sessions, ceding the floor to the specialist and regaining their position at the end of the class period. What we're describing here is the kind of teaching that occurs along a longer trajectory, one that involves significant collaboration on both assignment and course design. In these cases, the specialist with a graduate degree in text, geospatial, or data analysis is seconded to an instructor's course because that instructor has expressed interest in augmenting traditional methodological analysis with a digital approach. Most of the time these professionals are academic staff housed in the library or IT department. These specialists provide a bridge between the faculty member's content knowledge and the specialized methods of DH. At its most straightforward, the specialist collaborates with the instructor on assignment design, prepares datasets, tutorials, and participates in the instruction over several classes or weeks. At the other end of the continuum, the specialist is embedded in the course for the entire term, serving as a de facto co-instructor, involved in every phase of planning and implementation of the course. It is unusual for the specialist to participate in grading or student evaluation in a formal sense, although the instructor may ask for their input on certain aspects of a submitted assignment. Because the word "instruction" appears in the specialist's job description (although usually in reference to teaching workshops), this kind of labor is assumed to be a commitment on the part of the university to the instructor, and the (staff) specialist is not compensated for this adjunct work, which often requires incremental work hours beyond the established work week (in evenings and on weekends).

This kind of cooperative instruction can be highly effective, and we suspect that many students have had a more robust DH learning experience because of these multiple professionals providing different contexts for the work the students are doing. But the labor that is required by both the instructor of record and the specialist can be much more complex and intensive than for another type of course. For the instructor of record, class preparation and lesson plans take on extra levels of complexity because of the need to add time for hands-on work, for installing and troubleshooting software, and adjusting the syllabus for the inevitable need to plan for technical complications and student accommodations. For the specialist, a similar amount of preparation is required, and because the datasets and corpora must draw upon the subject matter of the class, there is rarely a "cookie cutter" solution to develop an assignment or ensure that the students' experiences are rigorous and satisfactory. There is no plug-and-play version of this kind of rich pedagogical intervention. The necessary collaboration between the instructor of record and the specialist constitutes a non-trivial amount of preparation on both their parts—incremental labor that they have to fit into already overloaded schedules. This is where the invisible labor inherent in co-instructed courses becomes evident. A faculty member may teach anywhere from one to five courses in a semester, all of which require intense and consistent planning over the entire semester. As we learned from our aforementioned survey, 24 percent of respondents indicated that they have worked as an embedded specialist in one or more classes. It is not uncommon for such a specialist to be committed to multiple courses, all of which require intense bursts of planning in courses across departments or disciplines. The instructor and the specialist need to trust one another so their work in the classroom is complementary and to be transparent about how their different forms of expertise complement one another. It takes an incredible amount of work to develop this kind of relationship, trust, shared vocabulary, and vision and to reach a consensus about how this collaboration will best serve students. But there is also a power dynamic at play in these relationships: if the university "provides" such human resources to faculty members, then it is understandable that faculty members assume that specialists are there to support them when and as needed. Faculty members express gratitude for the contributions of the specialists but do not necessarily understand the toll that commitment may take on the staff member. Likewise, the specialist may feel a heightened sense of value to the university's educational mission through this kind of engagement, and thus work harder to please the instructor through their level of commitment. More surreptitiously, the staff member may feel that they do not have the institutional capital to push back against high expectations of their time or expertise. For this relationship to work, both the instructor and the specialist need to recognize what each contributes, to make, in other words, the other's efforts visible.

PUBLISHING AND SHARING MATERIALS

It is a hallmark of DH that—wherever possible—we share our materials: we endeavor to make our projects and publications open access, our code open source, and our datasets open for others to test our hypotheses, run their own analyses, and take the work further and in different directions. So it should not come as a surprise that it has also been the practice of many who teach DH to share their materials—syllabi, lesson plans, rubrics—online, as well. In our survey, 61 percent of respondents indicated that they published assignments or syllabi for the for-credit courses they taught between July 2017 and July 2019. These numbers suggest that it is more common to share such work than it is to not do so within DH pedagogy. And, again, while we do not have data to say how frequently instructors outside of DH publish their teaching materials, our anecdotal experience makes clear that such sharing is far, far more prevalent in DH than it is in other fields.

Creating the stuff of teaching requires significant labor, a reality that is familiar to all teachers. Sharing that labor is, in turn, an additional and invisible burden since it is not accomplished as soon as the assignment or course schedule is drafted nor is it recognized by our institutions as, to use the language of the neoliberal university, a "value-add." Whether people share materials via Google Docs, an institutional or disciplinary repository, or host them on their own web server, there are costs—even if only the time of decision-making—that must be paid by the members of the DH teaching community.

Such sharing of materials often spurs additional invisible labor. Colleagues read our syllabi and assignments, either before or after the course has run, and they offer us feedback. The original designer then can review the feedback and make changes to their plans. These changes will likely improve the experience for the students, but they still cost the teacher—and their commenting colleagues—extra, unseen effort.

The reason for this sharing was originally, in large part, because the curriculum for teaching DH did not exist. During the authors' experiences as graduate students in literature programs, we were treated as apprentices—teaching assistants for faculty members in survey courses or as instructors in first-year writing courses. This apprenticeship, supervised by a faculty mentor, helped us to make sense of classroom mechanisms and course rhythms. Assignments we developed were modeled on those of our mentors and workshopped with others who were apprenticing alongside us. No such equivalent exists in DH.[1] Those of us who developed DH courses or assignments did so without that mentoring model. With no departmental store of ready-made syllabi or rubrics to draw from, we cobbled together new materials and road-tested them in our classrooms and then published them online to start building up that disciplinary cache. By sharing these materials, we collectively built a de facto primer not only on *what* to teach (tools, data) but also *how* to teach it (tutorials, assignment design). By uploading the materials we had developed for our own proto-DH courses to our personal websites; tweeting about rubrics; crowdsourcing syllabi at THATCamps; and curating our own pedagogical process, we became an open educational resource.

When contemplating this particular mode of invisible labor, we believe there is another turn of the screw to consider: *who* performs it. While there are always exceptions to a rule, we observed that much of the evolutionary and experimental DH curriculum development over the last decade was undertaken by colleagues who are either early in their careers (pre-tenure) or are in transient or insecure positions with regard to teaching (postdocs, adjuncts, alt-ac professionals), or completely off the tenure track. Put differently, those doing the invisible labor of making pedagogical labor public were those whose employment made them largely invisible to the academy. Some of these colleagues have since been promoted, but just as many have switched institutions and/or careers, a fact we each discovered while completing permissions work for our contributions to *Digital Pedagogy in the Humanities*.[2] In other words, and unlike other humanities disciplines, the people taking the biggest pedagogical risks are those with the most to lose. The invisibility of many of these colleagues and their labor became, in the end, total, and, taking the advice of Radiohead, they disappeared completely from the academy.

(IN)VISIBLE VALUES

Having articulated three broad categories of invisible labor that are, we believe, endemic to DH pedagogy, we want to turn our attention to the more foundational question: *why?* Why does DH teaching involve so much extra effort? What is it about either those who teach or the subject itself that leads people to go above and beyond in the classroom?

We believe that these labor practices owe a lot to the values of the DH community and the attempts of its members to live out those values. Perhaps the most succinct articulation of such can be found in Lisa Spiro's "'This is Why We Fight': Defining the Values of the Digital Humanities." In this influential essay, Spiro argues that we should define the field not with "particular methods or theoretical approaches" but via "a community that comes together around [shared] values" (2012, 16).[3] She encourages the community to develop such a statement through an "open, participatory, iterative, networked process" and explains that she is trying to "kick off the discussion" rather than prescribe a charter (18). In the years since Spiro's essay appeared, there has not, to our knowledge, been a concerted effort to draft such a statement. But Spiro's work continues to be cited as a clear and strong articulation of the values that DH practitioners tend to embrace. She may have wanted to start the conversation, but we believe she said her piece so well that it was simply greeted by a chorus of *Amen!*, and in truth she provides the framework that folks inclined toward DH pedagogy embrace.

Spiro presents five different ethical categories that she believes the "digital humanities community aspires to achieve" (2012, 24). These are openness, collaboration, collegiality and connectedness, diversity, and experimentation. She sees these values as intrinsic and essential to the work we do and the ways in which we do it, and yet they are not always recognized or made explicit enough to resonate with people in the broader academy. Openness is "a commitment to the open exchange of ideas, the development of open content and software, and transparency" (24). Collaboration is "essential to its [DH's] work and mission" which, among others, is "transforming how the humanities work," changing from a largely solo endeavor to drawing on diverse skills and perspectives (25). What's more, the DH community "acknowledges contributions by all involved, whether they are tenured faculty, graduate students, technologists, or librarians" (26). Collegiality and connectedness is closely connected to collaboration, whereby the community encourages contributory problem-solving. Diversity within our community of practitioners leads to a sense of vibrancy, in which "discussions are richer, and projects are stronger if multiple perspectives are represented" (28). The final value she suggests is experimentation, whereby we support "risk taking, entrepreneurship, and innovation" in our research and its pedagogy (28).

While the two of us know that Spiro did not intend to have her five suggested values become canonical, we think she cogently articulated what the two of us felt when we first began exploring the digital humanities. We were drawn to the field because it was open, collaborative, and so on, and we have continued to work—and, in particular, teach—in this space because we believe in these values. Still, we have noticed a disconnect over the years between the commitment so many of us have made to enacting these tenets and recognition of what this commitment requires of us. It is most evident in the ways our work is or is not acknowledged at an institutional or broader academic level, but surprisingly we have also noticed a kind of neglect within DH of the particular kinds of labor that are required to teach DH. As we began work on this essay, our instinct was to write something of an indictment, using Spiro's work as a foil. We would chronicle the DH community's fall from grace through reference to this ten-year-old declaration of its intentions!

But as we revisited Spiro's piece, we found a different explanation and so chose a different way in. Now, we believe that a key to understanding the invisible labor within DH pedagogy is in the attempts of teachers to live out these different values.

Why do DH teachers invite guest speakers to the classroom? As mentioned above, it's to help teach a skill, to respond to a reading, or to provide a methodological context. But teachers also extend these invitations because DH values diversity, connectedness, collaboration, and experimentation. We know that our class will be "richer" if we bring in other perspectives. We know that we do not have to be masters of all things DH and can draw on the expertise of colleagues. As we host guests, we can also perform connectedness for our students as we push back against academic hierarchies—in what other field does a "luminary" show up in your classroom? And yet we also know that introducing a new person into the classroom is a bit of a risk, but one that is worth the attempt if it leads to transforming how the humanities classroom works. Why do guest speakers accept our invitations? Because DH values collegiality and connectedness. We say "yes" when we are asked to appear in a colleague's classroom because we *are* collaborative. We want to offer help to those who need it, whether that assistance comes in the form of teaching XML or of responding to students in person (or virtually, but still present). We believe that making such an appearance in class—whether in person, traipsing from another corner of campus, or remotely—will enhance the experience of the students in the class.

Why do we co-teach, ask others to teach with us, or collectively support a complex pedagogical construction? Because the kind of radical experimentation inherent in good DH pedagogy requires multifaceted collaboration. As with guest lecturers, these new classroom paradigms heighten the sense of diversity that multiple instructors (and types of instructors) can provide our students. It is not just about teaching different types of tools, where one specialist introduces GIS and another TEI. Oftentimes a specialist will come to the classroom with a particular critical perspective on the method being taught or how it is applied in different contexts that changes how our students understand the material as well as the method. Perhaps it is simply that specialists lighten the load for the instructor of record, who end up believing they have the right to expect this labor of the specialists, the librarians, and other support staff. But we would like to think that the collegiality that DH prizes results in an acknowledgment that the commitment of labor by all involved should be acknowledged ... if not rewarded.

Why do we share our teaching materials online? Surely it is connected to the value the community places on openness. Much of the time our assignments and syllabi are developed in conversation with those of others who have made their work public, so it is only natural that we do the same thing. This open exchange of ideas and reliance on the work of others simultaneously enacts both the collegiality and collaboration that the DH community seeks to foster. Additionally, publishing teaching materials often draws on the ethos of experimentation. Although academia privileges the dissemination of research, it is much more unusual to make public what we do in the classroom (see Sample 2009). For a DH pedagogue, posting a syllabus may be an experiment that involves technology (learning a new platform, for example), but it is also just as likely to be an experiment in defying the norms of the academy, whose members tend to guard their artifacts as precious intellectual property. But this is where that hallmark of openness and generosity came to bear. Not only did we build those early DH courses from the ground up, we took the extra step of making them public. We broke with convention and summoned the extra energy to curate our own pedagogical process, inviting our colleagues to adapt and experiment in their own contexts.[4]

To state it clearly: we believe that the invisible labor that DH pedagogues commit themselves to is clearly connected to the values that the DH community espouses. At this point, however, a canny reader might raise an objection: if these teaching practices are spurred on by the community's values, then wouldn't the community, well, *value* this labor? Wouldn't the ethical commitments of DH practitioners to recognize the "contributions by all involved" lead to pedagogical work being appreciated and counted? To such a reader, we concede that DH does, in many ways, do more to acknowledge pedagogical work than many other fields. Journals like *Kairos* or the *Journal of Interactive Technology and Pedagogy* and collections like *Digital Humanities Pedagogy* (ed. Hirsch), *Teaching with Digital Humanities* (eds. Travis and DeSpain), or *Digital Pedagogy in the Humanities* (eds. Davis, Gold, Harris, and Sayers) regularly provide opportunities to discuss one's praxis in print.[5] If such DH-centric publications help shine a light on teaching, perhaps, such pedagogical labor is invisible when viewed within the context of the broader academic enterprise. While those in the DH community may see our efforts (whether on our public websites or as we appear in one another's classes), the institutions that employ us fail to recognize them. This should not be surprising since pedagogy is, as mentioned, an ill-favored stepsibling to "research" in the US universities' tripartite system of categorizing labor: scholarship, teaching, and service. To be very specific, Brian is on the tenure track (as "professional faculty" rather than "professorial"), and while his annual review wants to know about publications and invited or conference presentations, neither his publicly accessible syllabi and assignments nor his appearance in colleagues' classes meet the threshold of what the university considers a publication or presentation. Diane is in an alt-ac position, working within the library but affiliated with an academic program that allows her to teach as an instructor of record while she also supports faculty members in their own teaching. Her annual review process does not allow her to include any publication or presentation whatsoever. It also does not reflect upon any of the teaching she does in any context. The way in which we—both "we" the authors and "we" the DH pedagogy community—choose to teach DH is, in other words, inscrutable to our institutions. The efforts remain invisible not so much because the university *cannot* count them but because the university has *chosen* not to count them. In short, one reason that the work of DH pedagogy remains invisible is due to the fact that "some digital humanities values may clash with the norms of the academy" (Spiro 2012, 30).

But while DH, thanks to its values, does have avenues for recognizing such labor, we must also respond to our canny reader that these values concomitantly set up the expectation that one *will* participate in such praxes. After all, if you are a DH practitioner and you believe in these values, you would certainly want to do your part to advance them, right? Such an expectation is driven more by an individual's internal sense of collegiality (natch!) than it is imposed by the broader community. Nevertheless, this expectation for particular pedagogical behaviors means that when individuals engage in them, their efforts become less individually Herculean and more akin to muscle memory. And muscle memory, like the beating of our hearts or the contraction of our lungs, tends to escape our notice. The way in which the DH community's values end up enabling invisible labor—despite their clear intent to recognize such contributions—is perhaps manifest in Spiro's description of the value of openness: "a commitment to the open exchange of ideas, the development of open content and software, and *transparency*" (2012, 24, emphasis added). While we can understand "transparency" as the desire to make visible the workings of scholars and scholarship, teachers and teaching, it can simultaneously be understood as the making-invisible of those same subjects. Transparency within DH becomes a hinge: the openness that leads us to discuss our values and to enact them in our teaching becomes the expectation that we will—of course—do so, making transparent the cost it exacts on all of us.

CONCLUSION

In the end, it should not be surprising that there is invisible labor in digital humanities pedagogy since the structure of the university is such that pedagogical labor is always difficult to see. What sets DH teachers apart are the new categories of work that they have created for themselves in their efforts to teach as effectively as possible. Whether hosting guest speakers, taking part in a teaching cooperative, or sharing materials online, DH teachers participate in teaching activities that require even more effort than what tends to be expended in the classroom. What's more, we have found that a large number of those who teach DH engage in these practices.

We also believe that the shared values within the DH community help explain why so many DH practitioners spend the extra time to enhance their classes despite the fact that they will not be rewarded for doing so. When we teach collaboratively or share our materials with the broader world, we help build a more open, connected, diverse, and exploratory environment for our students and for our colleagues. Yet these values lead to expectations for this sort of labor, which in turn makes it more invisible in the end.

There is no easy solution to the problem of invisible labor within DH pedagogy. Academe as a whole chooses not to see pedagogical work. And while DH finds many ways to make it more public, it also increasingly produces a collective shrug, as we all continue to do things that would boggle the minds of those working in other fields but that seem just everyday occurrences as we try to live out our values. Perhaps the answer would be to add a new clause to the DH community's (imagined) value statement that explicitly states that these values should not be taken as edicts to be applied in all ways to all parts of our professional lives at all times. But who among us in DH is ready to ... just ... stop? How can we embrace and enact these values without it becoming the death of us?

In the end, this essay is our effort to make it possible to see what has been invisible. While we feel driven to encourage DH pedagogues to continue embracing the community's values, we also argue that these efforts be made as visible in as many ways as possible. Those who are in senior positions within the field should see and then advocate strongly for recognition of the extraordinary efforts DH teachers make. They should simultaneously encourage their junior colleagues to take a break once in a while. Those who are just starting out should not feel like they have to set up a website and publish a syllabus *and* invite guests AND create equal space for all instructors all at once. Try to take it one day at a time, but when that day is your annual review, insist on telling everyone what you've done!

NOTES

1. In an essay in our forthcoming *What We Teach When We Teach DH: Digital Humanities in the Classroom* (University of Minnesota Press), Catherine DeRose discusses efforts to train graduate students to teach DH in a manner that builds on the model within graduate disciplinary departments.

2. Many of the professionals who developed these learning objects were no longer employed by the universities for which they taught the courses referred to. Some were uncertain what rights they still had to those materials if they were housed on a university's servers to which they no longer had access. (In the middle of the *Digital Pedagogy in the Humanities* review process, a complex blogging assignment that Jakacki had designed while a postdoc at Georgia Tech was deleted from a server, and it could only be recovered through screenshots saved on the Internet Archive's *Wayback Machine*.) Some had left the academy altogether.
3. Spiro suggests that DH values come from the different communities that are blended together in DH practice. These include the cultures of academia, information technology professionals, librarianship, and the Internet (see Spiro 2012, 19–23).
4. Of course one should point out that this labor wasn't purely magnanimous: many who did this sharing were simultaneously on the job market trying to demonstrate what DH courses would look like to potential employers, a fact that might have contributed to these efforts.
5. While not explicitly about DH pedagogy, the contributors to *Digital Pedagogy in the Humanities* by and large come from that field and many of the collection's artifacts derive from DH courses.

REFERENCES

Croxall, Brian, and Diane K. Jakacki. 2019. "Who Teaches When We Teach DH? A Survey." https://hcommons.org/deposits/item/hc:27045/.

Croxall, Brian, and Diane K. Jakacki. 2020. "Who Teaches When We Teach DH? Some Answers and More Quesions [Sic]." https://hcommons.org/deposits/item/hc:31673/.

Davis, Rebecca Frost, Matthew K. Gold, Katherine D. Harris, and Jentery Sayers, eds. 2020. *Digital Pedagogy in the Humanities: Concepts, Models, and Experiments*. https://digitalpedagogy.hcommons.org/.

DeRose, Catherine. Forthcoming. "Pedagogy First: A Lab-Led Model for Preparing Graduate Students to Teach DH." In *What We Teach When We Teach DH: Digital Humanities in the Classroom*, edited by Brian Croxall and Diane K. Jakacki. Minneapolis, MN: University of Minnesota Press. https://dhdebates.gc.cuny.edu.

Hirsch, Brett D., ed. 2012. *Digital Humanities Pedagogy: Practices, Principles and Politics*. Cambridge: Open Book Publishers.

Radiohead. 2000. "How to Disappear Completely." *Kid A*. Parlophone.

Sample, Mark. 2009. "Transparency, Teaching, and Taking My Evaluations Public." *Samplereality* (blog). August 4. https://samplereality.com/2009/08/04/transparency-teaching-and-taking-my-evaluations-public/.

Spiro, Lisa. 2012. "'This Is Why We Fight': Defining the Values of the Digital Humanities." In *Debates in the Digital Humanities*, edited by Matthew K. Gold, 16–35. Minneapolis, MN: University of Minnesota Press. https://dhdebates.gc.cuny.edu.

Travis, Jennifer, and Jessica DeSpain, eds. 2018. *Teaching with Digital Humanities: Tools and Methods for Nineteenth-Century American Literature*. Illustrated edition. Urbana, IL: University of Illinois Press.

CHAPTER TWENTY-NINE

Building Digital Humanities Centers

MICHAEL PIDD
(DIGITAL HUMANITIES INSTITUTE, UNIVERSITY OF SHEFFIELD)

It is the early 1990s. In the Arts Tower at the University of Sheffield a new center has just been established called the Humanities Research Institute. It would one day take the name Digital Humanities Institute, but for now it is an office that provides desk space for two PhD students. One is studying the biblical *Book of Job* and the second is studying the eighteenth-century French *Encyclopédie*. There are also two research assistants on a new project called The Canterbury Tales Project. We are transcribing medieval manuscripts from microfilm printouts into Apple Mackintosh SE computers in order to reconstruct the early textual history of the poem. The project will produce scholarly CD-ROMs for Cambridge University Press. Every Friday a floppy disk containing the week's transcriptions is mailed to the project's head office in Oxford. We also share the office with a large desk that supports the weight of a huge machine called the Kurzweil Scanner. It is the size of an Austin Metro and we are told that it can be taught to read books. Nobody ever comes to use it. There is another office next door, unrelated to the Humanities Research Institute, whose occupants have been working away since 1989 on a project called The Hartlib Papers. The Canterbury Tales Project and the Hartlib Papers Project are examples of something new at Sheffield called *humanities computing*. But there is no director as such, no personnel directly employed by the institute, no budget, and no software engineers. Over the next twenty-seven years or so the Humanities Research Institute, the Hartlib Papers, and the term *humanities computing*—but especially scanners—will evolve into things that are incomprehensible in the early 1990s, in terms of what they do, what they mean, and how the institution uses them.

This chapter explores the institutionalization of digital humanities in the form of the "digital humanities center," drawing on my experience as director of the University of Sheffield's Digital Humanities Institute (DHI),[1] and the experiences of other center directors, past and present. It will not be representative of every digital humanities center, since there are many models, and it will not claim that the DHI's approach to digital humanities and its institutionalization is by any means the best. Further, we should acknowledge that digital humanities can be institutionalized in other ways: academic positions in digital humanities within a research-led teaching department, digitization units within university libraries, and dedicated "digital humanities support officers." Their institutional merits are beyond the scope of this chapter. Instead, we will look at how a center such as the DHI has evolved over time in response to its changing institutional role, the practical and logistical *dos* and *don'ts* of running a center, and what the future might hold for digital humanities centers. In particular, I will argue that the institutionalization process always requires digital humanities to transform into a broader subject domain in order to increase its relevance to its institutional stakeholders: management, colleagues, and of course students.

At the time of writing, the Digital Humanities Institute has collaborated on more than 120 research projects, worked with over a hundred institutions in the UK and internationally, and developed more than seventy research resources. It maintains an online hosting and publishing presence for the majority of these resources, some of which have been "live" for more than twenty years. It has sought financial sustainability with mixed success over the years, but all its staff have open-ended contracts (albeit on professional services contracts rather than academic contracts, which imposes some constraints around promotion and research recognition). The DHI has become embedded in the research services and academic research strategies of its home institution, and it is now transitioning towards a model that is more recognizably "departmental" in its academic function, institutional status, and aspirations. I refer to this type of journey as the *institutionalization* of the digital humanities center.

WHAT IS INSTITUTIONALIZATION?

The concept of institutionalization will be familiar to anyone who has attempted to build a center, in any subject. On a simple level, it is the desire to have one's center taken seriously by the university in which it resides. On a more complex level, it means that the center develops a business operation whose activities are considered to be in alignment with the strategic objectives of the wider university. It becomes strategically important at some level within the organization, and informs decision-making in areas where it can bring value through its knowledge and expertise. Digital humanities centers usually begin life as the brain-child of one or more individuals who have a shared desire to make the practice, study and/or teaching of digital humanities a strategic activity within their institution. Typically, this reflects their view that an institutional approach to digital humanities will yield academic value, whether it be measured in the form of increased research income, tuition fees, institutional ranking, or an internationally competitive research environment. Many centers in the arts and humanities, irrespective of the subject domain, appear to start life in an ad hoc way, poorly funded and informal. Occasionally a center will be established with a generous foundation grant. Irrespective of how the center is funded at the outset, its founders have been given a moment of agency by the institution, to establish the academic value of their idea. What this agency means in practice is that the center's founders have been given a period of time (its duration dependent on the initial funding) in which to develop and demonstrate a business model that will create academic value. If successful, the center and its way of thinking will be appropriated by the institution—the institution will adopt its ideas, knowledge, and way of thinking—and the founders will have to accept that the institution will also shape the center beyond its original plan (others within the institution will be given the agency to affect this). This is institutionalization. If unsuccessful, the center will be disbanded, dramatically or quietly, depending on how much investment the institution has lost.

The purpose of all centers is to facilitate interdisciplinary and cross-disciplinary research around a subject domain that cannot be adequately represented or addressed by an individual department or faculty. Digital humanities is one such domain, for it combines knowledge, expertise, methods, and concerns from across the arts, humanities, heritage, computer science, library and information science domains. As Patrick Svensson says, digital humanities is "an array of convergent practices" and "defies any precise definition" (2010). Historically, a center in the arts and humanities could take many forms: a web page, an occasional seminar series, a network of colleagues and projects, or an institutionally recognized unit which has an executive, staff, budget, and a physical space. Only the latter type has any possibility of becoming institutionalized, because it is the only format in which the founders are given the agency to unambiguously demonstrate academic and financial value to the parent institution through their activities and reporting. It was rare for the more informal types of centers (often termed "letterhead centers") to evolve and become institutionally recognized, and in recent years UK universities have largely forbidden their creation.

A SERVICE-ORIENTED BUSINESS MODEL

The DHI began its life in 1992 fully subsidized by the University of Sheffield. Each year it had to negotiate a small grant to cover its core costs. Initially the DHI was not concerned with digital humanities per se. Instead, it was a center intended to facilitate new ways of doing humanities research; thus the original name, Humanities Research Institute. There was a belief that these new types of projects, by being complex and process driven, necessitated a type of research that was relatively novel in the arts and humanities: collaboration between researchers. In fact, many of the achievements of the earliest digital humanities projects, such as The Canterbury Tales Project and the Electronic Beowulf Project, were as much about feats of collaboration as they were about technology in humanities research. For example, the former produced multi-authored works more resonant of scientific publications, and the imaging, technical, and editorial work of the latter was shared between transatlantic research teams.

The early years of the DHI also coincided with the creation of the Arts & Humanities Research Board (AHRB) in 1998, which would eventually become the Arts & Humanities Research Council (AHRC) in 2005. The AHRB enabled arts and humanities researchers for the first time to secure funds that were of a comparable scale to the sciences. It funded eight digital humanities projects at the DHI during these early years, including the André Gide Editions, Galdós Editions, Science in the Nineteenth-Century Periodical, and the Old Bailey Proceedings Online.[2] The DHI also benefited from the New Opportunities Fund (NOF), which co-funded Old Bailey Online. NOF was a government scheme to distribute National Lottery funds to health, education, and environmental projects. It funded 149 digitization projects between 2001 and 2004 at a cost of £50m. So it is perhaps no coincidence that the DHI evolved to focus on externally funded, collaborative research during its first ten years.

Over time it was the digital humanities component of the DHI that was able to develop a business model and an operational approach to replace its reliance on subsidization, in full or at least in part. The DHI sought to become largely self-funded by charging out its technical and project management support as a service on externally funded research grants led by academic investigators. The DHI's running costs were underwritten by the faculty, so any end-of-year surplus was reclaimed by the university, but the faculty also absorbed deficits in any lean years. Our costs were such that we could not rely on the funding success rates of academic investigators within the University of Sheffield alone, so we had to offer our services to colleagues in other institutions. The digital humanities support which we were offering was practical: grant writing, software programming, website development, data analytics, algorithm development, project management, publishing and long-term hosting of the digital outputs. In the DHI, digital humanities has always been a practice-led domain, focused on combining hard skills with arts and humanities research questions, methods, sources, and perspectives. Digital humanities is a "big tent," accommodating pedagogy, theorization, and reflexivity as well as practice. Focusing on the practice-led approach made our business model distinctive. We were offering a service that had increasing demand within arts and humanities research, but which had low supply in terms of personnel sufficiently qualified to deliver the service within their home institutions.

This business model had a distinct advantage over models that were focused on research excellence only. It demonstrated to the wider institution that the DHI was taking its financial sustainability seriously, and was able to evidence real payment-based services, business and outputs. The digital humanities element of any research award, in terms of the DHI's technical services, is typically 12 percent of the total value of the award. So, if a center such as the DHI is required to self-fund to the tune of, say, £240,000 annually in order to cover its core costs, it needs to be involved in successful grant awards totaling £2m. In practice this means that a faculty has a center that is driven to push up grant application numbers. But in reality this means working on grant applications that are being led by other institutions; over 50 percent of the DHI's business. Further, it resulted in the DHI being able to participate in a wider range of research topics than those presented by our colleagues at the University of Sheffield.

The DHI's status became what in UK universities is called a *research facility*: a resource consisting of personnel and equipment which is available to staff using a cost-recovery model. The chief advantage of this classification was that it enabled us to charge a day rate for the DHI's role on research projects; that is, a single price per day, including all overheads, irrespective of which member of staff would be working on the project. This facility model gave us the flexibility to assign staff to projects based on their current workload and expertise, because the budget was not tied to the salary of a named individual. The alternative, of having small percentages of a member of staff's time coded to different project accounts, was unmanageable from an operational and accountancy perspective; it made understanding the center's financial position at any given moment in time extremely difficult. Tedious though all of this sounds, the facility model made it easier to seek sustainability. As a research facility we became a recognized financial entity within the university (no soft money here or funds held in another department's accounts), and the day rate enabled us to charge and retain 100 percent of the income. Digital humanities centers never fail because they are poor at digital humanities; they tend to fail because they have a poor business model and an unsupportive institution.

Nowadays the DHI's service-oriented approach is termed a *digital lab* or *digital humanities lab*, and its personnel are frequently referred to as *research software engineers* in recognition of their specialism: engineering software solutions for researchers and, in the process of doing this, undertaking research themselves. The details of the model might vary: the services might be to support research or teaching; chargeable on a cost-recovery basis or free at the point of use; and for internal or external clients only. A notable implementation of the lab model is King's Digital Lab (KDL),[3] which has done important work to improve the practice of digital humanities project management, and to professionalize the role of research software engineers through career development pathways. If left unformalized, research centers can become the domain of fixed-term or casualized staff. Some of us will remember the days of the senior academic colleague and his digital humanities assistant: a highly talented software engineer who was retained from one month to the next on what was termed *soft money*. No doubt the arrangement still exists in some universities. A center must have a business model, or at least a strategy, that guarantees continuity of employment for its key staff. This improves institutional memory, retains skills and knowledge, and encourages long-term planning and a greater commitment to sustainability (of software and data but also business operation). Importantly it improves institutionalization, because staff are more likely to develop the deep collegial networks that help align the center with institutional cultures and priorities.

CHALLENGES WITH THE SERVICE-ORIENTED APPROACH

There are hard challenges with the DHI's business model, and perhaps with digital humanities labs in general, and over the years I have had colleagues in other universities tell me that such an approach with its emphasis on paid-for services would be unthinkable to them. They are no doubt fortunate. In England at least, where the government has removed a lot of state aid for universities and encouraged a more marketized approach through student fees, few universities would be willing to fund a research center in the arts and humanities for twenty-seven years without some form of financial return, either directly in the form of research income or indirectly through research activities that have made significant contributions to the institution's research environment. A center needs a business model that will enable it to align with the institution's own strategic priorities: world-leading research, but research that pays its own way. The emphasis towards one or the other tends to vary by institution, and even at the DHI we have always considered ourselves to be first and foremost digital humanities research practitioners but—out of necessity—with an eye on the bottom line as well.

The first problem with the DHI's service-oriented approach is the danger of only ever chasing the money, since this can result in undertaking a brand of digital humanities that is less about risk and innovation in digital research methods, and more about delivering safe, pedestrian software solutions. "Inevitably, the paradigm of funding shapes the kind of work that is done, and the tools and methods that are used" (Opel and Simone 2019). However, digital humanities is always at its best when framed as a practice within an arts and humanities subject domain, and technically pedestrian software solutions are often, nonetheless, innovative when applied to the subject domain in terms of the data-focused research that it facilitates or communicates. For example, in 2007 we published the *Sheffield Corpus of Chinese*.[4] Technically it is very pedestrian (albeit an early implementation of UTF-8), but every time we try to quietly retire the website we receive howls of protest from users around the world who maintain that their school dissertation, doctoral thesis, or peer-reviewed article cannot be completed without it. Reassuringly, content is still king, and the technically uncomplicated job of making high-quality data publicly accessible is still digital humanities' most important contribution to the academy. However, in the DHI we also try to counter the problem of pedestrianism by being proactive in project development, alongside our responsive services for colleagues: we write our own research proposals and lead or co-lead projects that have at their heart the scale of technical research and innovation that will hopefully make a research software engineer jump out of bed in the morning.

The second problem with the DHI's service-oriented approach is that it is a portfolio approach. The need to continually secure new project work, develop digital outputs (websites, apps, management systems and data), and then host the outputs over the long term (which is a key part of the service we offer) produces sustainability challenges: software and financial.

Software sustainability is concerned with hosting and maintaining (usually web-based) interfaces and data in a constantly changing technical environment. Jamie McLaughlin (2019) and Smithies et al. (2019) cover many of the challenges of software sustainability from their firsthand experience. If left unaddressed, software sustainability demands will grow at a constant rate until all the center's human resource is fully occupied in servicing existing outputs rather than working on new projects. The problem arises predominantly in relation to web interfaces rather than data, particularly legacy systems that were not written using modern software development frameworks. The DHI hosts and maintains more than seventy online resources, and so any upgrade to our servers, database software, or programming languages requires us to undertake a lengthy review of all our assets. However, from the perspective of the academy, although there is a local cost to sustaining these resources, it is more efficient and cost effective than if these resources were hosted by more than seventy individual institutions. We are confident that many of the resources would not be accessible today since most institutions lack the funding, expertise, and willpower to maintain and sustain them. Perhaps we should be lobbying more vociferously to have these costs covered. But until such a time when the academy is prepared to take these costs seriously, any new digital humanities center is strongly advised to consider the reality of sustaining its infrastructure and digital outputs over the long term, and to plan in from the outset how growth will be managed through good hosting practices, prudent technology decisions and a "take down" policy for retiring resources that have become too expensive to maintain.

Financial sustainability is concerned with maintaining a portfolio of research projects in a changing environment. The DHI's service-oriented approach is predicated on constantly acquiring new business. In UK higher education, digital humanities work is almost entirely project-based, and therefore relies almost exclusively on external funding from research councils and charitable trusts. There is funding to be had from commercial work, of course, and by re-orienting digital humanities research software engineering to capitalize on opportunities presented by industrial strategy challenges and knowledge exchange grants. However, whatever the source of funding, a center's financial planning is made challenging by the increased competition for reduced funding, long turnaround times (from bid development to award), unforeseen delays on the projects themselves, and the academy's overall reluctance to fund software development at realistic levels. I refer to the *academy* to avoid blaming funders, because as reviewers we all question "value for money" in funding bids, then complain that our own projects are not satisfactorily resourced. While the majority of time and financial resource on a digital humanities project is spent creating high-quality, open data, surprisingly little is spent on developing the interfaces that are intended to query and present the data—and yet this is the main value that a research software engineer brings to a digital humanities project, and it is often the main impediment to accessing the data in the long term.

Whether we are dealing with software sustainability or financial sustainability, the elephant in the room takes the form of a question: *is it sustainable?* The answer for both is the same: *no, not entirely*. If left unaddressed, both become impediments to the other: the need to acquire more projects produces more outputs to maintain, and more outputs need more project work to back-fund the maintenance. Both are at the mercy of changing environments: technology forces us to upgrade, whether we have the resources to do that or not, and the changing research-funding environment forces us to compete harder for work, because grant submissions are higher and awards are fewer. Both can be de-risked to a degree, and the DHI has been in the game for twenty-seven years at the time of writing, which would be impressive if it were an SME (small or medium-sized enterprise), but clearly a service-oriented model has to evolve and adapt to the center's changing environment if it is to remain sustainable. The research environment has changed a lot in twenty-seven years. None of this is to suggest that the DHI's service-oriented model is wrong. Instead, I believe that a successful service-oriented approach which can outlive any foundation grant is necessary—at the outset of a center's life—for developing the longer-term financial sustainability that then enables it to develop academic value. But a never-ending provision of paid-for services is not where a successful center is destined to end up. Instead, we should perhaps view digital humanities centers, whatever their business model, as the nurseries for digital humanities academics and departments.

ACADEMIC VALUE AND MONETIZING THE CENTER

An important driver in the institutionalization of a digital humanities center, or indeed any digital humanities role or activity, is the academic value that it brings to the institution, as opposed to financial value. Yet the academic value of digital humanities is surprisingly hard to sell. First, the term *digital humanities* does not resonate with most people, even colleagues within the arts and humanities, and especially senior leaders within our institutions. Some institutions and centers that deliver digital humanities programs describe how they have to teach digital humanities "by stealth," by calling it something more meaningful, such as cultural studies, media studies, cultural heritage studies, museology, computational linguistics, etc. In many respects it is a victim of being such a broad church. Second, the value of digital humanities is not immediately obvious using the measures that matter to most institutions: research rankings, student numbers, and income. Often there is no digital humanities department, no fee-paying digital humanities students, and no clear idea of what the employability value of digital humanities is. As Claire Warwick says, when digital humanities pursues a specific "service computing" model it can suffer from similar cultural problems to other university services that support academic research; they are "often undervalued by the very academics that use them, because they do not generate research in themselves" (Warwick 2012, 195). Those of us that find ourselves having to articulate the academic value of digital humanities have to rely on other, indirect measures: research grant awards in other departments that might not have been made had the project not included a sound and, where appropriate, innovative technical methodology; assets, e.g., open data resources that help establish reputation while also having commercial licensing value; research environment; skills and knowledge; impact and knowledge exchange, i.e., the assumption that digital humanities is inherently impactful through its collaborations with cultural institutions and creative industries, and its focus on disseminating outputs. If none of these are persuasive, we have to resort to presenting digital humanities as something that can make the arts and humanities appear almost STEM-like in the minds of the engineers and scientists that run most of our institutions. However, the demise of the UK's subject data centers for arts and humanities in the 2000s shows that institutions can maintain most of the above—once it has been established—without the expense of retaining an actual center. Ultimately, we have to rely on influential figures within the institution being favorable towards and supportive of digital humanities when articulating our academic value. The DHI has been extremely fortunate in this regard.

At the point when a service-oriented approach becomes less sustainable for a center, institutionalization comes into play more forcibly: has the center developed sufficient academic value to be monetized by the institution, or does the institution simply close it down? By *monetized* I mean changing the center's business model to re-acquire financial sustainability, but doing so by capitalizing on the assets (academic value) that the center has developed under the old business model. This process seems to be a trend for those digital humanities centers that started out with the service-oriented model, such as Humanities Advanced Technology and Information Institute (HATII) at the University of Glasgow, the Centre for Digital Humanities at King's College London, and the DHI. The trend can be seen with centers in other subject domains too, such as the Textile Conservation Centre (University of Glasgow)[5] and the Institute of Making (UCL).[6] What all of these centers have in common is that (unconsciously) they have used a service-oriented model for digital humanities research in order to establish themselves and develop academic value for the institution, in the form of research income, assets, research environment, skills, and knowledge. At a certain point in time the institution has then capitalized on the center's academic value by re-focusing its activities on delivering teaching and learning, which in the UK creates a much larger market for the center than research funding. But in doing so, the institution has also been required to broaden the center's subject domain beyond the scope of its founders in order to maximize its relevance.

MONETIZATION THROUGH PEDAGOGY REQUIRES DIGITAL HUMANITIES TO CHANGE

Finally, I want to explore how institutionalization favors digital humanities centers becoming more about pedagogy and less about research over the long term, and how this necessitates a digital humanities center transforming its subject domain into something that resonates more with management, staff, and especially students. It is perhaps not surprising that the older, established digital humanities centers now focus on research-led teaching, given the challenges of the research-funding environment and the opportunities afforded by domestic and international student markets: HATII, established in 1997, is now the Information School;[7] and the Centre for Digital Humanities at King's College London, established in 1991, is now the Department of Digital Humanities[8] (and not to be confused with the King's Digital Lab). The DHI has also followed this trend towards research-led teaching. Both Glasgow and King's College London have capitalized on their considerable, long-standing reputations as former centers for project-based digital humanities research when offering programs to students. And they have gone further, by becoming academic departments in their own right and delivering programs directly. Full institutionalization of the digital humanities center, in the UK at least, seems to necessitate departmentalization in order to fit in with HE's more streamlined, market-focused organizational structures. But in doing so, these centers have also changed what digital humanities is within their own institutions by broadening its scope and articulating its employability value. Both offer a Master's in Digital Humanities, but many of the programs on offer are now focused on topics that have greater resonance for industry and employability, such as media management, cultural heritage, and archives management. Their broadening of the subject domain (from digital humanities to information science with a focus on digital culture and society) in order to become an academic department and deepen the institutionalization is telling with respect to the future of digital humanities. Student registration data provided by HESA (Higher Education Statistics Agency) for 2019 shows that these types of "digital programs" recruit significantly more students than Digital Humanities MAs, across all UK institutions, not just at King's College London and the University of Glasgow.

This perhaps points towards how, today, digital humanities lies at the intersection of many subject domains that we would consider to be on trend, emerging, and distinctively twenty-first century: data science, big data, AI, digital media, digital society and culture, etc. In its more institutionalized form, digital humanities is being taught "by stealth," by addressing these more popular, employability driven topics from an arts and humanities perspective. For me this arts and humanities perspective revolves around a deep understanding of natural language (messy data), visual culture, the selection and use of sources, and the nature of meaning. These are all key skills and knowledge that are missing from most engineering and social science approaches. This prompts a question: if a digital humanities research center is to be fully institutionalized, must it always transition away from project-based research towards pedagogy (or at least research-led pedagogy) and broaden its domain? I think this is inevitable unless research funding becomes less competitive or we develop a new way of extracting academic value from digital humanities research.

IN PRAISE OF INSTITUTIONAL SUPPORT

In discussing the institutionalization of digital humanities centers, this chapter has perhaps spoken little about digital humanities itself. So the conclusion warrants a tale about a digital humanities project which, I hope, will demonstrate the importance of institutions sustaining centers and their staff, in whatever form, over the longer term.

It is now 1996. Although The Hartlib Papers Project[9] was never a "DHI project" as such, having predated the establishment of the center by a good seven years, we consider this to be our foundation project due to its ambition, invention, approach, and spirit. Samuel Hartlib (*c.* 1600–1662) was a great seventeenth-century "intelligencer" and man of science who set out to record all human knowledge and make it universally available for the education of all mankind. The project's objective was to create a searchable electronic edition with full-text transcriptions and facsimile images of all 25,000 pages of his correspondence. Hartlib would have loved the very idea of this project! I recall a busy office with large, boxy computer terminals, 5-inch floppy disks and an entire bookcase dedicated to storing the ring binders that contained the user manual for Microsoft Word for DOS. The transcriptions and facsimiles were accompanied by a powerful search tool and topic modeling engine called TOPIC by Verity Inc.[10] This seven-year project culminated in the publication in 1996 of two CD-ROMs by University Microfilms in Michigan. The CD-ROMs retailed for $4,995.[11] Advance orders had been made by some of the world's leading academic institutions and libraries.

The CD-ROMs, including the TOPIC software, were designed to run on the early versions of Microsoft's graphical user interface, Windows 3.x. In 1995 Microsoft released Windows 95, which was a significant upgrade for the operating system run by most of the world's PCs at that time. One year later, after the consumer boom in buying new PCs with Windows 95, the Hartlib Papers CD-ROM was published. Unfortunately the CD-ROMs, and particularly the TOPIC software, were incompatible with the new PCs. Verity reportedly declined to upgrade the software. Destroying the CD-ROMs and reprinting them was also expensive. So institutions that had made advanced purchases received CD-ROMs that were incompatible with their new suites of desktop PCs. The British Library reportedly needed to resurrect an old PC from storage in order to make its copy of the Hartlib Papers CD-ROM accessible to the public. Needless to say, in 1996 we all learnt about the dangers of relying on proprietary data formats (MS Word) and software (TOPIC), and the wisdom of open-source standards.

An enlarged edition, incorporating many Hartlib materials from libraries around the world, was published on CD-ROM by the DHI in July 2002. The *Hartlib Papers Second Edition* cost £1,570. Owners of the original CD-ROMs were given a free upgrade. Creating this new version was not a trivial process because for a long while we were unable to open or convert the MS Word for DOS documents using the current versions of Microsoft Word. Although some proprietary file formats, such as DOC, are used almost universally, we learnt that this is still no guarantee that the same company's product will work with our data files in the future. Eventually we converted the files into XML, and generated an HTML version for display through a web browser. The entire product ran off a CD-ROM disk, so we had to deploy a small, self-contained executable to serve as a search engine. This was a Java applet called JObjects QuestAgent. It relied on Microsoft's proprietary Java environment (JVM) which was discontinued in 2003 after a dispute over IP with Sun Microsystems (one year after publication of the new edition of Hartlib). Java applets were eventually deprecated entirely due to their security vulnerabilities. So in 2003, with another defunct CD-ROM, we learnt about the dangers of relying on proprietary web technologies and old publishing models.

In 2013 we published the third and current edition of the Hartlib Papers.[12] It uses entirely open data standards and web programming frameworks. It is published online and available for free. It now looks a little dated at the time of writing, but it still works perfectly and we have every confidence that it will continue to do so. Since 2013 the Hartlib data has been shared with other research groups and integrated into other, larger online research resources, most notably *Early Modern Letters Online*[13] and our own *Connected Histories* website.[14] Moving forwards, we plan to make the entire dataset more easily accessible for researchers under a Creative Commons license, via the DHI Data Service,[15] and no doubt we will revisit features such as topic modeling. By 2013 we had learnt that web services are far more economical, sustainable, and stress-free than hard media such as CD-ROM disks; servers, delivery frameworks, interfaces, and data can be maintained with minimal hindrance and at no cost to the end-user.

The journey of the Hartlib Papers as a publication has mirrored the intellectual journey of the DHI, as we have moved (and sometimes stumbled) from the old proprietary publishing model to a newer, open model in which even interfaces for data are becoming less important. It emphasizes the durability of data. Despite Hartlib's checkered publishing history, the quality and value of the transcriptions remains undiminished. The original investment in the staff who labored to produce full-text transcriptions of more than 25,000 pages of Hartlib's correspondence was sound, and many of the staff went on to other significant digital humanities projects, such as the Newton Project.[16] However, Hartlib's publishing history also emphasizes the importance of institutional support. A second edition was made possible because there was a digital humanities center with staff who remembered and understood the value of the data that the original project had placed in their custody. The third, online edition was made possible for the same reasons. We all know that digital humanities centers disappear as soon as the parent institution removes its support, whether that support is financial or purely political, and that the outputs (data, interfaces, etc.) will continue to be maintained only for as long as they present no cost. It is unlikely that Hartlib and many of the DHI's more than seventy other resources would be accessible today had the University of Sheffield not continued to support the center and its staff. Institutionalization is both an aim and a price that we have to pay if we want to build a successful digital humanities center. We need to be given agency and support to flourish and be successful, but in return the institution will want us to evolve in order to remain valuable within the changing academic landscape (or dare I say, market).

NOTES

1. https://www.dhi.ac.uk.
2. https://www.dhi.ac.uk/category/funders/ahrb/.
3. https://kdl.kcl.ac.uk/.
4. https://www.dhi.ac.uk/scc.
5. http://www.textileconservationcentre.org.uk/index.html.
6. https://www.ucl.ac.uk/ucl-east/academic-vision/institute-making.
7. https://www.gla.ac.uk/subjects/informationstudies/.
8. https://www.kcl.ac.uk/ddh.
9. https://www.dhi.ac.uk/projects/hartlib/.
10. https://web.archive.org/web/20210222235452/; http://www.fundinguniverse.com/company-histories/verity-inc-history/.
11. https://users.ox.ac.uk/~ctitext2/resguide/resources/h100.html.
12. https://www.dhi.ac.uk/hartlib.
13. http://emlo.bodleian.ox.ac.uk.
14. https://www.connectedhistories.org.
15. https://www.dhi.ac.uk/data.
16. https://www.history.ox.ac.uk/newton-project.

REFERENCES

McLaughlin, Jamie. 2019. "Strategies for Maximising the Maintainability of Digital Humanities Websites." In *Proceedings of the Digital Humanities Congress 2018*, edited by Lana Pitcher and Michael Pidd. Sheffield: The Digital Humanities Institute. https://www.dhi.ac.uk/openbook/chapter/dhc2018-mclaughlin.

Opel, Dawn and Michael Simeone. 2019. "The Invisible Work of the Digital Humanities Lab: Preparing Graduate Students for Emergent Intellectual and Professional Work." *Digital Humanities Quarterly* 13 (2).

Smithies, James, Carina Westling, Anna-Maria Sichani, Pam Mellen, and Arianna Ciula. 2019. "Managing 100 Digital Humanities Projects: Digital Scholarship & Archiving in King's Digital Lab." *Digital Humanities Quarterly* 13 (1). http://www.digitalhumanities.org/dhq/vol/13/1/000411/000411.html.

Svensson, Patrik. 2010. "The Landscape of Digital Humanities." *Digital Humanities Quarterly* 4 (1). http://digitalhumanities.org/dhq/vol/4/1/000080/000080.html.

Warwick, Claire. 2012. "Institutional Models for Digital Humanities." In *Digital Humanities in Practice*, edited by Claire Warwick, Melissa Terras and Julianne Nyhan, 193–216. London: Facet.

CHAPTER THIRTY

Embracing Decline in Digital Scholarship beyond Sustainability

ANNA-MARIA SICHANI
(SCHOOL OF ADVANCED STUDY, UNIVERSITY OF LONDON)

What kind of garbage will our digital work become?

—Sample 2019

Sustainability in Digital Humanities is nothing but a chimera. The flourishing culture of digital scholarship of the last decades has been eye-opening on the ways we can approach and understand anew human culture and knowledge while, at the same time, leaving us with a huge amount of digital (and born-digital) datasets, assets, and projects that we need to curate, migrate, and sustain—or abandon and forget altogether.

Sustainability has been originally introduced in the areas of environmental management and development to suggest that "it is possible to achieve economic growth and industrialization without environmental damage" or, in more vague terms, to describe "development that meets the needs of the present without compromising the ability of future generations to meet their own needs" (Adams 2006, 1). The core of mainstream sustainability thinking has become the idea of three dimensions, environmental, social, and economic, organically linked to the notions of progress, risk, and change, which have always been natural facets of human culture—and indeed life. Soon, the technological dimension was added to sustainability thinking, being the default area to face change, risk, and evolution.

The discussion towards sustainability emerged early on in the Digital Humanities community, almost hand in hand with the growth of the field as such. Since early 2000, Digital Humanities (DH) has been defining itself as an interdisciplinary academic field and a diverse community of practice through (not-always) funded project-based scholarship, cutting-edge computational frameworks, technologies and methodologies, institutional structures, and forums. As the field and its digital outputs started to expand significantly, the interest in sustainability continued to grow. From the very beginning, sustainability concerns and strategies about the longevity of DH work have been part of the planning of projects and institutions (Zorich 2003), policymakers' and funders' agendas alike (Maron and Pickle 2014; SSI 2018) to such an extent that through the years "sustainability" evolved from an area of practice to a substantive object of study and reflection in itself.

WHAT WE (DON'T) TALK ABOUT WHEN WE TALK ABOUT SUSTAINABILITY

Despite the many disparate uses of the term and the lack of consensus on what "sustainability" in DH actually is or how we can achieve it, sustainability in digital scholarship usually refers to the ability, alongside the necessary resources, to maintain the digital outputs of a project accessible and usable for the long-term against a set of challenges including technological shifts, staffing and funding constraints, or institutional and administrative changes.

From early on, sustainability approaches to fight technological obsolescence of platforms, formats, and software have been focusing on developing and using computational standards and open technologies (Pitti 2004), an ethos that led to the initial development of many technical standards in the field, such as the adoption of XML and the development of the Text Encoding Initiative. In a similar vein, concepts and practices of data curation and documentation, digital preservation, and web archiving have also been developed and associated with long-term sustainability of digital scholarship outputs. More recent sustainability approaches explore socio-technical factors on digital preservation strategies, through effective, iterative ongoing project management (Langmead et al. 2018) while others introduce Software Development Lifecycle (SDLC) approaches to DH development and archiving (Smithies et al. 2019). Technical sustainability is also often associated with DH projects' scale and development workflow, bringing into the discussion the benefits as well as the headaches of large-scale digital humanities infrastructure projects, aiming to offer a one-model-suits-all preservation layer to their outputs (Verhoeven and Burrows 2015), versus the small-scale, customized, and versatile microservices (van Zundert 2012).

Conceptualizing technical systems and infrastructure "as a human and community asset in need of maintenance and support, rather than a technical artefact in need of service management" (Smithies et al. 2019) helps us move beyond the technological prevalence in sustainability discussions. By calling attention to the human-intensive aspect of DH work, aspects of skills development, capacity building, and inclusivity as well as continuous user/community engagement and aspects of legacy management need also to be considered as key factors in sustainability planning (Edmond and Morselli 2020). Lately, concerns over the environmental impact and responsibility of our research endeavors have also been introduced in the mode of transdisciplinary eco-critical Digital Humanities (Posthumus and Sinclair 2014; Starling Gould 2021), through the works of the GO::DH Minimal Computing Working Group[1] as well as individual DH labs' initiatives (SHL 2020), in an attempt to model a more eco-conscious behavior for our DH praxis.

All in all, when we talk about sustainability in DH we are mainly referring to a dynamic network of technologies, infrastructures, people, financial, and managerial decisions that are performing a constant tug-of-war with one another while battling about the longevity of digital outputs. Depending on the case, each agent or stakeholder is focusing on a distinct set of priorities and employing different processes to accomplish their goal. In addition, different sustainability strategies, models, best practices, dos and don'ts, have been developed and applied to individual DH projects, in different stages and scales, manifesting, thus, the plurality of cases and approaches, from a step-by-step planning ending procedure for DH projects (Rockwell et al. 2014) to a set of off-the-shelf preservation and archiving solutions for DH outputs (Carlin 2018). In the end, sustainability in DH discourse and practice lies somewhere between reality checks, risk assessment, and forward-thinking.

In 2010, Jerome McGann brought a series of "hidden" parameters into the sustainability discussion by asking a couple of provocative rhetorical questions:

> Would the problems go away if we had access to a lot more money? Or technical support? Or perhaps if all our scholarly projects had well-crafted business plans? To think that they would is a fantasy we all, in our different ways and perspectives, have to reckon with. Of course funding and technical support are necessary, but to fixate there is to lose sight of the more difficult problems we're facing. These are primarily political and institutional.
>
> (McGann 2010, 5)

Indeed, sustainability for digital scholarship does not exist in isolation from the political and institutional contexts in which it is being implemented. In what follows, I will try to briefly map a few central concepts, practices, and contexts embedded within the DH culture, highlighting the political and institutional obstacles they raise towards facing or even conceptualizing how sustainability is defined, measured, and practiced in Digital Humanities.

"FOREVER" AND SUSTAINABILITY

The expectation that digital projects will (have to) live "forever" stands as an assumption inherited from the era of print materiality that all scholarly outputs, including digital ones, ought to last "for as long as books do." While printed books have been proven to be really durable material assets across the centuries, digital materiality is radically different: although sometimes invisible and thus easily overlooked, it's more fragile, dynamic, diverse, constantly evolving, and, thus, digital preservation and maintenance are far more complex processes than any paper or any analog-related procedure (Tzitzikas and Marketakis 2018). Furthermore, for the majority of technological systems we can't predict their exact life span and, even if we have an expected end-date for the project's funding, staffing, or its maintenance, for most of the cases data management plans might be easily and unexpectedly ruined along the way; using Jeff Rothenberg's famous words, "digital information lasts forever or five years, whichever comes first" (1995, 44).

Indeed, a lot of the physical, print-based functions and practices around knowledge production are being mimicked in the digital scholarship era, from the skeuomorphic user experience design of our most heavily used computer functionalities (take "folders" in any file browsing context) to citation mechanisms, but to try representing and achieving digital research outputs' longevity via mirroring their physical antecedents is not only anachronistic: it's futile and perilous. Let's also acknowledge that when we are talking today about active, as opposed to passive, print preservation and sustainability, we are mainly referring to mechanical (microfilming and photographic) or digital reproduction technologies, thus, external to the original. Similarly, for the digital outputs—and especially the born-digital ones—we might need to invent or even wait for a new, external technology to be employed for their maintenance and preservation.

"DONENESS" AND SUSTAINABILITY

Related to the notion of longevity is another concept inherited from the world of print-based scholarship and publication: the concept of a complete, finished end product, that needs to be maintained as such. Since 2009 Matthew Kirschenbaum has been questioning "the measure of 'completeness' in a medium where the prevailing wisdom is to celebrate the incomplete, the open-ended, and the extensible" (Kirschenbaum 2009). The concept of "doneness" circulates widely within the DH jargon, embellished with tasks, milestones, deliverables, end-products, and deadlines, contradicting the radically flexible and rapidly changing modes of digital scholarly work, collaboration, development, publication, and funding.

One can argue here that the core component of Digital Humanities, the "project," encompasses also this very contradiction between the closed and the open, the finished and the extensible: the term "project" refers to "'a planned or proposed undertaking; a scheme, a proposal; a purpose, an objective' into the foreseeable future, with gusto. Such an orientation is actually at odds with the definition of a project in relation to particular aims" (Brown et al. 2009). With the project being the currency of evaluating digital scholarship in academia, the pressure to deliver and maintain a finished and accessible digital output transforms any claim of experimentation and innovation to a quantifiable budget unit in a funding proposal and another line in academics' CVs. Not surprisingly, sustainability planning often comes up at the final stages of a project, while wrapping things up before its end-date.

"NEWNESS" AND SUSTAINABILITY

Digital scholarship programmatically tones down the desire for "completeness" and "closure," and, thus, by embracing experimentation, it might be better described by what Julia Flanders has called the "culture of the perpetual prototype."[2] This commitment to the proof-of-concept, the beta version, the constantly evolving output, alongside an identity commitment of many DH practitioners towards hacking and building things, makes digital scholarship work to look more closely to a startup company's ethos, biased towards "openness" and "newness."

With novelty and innovation being the flagship of DH work as well as what the funding agencies are actually supporting through "proof-of-concept," "start-up" and "seed" grant schemes, the undertaking towards longevity and sustainability for DH projects becomes a tough race and a silly joke at the same time. Despite the fact that sustainability and data management plans must be nowadays an integral part of research proposals in most funding schemes, and, somehow ironically, usually frame innovation statements in grant applications, let's be honest: we are getting "follow-on" grants to expand cutting-edge project work but there is not yet a funding scheme solely focusing on closing down a project.

CREDIT, EVALUATION, AND SUSTAINABILITY

The pursuit of sustaining a digital work as a final, closed research output corresponds to and supports the traditional institutional system of assigning credit in humanities scholarship, including the processes of evaluation for tenure and promotion. There is indeed a need to preserve and sustain digital outputs as the measure of scholarly success and the currency of professional reward and promotion for humanities scholars. However, the idea of a single-authored, completed, citable, successful, and always available-to-be-shown object has been challenged and nowadays replaced by a collaborative work, open and ongoing, that never managed to be perfect or finished, living probably in a (version-controlled) code repository.

This dynamic and open-ended aspect of digital scholarly work can be observed both on the production and the reception side and as such must be evaluated and maintained: as Bethany Nowviskie reminds us, "in many cases, the 'products' of digital scholarship are continually re-factored, remade, and extended by what we call expert communities [...] Sometimes, 'perpetual beta' is the point! Digital scholarship is rarely if ever 'singular' or 'done,' and that complicates immensely our notions of responsibility and authorship and readiness for assessment" (2012). Such insights call attention to our current academic modes of evaluation when it comes to collaborative and open digital outputs and force us to rethink assessment criteria for DH work that will value quality not in terms of final product but by focusing on the processes by which it is co-created and re-used by various audiences and communities. In such a framework, maintenance should be seen not as an end-stage task but as an ongoing, iterative process, both on the production and the reception side, especially in public Digital Humanities initiatives, where the active involvement of local communities secures the afterlife of the project.

FAILURE AND SUSTAINABILITY

While Digital Humanities have been traditionally favorable to experimentation, exploration, playfulness, "screwing around" (Ramsay 2010), and eventually errors and failure, project failings or failed projects never (or rarely) manage to be maintained. Failure, despite its many praises, still remains a strong taboo in digital scholarship and in the humanities more generally, favoring mainly glowing success stories, outputs that are worth every penny out of research funding and can land a tenure-track position. The invalid code snippet, the broken link, this error message, this now dead or not-so-much-visited web resource, this poor collaboration, all of them are living manifestations of our scholarship and we need to preserve and document them with forethought and rigor, like the brave story of Project Bamboo (Dombrowski 2014) and Shawn Graham's courageous narratives of "failing gloriously" (2019).

Also, we need to demystify the idea of "fail forwards"; not every failure does contain the seeds of future success. Most of the time a failure is just that, and not much more. The cost of failure, though, must be explicitly acknowledged, be it a blight upon the reproducibility and re-use of the digital outputs or related to brand/reputation management of the stakeholders, and this should be considered as the afterlife of the project, equally important as its life. So, in case we are still assessing/debating what to maintain and what to abandon from our digital scholarship, I will second here Unsworth's advice: "Our successes, should we have any, will perpetuate themselves, and though we may be concerned to be credited for them, we needn't worry about their survival: They will perpetuate themselves. Our failures are likely to be far more difficult to recover in the future, and far more valuable for future scholarship and research, than those successes" (1997).

BETWEEN MAINTENANCE AND BREAKING

So where are we heading through this constant effort to balance the need for innovation against the need for longevity and memory, the mandate towards flexibility, experimentation, and iterative improvements against the inescapable failure and unfulfilled maintenance in our everyday Digital

Humanities practice? Certainly not in prolonging the life of digital outputs, nor in necessarily improving the maintenance strategies we employ, nor in a more critical stance towards sustainability itself. Perhaps even discussing and agonizing too much about sustainability is acting as a brake on the fullest development of the (future-thinking of the) field.

After all, the decline or loss of (parts of) DH projects is an embedded part of their very unique identity as new modes of scholarship. So instead of endlessly searching for the best strategy to ensure the long-term maintenance of the digital outputs, perhaps we need to courageously embrace the very idea of change, failure, loss in the digital world and Digital Humanities, be it infrastructure failure, technical obsolescence, staff overturn, or lack of funding—or altogether. This is and will be our brand new data- and systems-saturated scholarly world; things will eventually break. Information and knowledge will be produced and lost in the blink of an eye.

Such an approach aiming to mitigate or even evaporate sustainability agonies from our DH concerns' repertoire has had its lonely yet brave supporters for a long time now. In 2010, Bethany Nowviskie and Dot Porter, after conducting a wide-ranging survey in the DH community with the eloquent title "Graceful Degradation: Managing Digital Projects in Times of Transition and Decline" proposed to start thinking about designing projects "in a way that takes periods of transition and possible decline into account from the very start" (Nowviskie and Porter 2010). More recently, in 2019, Mark Sample, while exploring this middle, slippery space between maintenance and breaking and referring to Steven Jackson's work on repair and breakdown, proposed a "speculative care" approach for DH projects:

> ... like the plane designed with its eventual crash in mind, speculative care builds new digital work from the ground up, assuming that it will eventually fail. Technology will change. Funding will stop. User needs evolve. It is ideal for DH projects to launch with sustainability strategies already in place; speculative care demands that there also be obsolescence strategies in place [and] embeds broken world thinking into the digital tool or project itself.
>
> (Sample 2019)

Reconceptualizing our workflows and starting a DH project not only with a data management plan or a sustainability strategy in place but with the deep knowledge that it will eventually degrade, this "broken world thinking" (Jackson 2014), enhances our DH practice, not with a deterministic set of concerns, but with a richer imaginary on how digital scholarship, alongside the world-as-we-know-it, can still be productive, useful, and hopeful when abandoned or broken.

Ultimately, though, as we move forward in the post-pandemic era, where we have all been smashed into the immediate and perpetual "now" and every risk management attempt has been proven not sufficient, we need to be pragmatic about our own expectations and choices for the years to come: we will carry on designing and developing digital projects following best practices and standards aiming for interoperability and access, while having in mind their future internal accident or end-date—sooner or later, it will eventually come. Through such a stance, we can cultivate a more comprehensive and timely ethics and practices of care in Digital Humanities (Parikka 2019), where care, concern, and responsibility, for our communities, for the wider environment around us, and for our systems and devices, will encompass an attempt less to look after and sustain than to understand, to have empathy for and to engage with.

NOTES

1. http://go-dh.github.io/mincomp/.
2. "The Long Now of Digital Humanities," Baker-Nord Center for the Humanities, Case Western Reserve University, https://case.edu/artsci/bakernord/events/past-events/2014-2015/long-now-digital-humanities.

REFERENCES

Adams, W. M. 2006. *The Future of Sustainability: Re-thinking Environment and Development in the Twenty-first Century*. Report of the IUCN Renowned Thinkers Meeting, January 229–31. https://portals.iucn.org/library/node/12635.

Brown, Susan, Patricia Clements, Isobel Grundy, Stan Ruecker, Jeffery Antoniuk, and Sharon Balazs. 2009. "Published yet Never Done: The Tension between Projection and Completion in Digital Humanities Research", *Digital Humanities Quarterly* 3 (2). http://www.digitalhumanities.org/dhq/vol/3/2/000040/000040.html.

Carlin, Claire. 2018. "Endings: Concluding, Archiving, and Preserving Digital Projects for Long-Term Usability." *KULA: Knowledge Creation, Dissemination, and Preservation Studies* 2 (1): 19. https://doi.org/10.5334/kula.35.

Dombrowski, Quinn. 2014. "What Ever Happened to Project Bamboo?" *Literary and Linguistic Computing* 29 (3) (September):326–39. https://doi.org/10.1093/llc/fqu026.

Edmond, Jennifer and Francesca Morselli. 2020. "Sustainability of Digital Humanities Projects as a Publication and Documentation Challenge." *Journal of Documentation* 76 (5): 1019–31. https://doi.org/10.1108/JD-12-2019-0232.

Graham, Shawn. 2019. *Failing Gloriously and Other Essays*. With a foreword by Eric Kansa and afterword by Neha Gupta. Grand Forks, ND: The Digital Press at the University of North Dakota.

Jackson, Stephen J. 2014. "Rethinking Repair." In *Media Technologies: Essays on Communication, Materiality and Society*, edited by Tarleton Gillespie, Pablo Boczkowski, and Kirsten Foot, 221–40. Cambridge, MA: MIT Press.

Kirschenbaum, Matthew G. 2009. "Done: Finishing Projects in the Digital Humanities." *Digital Humanities Quarterly* 3 (2). http://www.digitalhumanities.org/dhq/vol/3/2/000037/000037.html.

Langmead, Alison, Aisling Quigley, Chelsea Gunn, Jedd Hakimi, and Lindsay Decker. 2018. "Sustaining MedArt: The Impact of Socio-Technical Factors on Digital Preservation Strategies." Report of research funded by the National Endowment for the Humanities, Division of Preservation and Access, 2015–2018. https://sites.haa.pitt.edu/sustainabilityroadmap/wp-content/uploads/sites/10/2017/01/SustainingMedArt_FinalReport_Web.pdf.

Maron, Nancy L. and Sarah Pickle. 2014. *Sustainability Implementation Toolkit: Developing an Institutional Strategy for Supporting Digital Humanities Resources*. Ithaka S+R. http://sr.ithaka.org/?p=22853.

McGann, Jerome. 2010. "Sustainability: The Elephant in the Room." *Online Humanities Scholarship: The Shape of Things to Come*, 5–14. https://cnx.org/exports/3d5747d3-e943-4a39-acf9-beb086047378@1.3.pdf/online-humanities-scholarship-the-shape-of-things-to-come-1.3.pdf.

Nowviskie, Bethany. 2012. "Evaluating Collaborative Digital Scholarship (or, Where Credit is Due)." *Journal of Digital Humanities* 1 (4) (Fall). http://journalofdigitalhumanities.org/1-4/evaluating-collaborative-digital-scholarship-by-bethany-nowviskie/.

Nowviskie, Bethany and Dot Porter. 2010. "The Graceful Degradation Survey Findings: How Do We Manage Humanities Projects through Times of Transition and Decline?" Digital Humanities 2010, London. http://dh2010.cch.kcl.ac.uk/academic-programme/abstracts/papers/html/ab-722.html.

Parikka, Jussi. 2019. "A Care Worthy of Its Time." In *Debates in the Digital Humanities*, edited by Matthew K. Gold and Lauren F. Klein. University of Minnesota Press. https://dhdebates.gc.cuny.edu/read/untitled-f2acf72c-a469-49d8-be35-67f9ac1e3a60/section/38b93cc9-3b58-4bcf-a444-04bcdaf322ee#ch44.

Pitti, Daniel V. 2004. "Designing Sustainable Projects and Publications." In *A Companion to Digital Humanities*, edited by Susan Schreibman, Ray Siemens, and John Unsworth. Oxford: Blackwell. http://digitalhumanities.org/companion/view?docId=blackwell/9781405103213/9781405103213.xml&chunk.id=ss1-5-1.

Posthumus, Stephanie and Stéfan Sinclair. 2014. "Reading Environment(s): Digital Humanities Meets Ecocriticism." *Green Letters. Studies in Ecocriticism* 18 (3): 254–73.

Ramsay, Stephen. 2010. "The Hermeneutics of Screwing Around; or What You Do with a Million Books." https://libraries.uh.edu/wp-content/uploads/Ramsay-The-Hermeneutics-of-Screwing-Around.pdf.

Rockwell, Geoffrey, Shawn Day, Joyce Yu, and Maureen Engel. 2014. "Burying Dead Projects: Depositing the Globalization Compendium." *Digital Humanities Quarterly* 8 (2). http://www.digitalhumanities.org/dhq/vol/8/2/000179/000179.html.

Rothenberg, Jeff. 1995. "Ensuring the Longevity of Digital Documents." *Scientific American* 272 (1): 42–7.

Sample, Mark. 2019. "The Black Box and Speculative Care." In *Debates in the Digital Humanities*, edited by Matthew K. Gold and Lauren F. Klein. University of Minnesota Press. https://dhdebates.gc.cuny.edu/read/untitled-f2acf72c-a469-49d8-be35-67f9ac1e3a60/section/3aa0b4f4-bd72-410d-9935-366a895ca7a7#node-b3c2a783137fad8bf887d557464c1d3257d553da.

Smithies, James, Carina Westling, Anna-Maria Sichani, Pam Mellen, and Arianna Ciula. 2019. "Managing 100 Digital Humanities Projects: Digital Scholarship & Archiving in King's Digital Lab." *Digital Humanities Quarterly* 13 (1). http://digitalhumanities.org/dhq/vol/13/1/000411/000411.html.

SSI (Software Sustainability Institute). 2018. *Sustaining the Digital Humanities in the UK* (report). https://doi.org/10.5281/zenodo.4046266.

Starling Gould, Amanda. 2021. "Kitting the Digital Humanities for the Anthropocene: Digital Metabolism and Eco-Critical DH." In *Right Research: Modelling Sustainable Research Practices in the Anthropocene*, edited by Chelsea Miya, Oliver Rossier, Geoffrey Rockwell, 93–110. Cambridge: Open Book Publishers.

Sussex Humanities Lab (SHL) Carbon Use and Environmental Impact Working Group, Jo Walton, Alice Eldridge, James Baker, David Banks, and Tim Hitchcock. 2020. *The Sussex Humanities Lab Environmental Strategy* (Version 1.3). Zenodo. http://doi.org/10.5281/zenodo.3776161.

Tzitzikas, Yannis and Yannis Marketakis. 2018. *Cinderella's Stick – A Fairy Tale for Digital Preservation*. Basingstoke: Springer Nature.

Unsworth, John. 1997. "Documenting the Reinvention of Text: The Importance of Failure." *The Journal of Electronic Publishing* 3 (2). http://dx.doi.org/10.3998/3336451.0003.201.

Verhoeven, Deb and Toby Burrows. 2015. "Aggregating Cultural Heritage Data for Research Use: The Humanities Networked Infrastructure (HuNI)." In *Metadata and Semantics Research*, edited by Emmanouel Garoufallou, Richard J. Hartley, and Panorea Gaitanou, 417–23. Springer International Publishing [Communications in Computer and Information Science 544].

Zorich, Diane M. 2003. "A Survey of Digital Cultural Heritage Initiatives and Their Sustainability Concerns." *Council on Library & Information Resources (CLIR)*. https://www.clir.org/wp-content/uploads/sites/6/pub118.pdf

Zundert, Joris van. 2012. "If You Build It, Will We Come? Large Scale Digital Infrastructures as a Dead End for Digital Humanities." *Historical Social Research Journal* 37 (3): 165–86.

CHAPTER THIRTY-ONE

Libraries and the Problem of Digital Humanities Discovery

ROXANNE SHIRAZI
(THE GRADUATE CENTER, CITY UNIVERSITY OF NEW YORK)

Over the last decade, a new kind of librarian took shape in US academic libraries: the digital humanities librarian. In 2011, Jean Bauer was hired as Brown University's first DH librarian, marking the start of a trend in hiring that saw annual searches balloon from two to twenty-eight in just five years (Morgan and Williams 2016). The following year I helped launch *dh+lib*, a project designed to connect the work of LIS professionals and DH scholars, with the understanding that there was significant overlap between the two areas. Also in 2012, Micah Vandegrift published a piece called "What is Digital Humanities and What's It Doing in the Library?," the title playing on Matthew Kirschenbaum's now-classic musing on DH (2012). Vandegrift summed up his argument in a "TL/DR" (too long, didn't read) for the piece:

> Libraries and digital humanities have the same goals. Stop asking if the library has a role, or what it is, and start getting involved in digital projects that are already happening.
>
> Advocate for new expanded roles and responsibilities to be able to do this. Become producers/creators in collaboration with scholars rather than servants to them.
>
> (Vandegrift 2012)

Such bold statements reflect the tone of those years, when questions of "who's in and who's out" took center stage and DH was still reeling from its debut in the popular press. Academic librarians, it seemed, were so delighted to *discover* scholars who were doing work that looked like our work, who valued curatorial ways of knowing, who wanted to talk about controlled vocabularies and metadata that we jumped up to proclaim, "we do this too!" and suddenly the only way to engage with DH was to *do DH*.

The enthusiastic response from librarians looking to expand their roles and responsibilities was met with equal enthusiasm by library administrators who were thus presented with a quick fix for building capacity for digital scholarship (Posner 2013, 44). We speak of the digital humanities librarian as concierge, IT expert, researcher, curator, educator, mediator, interpreter, host, partner, advocate, consultant (Vedantham and Porter 2015, 184–6; Zhang, Liu, and Mathews 2015, 371–3). Bobby Smiley, reflecting on the evolution of DH librarianship, laments: "The nature of my work is saying 'yes' to almost everything" (2019, 416). In 2017, a Digital Library Federation Working Group formed under the tongue-in-cheek banner of "miracle workers," which—though met with some discomfort—gestured to the "ill-defined mandates" that digital humanities librarians encounter in shepherding institutional digital transformation with very little real power to do so (Bonds and Gil 2017; Morgan 2017).

This expansive view of what libraries and librarians have to offer digital humanities frames the library as lab, the library as producer, the library as platform. Yet with all of the attention paid to libraries and DH, in all of the calls for LIS practitioners to *do* DH and to work collaboratively with digital humanists (indeed, to recognize that librarian/scholar is not a dichotomy), we were perhaps a little too willing to shed the markers of traditional library work. We told ourselves that librarians needed reskilling to support DH. We needed to learn a new scholarly language: DHers create "tools," not software, they create "digital projects," not websites. DHers need help with Gephi, Mallet, Omeka; librarians needed to be conversant with digital humanities techniques and offer workshops steeped in digital humanities pedagogy. We eagerly translated the role of a subject librarian to DH librarian, proclaiming that instead of expertise with traditional sources, a DH librarian might have expertise with one or more "tools" and consult with researchers to pair their idea with an appropriate mode of analysis. Walking them through the process, helping them learn the tool. This was important work then, and it is important work now, but it shaped academic libraries' conception of how to support DH scholars within a limited framework of producer/creator. DH was about *building things*, so librarians focused on building too.

As libraries emphasized their crucial role in producing DH work as partners in developing, sustaining, and preserving digital resources, scant attention was paid to our role in description and discovery, our contribution to disciplinary formation that goes beyond our technology stacks and campus service models. Librarians, somewhat cavalierly, proclaim that "DH research output still lacks attention, integration, and sustainability" (Zhang, Liu, and Mathews 2015, 362), as though these qualities are birthed fully formed out of a developing field. The selectors, the catalogers, the bibliographers, those attending to the infrastructure of description and discovery, how might these "miracle workers" respond to DH? Perhaps in all of our asking about *the role of the library in DH*, we neglected to grapple with the many challenges presented by considering *the role of DH in the library*. By inverting the question that Vandegrift dismissed so long ago, we begin to grasp what has gone missing in the librarian drive to sustain DH knowledge production by becoming producers/creators of DH.

The discussion that follows is necessarily *one* area of libraries, *one* idea of librarianship, and *one* point of contention. The present conjecture depends on a choice between framing digital humanities as methodology, as a computational humanities; or digital humanities as presentational, as a mode of representation of scholarship. I have chosen the latter.

WHY IS IT SO HARD TO FIND DH PROJECTS?

When I teach DH to LIS students, I devote a class period early in the term to locating digital projects and the problem of DH discovery. Students work in pairs to complete a "DH Project Scavenger Hunt," in which they fill out a worksheet using an online search engine to find scholarly DH projects. Students are asked to find a project that meets various qualities, such as: employs a creative or unusual timeline; uses crowdsourced transcription; is not in English; uses twentieth-century texts or materials; includes historical reconstruction or modeling. They then choose one of the projects they've found and discuss these questions with their partner: Is it still active? How can you tell? How would you categorize the topic or discipline? How would you categorize the method or technologies used? (Is it evident what those are?) Why would you call this a DH project?

These are the kinds of questions that librarians (and faculty) encounter as DH students look for examples to analyze and critique, as they seek models for what's possible and what to incorporate in their own projects. Eventually, we settle on a list of techniques: check the NEH digital humanities grants archive, DH Commons, Around DH in 80 Days, the Black DH Google Doc, DHNow, *dh+lib*. Students are, understandably, mystified by this state of affairs. Why is it so hard to find DH projects?

A 2019 report found that "much of the work of DH discovery, according to those we spoke with, appears to rely upon a network of relationships" with one study participant noting, "you have to know who works with whom" (Hudson-Vitale et al. 2019, 17). At this late stage, why do we put up with such exclusivity? It is no wonder that your digital humanities looks nothing like my digital humanities, if we are traveling in closed circles with pockets of knowledge passed among ourselves. If DH is serious about moving towards equity and inclusion, we must ask: who benefits from a system that requires such insular knowledge as the price of entrance to a field?

Reading the literature in libraries around DH discoverability, one sees how easy it is to conflate findability with interoperability and re-use. Some take discoverability to mean an established metadata standard applied at the point of production to allow automated harvesting and aggregation, a dream of seamless access and magical description. Others conflate discovery with outreach, promotion, media attention, critique, and it becomes a problem of *dissemination*—again, a responsibility of the creator. In this discussion of discovery, I simply mean how and where researchers find and access digital humanities projects and whether they are collocated with other scholarly materials.

Alex Gil has admirably demonstrated that making DH projects visible and accessible does not require large institutional resources. In 2014, responding to the nescience of Anglophone scholars concerning non-Anglophone digital humanities projects, Gil spearheaded "Around Digital Humanities in 80 Days" to highlight the global nature of digital humanities practice. Working with Global Outlook::Digital Humanities (GO::DH), a special interest group of the Association of Digital Humanities Organizations (ADHO), Gil and colleagues collected DH project information in a crowdsourced Google Doc as the basis of an online exhibit that featured a different project each day for eighty days. Since its appearance the project has become a de facto directory of digital projects for those curious about what DH looks like in practice, and it is still sustained by GO::DH, which now identifies discovery as one of its core activities (GO::DH n.d.). Similarly, CariDiScho2020, "A Directory of Caribbean Digital Scholarship," draws on several crowdsourced lists to curate what is essentially an online reference work (CariDiScho2020 n.d.). Developed as part of a 2020 scholarly conference co-organized by Gil, Kaima L. Glover, and Kelly M. Josephs, the directory also includes an underlying Google Doc to allow for continual updating (the directory itself is static). While these initiatives were not reliant on grant funding or sustained institutional support, they do both rely on a social infrastructure of scholarly community, an identified need, and a collective response. Neither is attempting to create a master directory of *all* digital projects, nor are they harvesting and aggregating data on a large scale. These are community-curated lists, and they continue to serve their communities.

Yet, we—librarians—must acknowledge that projects like these are powered by individuals and their Google Docs. They are a lightweight, if temporary, response but they should not be mistaken for a solution to the problem of DH discovery. There must be a middle ground between back-of-the-napkin ideas and our dream of complete interoperability.

MAINTENANCE AND "THE DIRECTORY PROBLEM"

Anyone attempting to resolve the discovery problem by proposing a master database of DH projects will be met with skepticism, a knowing smirk, perhaps, and an outstretched hand to guide them through a history of failed projects that have come and gone. Quinn Dombrowski's insightful account of Project Bamboo and the lessons to be learned from amorphous, multi-institution, and overly ambitious infrastructure projects is one place to start (2014). If we consider Tara McPherson's proposition that "today's fantasies of total knowledge depend on the database," with a foundation of extreme abstraction and an underlying drive towards interchangeability, we might begin to understand the desire for an all-encompassing solution to the problem of DH discovery in the form of a directory (McPherson 2015, 487). Digital humanities practitioners may espouse the tenets of playful experimentation and ephemerality and engage in innovative, nimble responses but they are, at the same time, prone to the "no limits" dreams of totality that are endemic to our digital age.

Through the work of Dombrowski and others, however, DH has come to appreciate the work of maintenance and so we encounter cautionary tales to talk us down from our grand schemes. Shannon Mattern observes that "the world is being fixed all around us, every day" despite the obfuscatory economic discourse of innovation: "Maintenance at any particular site, or on any particular body or object, requires the maintenance of an entire ecology: attending to supply chains, instruments, protocols, social infrastructures, and environmental conditions" (2018). That is, maintenance and repair are not the sole province of any one group, and the negotiation over what gets maintained and who does the maintaining is necessarily fraught. Digital humanities as a community of practice that spans the full range of faculty and staff positions within and adjacent to academia must negotiate these issues as the field grows and matures.

The decision to let go of the DiRT Directory (Digital Research Tools), for example, was not for lack of use or need; indeed, tool directories are lauded for their role "providing recognition" for DH software work, a path towards valuation that takes place outside of peer-reviewed journals (Grant et al. 2020). There is an understanding that directories themselves are necessary, but an appreciation of the maintenance required to keep them current, and thereby useful, rightly dampens any enthusiasm for new initiatives. Without the technical and social infrastructures in place to sustain these projects, which are themselves providing a kind of infrastructure for DH research and scholarship, existing product owners face burnout and a lack of institutional recognition for the work. DiRT was ultimately merged with TAPoR (Text Analysis Portal for Research), but the post-mortem analysis identifies important challenges to those developing research infrastructure that is produced and maintained by disciplinary specialists.

Beyond maintenance, we may also be seeing a shift in our collective understanding of sustainability and preservation. Where once we lamented the burdensome challenges of digital preservation, and conveniently declared that ephemerality was a feature and not a bug, there is now an interest in reviving legacy digital projects, rescuing them not just for re-use but to re-produce them through a process of salvaging; not stripping for parts, exactly, but resurrecting, restoring and repurposing (Senseney et al. 2018; Antonini, Benatti, and King 2020).

A recent study concluded that digital humanities projects have a five-year "shelf life," or period of time a project can exist without updates, after which it can reasonably be considered "abandoned" (Meneses and Furuta 2019). For years I assumed that the loss created by such abandonment was purely epistemic, but now it is the wastefulness of DH projects left to wither on the vine that moves me. What was it all for?

COLLECTING DH

One of the unfortunate corollaries of second-wave DH is that making space for humanities scholars who *want* to present their work digitally, who want to re-present, re-mix, re-interpret, re-think through expansive mediums becomes understood as a call, a mandate, for everyone to do this to remain competitive professionally. Similarly, among academic libraries there grew calls to expand research support services, to expand a slate of workshops, and collectively expand our capacity for digital scholarship, to accommodate this new style of researcher.

Yet what does it really mean to make space for this expansive idea of scholarship in libraries? How do libraries appropriately value open and digital forms of knowledge production in the academy, beyond touting our role as facilitators and producers of this new knowledge? Academic libraries have spent most of the Internet age desperately updating our image, our rationale, our services when historically our strengths lie in notions of permanence, reliability, longevity. Like the phenomenon of habitual new media described by Wendy Hui Kyong Chun, twenty-first-century libraries have been frantically "updating to remain the same," fighting obsolescence even by "writing over" or destroying the very thing we wish to resuscitate (Chun 2016, 2). How often have we heard that libraries are no longer necessary because everything is online? So we (re)discover the library as lab, the library as producer, the library as platform. It is no wonder that the perennial future of libraries has been, like Elsevier, leaning towards "workflow" over content (Schonfeld 2017).

The irony is that to make space for the humanities, done digitally, libraries need to *collect* this work, and we need to figure out how to do so in the absence of any industry standards for freely available digital materials. Now, it is true that the sum of a library's contributions to academia goes beyond the maintenance of physical and digital collections, but to attend to our role in disciplinary formation we must acknowledge the legitimizing function of libraries. Our collections reflect our priorities, our values, and our sensibilities. We can't wait for grand schemes of interoperability and data re-use to materialize from the creators, or from the next four-year grant-funded initiative that investigates project sustainability. We need rogue collectors, working outside the rarefied world of special collections, looking not to preserve the entirety of a digital work but to record useful metadata and integrate it with our proprietary digital resources. We need to solve the directory problem using systems we already have.

What would it look like for a library to *collect* DH instead of *support* DH? Imagine selectors at research libraries scouring scholarly journals for digital project reviews, chasing project citations in published books and articles, attending conferences and noting works-in-progress to build a list of digital materials to "acquire." We might find institutions identifying their own collection strengths and weaknesses, drawing on the disciplinary knowledge of our producer/creator digital humanities librarians to fill the gaps. If this sounds absurd, consider that today's academic libraries already operate in a post-ownership environment for electronic resources. Many of our librarian colleagues already spend their days configuring (and battling) proprietary systems that offer little in the way of stable access to digital collections. Perhaps we should consider how libraries could approach the discovery issue for openly available digital humanities projects as a means of sustaining our scholarly communities. By collocating digital resources with print and audiovisual materials and providing access through *description*, even in the absence of ownership, we might tip the scale towards broader integration, attention, and sustainability.

Most DH projects are produced outside of any formal publishing apparatus, any profit motive to standardize the identification and exchange, any financial value ascribed to the object itself that warrants careful collection and preservation. Yet libraries have seen this before. In 1982, *Alternative Materials in Libraries* called on librarians to attend to the systemic barriers impeding the acquisition, cataloging, and use of "alternative publications, which include small press and other materials produced by non-standard, non-establishment groups or individuals" (Danky and Shore 1982). Librarians were (finally) responding to the explosion of counterculture literature and the ways in which our procurement and processing systems were simply not designed to accommodate non-standard, non-mainstream materials. The move to collect and catalog the alternative press and zines was not just another iteration of the canon wars; librarians needed to address the material conditions of our collecting and procurement processes in order to meet our obligation to our local communities, and that involved expanding our collection scope and integrating alternative materials into our existing catalog (Dodge 2008).

DH projects are cultural artifacts but unless libraries are responsible for full stewardship and preservation in partnership with the creators, we have no ownership. Without ownership, we have only a path to the work, which hinders our collective willingness to dedicate institutional resources to cataloging and integrating it into our discovery system. And yet, a path to a work starts to sound an awful lot like a citation. Imagine a practice of enumerative bibliography to record and document the steady growth of DH projects, to begin to identify where to allocate resources (and time) for deeper technical services work. What is stopping us? Blame the primacy of search, of Google; blame the fact that it's free, so we don't need to bother *collecting* it; blame the proliferation of LibGuides over stable and uniform reference guides, and the newer library services platforms that have turned libraries' attention from cataloging electronic resources to activating subscriptions and troubleshooting connections. The twenty-first-century traditional-digital scholarly publishing system is turning libraries into mere mechanisms for translating procurement funds into access points, and in our move to "write over" library collections we are finding new forms of enclosure surrounding the commons.

The open access (OA) movement has struggled with many of the same issues and a solution has not been forthcoming, but there is a commonality between the challenges of getting OA works into library discovery systems and the problems of DH discovery. "Mapping the Free Ebook Supply Chain," a grant-funded investigation into the mechanics of ebook discovery, concludes that "There is a clear need to make sure that OA books are included more rigorously in library catalogs" (Watkinson et al. 2017, 2). Report authors identified challenges that are all too familiar to DH: "systems and workflows designed to lock down not open up books," "questions about sustainability," and "lack of consistency in metadata" (Watkinson et al. 2017, 4). On institutional resources, they note: "Preparing books for deposit in platforms, including cataloging, incurs costs and it was unclear whether publishers or libraries would be willing to fund these fees" (4). Might libraries be willing to incur costs to catalog digital humanities projects? What if our producer/creator librarians, versed in DH tools and techniques, were able to leverage this knowledge in their libraries to design our own strategies instead of pushing it out to individual creators?

These are the problems that DH brings to the library, and by attending to the discovery and use of digital humanities projects we can approach the challenges collaboratively in a way that is impossible when working with commercial vendors. Libraries have already embraced, rightly or not, a post-ownership licensing model for digital materials; why not use some of our resources to tackle the OA discovery problem in a sphere where we are already working as partners in scholarly production? Archivists have developed a postcustodial model that works particularly well with participatory and community archives, in which an institution stewards materials that are not in their physical custody ("SAA Dictionary: Postcustodial" n.d.). Perhaps it's time for libraries to separate description from ownership and to mobilize our collective capacity and institutional resources towards developing and adapting our systems for the somewhat paradoxical "acquisition" of open access materials.

* * *

A library catalog record is evidence of a work's existence, somewhere. A citation is evidence of its existence, somewhere. A Google Doc is evidence of its existence, somewhere. And once you have that record of existence with a few key data points, you can begin the process of locating it, of tracing its path through production to reception to obsolescence and rediscovery. But to do this, you must have the initial trace, the scant evidence from which to begin. A library catalog is a surrogate for the library collection, and each record is designed to capture enough descriptive information such that you can easily assess whether the item itself is worth retrieving. Serials cataloging, for example, has intricate means of connecting publications across changes in name or publisher. These linking fields do more than describe the object at hand; they map relationships between objects and between records. This is painstaking work, some would call it unsustainable, and yet it is sustained by a system of paid professionals whose core concern is description and access through creating and maintaining records.

For a brief moment, in the early Internet age, libraries were enthusiastic about adding websites to library catalogs as just another information format, with an effort towards standards and training in the US being developed by CONSER, the Cooperative Online Serials Program administered by the Library of Congress's Program for Cooperative Cataloging (Gerhard 1997; Moyo 2002). Publications like the *Journal of Internet Cataloging* helped justify a move to expend institutional resources to describe freely available online sources that offered enduring research value (the publication has since been renamed the *Journal of Library Metadata*). Berkeley's Digital Library SunSITE articulated four "digital collecting levels" with corresponding preservation commitments: archived, served, mirrored, and linked ("Digital Library SunSITE Collection Policy" 1996). Similarly, the IFLA Working Group on Guidelines for National Libraries addressed the challenges of electronic resources for national bibliography and posed the question of whether users were better served by libraries recording what exists or recording only what is held (Žumer 2009). In the late 1990s and early 2000s, US catalogers had professional concerns about the sustainability of such efforts. They noted the importance of being format-agnostic in bringing quality sources into library catalogs, but faced a dilemma over efficiency and whether they should just "leave the job to search engines like Google" (Dong 2007). It's clear today which side of the argument won, though given what has transpired in the intervening years and the revelations surrounding Google's search algorithms, libraries may need to reassess our approach. Why not begin with digital humanities scholarship as a use case for selecting and cataloging freely available web content in research libraries?

Ultimately, we need to arrive at a point where non-host institutions can collect and preserve digital humanities projects. A recent Mellon-funded project called "Digits" explored software containerization as a possible path and identified technical and social barriers to the integration of DH research output. The group suggests that we need technical standards for creators if we are to solve the problem of portability, to allow scholars to deposit their work with a library and to move responsibility for low-stakes maintenance and preservation away from the individual creator (Burton et al. 2019). Regarding social barriers, DH scholars are beginning to see the value of reviewing digital humanities work in disciplinary publications, and in 2019, Roopika Risam and Jennifer Guiliano launched an experimental journal, *Reviews in DH*, designed to "foster critical discourse about digital scholarship in a format useful to other scholars" (Reviews in Digital Humanities n.d.). Such publications could also serve as a filtering mechanism for library selectors wishing to catalog DH at their libraries.

Knowing what we know now about the rise of Google, blogs, social media, and the systematic defunding of public institutions under neoliberalism, it is no surprise that cataloging websites was seen as a fool's errand. Yet the importance of *collecting* externally hosted online scholarly resources by *cataloging* them is better understood when presented with the limited use case of digital humanities researchers. We should also acknowledge that there are libraries who already catalog DH projects, but they are exceedingly rare, especially in relation to the number of libraries who have been expanding digital scholarship services over the last decade. The beauty of cooperative cataloging is that one library can create a record and another can enhance, refine, or correct it in a shared cataloging environment. The selection of online materials to be cataloged, if it's done at all, would typically be performed by subject specialists in the library, but since digital humanities librarians have reimagined that role away from traditional services it is very likely that no one is currently doing this work. I linger on these details because the maintenance of the system, the catalog, the record—this maintenance is the work of libraries, and it undergirds our ability to facilitate knowledge production. We are failing the digital humanities by not tending to our "traditional" library concerns while exhorting everyone to become a producer/creator of DH.

In discussing the challenges related to maintaining tool directories, Grant et al. note that, "we have experimental infrastructure, but it hasn't yet been woven into the practices of the field partly because the practices are still emerging" (2020). It may be that the practices of DH will have no use for the library catalog. But should the absence of a full-scale solution stop us from collecting and cataloging in piecemeal? We might find value in a multiplicity of approaches, weaving together a patchwork of library collections that are sustained and maintained as an alternative to our beloved Google Docs.

CONCLUSION

Cataloging digital humanities projects and tools might help make them legible as scholarship but it requires breaking the spell of *libraries doing DH*. If we can do that, if we can broaden the mandate of digital humanities librarianship away from the producer/creator model we can imagine bibliographies and indexes and compendiums and encyclopedias and all manner of reference works documenting the creative output of digital humanists. We can imagine libraries adjusting to a post-custodial mode of collecting, developing practices that align with the open access movement and breaking free from vendor solutions that move to enclose our digital commons.

Thinking about the role of DH in libraries causes us to wrestle with the challenge of an online, networked humanities. It causes us to wrestle with who is responsible for collecting "free" work, how we will curate those collections, and who will dedicate resources to protecting materials that are not valuable from a monetary standpoint, cannot (yet) be bought and sold as "rare" materials to a consumer collector. It forces us to question what is valuable about scholarship: the making? the doing? the representation?

Now, readers may interpret this provocation as a reprimand to librarians, a call to stop traveling amongst the producers and stay home tending their collections. I don't want to stop librarians at all, I want more librarians attending to DH in more ways than one. I've proposed here that the problem of DH discovery is intimately woven with librarians' discovery of DH. (Yes, librarians were involved in much earlier stages dating back to the days of humanities computing, but I am referring to the clear growth in positions identified earlier.) Recognizing the many ways that librarians can support and, indeed, *do digital humanities* beyond the producer/creator model is but one step towards the maturation of our professional engagement with the field. Consider this a call for a cadre of librarians attending to the work of description and access for openly available, online, digital resources, using digital humanities as a use case for what surely lies ahead.

Marisa Parham reminds us that "Working merely to map extant systems ... onto new kinds of scholarship not only assumes that such translation is possible; it also implies that we are in fact satisfied with what we already have" (2018, 683). We might not *want* to translate extant library systems to meet the challenge presented by digital humanities projects, but we don't have to be satisfied with mere translation. We can imagine alternatives and set our sights on building out the infrastructure of description and discovery for open materials on our own terms and in ways that meet the needs of digital humanists.

There may never be a single, correct, answer to the problem of DH discovery, but libraries already have the infrastructure to develop a network of local solutions and map connections between our multiple communities of practice. It's time for us to get started.

REFERENCES

Antonini, Alessio, Francesca Benatti, and Edmund King. 2020. "Restoration and Repurposing of DH Legacy Projects: The UK-RED Case." In *Book of Abstracts of DH2020*. Ottawa. https://dh2020.adho.org/wp-content/uploads/2020/07/138_RestorationandRepurposingofDHLegacyProjects.html.

Bonds, Leigh, and Alex Gil. 2017. "Introducing the 'Miracle Workers.'" *DLF*. December 15. https://www.diglib.org/introducing-miracle-workers-working-group/.

Burton, Matthew, Matthew J. Lavin, Jessica M. Otis, and Scott B. Weingart. 2019. "Digits: A New Unit of Publication." https://docs.google.com/presentation/d/e/2PACX-1vT0ksN1GnMilebHrmgGTeN5jhIK3nBURVQ4YBWevUoAYcANwqgSKXZ5MEV7wUh8Hz65IlzvnCS6eRlI/pub?start=false&loop=false&delayms=60000#slide=id.p.

CariDiScho2020. n.d. "Sources." CariDiScho2020. http://caribbeandigitalnyc.net/caridischo/.

Chun, Wendy Hui Kyong. 2016. *Updating to Remain the Same: Habitual New Media*. Cambridge, MA: MIT Press.

Danky, James Philip, and Elliott Shore. 1982. *Alternative Materials in Libraries*. Metuchen, NJ: Scarecrow Press.

"Digital Library SunSITE Collection and Preservation Policy." 1996. http://web.archive.org/web/20110628222622/http://sunsite.berkeley.edu/Admin/collection.html.

Dodge, Chris. 2008. "Collecting the Wretched Refuse: Lifting a Lamp to Zines, Military Newspapers, and Wisconsinalia." *Library Trends* 56 (3): 667–77. https://doi.org/10.1353/lib.2008.0013.

Dombrowski, Quinn. 2014. "What Ever Happened to Project Bamboo?" *Literary and Linguistic Computing* 29 (3): 326–39. https://doi.org/10.1093/llc/fqu026.

Dong, Elaine X. 2007. "Organizing Websites: A Dilemma for Libraries." *Journal of Internet Cataloging* 7 (3–4): 49–58.

Gerhard, Kristin H. 1997. "Cataloging Internet Resources." *The Serials Librarian* 32 (1–2): 123–37. https://doi.org/10.1300/J123v32n01_09.

GO::DH. n.d. "About." Global Outlook::Digital Humanities. http://www.globaloutlookdh.org/.

Grant, Kaitlyn, Quinn Dombrowski, Kamal Ranaweera, Omar Rodriguez-Arenas, Stéfan Sinclair, and Geoffrey Rockwell. 2020. "Absorbing DiRT: Tool Directories in the Digital Age." *Digital Studies/Le Champ Numérique* 10 (1). https://doi.org/10.16995/dscn.325.

Hudson-Vitale, Cynthia, Judy Ruttenberg, Matthew Harp, Rick Johnson, Joanne Paterson, and Jeffrey Spies. 2019. "Integrating Digital Humanities into the Web of Scholarship with SHARE." Association of Research Libraries. https://doi.org/10.29242/report.share2019.

Mattern, Shannon. 2018. "Maintenance and Care." *Places Journal*, November. https://doi.org/10.22269/181120.

McPherson, Tara. 2015. "Post-Archive: The Humanities, the Archive, and the Database." In *Between Humanities and the Digital*, edited by Patrik Svensson and David Theo Goldberg, 483–502. Cambridge, MA: MIT Press.

Meneses, Luis, and Richard Furuta. 2019. "Shelf Life: Identifying the Abandonment of Online Digital Humanities Projects." *Digital Scholarship in the Humanities* 34 (Supplement 1): i129–34. https://doi.org/10.1093/llc/fqy079.

Morgan, Paige. 2017. "Please Don't Call Me a Miracle Worker." December 17. http://blog.paigemorgan.net/articles/17/please-dont.html.

Morgan, Paige, and Helene Williams. 2016. "The Expansion and Development of DH/DS Librarian Roles: A Preliminary Look at the Data." Milwaukee, WI: OSF. https://osf.io/vu22f/.

Moyo, Lesley M. 2002. "Collections on the Web: Some Access and Navigation Issues." *Library Collections, Acquisitions, and Technical Services* 26 (1): 47–59. https://doi.org/10.1016/S1464-9055(01)00240-8.

Parham, Marisa. 2018. "Ninety-Nine Problems: Assessment, Inclusion, and Other Old-New Problems." *American Quarterly* 70 (3): 677–84. https://doi.org/10.1353/aq.2018.0052.

Posner, Miriam. 2013. "No Half Measures: Overcoming Common Challenges to Doing Digital Humanities in the Library." *Journal of Library Administration* 53 (1): 43–52. https://doi.org/10.1080/01930826.2013.756694.

Reviews in Digital Humanities. n.d. "About." Reviews in Digital Humanities. https://reviewsindh.pubpub.org/about.

"SAA Dictionary: Postcustodial." n.d. https://dictionary.archivists.org/entry/postcustodial.html.

Schonfeld, Roger. 2017. "Cobbling Together the Pieces to Build a Workflow Business." *The Scholarly Kitchen* (blog). February 9. https://scholarlykitchen.sspnet.org/2017/02/09/cobbling-together-workflow-businesses/.

Senseney, Megan Finn, Paige Morgan, Miriam Posner, Andrea Thomer, and Helene Williams. 2018. "Unanticipated Afterlives: Resurrecting Dead Projects and Research Data for Pedagogical Use." *DH2018* (blog). June 7. https://dh2018.adho.org/unanticipated-afterlives-resurrecting-dead-projects-and-research-data-for-pedagogical-use/.

Smiley, Bobby L. 2019. "From Humanities to Scholarship: Librarians, Labor, and the Digital." In *Debates in the Digital Humanities 2019*, edited by Matthew K. Gold and Lauren F. Klein, 413–20. Minneapolis, MN: University of Minnesota Press. doi: 10.5749/j.ctvg251hk.38.

Vandegrift, Micah. 2012. "What Is Digital Humanities and What's It Doing in the Library?" *In the Library with the Lead Pipe*, June. http://www.inthelibrarywiththeleadpipe.org/2012/dhandthelib/.

Vedantham, Anu, and Dot Porter. 2015. "Spaces, Skills, and Synthesis." *Digital Humanities in the Library: Challenges and Opportunities for Subject Specialists* (April): 177–98. https://repository.upenn.edu/library_papers/86.

Watkinson, Charles, Rebecca Welzenbach, Eric Hellman, Rupert Gatti, and Kristen Sonnenberg. 2017. "Mapping the Free Ebook Supply Chain: Final Report to the Andrew W. Mellon Foundation." University of Michigan. http://hdl.handle.net/2027.42/137638.

Zhang, Ying, Shu Liu, and Emilee Mathews. 2015. "Convergence of Digital Humanities and Digital Libraries." *Library Management* 36 (4/5): 362–77. http://dx.doi.org.ezproxy.gc.cuny.edu/10.1108/LM-09-2014-0116.

Žumer, Maja, ed. 2009. *National Bibliographies in the Digital Age: Guidelines and New Directions*. IFLA Series on Bibliographic Control 39. Munich: De Gruyter Saur.

CHAPTER THIRTY-TWO

Labor, Alienation, and the Digital Humanities

SHAWNA ROSS AND ANDREW PILSCH (TEXAS A&M UNIVERSITY)

In a piece dating from 1685, the philosopher Gottfried Leibniz describes a hand-cranked calculating machine that could overcome the limitations of scientific research (especially astronomy) imposed by "old geometric and astronomic tables." The existing tables lacked insight into the multitudinous "kinds of curves and figures" that can be measured through the calculus that he (simultaneously with but independently from Isaac Newton) discovered a decade before. Calculus, Leibniz exalts, opens up the measurement of a variety of new geometric figures, "not only for lines and polygons but also for ellipses, parabolas, and other figures of major importance, whether described by motion or by points," resulting in a panoply of formulas and tables useful for describing and predicting an unprecedented range of phenomena, including solar mechanics and the cycles of comets (Leibniz 1929, 180).

The calculation of such tables is extremely laborious, though comprising a voluminous number of arithmetic operations that, considered singly, are fairly simple. For Leibniz, this drudgery is necessary: having such values at hand will greatly accelerate scientific inquiry, but the hours of calculation distract from more creative, ideational work. Somewhat ironically, the Enlightenment polymath derided by Voltaire for his optimism elects not to seek a path out of this laborious double bind. Instead, he concludes "it is unworthy of excellent men to lose hours like slaves in the labor of calculation which could safely be relegated to anyone else if machines were used" (Leibniz 1929, 181). On the one hand, "excellent men" may avoid laboring "like slaves" at the mundane work of simple calculation, while "anyone else" could be made to work these calculations. That Leibniz later affirms the immorality of chattel slavery in his "Meditation on the Common Notion of Justice" makes it all the more disquieting that he casually consigns allegedly less excellent men to the operation of machines (see Jorati 2019). Here, the drudgery of mathematical tabulation is divided from the creative work of devising proofs or novel applications of theory. The kind of computing machines described by Leibniz do not reduce labor; instead, they make it possible to *redistribute* labor according to existing hierarchies of power.

Leibniz is amongst the first to recognize that computing machines enable a division of labor, but he is far from the last. Computers—which referred first to rooms full of relatively unskilled workers (often women or subalterns), then to the machines meant to replace the labor of said workers—divide inquiry into repetitive tasks and those deemed exciting by the mostly male, highly educated people who are often highlighted historically as the true instigators of scientific discovery. Understanding this division, we wish to consider digital humanities as, first and foremost, a labor formation and to evaluate the ways in which this labor formation reproduces the kind of labor practices that stem from this original hierarchical divide. We first trace a history of labor practices in computation that create hierarchies between skilled and unskilled laborers. We then consider these labor practices in the context of discussions of alienated labor in the humanities, especially the trend toward adjunctification, the collapse of the tenure-track job market, and the channeling of digital humanities PhDs into service-heavy "alt-ac" positions.

Focusing on the ways that digital humanities projects perpetuate the tech industry's alienating practices, we show how scholarly work continues to recapitulate the double bind originating from Leibniz. We offer two strands of suggestions for creating a better possible future, one that both 1) strategically decouples work from personal joy in the digital humanities, and 2) uses an altered model of DH that can intervene in the conditions of academic labor. This altered model privileges small-scale, short-range, and narrow-focused projects that are deeply embedded in humanistic and rhetorical practices from the past and present. Only by combining both of these strategies can we carefully cultivate both accountability and creativity in the digital humanities.

THE HISTORY OF COMPUTATION AND THE LABOR THEORY OF VALUE

In the spring of 1790, less than a year after the storming of the Bastille, the revolutionary government of France proposed to standardize measurements in the new republic. This ambitious project involved the production of the kind of chart whose absence so upset Leibniz. Chosen for the project was Gaspard Clair François Marie Riche de Prony, "a civil engineer with the country's elite Corps des Ponts et Chaussées" (Grier 2007, 33). Inspired by Adam Smith's account of the division of labor in *The Wealth of Nations*, de Prony devised a three-tier system that, he boasted, "could manufacture logarithms as easily as one manufactures pins" (qtd in Grier 2007, 36). The project was overseen by a "handful of accomplished scientists [who] had made interesting contributions to mathematical theory" who were not involved in the actual work of calculation; they instead "researched the appropriate formulas for the calculations and identified potential problems" (Grier 2007, 37). This creative work was strictly sequestered from the laborious process of calculating the detailed logarithms the revolutionary government of France hoped to offer as a testament to the rationality of the new republic.

For this purpose, De Prony created "two distinct classes of workers," both operating under the purview of the "accomplished scientists." These two groups "dealt only with the problems of practical mathematics." The first group, composed of semi-skilled mathematicians, "took the basic equations for trigonometric functions and reduced them to the fundamental operations of addition and subtraction." These reduced problems were passed to the second group, ninety workers who "were not trained in mathematics and had no special interest in science." This final group was largely composed of "former servants or wig dressers, who had lost their jobs" when their former employers, France's titled aristocracy, lost their heads during the Reign of Terror (Grier 2007, 36). De Prony's project was ambitious; his charts were detailed. After four years of computation, the astronomer Jérôme Lalande visited and marveled that the workers were "producing seven hundred results each day" (qtd in Grier 2007, 37). Even at this heady pace, the project took six years to complete, resulting in a nineteen-volume manuscript of logarithm tables.

De Prony organized his logarithm factory according to principles described by Adam Smith. Far from an aberration, the history of computing shows that the divide between creative and rote work is a perpetual feature, one that has been absorbed into many aspects of DH labor, as well. However, several more examples are provided to establish the link between revolutionary France and the present.

By the 1830s another prominent figure in the history of computing had taken notice of the labor practices that had supported de Prony's failed project (Grier 2007, 38). In *On the Economy of Machinery and Manufactures* (1832), Charles Babbage, the mathematician whose experiments with the difference engine and plans for the analytic engine are now viewed as precursors of the programmable computer, identified de Prony's model specifically as an extension of Adam Smith's theories of labor from manufacturing to the scientific professions. Babbage argues that "the division of labor can be applied with equal success to mental as to mechanical operations, and that it ensures in both the same economy of time" (1846, 191). He calls this the division of mental labor, thus recognizing that cognition has an emerging labor value (which had formerly been associated with physical exertion). While de Prony used Smith's concepts to structure his calculating office, Babbage uses the French calculating office to show that mental work can be studied, modeled, and controlled using the processes applied to physical labor (191–202).

The detailed analysis of the division of labor in practice and in theory, both mechanically and mentally, marks Babbage as an early innovator in operations management and the study of organizations (Lewis 2007). Joseph Shieber (2013) also argues that Babbage's account of mental labor is the first instance of socially distributed cognition—a cognitive-science theory in which thought is shared between groups of technical co-workers. Shieber also argues that Babbage's internalization of de Prony's tables and his realization of the complexity of the undertaking were what gave rise to the idea of programmable computation in the first place. In short, by formalizing de Prony's scheme as the way to structure technical endeavors, Babbage foregrounds and justifies computation as a labor practice: organizing, managing, and hierarchizing thought in the same ways bodies are organized, managed, and hierarchized in the factory.

Dividing labor into rote tasks (the basic addition and subtraction performed by de Prony's hairdressers) and creative tasks (the work of the select few scientists who manage the proceedings) allows for more sophisticated technical projects to be undertaken. As we have seen, it also creates hierarchies within knowledge work. As Babbage observes,

> [W]e avoid employing any part of the time of a man who can get eight or ten shillings a day by his skill in tempering needles, in turning a wheel, which can be done for sixpence a day; and we equally avoid the loss arising from the employment of an accomplished mathematician in performing the lowest processes of arithmetic.
>
> (1846, 201)

Thus, mental labor, like physical labor, becomes structured by a class hierarchy in which menial mental laborers are directed by those engaged in creative work.

In *Programmed Visions* (2011), Wendy Hui Kyong Chun documents how this division continued during the early days of the digital computer, where "[u]sers of the ENIAC usually were divided into pairs: one who knew the problem and one who knew the machine" (31–2). This division was often gendered. A male domain expert would be paired with a female programmer, whose job was often to crawl physically inside the massive machine and program it by reconnecting electronic components. The programmer had "first to break down the problem logically into a series of small yes/no decisions," performing both the work of the second and third class of workers in de Prony's original hierarchy (32). As this work became commonplace (but no less physically and mentally demanding), ENIAC operators, including Admiral Grace Hopper, began to imagine structured programming languages, which abstract the physical rewiring work into a series of repeatable, grammatical statements that could be keyed into the computer at distance. Chun argues this abstraction "both empowers the programmer and insists on his/her ignorance" by hiding the operation of the computer (37). However, it also grants "new creative abilities" (37). Chun's account of abstraction, coupled with the general shift to programming as an autonomous disciplinary pursuit no longer in service of other fields (physics, astronomy, ballistics, etc.), hybridizes this division between creativity and rote labor in the figure of the programmer: a laborer simultaneously engaged in the creative work of design and the drudgery of implementation. The modern programmer combines aspects of all three laboring classes in de Prony's logarithm computing office, which confounds—but does not erase—the hierarchization and exploitation endemic to computing as a form of labor.

This hybridization of job positions in computation-based industries has not freed workers from exploitation because, first, some repetitive tasks are simply outsourced to low-paid workers in India, Ukraine, and China, and second, the creative work allotted to more privileged workers is expected to compensate for the less humane expectations their employers have of them. Tracy Kidder's *The Soul of a New Machine* (1981) offers one of the earliest depictions of "crunch time," the practice of working extremely long hours as a product's impending launch approaches in order to finish on time. Documenting the creation of the Data General Eclipse MX/8000 computer during the late 1970s, Kidder describes the intense work pressures there as part of a general condition in the field: "short product cycles lend to many projects an atmosphere of crisis," an atmosphere that leads to "a long-term tiredness" affecting computer engineers too old to endure the physical and emotional strain of their work, usually occurring around the age of 30 (2000, 104). Even while discussing grueling work conditions, the engineers Kidder profiles reiterate their love of the job: the fun of solving technical problems, the rush of control when a machine works as intended, the sense of fascination when first taking apart a telephone as a child. Many work eighty hours a week. For instance, Josh Rosen, a burned-out engineer, slowly realized he had no life outside work because "he had spent nearly half of all the hours of all the days and nights of the last three years at work" (218). Rosen described that when he lost his fascination with computers, the work "was, all of a sudden, a job"—and not a healthy or agreeable one (219).

The people who work in this fashion often do not protest these long, uncompensated hours because they experience great joy in the minutiae of the computer's design. This is largely because, once computing emerged as a self-contained field, no longer tied to other disciplines, the distinction between rote and creative work blurred. Kidder describes seemingly mind-numbing work—a stack of notebooks full of handwritten CPU microinstructions, is one memorable example—as the height of interest and daring technical innovation (2000, 100–1). As computers become objects of technical fascination instead of means to other ends, the detail-oriented tasks that historically would have been allotted to human computers become the high-level design tasks normally reserved for Leibniz's "excellent men."

Long work hours and constant states of crisis have only intensified since the late 1970s. A pseudonymous blog post from 2004, titled "EA: The Human Story," revealed that, in large gaming companies such as Electronic Arts, the crunch conditions identified by Kidder have become standard practice. Work times increased from eight-hour days for six days a week, through twelve-hour days for six days a week, culminating in thirteen-hour days for seven days a week—all in service of projects that are "on schedule." Like Kidder, and like much of the subsequent coverage of crunch in the gaming press, the blog post assures readers that game developers tolerate crunch because of love for the job's creative aspects (ea_spouse 2004).

Crunch is an example of what Tiziana Terranova calls "free labor." "[S]imultaneously voluntarily given and unwaged, enjoyed and exploited," Terranova explains, "free labor on the net includes the activity of building websites, modifying software packages, reading and participating in mailing lists, and building virtual spaces." The Internet, she argues, is a "thriving place" because of freely given extra labor (2012, 68). Terranova argues that "the expansion of the Internet has given ideological and material support to contemporary trends toward increased flexibility of the workforce, continuous reskilling, freelance work, and the diffusion of practices such as 'supplementing' (bringing supplementary work home from the conventional office)" (69).

In the 1980s, Donna Haraway suggested that such free labor is a byproduct of cybernetics and is explicitly feminized:

> Work is being redefined as both literally female and feminized, whether performed by men or women. To be feminized means to be made extremely vulnerable; able to be disassembled, reassembled, exploited as a reserve labor force; seen less as workers than as servers; subjected to time arrangements on and off the paid job that make a mockery of a limited work day.
>
> (1990, 166)

Susan Leigh Star and Anselm Strauss respond to the feminized nature of digital labor by coining "invisible labor" as a term for the labor not traditionally counted as work. They first consider the attempt to gain pay for housework by the Women's Movement in the 1970s, then consider tasks like cleaning up after meetings, summarizing discussions, and otherwise working to get work back on track as examples of uncompensated work typically done out of love or duty by women (Star and Strauss 1999). Such accounting is crucial because, as work becomes increasingly mediated by digital technology, invisible labor such as "tuning, adjusting, and monitoring use and users grows in complexity" (Star and Strauss 1999, 10).

Invisible, free labor proliferates in digital spaces. Computing as a labor formation, then, operates as follows: divide work into rote and creative tasks, devalue the rote work, then create conditions by which that rote work is a labor of love or joy, done for free, as part of uncompensated crunch time. Feminist analyses of labor have linked this division and devaluation to a process of feminization that has specific ties to the undervaluing and undercompensating of work done by women. With these feminist approaches in mind, we turn to an analysis of these forces—division, devaluation, feminization—in the context of academic labor in the humanities.

FEMINIZED CRUNCH TIME IN DH

Division, devaluation, and feminization have characterized the digital humanities since its fabled inception with Roberto Busa's *Index Thomisticus*. Melissa Terras and Julianne Nyhan (2016) explain that this ambitious concordance of Thomas Aquinas's oeuvre relied on female punch card operators whose "contribution to the early days of humanities computing has been overlooked," before arguing that this gendered division of labor was also common in early humanities computing projects. Sharon M. Leon (2018) affirms that historical accounts of the digital humanities downplay women's contributions, while Barbara Bordaejo (2018) debunks "the general perception that digital humanities, as a discipline, is balanced in terms of gender" and observes that any recent gains are "most likely to apply to white women from affluent countries in the Northern Hemisphere."

One characteristic power imbalance in DH stems from its reliance on para-academics, who lack the security of a tenure-track job and struggle with financial and time constraints that stymie their own research. For instance, while Julia Flanders (2012) credits the rise of DH for making her labors visible as an intellectual endeavor (rather than a service obligation), she also recognizes the bulk of the everyday work done in DH is executed by "para-academic" employees who are treated as hourly wage-earners, not as permanent colleagues. Discussing how para-academics suffer from a "displacement of autonomy concerning what to work on when and how long to take," Flanders reflects,

> Our expectations of what work should be like are strongly colored by the cultural value and professional allure of research, and we expect to be valued for our individual contributions and expertise, not for our ability to contribute a seamless module to a work product. Our paradigm for professional output is authorship, even if actual authoring is something we rarely have enough time to accomplish. One would expect the result of this mismatch of training and job paradigm to be disappointment, and in some cases it is.
>
> (Flanders 2012)

This disappointment is frankly discussed by three other para-academics, Lisa Brundage, Karen Gregory, and Emily Sherwood (2018): "we often experience these labor conditions personally and viscerally, particularly if we still feel the tug of our scholarly training and our home disciplines," they note, adding that these "feelings of guilt and pressure" are not "personal" but instead "are illustrative of the bind" in which para-academics labor.

William Pannapacker (2013), links this double bind to the instrumentalization of love. Using language that recalls de Prony's labor hierarchy, Pannapacker recognizes that the rhetoric of love draws graduate students into degree programs that cannot promise job security:

> the rhetoric of "love" has an ambiguous meaning when it's applied to graduate school. It can be impossibly idealistic, and deeply rooted in powerful experiences that override economic self-interest. It also can be deeply cynical, a means of devaluing the work of some for the benefit of others. The transformation of higher education into a system of contingent labor—a reflection of larger transformations of the workplace—depends on "love" in both senses.
>
> (Pannapacker 2013)

Responding to Pannapacker's post, then-graduate student Jacqui Shine (2014) adds that the economic sacrifices of working-class students are acute and that for such students, affirming one's love for academia feels "like an entrance requirement," a performance demanded as a form of "gatekeeping." After graduation, the bind remains. Lee Skallerup Bessette (2019), describing her experiences in an institutional context of increasing adjunctification, documents how she redirected her love for knowledge work away from efforts at professionalization and towards acts of empathy toward her students.

This redirection of love—whereby the joy of self-directed academic research transforms into empathy for one's students and co-workers, both inside and outside the tenure track—is the best solution that the digital humanities currently offers for our own de Pronyian bind. It powerfully informs the most recent iteration of *Debates in the Digital Humanities*, the series of essay collections edited by Matthew K. Gold and Lauren Klein. Each successive volume pays more attention to labor, with the 2019 installment containing an entire forum on the topic, "Ethics, Theories, and Practices of Care." This gathering of essays shifts the object of scholarly love from work itself to one's fellow workers. It is certainly laudable to privilege people over projects, and it bears repeating that we must respect, protect, and support our fellow workers and students. Yet proposing care as a solution will increase labor—especially those of female, queer, and non-white scholars—unless (at least) two conditions are met. First, these additional tasks need to be neutralized by reducing expectations for other tasks and by recognizing care work formally as visible, billable labor. Second, we must overthrow the labor practices that make heroic acts of care seem necessary.

This requires us acknowledging that "crunch time" exists in our field, that we *ourselves* reproduce the conditions of labor like those in the software and gaming industries. What will not work is to outsource tasks to workers who do not benefit intellectually from their participation, to repeat never-fulfilled utopian promises about extensibility or interoperability, to invoke "collaboration" without formalizing fair distributions of labor and remuneration, or to seek productivity tools and "hacks" instead of reducing real workloads. These examples illustrate how DH tends to combine the worst aspects of labor in computing and in the humanities: we perpetuate both the rote/creative division of early computational labor and the rhetoric of love present in academia, combining these two labor structures to bind workers asked to do the less interesting tasks on behalf of Leibnizian "men of brilliance." Consider, for example, DH projects that follow a "PI model," in which an academic with an idea for a project rooted in their *own* research agenda serves as a "principal investigator" who assigns rote, repetitive tasks to less visible workers. Their contributions are necessary because the projects (particularly grant-based ones) are designed at a scale that would be impossible for a single individual to perform (particularly those who are balancing research, teaching, and service responsibilities).

Although decisions made by individual managers can be made with the best of intentions to secure equitable working conditions and adequate compensation, this cannot compensate for the fact that academic researchers in all positions have been trained to expect a certain degree of autonomy and personal investment in their work. Even heroic acts of care cannot compensate for the non-distributed nature of passion in DH projects. Unless we agree to concede that academia can no longer support researchers following wherever their curiosity leads, DH will be morally compromised by its foundational paradox: its application of tools designed to maximize the scale of its calculations to investigate fundamentally non-scalable research questions and produce non-scalable research outputs. It is motivated by research agendas whose value to the scholarly community resides in their uniqueness and novelty, yet it is enabled by technologies whose power resides in the sheer repetition of calculations iterated both by computers and the workers operating them.

Humanities research resists scalability—or rather, it becomes easily scalable only with exploitation. In their essay "Against Cleaning," Katie Rawson and Trevor Muñoz (2019) consider the question of scale in DH. Building on Anna Tsing, they theorize that all digital projects include both scalable and non-scalable elements and conclude that some forms of data diversity need to be preserved. In the same volume, Andrew Gomez (2018) summarizes a decade of arguments that DH has reproduced narrow canons (usually populated by white figures, especially men) because "digital humanities thrives on abundance," on subjects whose textual records are abundant, repeatable, and homogeneous units. Like Rawson, Muñoz, and Gomez, we are concerned that the erasures necessary for large-scale projects threaten to erode progress made to diversify and decolonize the humanities. Here, we want to expand these theories of nonscalability to labor—to the *work* being done to the objects under inquiry, in addition to those objects themselves—to build fairer working conditions in DH.

UNSCALED DH: A FUTURE FOR FAIR ACADEMIC LABOR

An unscaled DH will refuse to reproduce exploitative labor practices from software industries, which make use of "crunch time" or perpetuate unfair divisions of labor, dividing the haves from the have-nots of mental labor. It will vigorously support and market projects that are not "big data" projects. When it uses "scalable" data, it will not exploit workers to do so. It will ameliorate or neutralize unfair practices already within academia, where underpaid or unacknowledged labor has always been a norm, from wives proofreading their husbands' manuscripts to graduate students creating bibliographies and workers in the Global South entering data in reference databases. Because DH inherits both of these strands of unfair labor practices, it will be complicated to upend the labor patterns here labeled exploitative. We therefore offer two overarching axioms. *All workers should be hybrid workers. Do not create crunch conditions.*

These axioms can serve as a broad mantra for DH project design, as well as suggest concrete guidelines for project management. To address the former first, make sure that all workers are assigned both creative and rote tasks. The simplest way to avoid crunch is to design projects to require sustainable levels of work in the first place, thereby obviating the risk of creating crunch conditions. We can lean into the fundamental nonscalability of humanities data by focusing on smaller projects. Right-size projects by tailoring scope to the fund of financial and human resources that you can likely devote to it. Such "self-sustaining" and "modest" efforts, to draw from Roh (2019), help diversify the currently narrow range of established, prominent DH projects by highlighting small, incomplete, or nontraditional archives.

Once a project is designed, crunch conditions must be avoided through careful project management. Do not exceed set work hours for workers. Do not practice supplementing, the practice of asking work to be taken home, discussed by Terranova (2012). Do not treat nights and weekends as if they belong to the project. These toxic practices emerge from a shared sense of passion, from a common desire to build and create. But they are also exploitative. Too often, heroic effort is normalized or even becomes compulsory. As David S. Roh (2019) has established, startup culture has infected DH, making it overemphasize new projects that initially enjoy a great deal of traction but rely too much on soft money—a pattern that necessitates us to develop "a critical awareness of, and resistance to, a startup logic demanding performativity." While many projects, especially early on, run primarily on the personal passion felt by their originators, this passion is not something that should, or even could, scale as projects grow. Rather than leverage passion and joy as an unconscious cover for exploitation, remember that others, especially the alt-ac and student labor marshaled into DH projects, do not—and need not—share the same passion for the project's creation.

As projects grow, the commitment to banish crunch will need to be renewed at every level. Refuse to exploit faculty, students, and staff in IT, in DH centers, and the libraries. Rather than assume that everyone knows how to operate computers, seek opportunities for ourselves and our collaborators to be trained in the rudiments of computer usage—including file management, intellectual property, information security, and the history of computing—rather than invest solely in learning a single platform or workflow. Being more inclusive in your training will make your initiative more useful for the people you draw into its sphere. If you manage grant funds, as Cole et al. (2019) remind us, disburse them in an accountable, equitable, feminist manner. If you work with graduate students, devote time to mentoring, including conveying a clear understanding of the job market and helping them explore a variety of career trajectories. Above all, do not allow your project to encroach on their research time or impede their progress to degree.

We must do more than renovate our management practices: we must consciously design our software, projects, classes, and workshops to respond to what we really need, not what our available equipment and previous training nudge us toward. Machine-first decision-making is the result of privileging outputs over experiences. We must give more than just lip service to the oft-repeated truism that DH values process over product. We must privilege objectives over objects. What are you trying to learn and teach *along the way to* your deliverable? This question should be used to decide what projects need to exist and which features are non-negotiable. What will always matter for the humanities are our foundational methods—both those that unite the humanities and those that are discipline-specific—suggesting that we should privilege projects that advance current methodologies or create new ones. For example, within our home fields of rhetoric and literature, projects could advance the foundational methods of bibliography and book history, intervene in existing debates in particular subfields (Ross and O'Sullivan 2016; Ross 2018), or create links to external fields aligned with textuality, such as new media studies.

All of these strategies can ensure that your project truly benefits your collaborators and your audience. Perhaps many of the projects in the "graveyard of digital empires" (Graham 2019) were buried because they did not find an audience. Even allowing that any DH project must survive a harsh environment characterized by dwindling grant money, high platform turnover, and overburdened laborers, there must be a reason why some projects survive while others do not. And the first question we, like any good detective, should ask is *Cui bono*? ("Who benefits?") Work that helps other people work or reaches a public audience will be favored in this harsh environment. Boutique projects will only survive if you have the ability to put in the time yourself. This devotion will be rewarded if you reserve some effort to the task of gaining recognition for it, whether by having it reviewed, applying for awards, or developing spinoff projects. In addition, contributing a new digital object when it is a byproduct of your study—a web companion for a book or a digital exhibition of materials from an uncatalogued box at an archive consulted for an article—will naturally emphasize process over product by embedding outcomes within a larger mission of learning and educating.

Whereas these recommendations focus on how to encourage fair labor practices for niche initiatives based on your particular research specialties, at the other end of the spectrum are projects rooted in public service. Asking scholars to foreground such projects, Alan Liu (2012) writes,

> At a minimum, digital humanists—perhaps in alliance with social scientists who study Internet social activism—might facilitate innovative uses of new media for such traditional forms of advocacy as essays, editorials, petitions, letter-writing campaigns, and so on. But really, digital humanists should create technologies that fundamentally reimagine humanities advocacy.

Thinking about labor in computing and DH allows us to come at Liu's suggestions from a different direction. Not only has DH been, historically, a call to reimagine humanistic scholarly practice, but it can also serve as a means of rearticulating computing practices through the insights earned by humanities fields: namely, issues of social justice, equity, criticality, and mutuality. If we are to "fundamentally reimagine humanities advocacy" as digital humanists, we cannot found it on the back of labor practices we would decry if we were to encounter them as objects of study.

REFERENCES

Babbage, Charles. 1846. *On the Economy of Machinery and Manufactures*. London: J. Murray. http://archive.org/details/oneconomyofmachi00babbrich.

Bessette, Lee Skallerup. 2019. "Contingency, Staff, Anxious Pedagogy—and Love." *Pedagogy* 19 (3): 525–9.

Bordalejo, Barbara. 2018. "Minority Report: The Myth of Equality in the Digital Humanities." In *Bodies of Information*, edited by Jacqueline Wernimont and Elizabeth Losh. Minneapolis, MN: University of Minnesota Press. https://dhdebates.gc.cuny.edu/read/untitled-4e08b137-aec5-49a4-83c0-38258425f145/section/46cde2dd-e524-4cbf-8c70-e76e2c98f015#ch18.

Brundage, Lisa, Karen Gregory, and Emily Sherwood. 2018. "Working Nine to Five: What a Way to Make an Academic Living?" In *Bodies of Information*, edited by Jacqueline Wernimont and Elizabeth Losh. Minneapolis, MN: University of Minnesota Press. https://dhdebates.gc.cuny.edu/read/untitled-4e08b137-aec5-49a4-83c0-38258425f145/section/db658807-141d-4911-b2d6-8e5682c8d08f#ch17.

Chun, Wendy Hui Kyong. 2011. *Programmed Visions: Software and Memory*. Cambridge, MA: MIT Press.

Cole, Danielle et al. 2019. "Accounting and Accountability: Feminist Grant Administration and Coalitional Fair Finance." In *Debates in the Digital Humanities 2019*, edited by Matthew K. Gold and Lauren Klein. Minneapolis, MN: University of Minnesota Press. https://dhdebates.gc.cuny.edu/read/untitled-4e08b137-aec5-49a4-83c0-38258425f145/section/91e872e4-947a-4bfa-bc12-c82d3521b8cd#ch04.

ea_spouse [pseud.]. 2004. "EA: The Human Cost." *Ea_spouse* (blog). November 10. https://ea-spouse.livejournal.com/274.html.

Flanders, Julia. 2012. "Time, Labor, and 'Alternate Careers' in Digital Humanities Knowledge Work." https://dhdebates.gc.cuny.edu/read/untitled-88c11800-9446-469b-a3be-3fdb36bfbd1e/section/769e1bc9-25c1-49c0-89ed-8580290b7695.

Gold, Matthew K. and Lauren Klein, eds. 2016. *Debates in the Digital Humanities 2016*. Minneapolis, MN: University of Minnesota.

Gomez, Andrew. 2018. "The Making of the Digital Working Class: Social History, Digital Humanities, and Its Sources." In *Debates in the Digital Humanities 2019*, edited by Matthew K. Gold and Lauren Klein. Minneapolis, MN: University of Minnesota Press. https://dhdebates.gc.cuny.edu/read/untitled-f2acf72c-a469-49d8-be35-67f9ac1e3a60/section/3788efb8-3471-4c45-9581-55b8a541364b#ch33.

Graham, Elyse. 2019. "Joyce and the Graveyard of Digital Empires." In *Debates in the Digital Humanities 2019*, edited by Matthew K. Gold and Lauren Klein. Minneapolis, MN: University of Minnesota Press. https://dhdebates.gc.cuny.edu/read/untitled-f2acf72c-a469-49d8-be35-67f9ac1e3a60/section/38d72a93-4bad-48c2-b709-59145658dc98.

Grier, David Alan. 2007. *When Computers Were Human*. Princeton, NJ: Princeton University Press.
Haraway, Donna J. 1990. "A Manifesto for Cyborgs." In *Simians, Cyborgs, and Women: The Reinvention of Nature*, 1st edn, 149–81. New York, NY: Routledge.
Jorati, Julia. "Leibniz on Slavery and the Ownership of Human Beings." *Journal of Modern Philosophy* 1 (1) (December 4): 10. https://doi.org/10.32881/jomp.45.
Kidder, Tracy. 2000. *The Soul of a New Machine*. Boston, MA: Back Bay Books.
Leibniz, Gottfried. 1929. "On His Calculating Machine." In *A Source Book in Mathematics*, edited by David Eugene Smith, translated by Mark Kormes, 173–81. New York, NY: McGraw Hill.
Leon, Sharon M. 2018. "Complicating a 'Great Man' Narrative of Digital History in the United States." In *Bodies of Information*, edited by Jacqueline Wernimont and Elizabeth Losh. Minneapolis, MN: University of Minnesota Press. https://dhdebates.gc.cuny.edu/read/untitled-4e08b137-aec5-49a4-83c0-38258425f145/section/53838061-eb08-4f46-ace0-e6b15e4bf5bf#ch19.
Lewis, Michael A. 2007. "Charles Babbage: Reclaiming an Operations Management Pioneer." *Journal of Operations Management* 25 (2): 248–59. https://doi.org/10.1016/j.jom.2006.08.001.
Liu, Alan. 2012. "Where Is Cultural Criticism in the Digital Humanities?" In *Debates in the Digital Humanities*, edited by Matthew K. Gold. Minneapolis, MN: University of Minnesota Press. https://dhdebates.gc.cuny.edu/read/untitled-88c11800-9446-469b-a3be-3fdb36bfbd1e/section/896742e7-5218-42c5-89b0-0c3c75682a2f.
Pannapacker, William. 2013. "On Graduate School and 'Love.'" *The Chronicle of Higher Education*. January 1. https://www.chronicle.com/article/on-graduate-school-and-love/.
Rawson, Katie and Trevor Muñoz. 2019. "Against Cleaning." In *Debates in the Digital Humanities 2019*, edited by Matthew K. Gold and Lauren Klein. Minneapolis, MN: University of Minnesota Press. https://dhdebates.gc.cuny.edu/read/untitled-f2acf72c-a469-49d8-be35-67f9ac1e3a60/section/07154de9-4903-428e-9c61-7a92a6f22e51#ch23.
Roh, David S. 2019. "The DH Bubble: Startup Logic, Sustainability, and Performativity." In *Debates in the Digital Humanities 2019*, edited by Matthew K. Gold and Lauren Klein. Minneapolis, MN: University of Minnesota Press. https://dhdebates.gc.cuny.edu/read/untitled-f2acf72c-a469-49d8-be35-67f9ac1e3a60/section/b5b9516b-e736-4d23-bf5c-7426e7a9de2d.
Ross, Shawna. 2018. "From Practice to Theory: A Forum on the Future of Modernist Digital Humanities." *Modernism/modernity* Print+ 3 (2) (Summer). https://modernismmodernity.org/forums/practice-theory-forum.
Ross, Shawna and James O'Sullivan. 2016. *Reading Modernism with Machines: Digital Humanities and Modernism Literature*. Basingstoke: Palgrave.
Shieber, Joseph. 2013. "Toward a Truly Social Epistemology: Babbage, the Division of Mental Labor, and the Possibility of Socially Distributed Warrant." *Philosophy and Phenomenological Research* 86 (2): 266–94. https://doi.org/10.1111/j.1933-1592.2011.00524.x.
Shine, Jacqui. 2014. "Love and Other Secondhand Emotions." *The Chronicle of Higher Education Blogs*, February 3. https://web.archive.org/web/20140301001701/https://chroniclevitae.com/news/309-love-and-other-secondhand-emotions.
Star, Susan Leigh and Anselm Strauss. 1999. "Layers of Silence, Arenas of Voice: The Ecology of Visible and Invisible Work." *Computer Supported Cooperative Work (CSCW)* 8 (1): 9–30. https://doi.org/10.1023/A:1008651105359.
Terranova, Tiziana. 2012. "Free Labor." In *Digital Labor: The Internet As Playground and Factory*, edited by Trebor Scholz, 67–114. London: Routledge. http://ebookcentral.proquest.com/lib/tamucs/detail.action?docID=1047015.
Terras, Melissa and Julianne Nyhan. 2016. "Father Busa's Female Punch Card Operatives." In *Debates in the Digital Humanities 2016*, edited by Matthew K. Gold and Lauren Klein. Minneapolis, MN: University of Minnesota Press. https://dhdebates.gc.cuny.edu/read/untitled/section/1e57217b-f262-4f25-806b-4fcf1548beb5#ch06.

CHAPTER THIRTY-THREE

Digital Humanities at Work in the World

SARAH RUTH JACOBS (CITY UNIVERSITY OF NEW YORK)

The digital humanities is commonly understood as a field of interrelated academic disciplines, albeit a field that pushes beyond the boundaries of traditional scholarly inquiry and into realms of activism, public engagement, and maker culture (Nyhan, Terras, and Vanhoutte 2016, 2). However, in recent years, scholars have increasingly suggested including work done outside of an academic environment under the digital humanities umbrella. In their introduction to *Debates in the Digital Humanities 2019*, Matthew K. Gold and Lauren Klein discuss the value of including "work outside of digital humanities and outside of the academy" within digital humanities (2019, xii). The examples that Gold and Klein give for this "expanded field" are of an African American History project at the University of Maryland and two professional conferences (2019, xii). Yet how are we to categorize for-profit, activist, partisan, corporate, and independent work that uses digital technology to speak to or represent humanistic interests? If the definition of the digital humanities were to be extended beyond the realm of scholarly work to include all humanistic thinking or work that utilizes digital spaces or tools, as this chapter will endeavor to do, then digital humanistic work would be acknowledged as being inescapable, vibrant, and thriving. Why can't teenage climate activist Greta Thunberg, for example, be considered a digital humanist? Her social media feed employs statistics, images, and appeals in different languages as a means to enact global unity and mobilize a planet against global warming. Yet she has no advanced degree, and in order to pursue her activist work, she took a year off from formal schooling (Associated Press 2020). Many news organizations and commentators earn millions of dollars in revenue by engaging in partisan social commentary that is transmitted through social media, satellite radio, digital television, and even memes. Every day this commentary helps people frame their worldviews, define their identities, and locate the sources of their problems. Why not consider these news producers digital humanists? In fact, the hesitancy of academic disciplines, or at least academic job and tenure guidelines, to consider the kind of work that is done by untrained, activist, for-profit, government, and corporate entities as existing within the same universe as academic work in the humanities has weakened the potential position and relevance of scholarship.

Furthermore, what if anti-humanistic work in the digital realm was also considered as existing within a spectrum of digital humanism—in this case, digital anti-humanism? The Chinese government's blocking of certain websites and its planned social credit system (Kobie 2019) are clearly using the digital realm to restrict, surveil, and arguably to dehumanize people. Russia's information warfare tactics are using disingenuous comments and posts in order to divide people and warp minds (Shane 2018). In the corporate realm, a Facebook algorithm "promotes Holocaust denial" (Townsend 2020). If these actions are not recognized as a form of digital humanism, then the tactics needed to fight and resist them will likewise be seen as out of the purview of digital humanities practitioners. This chapter will discuss the potentialities that could be enacted through such a radically expanded definition of digital humanities, and, by extension, an expanded understanding of what constitutes scholarly work. Digital humanities, understood broadly, is ubiquitously at work in the world, for better and for worse. Scholars need to meet it where it is and not just observe it but have the courage to alter it.

The reason that scholars feel pressed to discuss the role of the digital humanities in the world is that scholarship, not to mention overworked faculty, still often struggles to bridge the gap between theory and real-world application. In 2019 scholars were still wondering "how ... digital humanists [can] ally themselves with the activists, organizers, and others who are working to empower those most threatened by" the current, politically "charged environment" (Gold and Klein 2019, ix). Prescott (2016) eloquently states that the very form of institutionalized learning gets in the way of worldly action: "... universities are generally conservative bureaucratic environments ... If flexibility and openness are preconditions for success in the digital world, can this be achieved by digital humanities units in a university environment?" (463).

To answer Prescott's question, there are many examples of digital humanists and scholarly projects that do engage in activism and affect non-academic publics. As a prime example, Dr. Roopika Risam at Salem State University worked as part of a team of scholars to map the locations of immigrant children who had been separated from their families at the US and Mexican border (McKenzie 2018). Such group work, which is often required by the scope of digital humanities projects, may not get the recognition it deserves in the hiring or tenure processes because of the privileging of the order of authors. Here and there, scholarly collectives, centers, and initiatives are likewise showing promise in terms of worldly engagement. Columbia University's XPMethod is a group "for experimental methods in humanities research" that has "led efforts to map areas of Puerto Rico affected by Hurricane Maria, and to teach coding to inmates at Rikers Island correctional facility" (McKenzie 2018). There are also professors who partner with non-academic institutions, such as Dr. Jacqueline Wernimont at Dartmouth College who previously directed the Nexus Lab, a cooperative group of scholars, activists, and artists at Arizona State University (Fenton 2017). Dr. Wernimont worked with the Global Security Initiative, which is a partnership between Arizona State University and US defense and diplomatic stakeholders, as well as the Center for Solutions to Online Violence (Fenton 2017). The excitement and relevance of digital humanities work has certainly propelled many major colleges and universities to establish centers and labs for DH work. Yet much scholarly and institutional revisioning remains to encourage and enable DH work that might strike out beyond traditional scholarly frameworks or models.

Digital humanists have already succeeded in pushing the boundaries of academia. The move toward open access academic journals and the embrace of free open-source software by many prominent digital humanists have demonstrated formidable resistance to the walled garden model espoused by traditional scholarly work. The problem is not so much in the approach of digital humanists, but the limitations imposed by the job descriptions and tenure requirements set forth by colleges. The need for academia to embrace broader definitions of scholarship has for at least ten years been pointed out by digital humanists (Ramsay and Rockwell 2012), and to be fair some academic job listings in the United States have recently prioritized public scholarship. Yet the monolithic and intractable nature of endowed, accredited, and administratively bloated (Ginsberg 2011) institutions in higher education continues to reject broad and significant changes in job responsibilities and tenure requirements for faculty.

Rather than being encouraged to simply work in allegiance with outside entities, academic digital humanists must be afforded the spaces, course releases, and altered tenure guidelines needed for them to *be or become* activists, organizers, founders, and even for-profit and nonprofit business workers. As such, the mission of digital humanists in many cases will need to change and expand, to shift from walled-in or even open access scholarship to non-scholarly work done for more general and non-academic citizens. Paradoxically, in order for many humanistic departments to thrive, the disciplinary and departmental models must to an extent give way to career-based, skill-based, or alternative degree paths that may draw from courses in multiple departments. Likewise, digital humanities centers can help parlay members' skills and research into ethical partnerships with for-profit and nonprofit companies. Students, particularly graduate students, can be encouraged and given the opportunity to engage in entrepreneurial pursuits. University presses can expand into popular mediums, from the radio to websites and television, aiming for new audiences. Ultimately, if imperiled humanities departments cannot redefine their mission, change their output, and ethically co-create alongside for-profit and nonprofit organizations, then they will likely instead be shaped, cut, and distorted by the forces of corporatism, austerity, and political power.

THE EXAGGERATED DECLINE OF THE HUMANITIES

There seems to be an unending panic concerning how the humanities are in crisis and decline (Reitter and Wellmon 2021). More alarming than the sharp decline in US English majors (National Center for Education Statistics 2019) or the more modest decline in humanities study in OECD countries (with the exception of community colleges) (American Academy 2021) is the worldwide decline of democracies and the extreme rise in countries that are moving toward autocracy (Alizada et al. 2021). Again, in asking the aforementioned questions about how the digital humanities can expand its boundaries and engage with outside activists and institutions, there is an acknowledgment that humanistic education has failed in some respects, and that the academy must expand its mission and begin to properly engage with worldly problems as and where they arise.

Certainly, a large part of the interest in pursuing digital humanities in the world comes from a sense that humanities study has lost its relevance. Much has been written about the "decline and fall of the English major" (Klinkenborg 2013) and how "the humanities have a marketing problem" (Hayot 2021).

Yet not all humanities majors are in decline. In the US, where major-specific data is the most available, the number of communication and journalism majors increased by 11 percent between 2010 and 2018, which is less than the overall increase in degrees awarded, but still shows some growth (National Center for Education Statistics 2019). At 92,528 graduates in 2018–2019, communication and journalism majors greatly eclipsed English majors, who had 39,335 graduates that year (National Center for Education Statistics 2019). The number of multi- or interdisciplinary studies degrees increased by 25 percent between 2010 and 2018 (National Center for Educational Statistics 2019). Similarly, in OECD countries, communication majors have increased 10 percent between 2012 and 2018 while English majors have declined at least 25 percent over the same period (American Academy 2021). Over thirty digital humanities programs were founded between 2006 and 2016, though only about a third of these programs are for undergraduates (Sula, Hackney, and Cunningham 2017). Many humanities fields, especially some emerging fields, are growing and relevant to students.

The number of English majors has fallen because students do not see the field as being as "relevant" or "promising" as other majors in both the humanities and the sciences. Simply excoriating students because they don't "know how to write," praising the beauty of a "lifelong engagement with literature" (Klinkenborg 2013), or pointing out that English major salaries are "only" $15,000 less than STEM salaries upon graduation (Deming 2019) is not going to bring students back to the study of literature. Maybe it is time to stop defending the vitality and relevance of declining undergraduate departments, and even time to stop defending the humanities doctorate as being worthwhile in a practical, career-conscious way for anyone other than those who already have great privilege. Klinkenborg (2013) suggests that, apart from some "shifting" in the "canon," English literature is timeless and eternally relevant. Yet it is this kind of attitude in which the humanities are seen as gloriously impractical and siloed from contemporary culture that creates such a separation between academic study and real-world applications, scaring away potential majors. Other thinkers, at least, are willing to admit that in many cases "the curriculums are stale. The majors are stale. Neither of them represents the best of what the humanities can be, or foregrounds the power of humanist reason and humanist work" (Hayot 2021). As much as academics would like to pretend that humanities departments are bastions of contemporary thought and self-interrogating, the case is quite the opposite. English, and the insistence on "proper" English, has long been political, discriminatory, and authoritarian. Proper English used to be a way to maintain the status quo, and to a great extent, it still is. The fact that many college students no longer consider English "enough" to ensure their success in comparison to the less educated is likely something to be celebrated more than mourned. For the digital humanities or "threatened" humanities disciplines to be applicable outside of the academy, curricula must be forward-looking and considerate of social and workforce trends. Most faculty in higher education who are granted a modicum of academic freedom are in the unique position of being able to approach skills training and career-conscious education with a critical and ethical lens. If colleges with strong histories of academic freedom do not take on the work of both teaching career-based and vocational skills and having students question how such skills can be utilized in an ethical and socially just way, then the work will be taken on by less ethical, intellectually grounded, or scrupulous institutions, such as for-profit colleges and certificate programs. While this seems like an obvious observation, many humanities departments continue to dismiss emerging competencies and industry trends at their peril.

EXPANDING MISSIONS AND AUDIENCES

In order for digital humanities work to reach its full potential outside of the academy, the missions, forms, and intended audiences of DH need in many cases to be radically expanded. For example, with regards to the work of academic digital humanists, is politically motivated misinformation an interesting and unfortunate phenomenon that should be studied and discussed in college courses and scholarly journals, or is it the result of a crisis in literacy education that should be actively dealt with in the public sphere as well as in all levels of educational instruction? The answer to this question relies on the defined scope and mission of digital humanities work.

DISCIPLINARY BOUNDARIES AND FOCUSED TRACKS

In recent years many scholarly voices have come together to call for the blurring and eradication of disciplinary boundaries (Klinkenborg 2013). Allowing disciplinary crossover will magnify the influence of the digital humanities in potential real-world applications. For example, economists and financial workers would be able to make more humane and compassionate decisions if they had taken a digital humanities course on how to visualize, represent, and improve the labor and sustainability standards of select global trade networks. Whether or not such cross-pollination will help vulnerable humanities departments should not be the main consideration, though if computer science students were required to take a course in the ethics of digital surveillance, that could only strengthen the real-world impact of digital humanities.

Professors such as Klinkenborg (2013) express concern about how remaking disciplines with contemporary career paths in mind can discourage students from exploring and pursuing knowledge for personal enrichment. Yet, allowing students both the room to explore and then, later, to settle upon a practical area or areas of specialization can allow for both personal and professional fulfillment. Instead of the English major, students interested in the humanities might take a wider variety of courses (with the requirements loosened and expanded into other fields) and complete one or two relatively narrow "practical" areas of specialty, such as artificial intelligence computer programming, project management, or database development with MySQL. Each specialty could involve a sequence of three or four courses. Students would emerge from college with both a variety of self-directed areas of knowledge in the humanities (and other disciplines) as well as specific, marketable skills. Katina L. Rogers suggests similar competency-based courses in areas like "collaboration, project management, interpersonal skills, and technical skills" (2020, 93). Hayot (2021) argues in favor of abandoning disciplines for conceptual "four–course modules" like "social justice" and "migration studies":

> What if, then, we reorganized the undergraduate curriculum around a set of concepts that instead of foregrounding training in the graduate disciplines, foregrounded topics, skills, and ideas central to humanistic work and central to the interests of students? What if the humanities were marketed within the academy by the names of their best and most important ideas, and not by the names of their calcifying disciplinary formats?

Hayot (2021) here is arguing for replacing majors with topic areas, and for training undergraduates to apply humanities values in their future career and community, as opposed to training undergraduates to become grad students.

FOUNDERS, INCUBATORS, AND INDUSTRY COLLABORATORS

For many technology founders, such as Mark Zuckerberg and Sergey Brin, dropping out of Ivy League schools seems to be a badge of honor. Stanford has arguably become more prestigious given the number of highly successful dropouts who have passed through its halls. What, other than its proximity to Silicon Valley, makes Stanford such a fount of new tech companies? Well, in short, it supports and actively helps students find funders and collaborators for new tech startups (Walker 2013). Students go to Stanford with ideas for startups, and there their professors connect them with other students who can join the startup as well as mentors and investors in the tech industry (Walker 2013). In contrast, at most schools professors don't do anything to help students find a job in industry, nor are most professors actively connected to industry professionals who could make a student idea become a reality. Katina L. Rogers suggests that "Graduate programs could include opportunities to learn and apply [practical and collaborative] skills by partnering with organizations willing to host interns, or by simulating a work environment through collaborative projects with public outcomes" (2020, 93). If digital humanities concepts are ever to be applied in non-academic spheres, then digital humanities programs need to be places that are full of industry connections, and where it is glamorous to drop out.

In some cases, and although this is not a common model, a college might consider holding a stake or investing in a proprietary product, like software; a for-profit company; or a nonprofit organization. Though most digital humanities projects are open access and 100 percent grant-funded, such projects are difficult to sustain without steady internal funding. If a digital humanities department managed to hold a large enough stake in a company like Facebook, then the privacy and algorithmic manipulation of users might have been minimized enough to change something as big as the direction of the country. Furthermore, with the public funding of colleges vulnerable to economic downturns, partnerships with industry could give departments and institutions more security and stability.

While partnerships between higher education institutions and private companies could seem to invite corporate influence into the curriculum, such influence seems to be inevitable. Whether it is politicians who are advocating for defunding the humanities (Cohen 2016) or a university's own board or president who is valuing departments on the basis of their profitability (Bardia et al. 2020), the wolves are already amongst the flock. At least if a private enterprise were to begin in the college or digital humanities setting, it would have a higher chance of being developed ethically and humanely. Collaborative partnerships between industrial and academic entities would likewise hopefully foment the adoption of humanistic values in industry.

PROFESSORS IN INDUSTRY

Humanities professors should be advising government, corporate, and nonprofit entities. Sharing concepts with students is valuable, but it will in most cases be a decade or more before most humanities students have the power and positioning to shape policy. By then, they will likely have already been shaped more by their work environment and industry norms than by their undergraduate learning. Such advisory positions are not unheard of in science.

In addition to advisory positions, it would be ideal for professors to have some industry experience, with a master's program and five years of humanities-related industry experience being assigned the same value as a PhD. The most efficient way for digital humanities to effect change outside of higher education is to have instructors who have strong industry connections, a past in industry, or even a current industry tie.

MODELS

There are already many inspiring and innovative examples of digital humanities projects that exist partly or entirely outside of higher education. Patrik Svensson (2016) lists some of the different iterations of digital humanities organizations and partnerships that extend outside the academy:

> Work with other organizations and fields in order to manifest and sustain digital humanities as a key platform for engaging with the humanities and the digital: memory institutions, all humanities disciplines, other platforms for the humanities (such as humanities centers and the 4humanities initiative), some interpretative social science institutions, technology and science fields, intersectional fields such as gender studies, and organizations such as HASTAC. Double or triple affiliation can be a very useful institutional strategy. People are not restricted to one identity in any context ... Actual institutional configurations and possibilities vary considerably, but the basic idea of multiple affiliations and being a contact zone can be implemented in very different ways. Also, there can be a rich collaboration with individuals who are based elsewhere, but do not have a formal affiliation with a digital humanities initiative.
>
> (Svensson 2016, 488)

An example of such a "contact zone" would be the collaborative effort that occurred between the Google Books team, publishing professionals, and Harvard researchers from a variety of disciplines to develop the Google Ngram Viewer, with the project idea originating within a digital humanities center at Harvard (Michel et al. 2011). The Ngram Viewer is a tool to trace linguistic development, or controversially to consider cultural trends by looking at the frequency of certain phrases in vast quantities of books (Michel et al. 2011). The initiative needed for researchers to contact potential collaborators in private tech companies and in other disciplines should be an essential value and lesson of digital humanities work.

Another model of collaboration between private industry, government, and higher education institutions would be mission-driven organizations. For example, Cornell University, the City University of New York, Verizon, Pivotal Ventures (an investment firm), and Cognizant U.S. Foundation (a private consulting company) members have combined forces to oversee Break Through Tech, a training initiative that "is committed to propelling more high school and college women into tech careers through education, work experience, and community building" (Cornell Tech n.d.). Such models may be, at less scrupulous institutions, open for a corporate takeover of curriculum or a very top-down controlling of a student's career path. Thus with any type of institutional partnership it is essential that every step of approval be authorized through strong, veto-empowered, faculty governance.

FINANCIAL INDEPENDENCE

Just as humanities departments at public and private colleges alike are vulnerable to reductions in tax revenue, public funding cuts, or shrinking enrollments, digital humanities centers and grant-funded initiatives are likewise often on the verge of losing funding, with staff being forced to endlessly invent new projects to apply to grants (Prescott 2016), unless they are granted a steady stream of institutional support and grouped with library or IT services, as scholars such as Sula (2013) and Prescott (2016) strongly suggest is vital for DH centers' success. Many of the concepts and examples mentioned above can allow for colleges and digital humanities centers to gain some degree of independence through long-term outside revenue or investment.

Many digital humanities values seem to reject monetary gain for DH work. For example, open access scholarship rejects the subscription revenue model, thereby undercutting the financial sustenance of humanities work: a price is paid, so to speak, for there being no price. Most academic presses are not self-sustaining and must rely on institutional support (Somin 2019).

The digital humanities believe in open access and public-facing work. DH is not going to make money from targeted advertising, manipulative algorithms, or online surveillance. The question of how the digital humanities can be both ethical and potentially profitable or at least self-sustaining is a vital one to consider. One model is contracts and partnerships with the government. Digital humanists can help provide training and enrichment programs for underserved communities, and this could be accomplished through partnerships with local, state, and federal governments.

Academic presses could learn from the success of "Poem-a-Day" and other nonprofit or for-profit media producers that are voluntarily supported by their readers and through local and national funding (Academy of American Poets n.d.). Many academic presses are in the business of publishing some books that are aimed at general audiences, though Poe (2012) suggests that academic presses should go further and reach potential readers through blogs and other digital media. Were they permitted by their institutions, academic presses could produce TV shows, radio, podcasts, popular articles and other media, and they could get revenue from ads on their website or their videos. Not only could academic presses have public-facing content, but there is also a potential for academic-led think tanks, academic-connected political action committees, or academic-led nonprofits.

CONCLUSION

As Gold and Klein (2019) point out, calls for academics to embrace putting their knowledge into practice, solve worldly problems, and address new audiences have been surfacing for almost ten years (the example they give of this call is of Tara McPherson's work in 2012). Unfortunately, while there have been many success stories, the resistance of academic institutions, disciplinary norms, tenure expectations, and accreditation guidelines can be overwhelming and delay progress or even the possibility of progress.

What may need to happen to allow radical curricular and structural changes is the success of and competition from certain model institutions. A transformation of college learning is already in the works. Existing colleges are already competing with accelerated, skills- or career-based, and often non-accredited programs. Given that college enrollments will likely drop in the US by 15 percent in the next decade (Kline 2019), there will be increasing pressure for colleges to compete for and attract students. With this pressure, colleges and departments will have to appeal to what students want.

Not every college instructor can be embedded in industry or engaged in activism or media production. The suggestions in this chapter are intended for a select number of activist and industry-involved scholars. There will always be a place for theoretical discussion, critical theory, scholarly work, and informed analysis. As Katina L. Rogers (2020) suggests, both career-oriented curriculum planning and traditional theory have their places and their values. Of course, many changes can be incremental and be arranged without much fuss on the part of the administration. Thoughtful and symbiotic partnerships with private companies can be arranged. Some meaningful public editorials can be written or activist work can be done. New curricular offerings that train students in high-demand skills can be approved.

Many digital humanists are involved in direct activism, but they often do not get the deserved credit for their work, especially when it is intangible or in untraditional formats. Graduate students in digital humanities programs have found novel ways to create tools that can be used by specific publics. While the job market seems to be ever more abysmal, there are promising, albeit temporary programs that encourage the worldly application of digital humanities competencies. These programs include the Mellon/ACLS Public Fellows Program for Recent PhDs, which matches postdocs with public-serving institutions; the ACLS Leading Edge Fellowship, which emphasizes using humanistic and social science skills to "promote social justice" in "communities" (ACLS 2021); and a postdoctoral position at the PublicsLab at the Graduate Center of the City University of New York that asks prospective applicants to propose and begin a business or organization that utilizes skills garnered from their doctoral study (Beauchamp 2021). Outside of the US, many forward-thinking fellowships are available, such as the global call for digital humanists and other scholars to attend "research and training" at the Ca' Foscari University of Venice that would "[bridge] the gap between academic and applied research, and between research and market" related to a "global [challenge]" (Ca' Foscari University of Venice 2021). Unfortunately, there are fewer opportunities for students outside of Europe and North America, and Western countries and organizations should look to do more to support and empower foreign scholars and institutions.

If the academy does not unmake and reimagine itself in ways that give students the frameworks and skill sets to address real-world problems, there are plenty of private actors who are ready to "get in" on education, and they do not always have the best intentions. Students are already choosing between a four-year degree that will put them into substantial debt and not guarantee a job versus career-targeted training designed by Microsoft and delivered on LinkedIn (Nellis 2021).

REFERENCES

Academy of American Poets. n.d. "About Us." https://poets.org/academy-american-poets/about-us#7.

ACLS. 2021. "ACLS Leading Edge Fellowships." https://www.acls.org/Competitions-and-Deadlines/Leading-Edge-Fellowships.

Alizada, Nazifa, Rowan Cole, Lisa Gastaldi, Sandra Grahn, Sebastian Hellmeier, Palina Kolvani, Jean Lachapelle et al. 2021. *Autocratization Turns Viral: Democracy Report 2021*. University of Gothenburg: V-Dem Institute. https://www.v-dem.net/static/website/files/dr/dr_2021.pdf.

American Academy of Arts and Sciences. 2021. "Humanities Degrees Declining Worldwide Except at Community Colleges." *American Academy of Arts and Sciences*, June 14. https://www.amacad.org/news/humanities-degrees-declining-worldwide-except-community-colleges.

Associated Press. 2020. "Greta Thunberg Returns to School in Sweden after a Year Off." *The Washington Post*, August 25. https://www.washingtonpost.com/world/national-security/greta-thunberg-returns-to-school-in-sweden-after-y.ear-off/2020/08/25/905d0584-e6b0-11ea-bf44-0d31c85838a5_story.html.

Bardia, Aman, Srishti Yadav, Kalpa Rajapaksha, and Akhil SG. 2020. "The New School is in Crisis." *Jacobin Mag*, December 2. https://jacobinmag.com/2020/12/the-new-school-nssr-austerity-covid-neoliberalism.

Beauchamp, Justin. 2021. "Postdoctoral Fellowship in Humanities Entrepreneurship, 2021–2022." *PublicsLab*, March 16. https://publicslab.gc.cuny.edu/tag/fellowship/.

Ca' Foscari University of Venice. 2021. "Fellowships: Global@Venice." *Institute for Global Challenges*. https://www.unive.it/pag/40610/.

Cohen, Patricia. 2016. "A Rising Call to Promote STEM Education and Cut Liberal Arts Funding." *The New York Times*, February 21. https://www.nytimes.com/2016/02/22/business/a-rising-call-to-promote-stem-education-and-cut-liberal-arts-funding.html.

Cornell Tech. n.d. "Break Through Tech." https://tech.cornell.edu/impact/break-through-tech/.

Deming, David. 2019. "In the Salary Race, Engineers Sprint but English Majors Endure." *The New York Times*, September 20. https://www.nytimes.com/2019/09/20/business/liberal-arts-stem-salaries.html.

Fenton, Will. 2017 "The New Wave in Digital Humanities." *Inside Higher Ed*, August 2. https://www.insidehighered.com/digital-learning/article/2017/08/02/rising-stars-digital-humanities.

Ginsberg, Benjamin. 2011. *The Fall of the Faculty*. Oxford: Oxford University Press.

Gold, Matthew K. and Lauren F. Klein. 2019. "A DH That Matters." In *Debates in the Digital Humanities 2019*, edited by Matthew K. Gold and Lauren F. Klein, ix–xiv. Minneapolis, MN: University of Minnesota Press.

Hayot, Eric. 2021. "The Humanities Have a Marketing Problem." *The Chronicle of Higher Education*, March 21. https://www.chronicle.com/article/the-humanities-have-a-marketing-problem.

Kline, Missy. 2019. "The Looming Higher Ed Enrollment Cliff." *Higher Ed HR Magazine*. https://www.cupahr.org/issue/feature/higher-ed-enrollment-cliff/.

Klinkenborg, Verlyn. 2013. "The Decline and Fall of the English Major." *The New York Times*, June 22. https://www.nytimes.com/2013/06/23/opinion/sunday/the-decline-and-fall-of-the-english-major.html.

Kobie, Nicole. 2019. "The Complicated Truth About China's Social Credit System." *Wired*, June 7. https://www.wired.co.uk/article/china-social-credit-system-explained.

McKenzie, Lindsay. 2018. "Digital Humanities for Social Good." *Inside Higher Ed*, July 9. https://www.insidehighered.com/news/2018/07/09/when-digital-humanities-meets-activism.

McPherson, Tara. 2012. "Why Are the Digital Humanities so White? or Thinking the Histories of Race and Computation." In *Debates in the Digital Humanities*, edited by Matthew K. Gold, 139–60. Minneapolis, MN: University of Minnesota Press.

Michel, Jean-Baptiste et al. 2011. "Quantitative Analysis of Culture Using Millions of Digitized Books." *Science* 331 (6014) (January): 176–82. doi: 10.1126/science.1199644.

National Center for Educational Statistics. 2019. "Bachelor's Degrees Conferred by Postsecondary Institutions, by Field of Study." *Digest of Education Statistics*. https://nces.ed.gov/programs/digest/d19/tables/dt19_322.10.asp.

Nellis, Stephen. 2021. "Microsoft Targets 50,000 Jobs with LinkedIn 'Re-skilling' Effort." *Reuters*, March 30. https://www.reuters.com/article/us-microsoft-jobs/microsoft-targets-50000-jobs-with-linkedin-re-skilling-effort-idUSKBN2BM1S2?il=0.

Nyhan, Julianne, Melissa Terras, and Edward Vanhoutte. 2016. "Introduction." In *Defining the Digital Humanities*, edited by Julianne Nyhan, Melissa Terras, and Edward Vanhoutte, 1–12. Farnham, UK and Burlington, VT: Ashgate Publishing.

Poe, Marshall. 2012. "What Can University Presses Do?" *Inside Higher Ed*, July 9. https://www.insidehighered.com/views/2012/07/09/essay-what-university-presses-should-do.

Prescott, Andrew. 2016. "Beyond the Digital Humanities Center: The Administrative Landscapes of the Digital Humanities." In *A New Companion to Digital Humanities*, edited by Susan Schreibman, RaySiemens, and John Unsworth, 459–75. Chichester: Wiley-Blackwell.

Ramsay, Stephen and Geoffrey Rockwell. 2012. "Developing Things: Notes toward an Epistemology of Building in the Digital Humanities." In *Debates in the Digital Humanities*, edited by Matthew K. Gold, 75–84. Minneapolis, MN: University of Minnesota Press. http://www.jstor.org/stable/10.5749/j.ctttv8hq.8.

Reitter, Paul and Chad Wellmon. 2021. *Permanent Crisis: The Humanities in a Disenchanted Age*. Chicago, IL: University of Chicago Press.

Rogers, Katina L. 2020. *Putting the Humanities PhD to Work*. Durham, NC and London: Duke University Press.

Shane, Scott. 2018. "Some of the Popular Images and Themes the Russians Posted on Social Media." *The New York Times*, December 17. https://www.nytimes.com/2018/12/17/us/russian-social-media-posts.html.

Somin, Ilya. 2019. "University Presses Shouldn't Have to Make a Profit." *The Atlantic*, May 11. https://www.theatlantic.com/ideas/archive/2019/05/why-cuts-stanford-university-press-are-wrong/589219/.

Sula, Chris. 2013. "Digital Humanities and Libraries: A Conceptual Model." *Journal of Library Administration* 53 (1) (January): 10–26. doi: 10.1080/01930826.2013.756680.

Sula, Chris Alen, S. E. Hackney, and Phillip Cunningham. 2017. "A Survey of Digital Humanities Programs." *The Journal of Interactive Technology and Pedagogy* (11). https://jitp.commons.gc.cuny.edu/a-survey-of-digital-humanities-programs/.

Svensson, Patrik. 2016. "Sorting out the Digital Humanities." In *A New Companion to Digital Humanities*, edited by Susan Schreibman, Ray Siemens, and John Unsworth, 476–92. Chichester: Wiley-Blackwell.

Townsend, Mark. 2020. "Facebook Algorithm Found to 'Actively Promote' Holocaust Denial." *The Guardian*, August 16. https://www.theguardian.com/world/2020/aug/16/facebook-algorithm-found-to-actively-promote-holocaust-denial.

Walker, Tim. 2013. "The Billionaire Factory: Why Stanford University Produces so Many Celebrated Web Entrepreneurs." *The Independent*, July 15. https://www.independent.co.uk/student/news/billionaire-factory-why-stanford-university-produces-so-many-celebrated-web-entrepreneurs-8706573.html.

PART FIVE

DH Futures

CHAPTER THIRTY-FOUR

Datawork and the Future of Digital Humanities

RAFAEL ALVARADO (UNIVERSITY OF VIRGINIA)

According to the late sociologist of religion and knowledge Peter Berger, religions, in order to survive, must make the transition from "world building" to "world maintenance" (Berger 1990). In the period of world building, often triggered by an existential threat or social disruption, religious movements are sustained by charismatic leaders who promise a new order that will transform the myriad ills faced by humans in their collective project of survival into a world of lasting prosperity and justice. In the period of world maintenance, the expectation of change that brought the movement into being in the first place must be reconciled with the inevitable failure to fully realize the envisioned world. Where once prophets and mediums inspired change, scholarly priests emerge, often in collaboration with a warrior class, and build a set of institutions that can reproduce themselves, through an exchange of ritual service for livelihood and protection.[1]

In many respects, the digital humanities (DH) may be characterized as a movement in this sense.[2] Although rooted in the past—DH people often trace their ancestry to Busa's machine-generated concordance in 1951—it began in a period of rapid cultural change, prompted by the emergence of a global, networked media ecosystem that reconfigured the public sphere and accelerated the neoliberal transformation of the university. In contrast to the measured scholarship and philosophical concerns of humanities computing, DH adopted a revolutionary posture toward traditional academia that would address both epistemic opportunities and professional crises. Although it had no major prophets, it had many shamans and devotees who gleefully espoused an ethos of play and egalitarianism, all of whom eschewed serious attempts to define the field, an act perceived as either irrelevant or in bad taste. Audrey Watters rightly characterizes the movement as millennialist, a property she associates with the ideology of economic disruption (2013).[3] A decade ago, I characterized DH as a *situation*, in Goffman's sense, a metastable social arrangement in which the play of digital representation is a central value (Alvarado 2012). This revolutionary sensibility continues today as DH positions itself as a form of social activism on a variety of fronts, and expands globally as part of a general trend toward decolonization. The period of world building continues.

It is clear, however, that the transition to world maintenance has also been under way. This is evident in the establishment of DH centers and programs across the United States, although the formation of departments and postgraduate programs has not taken root to the extent that it has in Canada, the UK, and Europe. The difference is instructive; we see the emergence of two very different models for world maintenance. Although this difference appears to be the result of local adaptations to different institutional environments, it has consequences at the epistemic level. And this is where the heart of our question lies: Whether or not DH survives, and in what form, depends to a large extent on the institutional vehicle it takes as it transitions from building to maintenance.

The point can be illustrated by an anecdote. At the 2015 European Summer University in Digital Humanities at the University of Leipzig in Germany, there was a panel on the topic of "Digital Humanities: Towards Theory and Applications" organized by Laszlo Hunyadi. Panelists were asked to respond to the following prompt:

> Several fields of science (among them mathematics, physics, chemistry) are subcategorised into two, essentially independent but also interconnected branches, theoretical and applied: the first is aimed at discovering new knowledge about the given field, the second at applying some of these discoveries to other areas, sometimes well beyond its origin. One can wonder if the emerging field of Digital Humanities bears any of these characteristics; whether it is developing into a similar scientific field with its own independent properties. We witness as one of the strengths of DH to be the building of applications to serve wide ranges of interest across cultures, languages and technologies. One may ask: are these applications based on knowledge simply borrowed from other sciences or is DH creating its own theoretical foundations for these new developments? And going even further: can DH have an impact on the development of other (theoretical or applied) scientific fields?
>
> (Hunyadi 2015)[4]

In the lively conversation that followed, two facts emerged. First, there was a consensus among the panelists (it seemed to me) to regard DH as a science in its own right, at least in principle. Second, it was conceded, after some pushback, that one of the reasons for defining DH in this way is to secure funding within the European educational systems, where, to achieve the status of maintenance, a field must become a science. Not science in the broad, ecumenical sense of any system of knowledge, but in the historically specific and socially embedded sense with which we are familiar today—empirical, quantitative, publication driven, and grant-funded. Think CERN, not Leibniz.

This admission was surprising for its transparent opportunism, but it's clear that here necessity and truth do converge. In Europe, in contrast to the US, DH is more closely associated with linguistics, in particular corpus linguistics and stylistics, and linguistics is certainly a science in the modern sense—indeed, as Lévi-Strauss argued, it is the paradigmatic human science that showed the way for all others. Still, the classification of DH as a science must be rejected in principle if not in practice. If it is a science in the sense proposed, then it is nothing more than a rebranded form of linguistics. If so, what do we make of other strongly represented concerns that have no direct connection to linguistics, such as mapping, visualization, and augmented reality—not to mention the activist practices found in the US and elsewhere? At best, one might argue that DH has been very successful in the domain of linguistics, where the application of digital methods and the digitization of primary sources has borne the most fruit. And DH may indeed become an umbrella field within which a variety of conceptually related but institutionally balkanized fields might unite under a common head. Such an arrangement would do justice to the precarious status that linguistics has had over the years, especially in the US.

If there is a will to brand DH as a science in Europe, in the US we see movements in the opposite direction, to delegitimize DH for masquerading as a science, or for adopting its methods as it pursues another goal, that of becoming a new form of literary criticism and historical research. These come from internal critics of DH who, in effect, would rather the field become a kind of media studies, or perhaps science, technology, and society studies (STS) (e.g., Koh 2015). It also comes from unaffiliated scholars in the humanities—mainly from departments of English—who are dismayed and annoyed by the arrival of a foreign set of methods into their field, and who are incredulous that quantitative approaches can add anything to the project of interpretation (Da 2019; S. Fish 2019). An extreme element of this group regards DH as irretrievably tainted by its embrace of technology, associated with both the supposedly dehumanizing forms of rationality and the neoliberal dismantling of the academy (Golumbia, Allington, and Brouillette 2016).

The most trenchant of these criticisms are those that make the strong epistemic claim that numbers or computers can never produce anything of value to the work of interpretation. If true, then the idea of a scientific DH is a non-starter—at least in its US variant, where the focus has been on literary studies as opposed to linguistics—and DH therefore loses this institutional vehicle to move toward world maintenance. And it is clear that the end game of DH's critics is to prevent this from happening, to block what is perceived to be the annexation of their land. As with advocates of DH-as-science in Europe, its US critics are motivated by the observed connection between being a science and acquiring funding and prestige. Given the zero-sum politics of the academy, and the struggle in the public sphere to remain relevant when higher education is regarded by many as a bubble about to burst, we may surmise that the main beef that critics have with DH is not its epistemic status but its institutional one, even if their arguments focus on the former. Rather than dismiss these arguments as specimens of motivated reasoning, I believe they concern a—even the—fundamental question that DH must answer in order to survive. It therefore must be addressed directly.

Although narrowly focused on what she calls computational literary studies (CLS), "the project of running computer programs on large (or usually not so large) corpora of literary texts to yield quantitative results," Da's arguments can be generalized to apply to any effort to study literature and culture quantitatively (Da 2019, 601). The claim, for example, that "human and literary phenomena are irreducible to numbers" is a charge against any attempt to apply digital representations to the material of the humanities, to the extent that the former involve any form of measurement or discrete and countable classification of literary phenomena, not just the works of those she addresses. Given that science is rooted in measurement, her verdict is as universal as Fish's declaration that "[t]he desire to generate insight into human expression by 'scientific' means is futile" (2019, 333).

A less absolute framing of Da's argument is the following. If DH is to embrace the methods of science in exchange for institutional stability, then it must accept the terms of this bargain. Its practitioners—in CLS or any other area that claims to have discovered or know social or psychological facts, either explicitly in their theories or implicitly in their activism—must conform to some conventions of rigorous and intersubjective proof. But this is where things get interesting: these conventions are by no means clear, even if we accept some basic ground rules about the definition of science, such as the essential roles played by measurement and mathematics. As Underwood points out, Da appears to believe that conventions of validity are to be found in traditional, frequentist statistics (Underwood 2020). For example, in her criticism of literary text-mining, Da asserts that

> Quantitative analysis, *in the strictest sense of that concept*, is usually absent in this work. Hypothesis testing with the use of statistical tools to try to show causation (or at least idiosyncratic correlation) and the explanation of said causation/correlation through fundamental literary theoretical principles are usually absent as well.
>
> (2019, 605; emphasis added)[5]

The implied identification of quantitative analysis with hypothesis testing is telling. In point of fact, the former may have nothing to with the latter, and many areas of science have moved away from it. For example, in the area of machine learning, a subfield of the science of artificial intelligence, the Neyman–Pearson paradigm (to which one imagines Da refers) has been replaced by a set of performance measures that do not rely on p-values. In the areas of natural language processing and text analytics, an entirely different set of criteria is employed, based on the concept of entropy and its variant, perplexity. Da mentions these concepts in her critique, but does not grasp their methodological significance. More important, in these cases interest in causality (or even correlation) is not the norm; instead, the goal is often exploratory pattern discovery, through measures such as mutual information. Whether these results are compelling or not is a matter of perspective, but it's always easy to regard something as obvious in the perfect vision of hindsight.

Regarding the issue of genuine causal inference, hypothesis testing is not the gold standard of scientific verification that many seem to think it is. Judea Pearl has argued for decades that statistical associations expressed in the language of probability (on which hypothesis testing builds) can never be used to infer, or even represent, causal relations. For these, one must introduce exogenous, structural models that specify entities and their relations, and these usually are found in the heads of the scientists who employ statistical models, not in the latter's algebraic representations (Pearl 2009, 5). There are formalisms for expressing these models, such as structural causal models, but these are not what Da has in mind. (Nor are they required for rigorous scientific description of literature and culture.) More recently, Jeffrey Blume has argued convincingly that hypothesis testing with p-values is needlessly punitive of negative results and, in addition to the well-known problem of p-value hacking, is a source of the reproducibility crisis in the human sciences, since its use throws away meaningful results—what Da cavalierly dismisses as "non-results" (Blume 2017; Blume et al. 2019).[6] To this mathematical critique, we might add the speculation that p-values have been so successful because they provide the necessary binary filter required for gatekeeping in the sciences, which in turn reinforces research that succeeds within its terms.

Given the high bar that critics set for the work of CLS, and the absolute manner in which its project is dismissed, one wonders whether traditional literary studies (TLS) are held to a similar standard. Are the methods of close reading or reader-response theory any better at producing significant conclusions? Has TLS ever provided effective ways to check criticism, other than by appeals to authority? Has anyone, for example, ever really validated, much less clearly defined, Fish's concept of the "interpretive community," or his many claims regarding how this mysterious agent brings all things into being? Are there any non-non-results from *this* field? We have reason to be skeptical. The truth is that practitioners of TLS on historical and literary texts are forced to invent the pre-understanding that must be shared between writer and reader required for communication to take place.[7] To be sure, this is accomplished by the critic's significant learning about the context and culture in which texts are embedded—but here, too, one encounters the same limitations introduced by what Ricoeur calls the fixation and distantiation of discourse introduced by writing (1973). Given this, we might conclude that TLS has never been able to avoid the problem of projecting readers' ontologies, interests, and biases onto the work they interpret—a scandal celebrated by reader-response theory. In fact, the main problem with TLS from the perspective of CLS is not, as Moretti argues, that the critic is unable to read everything in her subject domain—a charge that only makes sense if one adopts the frequentist metaphysics of the long run, and the populism associated with it. It is that literary criticism, especially the best of it—Auerbach, Gilbert and Gubar, Said—has always been a form of high ventriloquism, in which the reviewed corpus is used to voice and legitimate the critic's own ideas, often with the purpose of advancing an agenda, laudable as that may be, or to become more celebrated than the authors they interpret. However naive or undeveloped current examples of CLS may be, their practitioners are openly and fundamentally engaged with the vexing problem of interpretive validity. Their efforts to produce an operational hermeneutics, whose "logics" are laid bare in code, and therefore open to the kind of investigation that Da pursues, are precisely the point.

It should be clear that statistical significance is not the essential issue here, as ripe as that topic is for critical review, and concern for the investigation of causality is, at least at this point, misplaced. What the eminent sociologist Paul DiMaggio wrote regarding social science applies equally well to the humanities:

> Engagement with computational text analysis entails more than adapting new methods to social science research questions. It also requires social scientists to relax some of our own disciplinary biases, such as our preoccupation with causality, our assumption that there is always a best-fitting solution, and our tendency to bring habits of thought based on causal modeling of population samples to interpretive modeling of complete populations. If we do, the encounter with machine learning may pay off not just by providing tools for text analysis, but also by improving the way we use more conventional methods.
>
> (DiMaggio 2015, 4)

The point is echoed by Underwood who argues that critics—and, I would add, many advocates—of a scientific DH miss the exploratory nature of CLS and other applications of quantitative and computational methods to the study of literature and culture. This work is exploratory at two levels. First, it is engaged with exploring patterns that may be induced from culturally and historically variant corpora through the application of novel methods of computer-aided reading. As many have noted before, these methods constitute a kind of scientific instrument, something like the microscope or telescope. Recall that when these were invented, they were employed with playful curiosity, which yielded substantive discoveries—the existence of cells, the moons of Jupiter. Second, and more profound, CLS is concerned with exploring the relationship between quantitative and interpretive modes of reasoning, and between the linguistic and the literary levels of description; in other words, the epistemic opportunity opened up by the fertile intersection of the humanities and the sciences. In this regard, CLS is quintessential DH—it exemplifies the serious play of representation that attends the acts of encoding, transforming, modeling, visualizing, and interpreting texts and their analytic results. Da dismisses these as "basic data work," an aspersion based on a misrecognition of the hermeneutic nature of these processes.[8]

Let us embrace the term *datawork*, as it captures an essential quality of CLS, and DH more generally: the conscious grounding of knowledge in the actual work of representation, associated with activities ranging from the operationalization of a received theory to the cartographic mapping of geographic references in a work of literature. What do we learn from datawork in this sense? For one, it allows us to examine, empirically and phenomenologically, broad claims about what it is possible to know and not know, and how our instruments and media of knowledge—from index cards and concordances to text encoding schema and vector spaces—shape our understanding of the observed. One such claim is that it is impossible to arrive at new, interesting, or useful literary knowledge about texts from counting their words. This claim is made repeatedly by Fish and Da in one form or another and is axiomatic to their rejection of CLS and DH. The idea is not new; it is a dogma of TLS, inherited from the nineteenth-century distinctions between explanation and understanding, *naturwissenschaften* and *geisteswissenschaften*, and inscribed in the organization of the university, both in the taxonomy of schools and departments and in the epistemic fissures that internally divide departments, such as philosophy and anthropology.[9] Fish inherits this distinction, but he amplifies the difference: he places an uncrossable ocean between these two realms by insisting on a collectivist idealism that locates all agency in the encompassing will of the community. It is not surprising that he would discount that which CLS counts since, in his framing, texts as empirical objects literally do not exist but are magically instantiated when read. But not all theories of culture and collective representations need be so romantically idealist. To be sure, the distinction does receive support from computational quarters. Shannon's mathematical theory of communication, which is foundational to the computational study of text, begins by separating out concern for meaning from the engineering problem of efficiently transmitting messages through signal media. Nevertheless, quantitative approaches to the study of meaning are not only possible but quite fruitful, as for example the study of lexical semantics by means of vector space and geometric models of text. Examples of this approach include latent semantic analysis, topic modeling, and word embedding. These approaches exemplify what Turney and Patel call the statistical semantics hypothesis (Turney and Pantel 2010, 146), which includes Zellig Harris's distributional hypothesis, an operationalized version of Wittgenstein's later theory of meaning (Harris 1954).

Now, I believe it is not possible to fully grasp the relationship between the statistical and the semantic without actually doing, or tracking, the datawork required to transform a text into encoded form, and then into a vector space representation (of which there are many types beside a bag-of-words), and then to apply, for example, matrix factorizations or kernel density estimates or probabilistic generative models to this representation to produce the results with which we are familiar—e.g., low-dimensional visual representations of high-dimensional spaces. This process, which is precisely deformative and performative in McGann's sense (McGann 2003, 10), forces one to make a series of interpretive decisions that both depend on and help develop a model of text at the level of what may be called its primary discursive structure.[10] What becomes apparent in this process is that one never simply counts words (or their proxies as delimited strings); one is always also defining and redefining a geometric structure of containers and connections—the text model—within which the counting makes sense. And it is this geometry that connects counting to meaning, through observations of co-occurrence, distance, entropy, and other measures.

In other words, CLS—and more broadly what I call exploratory text analytics (ETA)—is never just about counting, and so Da's argument collapses, even if some of the examples she discusses are equally silent on this point.[11] It is always about counting in the context of structures from which higher order patterns emerge. In Bateson's language, it's about discovering "the pattern which connects" the statistical patterns which, by themselves, indeed amount to an impoverished understanding of text (Bateson 2002). I have no reservation in calling some of these *patterns of culture*, in the sense meant by Ruth Benedict and cultural anthropologists since then, such as Benjamin Colby in his early work with the General Inquirer (Benedict 1934; Colby 1966). Nor do I hesitate to agree with Bill Benzon when he invites us to consider some of these patterns as representing "pathways of mind" and "historical process" (Benzon 2018, 3).

Fish's argument is harder to dismiss, since the process I describe supports his theory that the text is constructed in the process of its being read. But the structures that one discovers through working with the text, at least at the level of description indicated, namely the primary discursive structure of the text, are not put there by the analyst. They are properties of the text as real as the bones in my body. One does not need to succumb to the "hegemony of formalism" to believe this (Fish 1980, 8). Believing in the existence of something does not require one to subscribe to a monolithic theory built on the premise that it is only that thing exists or matters. (Such theories are more about building schools of thought than they are about understanding the world.)

Let us borrow the language of Fish and say that practitioners of CLS and ETA constitute a community of interpretation whose members share experience in this core process of datawork, a special case of the general DH situation of representational play. As a participant-observer, I believe this community believes that although readers' responses to texts are authentic and not to be dismissed, they are not the only interesting things that can be said about or done with texts. They also believe that although close reading—whose condition of possibility is the print-based media infrastructure of dictionaries, concordances, critical editions, archives, and libraries—is effective and fruitful as far as it goes, it is not the only approach to making sense of texts, especially given the conditions of mass-digitization and encoding of texts in which we are embedded. In response to Crane's question—What do you do with a million books?—this community has responded by combining the quantitative methods necessary to perceive them as such with the broad hermeneutic goals shared by their fields of origin. I believe this engagement has produced what I have called an operational hermeneutics based on a shared understanding of text that, although evolving, is remarkably consistent with the work of humanities computing as well as the deeper, pre-computational hermeneutic tradition of the interpretive human sciences. It has been articulated in pieces in various places, recently in Benzon's theory of the corpus (Benzon 2018). For our purposes, it is worth summarizing and expanding on these views.

What is this understanding of text? To begin with, texts *qua* texts are physical (but not material) objects with statistical and structural properties resulting from the matter-energy exchanges that attend their production—through the transductive acts of recording, writing, editing, publication, and other forms of entextualization—and transmission—via documents, reproduction, signal transmission, etc.—prior to and independently of their consumption through reading, close or distant.[12] In this view, a text is just a finite but unfixed series of symbols, drawn (with replacement) from a finite but fixed physical (so-called material) symbol system. A text in this sense is a property of a message that survives the several transductions it goes through as it passes from production to consumption. Typically, a message is produced by a speaker or writer with the intent to convey information or to influence the behavior of another person or group of people. Of course, other kinds of messages exist, such as those that seek to express one's experience or understanding of the world to another (or to one's self). The text of a message is that which is copiable, and in principle perfectly copiable, from one medium to another, for example from dictation to writing to printing to reading. Something survives that process, something is transmitted through that process, and that something is describable as a series of characters. This is true even when, as typically happens, that message is distorted, through loss or addition, in the process. Whether the characters of the message are produced by a Mayan scribe, a medieval Christian monk, or philosopher at a typewriter, the same thing is true. What differs is the shared symbol system required for communication to take place, and which indeed is the product of culture, as is the specific symbol system. This shared system includes rules of combination and a world in which meanings emerge. From this perspective, texts are like the fossils and settlement patterns studied by archaeologists who have long used quantitative methods to infer cultural and social facts from these data.

Now, what critics such as Fish and Da mean by a text is the elaborate structure of meaning, the cognitive and affective product, the fully realized understanding of a message that emerges from the acts of reading and interpretation, especially when the message is a poem or other work of literature. It is the intellectual content and emotional effect of a message fully considered, the radiant network of rich symbols, meanings, and implications evoked by the text in the reader that results from the work of reading and interpretation. For his part, Fish inherits this definition from the formalists, who conflated these two meanings of text and projected the latter onto the former, thereby fetishizing the text and its supposed autonomy from the intersubjectivity of the author and the reader.

In effect, we have before us two very different meanings of "text," which, when viewed together, amount to parts of the proverbial elephant. As parts of a system, we may think of the two meanings, the physical and the mental variants, as constituting two phases of a single, larger process. At issue in understanding the relationship between these two kinds of text is what is often called the location of meaning, a proxy for one's understanding of the nature of meaning. Clearly, none of the mental text is "in" the physical text; the physical text does not "contain" meaning.[13] The reader and the writer, the consumer and the producer, of a message must share a pre-understanding that includes linguistic competence but also an extra-linguistic understanding of the world.[14] Typically, this shared understanding is the product of common experience and education, through ritual processes in shared situations that synchronize cognition and affect in communities. So, yes, readers must assign meanings to the text, and these assignments take place within the context of the cultural field—the text in Fish's sense is constituted through the act of reading. But this does not free the reader from the obligation of making sense of the text as a message sent by another, especially when the reader knows that the message was sent over great cultural distance, removed by historical time or geographic space. Readers do the work of inferring meanings in messages that they, in good faith, assume to have existed in the minds of their authors. The fact that this relationship may be disrupted by the alienation of a text from its productive situation, is beside the point. None of this eliminates the reality of the physical text.

But the point is not simply that we can assert the existence of a text independently of its reading. It is that physical texts, even though they do not contain meanings—even though it makes no sense to think of meaning and text in this way—are nonetheless essential to the production and reproduction of mental texts. And here forgive me if I slip into the language of data analysis to observe that this process is not unlike that of Bayesian inference, an observation that has been made by others. For what I mean by physical text is simply that text is a form of data D—its quintessential form, in fact—and what I mean by mental text, the interpretive product, is simply the reader's hypothesis θ, broadly speaking, of what an author or a source speech community means or intends.[15] (The reader may, of course, be toyed with: in the case of non-intentionally created texts, for example those created by the cut-up method or drawn from the Library of Babel, which are subject to the effects of distantiation, the reader is sent on a wild goose chase.)[16]

Now, given this understanding of text, where does DH belong? Among the linguistic scientists or the literary critics? I believe neither. For one thing, DH is much broader than CTS, and would be impoverished by too close an alignment with either field. But more important, DH has created a new subfield in the still liminal space between the science of linguistics and the scholarship of criticism, where new forms of reasoning and sense making are being shaped in response to an emerging assemblage of media that, subject to the laws of media, have opened new opportunities for the study of mind, culture, and society. These forms are neither those of frequentist statistics nor of readerly literary criticism, and it only distorts the work of DH, CLS, and ETA to view them in these lights. But the question of an academic home, a vehicle for world maintenance, remains.

NOTES

1. I've taken some liberties in summarizing Berger, based on my own familiarity with the anthropology of religion.
2. Whatever your view of religion, it is difficult not to see parallels between it and other social movements and institutions, including those of academia. Especially academia. We are the inheritors of religion, as is evident in the history and architecture of so many colleges and universities. Even a place like the University of Virginia, whose founder explicitly renounced any connection to established religion, nevertheless maintained that connection through systematic opposition—the orientation of the Lawn, which runs North and South, contradicts the ecclesiastic convention to orient (it's even in the word) space along the path of the sun, and so reproduces that concern.
3. This view is consistent with the Web 2.0 ethos and aesthetic adopted by many in DH at the time. The phrase Web 2.0, whose appearance coincides historically with the rebranding of humanities computing as DH (Watters 2013), expressed the utopian hope for a new world opened up by the participatory web, in which individualism would be replaced by community, labor by creativity, and money by purpose. The viral success of the term must owe to its combination of a traditional religious belief—the Second Coming—with the banal inevitability of a software upgrade (Alvarado 2011).
4. The questions asked reflect the spirit of the program, whose director, Elizabeth Burr, has described DH as an epistemology, a view to which I am sympathetic.
5. Da claims to be making a "computational case" against CLS, but at no point does her argument employ computational ideas. Instead, she confounds computer science and statistics under the term "quantitative," which, far from having a strict sense, functions as a catch-all term without a clear meaning.
6. As a more useful measure of significance, Blume has defined "second-generation p-values," p_δ (SGPV).
7. This is a principle asserted from Schleiermacher to Shannon, by both traditional hermeneutics and the mathematical theory of communication. A shared grammar and understanding of the world is a condition of possibility for communication.
8. The dismissal of datawork also reflects a leisure class condescension towards manual labor. Later, Da writes of datawork, as in busywork, resorting to what might be called an *ad laborem* argument. As in: these arguments may be dismissed by the nature of the work they represent (Da 2020). It is a move similar to that made in previous decades against the use of the computer for literary purposes by referring to its use as mere typing.
9. Think of the not always amicable separation between analytical and continental philosophy, or of the department of anthropology at Stanford, which was at one time split into two, one to serve its scientific subfields, the other its humanistic ones.
10. I have in mind the model of text as understood by humanities computing in the 1990s (Renear, Mylonas, and Durand 1993; Renear, McGann, and Hockey 1999), but also in the hermeneutic sense,

which describes text at the level of what Ricoeur called "the linguistics of discourse" (1973, 92). To the extent that the model of text developed by humanities computing is preserved by the Text Encoding Initiative's collection of schema (TEI), the value of this markup for subsequent text analysis is invaluable.

11. The point remains even though many text analysts miss the connection between the work of text encoders and their own.
12. Reference to the immateriality of texts raises the sticky issue of hylomorphism, which will raise the hackles of some. I simply mean to assert that texts and documents are ontologically distinct, and that the former are analytically separable from the latter. Texts are physical in precisely the sense that information is physical, that they are subject to measurement but have no mass. In other words, I follow Wiener's dictum that "[i]nformation is information, not matter or energy" (Wiener 1948, 132). This perspective complements the materialism of text criticism and bibliography.
13. That texts do not contain meanings is a well-established point made by Wittgenstein, Shannon, and later by Reddy in his discussion of the conduit metaphor (Reddy 1979).
14. By extra-linguistic, I do not mean language-independent, but rather mental phenomena that are not in themselves linguistic but which are connected to the larger cognitive system in which language is central. One example of such an extra-linguistic component is what Hanks calls the deictic field, within which ostensive reference is made possible even when speakers are co-present and share a situation (Hanks 1987, 2005).
15. If we think of interpretation in a probabilistic framework, we may express the hermeneutic relationship between physical symbols (*Sr*) and mental meanings (*Sd*) as follows:

$$P(Sd|Sr) = \frac{P(Sd)P(Sd|Sr)}{P(Sr)}$$

Inasmuch as *Sr* and *Sd* represent what Schleiermacher called the "linguistic" and "psychological" aspects of interpretation, the formula also expresses the logic of the hermeneutic circle as a matter of updating the prior and recomputing the posterior (Palmer 1969). The analogy between the Bayesian approach to causality and the hermeneutic approach to meaning has been noted by others (Frank and Goodman 2012; Friston and Frith 2015; Ma 2015; Groves 2018; Reason and Rutten 2020).

16. An obvious criticism of my understanding of text is that it is suspiciously convenient for the work of CLS and ETA. And it is. But from the fact that one selects various aspects of an object to study, it does not follow that one has created those aspects. More generally, from the fact that one is put into a position to observe something, it does not follow that that which is observed has been created by the act of observing. If the reader is now reaching for their quantum physics, realize the phenomena in that domain are far removed from what we were talking about here. If we are going to deny the existence of physical text on the grounds of the uncertainty principle, you'll have to do the same with other things, like viruses and climates.

REFERENCES

Alvarado, R. 2011. "Science Fiction as Myth: Cultural Logic in Gibson's Neuromancer." *Science Fiction and Computing. Essays on Interlinked Domains*: 205–13.

Alvarado, Rafael. 2012. "The Digital Humanities Situation." In *Debates in the Digital Humanities*, edited by Matthew R. Gold, 50–5. Minneapolis, MN: University of Minnesota Press.

Bateson, Gregory. 2002. *Mind and Nature: A Necessary Unity*. New York, NY: Hampton Press.

Benedict, Ruth. 1934. *Patterns of Culture*. Boston, MI and New York, NY: Houghton Mifflin Company.

Benzon, Bill. 2018. "Toward a Theory of the Corpus." https://www.academia.edu/38066424/Toward_a_Theory_of_the_Corpus.

Berger, Peter L. 1990. *The Sacred Canopy: Elements of a Sociological Theory of Religion*. Illustrated edition. New York, NY: Anchor.

Blume, Jeffrey D. 2017. "An Introduction to Second-Generation p-Values." Statistical Evidence. https://www.statisticalevidence.com/second-generation-p-values.

Blume, Jeffrey D., Robert A. Greevy, Valerie F. Welty, Jeffrey R. Smith, and William D. Dupont. 2019. "An Introduction to Second-Generation p-Values." *The American Statistician* 73 (supplement 1): 157–67. https://doi.org/10.1080/00031305.2018.1537893.

Colby, Benjamin N. 1966. "Cultural Patterns in Narrative." *Science* 151 (3712): 793–98. http://www.jstor.org/stable/1718617.

Da, Nan Z. 2019. "The Computational Case against Computational Literary Studies." *Critical Inquiry*, March. https://doi.org/10.1086/702594.

Da, Nan Z. 2020. "Critical Response III. On EDA, Complexity, and Redundancy: A Response to Underwood and Weatherby." *Critical Inquiry* 46 (4): 913–24. https://doi.org/10.1086/709230.

DiMaggio, Paul. 2015. "Adapting Computational Text Analysis to Social Science (and Vice Versa)." *Big Data & Society* 2 (2): 2053951715602908. https://doi.org/10.1177/2053951715602908.

Fish, Stanley Eugene. 1980. *Is There a Text in This Class?: The Authority of Interpretive Communities.* Cambridge, MA: Harvard University Press.

Fish, Stanley. 2019. "If You Count It, They Will Come." *N.Y.U. J.L. & Liberty* 12 (January): 333.

Frank, Michael C. and Noah D. Goodman. 2012. "Predicting Pragmatic Reasoning in Language Games." *Science* 336 (6084): 998. https://doi.org/10.1126/science.1218633.

Friston, Karl J. and Christopher D. Frith. 2015. "Active Inference, Communication and Hermeneutics." *Cortex: A Journal Devoted to the Study of the Nervous System and Behavior* 68 (July): 129–43. https://doi.org/10.1016/j.cortex.2015.03.025.

Golumbia, David, Daniel Allington, and Sarah Brouillette. 2016. "Neoliberal Tools (and Archives): A Political History of Digital Humanities." *Los Angeles Review of Books*, May 1. https://lareviewofbooks.org/article/neoliberal-tools-archives-political-history-digital-humanities/.

Groves, Robert. 2018. "Hermeneutics and Bayesian Statistics." *The Provost's Blog* (blog). January 10. https://blog.provost.georgetown.edu/hermeneutics-and-bayesian-statistics/.

Hanks, William F. 1987. "Discourse Genres in a Theory of Practice." *American Ethnologist* 14 (4): 668–92.

Hanks, William F. 2005. "Explorations in the Deictic Field." *Current Anthropology* 46 (2): 191–220. https://doi.org/10.1086/427120.

Harris, Zellig S. 1954. "Distributional Structure." *Word* 10 (2–3): 146–62.

Hunyadi, Laszlo. 2015. "Digital Humanities: Towards Theory and Applications." "Culture & Technology" – The European Summer School in Digital Humanities. https://esu.culintec.de/?q=node/609.

Koh, Adeline. 2015. "A Letter to the Humanities: DH Will Not Save You." *Hybrid Pedagogy* 19.

Ma, Rong. 2015. "Schleiermacher's Hermeneutic Circle and Bayesian Statistics." *RongMa@Penn* (blog). November 28. https://rongma.wordpress.com/2015/11/28/schleiermachers-hermeneutic-circle-and-bayesian-statistics/.

McGann, Jerome. 2003. "Texts in N-Dimensions and Interpretation in a New Key." Special Issue on the Ivanhoe Game. *Text Technology* 12 (2003): 1–18.

Palmer, Richard E. 1969. *Hermeneutics: Interpretation Theory in Schleiermacher, Dilthey, Heidegger, and Gadamer*. Evanston, IL: Northwestern University Press.

Pearl, Judea. 2009. "An Introduction to Causal Inference." *Statistics Surveys* 3: 96–146. https://projecteuclid.org/euclid.ssu/1255440554.

Reason, Suspended and Tom Rutten. 2020. "Predictive Hermeneutics." *PsyArXiv*, August 11. https://psyarxiv.com/tg8ym/.

Reddy, Michael J. 1979. "The Conduit Metaphor: A Case of Frame Conflict in Our Language about Language." *Metaphor and Thought* 2: 164–201.

Renear, Allen, Jerome McGann, and Susan Hockey. 1999. "What Is Text? A Debate on the Philosophical and Epistemological Nature of Text in the Light of Humanities Computing Research." Panel presented at ACH–ALLC 2000 International Humanities Computing Conference, University of Virginia, June 10. https://web.archive.org/web/20040624214640/http://www.humanities.ualberta.ca:80/susan_hockey/ACHALLC99.htm.

Renear, Allen, Elli Mylonas, and David Durand. 1993. "Refining Our Notion of What Text Really Is." http://www.stg.brown.edu/resources/stg/monographs/ohco.html.
Ricoeur, Paul. 1973. "The Model of the Text: Meaningful Action Considered as a Text." *New Literary History* 5 (1): 91–117. https://doi.org/10.2307/468410.
Turney, Peter D. and Patrick Pantel. 2010. "From Frequency to Meaning: Vector Space Models of Semantics." *Journal of Artificial Intelligence Research* 37: 141–88.
Underwood, Ted. 2020. "Critical Response II. The Theoretical Divide Driving Debates about Computation." *Critical Inquiry* 46 (4): 900–912. https://doi.org/10.1086/709229.
Watters, Audrey. 2013. "The Myth and the Millennialism of 'Disruptive Innovation.'" *Hack Education* 24.
Wiener, Norbert. 1948. *Cybernetics: Or Control and Communication in the Animal and the Machine*. Paris: Librairie Hermann & Cie.

CHAPTER THIRTY-FIVE

The Place of Computation in the Study of Culture

DANIEL ALLINGTON (KING'S COLLEGE LONDON)

HERMENEUTIC AND EMPIRICAL CULTURES OF KNOWLEDGE PRODUCTION

In the course of a somewhat weary trek through more departments and disciplines than I ever intended, it gradually dawned on me that, while there isn't really any difference between that which is generally called "Social Science" and that which is sometimes called "the Humanities" and sometimes "the Arts," there is a very real distinction to be drawn between two cultures of knowledge production, each of which is to be found in both of the aforementioned. There is a place for computation—that is, for mathematical calculations, executed with or without the assistance of a machine—in one of these cultures, but in the other, there is not—and I submit that wider acknowledgment of this point might be of help to that which refers to itself as "the Digital Humanities," as its relationship to the rest of the academy continues to develop and evolve. In order to draw a line between my argument and that of C. P. Snow (1959), I shall refer to the two cultures that I have in mind as the "empirical" and the "hermeneutic."

Under the empirical culture of knowledge production, one answers research questions that are fundamentally *about the world*. Some (although not all) such questions are susceptible to approach through the use of what are conventionally known as "quantitative methods"—which is to say, through the use of methods wherein the analytic stage involves computation. However, under the hermeneutic culture of knowledge production, one answers research questions that are, by contrast, fundamentally about *meaning*. Although there have been attempts to recast hermeneutic questions in apparently objective terms—most famously, the intentionalist argument that the only meaning that can validly be ascribed to a text is the meaning that its author demonstrably intended (see especially Hirsch 1967), and the structuralist argument that the meaning of a text is a function of its relationship to a system that can potentially be described in its entirety (see above all Culler 1975)—it remains the case that the only measure of success for an interpretation of *Hamlet*, the *Theses on Feuerbach*, or the Sermon on the Mount is its intersubjective acceptability for an audience of the interpreter's peers. And computation has nothing to contribute under such a paradigm—unless we mean those computations which go on, unnoticed, in the background, incessantly, so that emails can take the place of the postal service and a word processor can take the place of a typewriter. But almost nobody cares about those.

That's just the way it is, and I'm not going to waste anybody's time by trying to change it. I'm not talking about a revolution.

QUANTIFICATION AND THE HERMENEUTIC CULTURE OF KNOWLEDGE PRODUCTION

Both cultures are to be found across the contemporary university. However, there are some disciplines—such as literature, philosophy, and theology—which are *essentially* hermeneutic. These may have adjunct empirical disciplines, such as the history of books, the history of concepts, and the history of religions, and the august figures who rule over the hermeneutic parent disciplines may sometimes allow representatives of these quasi-legitimate offspring to hold posts in their departments, but, at their core, those departments nonetheless continue to operate by rules alien to the empirical culture of knowledge production. This is to say that, by design, the critic, philosopher, or theologian addresses questions which are not susceptible to computation; moreover, he or she addresses them in relation to objects few enough in number to make computation redundant. If the scope of one's research is Sonnet 116 or a fragment of *On Nature* or a hadith, then qualitative methods are sufficient because each of those objects is unique. Hermeneutic scholars may write of relatively large classes of objects, such as "Elizabethan poetry," or "pre-Socratic philosophy," or "the sayings of the Prophet." But one of the key things taught to undergraduate students of literature, philosophy, and theology is to avoid making generalizations that would be impossible to support through careful deployment of quotations from irreducibly unique individual texts.

All this is fine. But the value and interest of quantitative methods lies in their potential to generate evidence to support—or, depending on the outcome, to *refute*—precisely those kinds of generalizations that students of essentially hermeneutic disciplines are trained to avoid. Quantification is inherently a form of generalization. In order to count something, one must first adopt the assumption that all examples of that thing can be treated as equivalent—at least from the point of view of the count. This involves a sort of anonymization of experience, treating everything that is to be counted as an instance of a type.

Types do not have to be reified, and may of course be adopted only provisionally, for the purposes of a single research question. But, despite this, it remains true that to think in such terms goes against every instinct of the hermeneutic scholar. A major part of disciplinary training in literary studies, art history, film studies, media studies, cultural studies, and multiple forms of qualitative social science consists in learning to see everything as unique, and in learning to focus most closely on that which distinguishes any given object or occurrence from all else that might be compared to it. The same word may have very different connotations in two different sentences. The same sentence may have very different implications in two different conversations, and even in two different parts of the same conversation. But if everything is unique, then everything is its own category, and every category has already been counted, and the total for every category is 1.

This is not, I must emphasize, a matter of the humanities versus the social sciences. The following, for example, appears in a sociology professor's response to an article promoting certain forms of quantitative analysis:

> Pierre Bourdieu in *Distinction* inadvertently confirms how quantification relaxes testing … As we know, Bourdieu compares the frequency with which persons agree that a sunset over the sea is suitable material for a beautiful photograph …. Persons with low occupational and educational rankings are more likely to affirm this subject matter as promising than are persons

> with high occupational and educational rankings. Supposedly this result confirms that persons with more education deploy a distinctive habitus by which they distance themselves from the trite and commonplace. The numbers confirm a differential, to be sure, but ... [b]y formalising people's aesthetic appreciation into countable agree/disagree results, Bourdieu made it facile to confirm his hypothesis.
>
> (Biernacki 2015, 343)

It's important to recognize that the above is not simply a critique of Bourdieu, whose theories are controversial even among those who do not reject his methods. It is, rather, a dismissal of the very idea of quantification, and it forms part of an assertion of the superiority of hermeneutics over empiricism: elsewhere in the same essay, it is asserted that "[h]umanistic method preserves arcane detail because none of it in a text is automatically privileged, which is to say, [because] almost any of it can become a cornerstone thanks to shifting the frame of reference" and that "[t]o maximise the potential of 'surprise' for theorising ... we must preserve the minutia as pivots of interpretation and therefore as source points for competing guesses" (2015, 342). Although made in this case by a sociologist, the same points could have been made in the same way by many a literary scholar, outraged at some of the stronger claims which have been made for the Digital Humanities. And those points are, moreover, completely valid: every cultural production and every experience *is* unique, if you only look at it closely enough, and so is every historical event, and much is lost along with the "arcane detail" that must be generalized away before computation can begin.

And yet to admit empiricism to a discipline is to open up a space within it for quantification.

QUANTIFICATION AND THE EMPIRICAL CULTURE OF KNOWLEDGE PRODUCTION

To quantify means to establish the numerical quantity of—which is to say, to measure or to count, or simply to express as a number. Quantification, then, involves temporarily laying aside the knowledge that everything is unique, and instead provisionally adopting the contrary assumption that it may be—for particular purposes—useful to treat particular things as *generic*. This is the opposite of what one learns to do when one learns to read poetry and scripture, and also of what one learns to do when one learns to read movies and videogames as if they were poetry and scripture. By analogy with close reading, quantitative methods have sometimes been sold to humanities scholars as "distant reading" (e.g., Moretti 2013)—which is a clever piece of marketing, but ultimately rather misleading, because quantification is not reading at all. Reading involves grasping the particularities of some text as a whole, and this is the case even when one does *not* read closely—for example, when leafing through a tedious novel to find out how it ends, or when skimming an academic article in order to pick out the major findings or to discern the overall argument.

By contrast, as Stephen Stigler (2016) has argued, statistics—the science of numerical analysis—is effectively founded on the realization that the informativeness of data is increased through aggregation, even though (and in fact, *because*) this means discarding a particular kind of information that might on the face of it seem rather valuable, i.e., information relating to the unique circumstances and characteristics of each observation or phenomenon. This is what happens whenever one takes the mean of a series of numbers (for example, describing a group of texts, or people) and henceforth uses the resulting number to stand for the whole series, often for purposes of comparison with some other series (for example, describing another group of people or texts).

Does this all sound rather reductive? Naturally it does! The ultimate point is a) to end up with a yes or a no, such that the generalization in question can be more or less confidently added to humanity's store of things that are or are not considered to be true, where this confidence comes from the fact that b) one arrived at the yes or at the no through a truth-determining procedure so anonymous that anyone could theoretically have followed it and arrived at a similar result. Quantification is reductive *by design*; it is reductiveness raised to a virtue. The idea of treating cultural artifacts in such a way may appear scandalous, because it is precisely the opposite of what critical interpretation entails. But—paradoxical as it may seem today—such resistance to aggregation was once commonplace even in a discipline such as economics:

> When, in the 1860s, William Stanley Jevons proposed measuring changes in price level by an index number that was essentially an average of the percent changes in different commodities, critics considered it absurd to average data on pig iron and pepper. And once the discourse shifted to individual commodities, those investigators with detailed historical knowledge were tempted to think they could 'explain' every movement, every fluctuation, with some story of why that particular event had gone the way it did. … It was not that the stories [they] told about the data were false; it was that [those stories] (and the individual peculiarities in the separate observations) had to be pushed into the background. If general tendencies were to be revealed, the observations must be taken as a set; they must be combined.
>
> (Stigler 2016, 15)

As with commodity prices, so with poetry, and music, and everything else: taking things in aggregate prevents us from explaining their individuality through stories founded on our detailed knowledge of the particularities of the things themselves and their histories. What it enables us to do instead is to observe general tendencies, and then to test our explanations of those general tendencies. It's never going to work quite as well when the generalizations being tested relate to social or cultural objects of knowledge, because the objects themselves are messier than physical objects of knowledge such as molecules or bacteria and are substantially more difficult (sometimes impossible) to consider in isolation. But that doesn't mean that we have to give up, nor even that it would be morally justifiable to do so. It only means that, if we are going to apply quantitative methods to such objects, we will have to proceed with exceptional caution and care.

This is unlikely to generate much excitement from scholars committed to the hermeneutic culture of knowledge production, because it isn't what they went into academia to do. On the contrary, they came to gain detailed knowledge of particularities and then to use it to tell enlightening stories. Which is fine: it is; honestly, it is. Better a single erudite monograph than a thousand badly designed experiments. But if we can design our quantitative studies carefully and well, and carry them out scrupulously, and demonstrate the contribution that the ensuing calculations make to existing knowledge—however small, however provisional, however apparently banal—then perhaps we may slowly begin to demonstrate a different kind of value, as those contributions accumulate.

THE TEMPTATIONS OF PSEUDOSCIENCE

But *will* those contributions accumulate? There is no guarantee. In a famous 1974 lecture, the Nobel prizewinning physicist Richard Feynman observed that much research in the social sciences was analogous to the ritual practices of "cargo cults," as they were then understood by Western anthropologists. The term "cargo cult," which covers "a range of millenarian ideas, cults, and movements that originated [in Melanesia] in the wake of Western colonisation and, more often than not, involved a strong concern with the acquisition of Western goods" (Otto 2009, 82–3), is now widely recognized as problematic: Lamont Lindstrom (1993) argues that it refers essentially to a Western myth, and Elfriede Hermann observes that residents of areas in which supposed "cargo cults" have been observed came to recognize the term "cargo cult" as "a catchphrase" used by colonial authorities and Melanesian elites alike to "degrade their culture" (1992, 56, 68). But if we understand what Feynman termed "Cargo Cult Science" as a metaphor whose vehicle is not an actual Melanesian religious practice but a Western myth or fable of "people at the margins of capitalised society who are conspicuously obsessed with getting hold of industrial products but are using the wrong means to do so" (Otto 2009, 93), it provides a useful means by which to understand the equally Western practice of *pseudoscience*: the natural sciences yield useful and reliable findings, like cargo planes coming in to land; not understanding the vast infrastructure necessary to produce this end result, some researchers from outside the natural sciences build runways and control towers of their own, and vainly hope for a drop-off. The runways and control towers are methods of numerical data collection and statistical analysis whose power is that "[w]hen [they] are employed correctly, they usually serve as a self-correcting system of checks and balances" (Cokley and Awad 2013, 27). But very often, they are not employed correctly—in part, because it is simply harder to employ them with regard to the more nebulous objects of humanities and social science scholarship, and in part, because of the lack of a culture in which "[o]ther experimenters will repeat your experiment and find out whether you were wrong or right" (Feynman 1974, 11). As a result, "the planes don't land," even though "all the apparent precepts and forms of scientific investigation" seemed to have been followed (Feynman 1974, 11). The fundamental problem is that all the statistical procedures in the world can't make up for the absence of the *collective* practice of formulating hypotheses and then using empirical evidence to test those hypotheses to destruction.

Unfortunately, much Digital Humanities advocacy reads as a plea to be allowed to establish a pseudoscience in peace. For example, the following argument from a leading Digital Humanities scholar was widely circulated in a number of different forums:

> Eventually, digital humanities must make arguments. It has to answer questions. But yet? Like eighteenth-century natural philosophers confronted with a deluge of strange new tools like microscopes, air pumps, and electrical machines, maybe we need time to articulate our digital apparatus, to produce new phenomena that we can neither anticipate, nor explain immediately. (Scheinfeldt 2013, 58–9)

The fact that some scientific progress was made before the formalization of the scientific method as understood today does not mean that it would be advantageous, or even ethically acceptable, to suspend that method's usual requirements for a discipline that is "young." (It should also be emphasized that the Digital Humanities are not so young as all that: the term "Digital Humanities" has been around since the turn of the Millennium, and the Department of Digital Humanities in which I work is much older than that, albeit that it originally did business under another name.) If a comparison may be drawn between Digital Humanities and Computational Biology, the origins of the latter field are usually traced back to the publication of Alan Turing's article, "The chemical basis of morphogenesis" (1952), which, far from aiming to "produce new phenomena" that cannot immediately be explained, instead proposed an explanation of known facts in terms of established physical laws. One wonders whether computational biology would have reached the prominence that it enjoys today had its founders given themselves a pass on answering research questions on the grounds that a new discipline needs time to find its feet and, after all, "[t]he eighteenth-century electrical machine was a parlour trick" (Scheinfeldt 2013, 57).

I don't mean, of course, to imply that Digital Humanities is inherently pseudoscientific, nor that Digital Humanities research has added nothing to the store of human knowledge. For a contrary example to these strawman positions, one only has to think of the stupendous contribution to scholarship which has already been made by the incomplete, fully digital third edition of *The Oxford English Dictionary*; if one wishes for something more numerical, there are many impressive examples, such as Ted Underwood and Jordan Sellers's (2016) analysis of diction as a predictor of whether or not a given work of nineteenth-century literature would be reviewed in a prestige venue. But these examples have nothing in common with parlor tricks: the former evolved directly from existing lexicographic practice, iteratively extended through application and development of forms of technology with which the institution responsible had been working for many years, while the latter was designed in order to test a specific hypothesis emerging from existing, non-digital literary scholarship (i.e., "that a widely discussed 'great divide' between elite literary culture and the rest of the literary field started to open in the late nineteenth century," Underwood and Sellers 2016, 323), and as such explicitly adopted the most straightforward form of what is known as the scientific method. In other words, where Digital Humanities research succeeds most impressively, it is by adopting the standards expected of researchers outside the Digital Humanities, and by attempting to contribute to knowledge in a very clearly defined way.

This may explain why non-computational traditions of Digital Humanities research have often achieved more convincing results than the computational traditions on which this chapter's critical eye is primarily focused. When textual critics turned to digital technology in hopes of extending their existing scholarly practice beyond what was possible given the limitations of print as a medium in which to embody it, they knew what they were about (see Bode 2017, 94–102). By contrast, computational methods in the Digital Humanities have too often been justified either as works-in-progress that may eventually become rigorous modes of scientific enquiry at some unspecified point in the future, or otherwise as unpredictable procedures for the generation of ideas which may or may not prove fruitful as starting points for conventional (and often hermeneutic) scholarship. An example of the former is provided by the following piece on topic modeling, written for a Digital Humanities audience by one of the computer scientists who had pioneered the technique a decade previously:

> Here is the rosy vision. A humanist imagines the kind of hidden structure that she wants to discover and embeds it in a model that generates her archive. The form of the structure is influenced by her theories and knowledge — time and geography, linguistic theory, literary theory, gender, author, politics, culture, history. With the model and the archive in place, she

> then runs an algorithm to estimate how the imagined hidden structure is realized in actual texts. Finally, she uses those estimates in subsequent study, trying to confirm her theories, forming new theories, and using the discovered structure as a lens for exploration. She discovers that her model falls short in several ways. She revises and repeats.
>
> ...
>
> The research process described above ... will be possible as this field matures.
>
> (Blei 2012)

As an iterative process of hypothesis formation and confirmatory testing, this sounds a lot like the scientific method. But it is, by the author's own admission, only a "rosy vision" which—it is suggested—may become possible to realize at some point in the future, given unspecified technological developments. By contrast, the other mode of presenting a Digital Humanities methodology is evident in the following, written by a Digital Humanities scholar and published in the same special issue of the same Digital Humanities journal:

> How the actual topic modelling programs [work] is determined by mathematics. Many topic modelling articles include equations to explain the mathematics, but I personally cannot parse them.
>
> ...
>
> Topic modelling output is not entirely human readable. One way to understand what the program is telling you is through a visualisation, but be sure that you know how to understand what the visualisation is telling you. Topic modelling tools are fallible, and if the algorithm isn't right, they can return some bizarre results.
>
> ...
>
> Topic modelling is not necessarily useful as evidence but it makes an excellent tool for discovery.
>
> ...
>
> Topic modelling is complicated and potentially messy but useful and even fun. The best way to understand how it works is to try it. Don't be afraid to fail or to get bad results, because those will help you find the settings which give you good results. Plug in some data and see what happens.
>
> (Brett 2012)

Unlike the previously quoted piece, this is a description of a Digital Humanities practice already possible in the here and now. Tellingly, there is no discussion of what might constitute "good results" or "bad results," nor of what features of a topic model might be understood through which kind of visualization, nor of how to recognize "fail[ure]" in topic modeling, nor even of what it means for a method such as topic modeling to be "fallible," nor of what one can really hope to achieve by "[p]lug[ging] in some data"—which data?—to a process whose details one "cannot parse." It is certainly conceivable that a hermeneutic scholar might find inspiration in the output of such a process, much as Surrealist poets found inspiration by cutting up printed texts and rearranging the words at random. Well, fair enough. But, to return for a moment to Feynman's metaphor (with an acknowledgment, once more, that its vehicle is not a Melanesian reality but a Western myth), no one should be surprised if the planes don't land.

Which of the two visions above provides a better description of topic modeling as it is *generally* taught and practiced in Digital Humanities? In a piece written four years later under the title of "Topic modeling: what humanists actually do with it," one scholar surveyed numerous examples before concluding that topic modeling "allows us to chart new paths through familiar terrain by drawing ideas together in unexpected or challenging ways" that "consistently call us back to close reading" (Roland 2016). Computer as inspiration for hermeneutic scholarship, in other words.

The author of that later piece appears not to have meant his comments as a criticism. But not everyone in the Digital Humanities has been so relaxed. For example, while others were advocating for topic modeling as the next big thing in humanities methodology, Benjamin Schmidt was cautioning that humanities scholars "are too likely to take ... [topic models] as 'magic', rather than interpreting [them] as the output of one clustering technique among many," and observing that those humanities scholars who already use topic modeling tend to approach it in a way that "ignores most of the model output" (2012a), employing a research practice which "ensures that they will never see the ways [in which] their assumptions fail them," and in the process "creat[ing] an enormous potential for groundless—or even misleading—'insights'" (2012b). That critiques such as this have emerged within Digital Humanities draws attention once again to the heterogeneity of the field, which is, as I have emphasized above, hospitable not only to under-theorized and often aimless tool use but also to the most rigorous enquiry. But the question must arise: if the former is going to be treated as *good enough*, then why put in the effort to carry out the latter?

There is nothing wrong, of course, with playing with tools as though they were toys, nor with being pleasantly surprised with intriguing "black box" computer output that one cannot fully understand. But Digital Humanities pedagogy is forever at risk of being overwhelmed by the confusion of such entertaining pastimes with the pursuit of knowledge. My personal experience is that, because students must be assumed to come to Digital Humanities with no prior experience of computer programming or of statistics, one must either take the slow approach of building up students' knowledge firstly of statistics and secondarily of programming to the point where they can grasp at least *some* approaches to analysis not as magical incantations but as a form of reasoning—or otherwise teach the *application* of a wide range of techniques without any real exploration of their theoretical and algorithmic underpinnings, thus creating the perfect conditions for pseudoscience to flourish.

While the former approach is (in my view) more intellectually defensible, and also happens to furnish students with knowledge and skills more readily transferable to other domains, including those of the workplace—the demand for workers with a little knowledge of Python but none of mathematics has often been exaggerated—it is also more challenging for students, especially given the tight time constraints imposed by a modern modular degree program, which is likely to require all of these things to be taught within a single course no longer nor more intensive than the several other parallel courses on more "theoretical" topics which students are likely to be studying simultaneously. And this is before we have even touched on questions such as data collection and research design: questions which are fundamentally part of both undergraduate and postgraduate teaching in subjects such as psychology or demography, and yet which struggle for space in a typical one-year postgraduate program in the Digital Humanities. In other words, while Andrew Goldstone is right to argue that "[c]ultivating technical facility with computer tools – including programming languages – should receive less attention [in Digital Humanities pedagogy] than methodologies for analysing quantitative or aggregative evidence" (2019), the reality is not only that it typically receives less but also that there is usually not enough time for *either*. Under such conditions, it is easier to teach students *just enough* technical facility to be able to create an attractive visualization and then display it as an end result—as Goldstone further noted, Digital Humanities as a field "is enchanted with visualisation, to such an extent that many people in the humanities identify the whole of data analysis with the production of visualisations" (Goldstone 2017)—or to create a topic model and then to interpret the individual topics as though they were a sort of haiku.

NURTURING THE DEVELOPMENT OF CULTURES OF QUANTIFICATION

Quantification—and therefore also computation—can only contribute to knowledge within an empirical, as opposed to a hermeneutic, framework, and such a framework does not exist, or exists only at the margins, in many of the disciplines that are regarded as the natural home of research questions concerning cultural artifacts. This means that, realistically, there is scant prospect for the development of quantitative methods in primarily hermeneutic disciplines such as literature, philosophy, or theology. I therefore conclude this chapter by proposing that the most promising future for computation in the study of culture may be through integration into primarily empirical disciplines such as history and anthropology, and into what I have referred to above as the empirical "adjunct disciplines" of a typical Faculty of Arts and Humanities' hermeneutic star attractions: that is, through engagement with disciplines such as the history of books (or concepts, or religions) in a deep enough way to furnish research questions that scholars working within those disciplines find meaningful with answers which those same scholars are likely to find credible, comprehensible, and useful. This is unlikely to mean walking into unknown intellectual territory armed with whizz-bang computational methods whose underlying mathematics almost nobody can understand.

There are three models on which I contend that the sort of engagement that I am arguing for is more likely to be successful. The first is that of immersing oneself completely in a given discipline for a substantial period of time and coming to understand it intimately, so that one eventually becomes able to start from first principles in designing methods and projects that will truly address that discipline's core concerns (both methodological and substantive). The second is that of working in partnership with researchers from a given discipline in order to develop mutually comprehensible approaches to the extension of their existing practice through judicious application of well-understood methods for dealing with numerical information (or at least with information that can be meaningfully represented in numerical form). And the third is that of apprenticing oneself—so to speak—to quantitative currents which may already have emerged within the discipline in question: history, for example, has a vigorous—albeit sadly marginalized—quantitative tradition which predates the Digital Humanities by several decades (see e.g., Termin 1973).

Each of these models represents a radical rejection of the bombast and theoretical naivety which has unfortunately characterized some of the highest-profile examples of Digital Humanities "macroanalysis" (see Bode 2017, 77–94). Moreover, each takes for its starting point the recognition that, because the computational researcher's *existing* skill set will be of uncertain value in relation to any given scholarly objective, the most important traits for him or her to develop are adaptability and a willingness to learn. Each model can thus be thought of as a different inflection upon a single program.

Central to that program is recognition that the most superficially impressive methods—the ones which most obviously wear the trappings of *science*—are unlikely to be the most useful ones. The ability to process large volumes of digital text, for example, is likely to be useful only in exceptional circumstances, and tests of statistical significance will often be strictly invalid. On the other hand, the world's archives (now and for the foreseeable future, mostly un-digitized) contain vast stores of commercial and administrative data, ripe for systematic analysis by those who have both the ability and the imagination. Such data may at first sight seem dry, but their potential is immense, as we see from William St Clair's (2004) economic history of the British publishing industry, or from Fabien Accominotti and colleagues' (2018) study of the changing class composition of audiences at the New York Philharmonic. Moreover, as Kevin Cokley and Germine Awad emphasize, there is an urgent need for the development of "empirical constructs which are rooted in the values and realities of ... marginalised and/or oppressed population[s]" in order to "help empirical researchers [to] conduct more culturally competent research which will be responsive to the unique needs of ... [those] populations" (2013, 30): a task for which scholars trained both in quantitative methods and in cultural critique should be ideally prepared.

Perhaps these ambitions will strike some as dispiritingly modest: rather than learning from and seeking to contribute to other disciplines, should not the Digital Humanities be blazing a trail for them to follow? That was, after all, the promise with which they burst onto the scene and were proclaimed the next big thing. But that was long ago: today, the Digital Humanities are no longer new, and have not been new since before the majority of current undergraduate students were born. And I hope that the Digital Humanities preserve enough institutional memory of their origins in various academic support services—my own department began long ago as a Centre for Computing in the Humanities, for example, and had no students of its own until comparatively recently—to be able to embrace a new service role, this time on an equal footing, with computational methods specialists working in partnership with scholars from across the humanities and social sciences to enrich each of the disciplines that they enter into by nurturing the development within it of a culture of quantification local and specific to the discipline itself. (And also by training their students to do the same—and not only within but also without the academy, where they may contribute to institutions other than disciplines in much the same way.)

A bigger problem, then, is that the ambitions that I have laid out are so difficult to realize: most of us are not yet ready to serve, in effect, as statistical consultants to the empirical humanities. In this connection, there are three observations that I should make. The first is that it's never too late to go back to school. The second is that "going back to school" means learning at the foot not only of statisticians but also of those who have spent their careers applying quantitative methods in empirical humanities disciplines independently of the developments associated with the Digital Humanities. And the third is that, with regard to the sort of program that I have outlined above, it will often be the older and the simpler methods that are most genuinely productive—in large part, because they are so much more straightforward to reason about. Some of those methods would not be taught on a typical statistics course. For example, a typical history project will probably benefit less from—say—spline regression than from an appreciation, grounded both theoretically and in research practice, of—say—the difficulties involved in aggregating quantitative data from different historical sources. And this is the sort of expertise that a historian with an interest in statistics is as well placed to develop as a statistician with an interest in history. Were it to become the foundation of undergraduate and postgraduate training in the Digital Humanities, I think that we might start to see widespread progress.

Alexander Pope's famous warning still applies: *Drink deep, or taste not the Pierian Spring*. No, a little statistical learning is not "enough," any more than a little historical learning could be. And no—despite years of chipping away at my ignorance—I cannot claim to possess more than a very little of either. But the knowledge is there to be made use of, if only we can make the effort—and there are many, many places where it is needed. Not having yet learnt *enough* is no excuse for not beginning the work, in whatever small way—provided that one can remain forever mindful of one's own limitations, of course. To switch momentarily to a new metaphor: I may not have what it takes to become a doctor, but that would be no reason not to take a course in first aid, and no excuse at all for refraining from applying the knowledge gained on such a course in the event that I found myself in a situation where it could be of benefit.

First aiders to those that don't yet know any relevant statistics, and first aiders to those that do, but don't yet know much about the complexities involved in collecting and working with cultural and historical data: as ambitions go, I've heard worse. Again, I'm not talking about a revolution. I'm only talking about how much good might be done in nurturing local cultures of quantification across the so-called humanities and the so-called social sciences—at least where there exists the potential for such cultures to grow.

Where there doesn't, the smart thing to do is probably to walk on. There's too much work to do anyway.

ACKNOWLEDGMENTS

I would like to thank James Smithies for offering reflections and comments on this chapter. All errors of fact and failures of interpretation rest with the author.

REFERENCES

Accominotti, Fabien, Adam Storer, and Shamus R. Khan. 2018. "How Cultural Capital Emerged in Gilded Age America: Musical Purification and Cross-Class Inclusion at the New York Philharmonic." *American Journal of Sociology* 123 (6): 1743–83.

Biernacki, Richard. 2015. "How to do things with historical texts." *American Journal of Cultural Sociology* 3 (3): 82–102.

Blei, David M. 2012. "Topic Modelling and Digital Humanities." *Journal of Digital Humanities* 2 (1).

Bode, Katherine. 2017. "The Equivalence of "Close" and "Distant" Reading; or, Toward a New Object for Data-Rich Literary History." *Modern Language Quarterly* 78 (1): 77–106.

Brett, Megan R. 2012. "Topic Modelling: A Basic Introduction." *Journal of Digital Humanities* 2 (1).

Cokley, Kevin and Germine H. Awad. 2013. "In Defense of Quantitative Methods: Using the 'Master's Tools' to Promote Social Justice." *Journal for Social Action in Counselling and Psychology* 5 (2): 26–41.

Culler, Jonathan. 1975. *Structuralist Poetics: Structuralism, Linguistics, and the Study of Literature*. London: Routledge and Kegan Paul.

Feynman, Richard P. 1974. "Cargo Cult Science: Some Remarks on Science, Pseudoscience, and Learning How to Not Fool Yourself." *Engineering and Science* 37 (7): 10–13.

Hermann, Elfriede. 1992. "The Yali Movement in Retrospect: Rewriting History, Redefining 'cargo cult'." *Oceania* 63 (1): 55–71.

Hirsch, E. D. Jr. 1967. *Validity in Interpretation*. New Haven, CT and London: Yale University Press.

Lindstrom, Lamont. 1993. *Cargo Cult: Strange Stories of Desire from Melanesia and Beyond*. Honolulu: University of Hawai'i Press.

Moretti, Franco. 2013. *Distant Reading*. London and Brooklyn: Verso.

Otto, Ton. 2009. "What Happened to Cargo Cults? Material Religions in Melanesia and the West." *Social Analysis* 53 (1): 82–102.

Roland, Teddy. 2016. "Topic Modeling: What Humanists Actually Do With It." *Digital Humanities at Berkeley*, July 14. https://digitalhumanities.berkeley.edu/blog/16/07/14/topic-modeling-what-humanists-actually-do-it-guest-post-teddy-roland-university.

Schmidt, Ben. 2012a. "When You Have a MALLET, Everything Looks like a Nail." *Sapping Attention*, November 2. http://sappingattention.blogspot.com/2012/11/when-you-have-mallet-everything-looks.html.

Schmidt, Ben. 2012b. "Words Alone: Dismantling Topic Models in the Humanities." *Journal of Digital Humanities* 2 (1).

Snow, C. P. 1959. *The Two Cultures*. Oxford: Oxford University Press.

Stigler, Stephen M. 2016. *The Seven Pillars of Statistical Wisdom*. Cambridge, MA: Harvard University Press.

St Clair, William. 2004. *The Reading Nation in the Romantic Period*. Cambridge: Cambridge University Press.

Termin, Peter. 1973. *New Economic History: Selected Readings*. Harmondsworth: Penguin.

Turing, Alan M. 1952. "The Chemical Basis of Morphogenesis." *Philosophical Transactions of the Royal Society B*, 237: 37–72.

CHAPTER THIRTY-SIX

The Grand Challenges of Digital Humanities

ANDREW PRESCOTT (UNIVERSITY OF GLASGOW)

Digital humanities is shaped by funding opportunities. This is sad, but true. It has been claimed that the project, as the basic unit of digital humanities activity, reshapes humanities research because of its team-based, iterative, and collaborative nature (Burdick, Drucker, Lunenfeld, Presner and Schnapp 2012, 124–6). But the requirement for highly structured, carefully planned projects derives from research funders and is not confined to digital research. A project is a convenient bureaucratic funding unit, not a new vision of how scholarship might be structured. Likewise, the preponderance of the digital humanities center reflects the fondness of universities for funding models in which they provide initial seed corn capital funding for a new center, in the pious hope that sustainability can be achieved from future research income. The digital humanities center reflects financial exigencies, not new intellectual models (Prescott 2015).

The future of digital humanities will be chiefly shaped by the way funding priorities develop and shift. Thus, recent enthusiasm for developing masters' courses in digital humanities is due not to increasing intellectual maturity in the subject area but rather to the drive to secure lucrative fees from international students. One notable trend in research funding over the past twenty years has been the emphasis on addressing Grand Challenges. A Grand Challenge is a major barrier to progress which if removed would help solve an important societal problem and have a global impact. Grand Challenges are often technological or social in nature. They may consequently seem far removed from traditional humanities research. Indeed, it may be felt undesirable that humanities research should address such political and social agendas and that pursuing Grand Challenges risks debasing and instrumentalizing the humanities. However, it is equally possible that the humanities by encouraging a more collective and less technological approach to Grand Challenges could address shortcomings in the Grand Challenge concept. This essay explores the ways in which digital humanities and the humanities more widely might engage with the Grand Challenge agenda. It will be suggested that Grand Challenges around such areas as the health of the Internet and bias in artificial intelligence may provide fruitful areas for future development of the digital humanities.

The idea of organizing research around Grand Challenges originated with the famous German mathematician David Hilbert, who in 1900 listed a set of twenty-three problems in mathematics. Hilbert argued that solutions to these problems would generate breakthroughs across many scientific fields (George, Howard-Grenville, Joshi, Tihanyi 2016). The concept of the Grand Challenge was revitalized by the Bill and Melinda Gates Foundation which in 2003 identified fourteen Grand Challenges in global health (gcgh.grandchallenges.org). Both the United States and Canada subsequently developed funding programs using the Grand Challenges approach and, by 2011, fifty individual challenges in such fields as global health, chronic non-communicable disease, and engineering had been identified (Grand Challenges Canada/Grand Défis Canada 2011).

Another high-profile expression of the Grand Challenges method are the sustainable development goals which were adopted by the United Nations in 2015 and aim to end poverty, protect the planet, and ensure prosperity for all.[1] The UN sustainable development goals form the basis of such research-funding programs as the Global Challenges Research Fund (GCRF) administered by United Kingdom Research and Innovation (UKRI) which has funded research projects directly related to the sustainable development goals.[2] The use of societal challenges as a focus for research funding was also prominent in the European Union's Horizon 2020 program.[3]

The benefit of a Grand Challenge structure is that it provides a framework for coordinating and sharing sustained input from a variety of participants across different sectors and disciplines towards a clear and concrete goal. The core features of a Grand Challenge are that it is a specific bottleneck or barrier which stands in the way of solving pressing global problems. The barrier is difficult but not impossible to overcome and, once removed, will have global impacts. One definition of a Grand Challenge, based on the Canada Grand Challenges program, is:

> specific critical barrier(s) that, if removed, would help solve an important societal problem with a high likelihood of global impact through widespread implementation.
>
> (George, Howard-Grenville, Joshi, and Tihanyi 2016, 1881)

However, Grand Challenges are also by their nature complex. They interact across many social, cultural, economic, and technological areas, and the resulting interrelationships can be multifarious. They cut across existing disciplinary and organizational structures and boundaries in an elaborate fashion, posing formidable logistical challenges. Above all, Grand Challenges inhabit a world of radical uncertainty. We cannot easily tell how the future will map out if Grand Challenges around such areas as climate change are not solved. It is tempting to imagine that Grand Challenges can be solved by heroic managerial intervention, marshaling straightforward management panaceas, but the complexity of structure of Grand Challenges means that participative, collaborative, and exploratory approaches are more likely to be productive (Ferraro, Etzion, and Gehman 2015).

The Covid-19 pandemic has fostered enthusiasm for Grand Challenge approaches (Gümüsay and Haack 2020; Howard-Grenville 2020). The speed with which vaccines for Covid-19 were made available generated political enthusiasm for research programs with a strong clear focus whose output offers a clear social benefit. Although the experience of the pandemic has so far illustrated at every turn the unpredictability of such major global events, nevertheless the pandemic has encouraged a political preference across the world for focused programs of research confronting clear-cut challenges. The memory of the pandemic is likely to encourage in the future the use of such funding models for all disciplines, including digital humanities.

In March 2021, the UK government announced the creation of the Advanced Research and Invention Agency (ARIA), which is intended to supplement the existing research-funding infrastructure coordinated by UK Research and Innovation (UKRI). ARIA will provide flexible rapid funding programs with a minimum of bureaucratic intervention for transformational research in areas which present global challenges.[4] The government announcement of the creation of ARIA explicitly cites the UK's coronavirus response, particularly the Vaccines Taskforce and the Rapid Response Unit, as exemplars of the importance of agility in funding and decision-making. ARIA is modeled on the United States's Defense Advanced Research Projects Agency (DARPA) which has also inspired Japan's Moonshot R&D and Germany's SPRIN-D programs. DARPA's early work on mRNA vaccines and antibody therapy was cited as one of the justifications for the creation of ARIA.

The idea of Grand Challenge funding is here to stay, at least for the foreseeable future. There are of course problems with the political enthusiasm for robust assaults directed at Grand Challenges. For example, while on the one hand creating ARIA, the UK government has on the other been criticized for providing insufficient funding for international collaborative programs such as Horizon Europe which are more likely to generate progress on global issues (Patten 2021). Notwithstanding such misgivings, if the arts and humanities are to fully access and exploit available research funding, it is inevitable that they engage with the Grand Challenge agenda.

Ways in which the arts and humanities might contribute to research on global challenges are illustrated by the GCRF in the UK. The challenge portfolios for GCRF cover themes such as global health, food systems, conflict, resilience, education, and sustainable cities.[5] The arts and humanities have a great deal to contribute in addressing these challenges and many fascinating projects have been funded, encompassing such varied activities as the use of artworks, performance, and storytelling to document healthcare experiences in the poorest part of the Philippines,[6] the creation of a large-scale global research network on parliaments and people in order to strengthen democracy in politically fragile countries,[7] and the Quipu project which has created an archive and online interactive documentary telling the story of the forced sterilization of 272,000 women and 21,000 men as part of a government program in Peru in the late 1990s.[8]

Because the GCRF program has formed part of the UK overseas aid program, GCRF has made funding available for arts and humanities research on a scale not previously seen in Britain. The former Chief Executive of the UK's Arts and Humanities Research Council has emphasized the immense benefits the GCRF program has brought to research worldwide and argued against threats to reduce funding to the program as a result of UK government cuts in overseas aid:

> Amid what may be the fastest technological change in history, what is the economy of tomorrow into which governments are trying to move their citizens? In a resource-constrained world, what do human-centred economies look like? Over the past five years, the UK's pioneering Global Challenges Research Fund (GCRF) has wrestled with these questions by reconnecting economic with social and environmental issues, and by forging new, more equitable partnerships between northern and southern researchers.
>
> (Thompson 2021)

In a post-pandemic world, these questions, and the potential contribution of the GCRF, seem even more pressing. But nevertheless a niggling doubt remains about the role of the arts and humanities in identifying these global challenges. Do these Grand Challenges address major barriers in arts and humanities scholarship? Do such challenges as resilience, food systems, and global health reflect priorities in the arts and humanities research or are they instead generated by STEM and social science concerns?

These questions reflect wider criticisms of the Grand Challenges approach which have been current ever since the Bill and Melinda Gates Foundation announced its challenges in 2003.

Anne-Emanuelle Birn (2005) in an influential critique of the approach of the Gates Foundation accused the Grand Challenges of focusing excessively on technological interventions and giving insufficient weight to the importance of addressing poverty and inequitable access to health systems:

> Individually, and as a whole, the 14 Grand Challenges share an assumption that scientific and technical aspects of health improvement can be separated from political, social, and economic aspects. Indeed, the Grand Challenges initiative has made this division explicit by excluding the problems of "poverty, access to health interventions, and delivery systems" from this competition. This is not simply an argument of economics versus medicine: if isolated technical approaches may result in few sustained health gains, historical evidence suggests that laissez-faire economic growth, with no focus on wealth distribution is likely to provoke "the four D's" of disruption, deprivation, disease, and death.
>
> (Birn 2005, 516)

Birn's criticisms of the Gates Foundation's technologically driven approach, with its strong emphasis on vaccination and failure to address issues of health inequality, seems particularly pertinent in the wake of the Covid-19 pandemic and the enthusiastic promotion of vaccination as a magic bullet solution by many governments. Similar doubts have been expressed about Grand Challenges in other fields. Erin Cech (2012) has criticized the US National Academy for Engineering's *Grand Challenges for Engineering* report for, among other things, its technological determinism and separating "technical" matters from the social, political, ethical, and cultural realms of these challenges. Desalination is presented as a solution to water shortages, without any reference to the complex social, political, and legal problems involved. Cech criticizes engineers for failing to engage with other professionals such as "political scientists, sociologists, historians, science and technology studies scholars, ethicists, etc." (2012, 91).

The kind of problems that critics like Birn and Cech have flagged suggest that there is an urgent need for humanities to become more involved with the Grand Challenge agenda in order to create a less technologically driven and more socially responsive approach. The humanities can provide stronger historical perspectives, develop more nuanced ethical frameworks, and contribute to the creation of more multidisciplinary and collaborative methodologies. There are many other areas where the humanities—and particularly digital humanities—are well placed to help generate a less technologically deterministic view of Grand Challenges. This makes it all the more surprising that there has been no concerted attempt to articulate a set of humanities Grand Challenges. There have been some calls by individual scholars for humanities Grand Challenges (Bostic 2016; Finnemann 2016), but never any campaign across the community of humanities scholars to identify Grand Challenges in the way we have seen for global health or engineering. Should there be a set of humanities Grand Challenges? Is the idea of the Grand Challenge appropriate in the humanities or is the humanities too diffuse and too varied in its perspectives to identify concrete challenges in the way STEM subjects have?

There clearly are potential Grand Challenges in the humanities. One need only think of such subject areas as religion or language to realize the fundamental role that the study of the humanities can play in addressing the world's problems. Humanities scholars in the past have addressed issues which can be considered Grand Challenges—Lord Acton's ill-fated attempt to write the history of liberty can be considered a Grand Challenge. But perhaps there are methodological issues which prevent the humanities readily embracing the idea of a Grand Challenge. Humanities research rarely resolves itself into a concrete problem which can be "resolved" or "overcome." In his discussion of the idea of the humanities, published in 1967, the literary scholar Ronald Crane observed that the sciences are most successful when they move forward from a variety of observations towards as high a degree of unity, uniformity, and simplicity as their material permit. By contrast, according to Crane, the humanities

> are most alive when they reverse this process, and look for devices of explanation and appreciation that will enable them to preserve as much as possible the variety, the uniqueness, the unexpectedness, the complexity, the originality that distinguish what men [sic] are capable of doing at their best from what they must do, or tend generally to do, as biological organisms or members of a community.
>
> (Crane 1967, 12)

Crane's view of the humanities appears nowadays extremely retrograde (not least in its casual sexism), and is chiefly of interest as a historical document illustrating the traditions which shaped the humanities, but nevertheless his distinction between a tendency in the sciences to look for focused concrete findings against studies in the humanities which emphasize complexity and diversity may partly explain why the idea of Grand Challenges has been less prominent in the humanities than the sciences.

The humanities reticence about Grand Challenges may, however, have deeper roots and reflect diffidence about the role of technology in the humanities. Willard McCarty has memorably described the timidity with which literary scholars in the 1960s viewed the first experiments with the use of mainframe computers in literary and historical studies:

> Fear was variously expressed in the professional literature of digital humanities: fear of the distortions computing would work on the humanities if taken seriously, evinced by the work and words of those who did take it seriously; fear of its mechanization of scholarship, parallel to the mechanization of which public intellectuals had been warning; fear of its revolutionary force, threatening to cast aside old-fashioned ways of thinking, as literary scholar Stephen Parrish declared was about to happen; and fear expressed in reassurances, such as literary critic Alan Markman's, that the computer is no threat to scholarship or a dehumanizing machine to be feared.
>
> (2014, 290)

This reassurance that the humanities were not about to be enslaved by technology was expressed in Alan Markman's insistence that "Man made the machine, and men will use it as a tool" (1965, 79) and this affirmation that the computer is a tool and that the important thing is the quality of the humanities scholarship it enables has been a recurrent theme ever since the time of Busa.

McCarty links some of these anxieties to the emergence of the computer during the early phase of the Cold War, but we can take this humanities anxiety about technology much further back. The modern study of the humanities, originating with the Romantics in the early nineteenth century, was a reaction to industrialization and the rise of a mechanistic utilitarian society (Prescott 2012). Samuel Taylor Coleridge called for "a general revolution in the modes of developing and disciplining the human mind by the substitution of life and intelligence for the philosophy of mechanism which, in everything that is most worthy of the human intellect, strikes *Death*" (1815).

Coleridge proposed the establishment of the Clerisy, an intellectual class paid by the state, who would be "the fountainhead of the humanities, in cultivating and enlarging the knowledge already possessed and in watching over the interests of physical and moral science" (Williams 1961, 78). Coleridge's concerns about the nature of the emerging industrial society and its effects on the population were shared by his contemporaries. In Wordsworth's poems, we encounter beggars whose anxious movements reflect the pointless and repetitive movements associated with the introduction of machines (Simpson 2009). Cobbett expressed his horror at the way the cash nexus was becoming all pervasive: "We are daily advancing to the state in which there are but two classes of men, masters and abject dependents" (Williams 1961, 33).

The humanities as we currently understand them were constructed as a defense mechanism against the effects of technology and industrialization, using culture to keep a mechanistic reductionist science at bay. The terms "art" and "artist" first began to be used with reference to the imaginative and creative arts during the period 1760 to 1830 and the redefinition of these terms was a means of affirming spiritual values in the face of the dehumanizing effects of the industrial and agricultural revolutions (Williams 1961, 15; 1976, 27–35). In his seminal work *Culture and Society* (Williams 1961), Raymond Williams traces the way in which culture similarly came to be constructed as a bulwark against the effects of industrialization. Williams analyzed what F. R. Leavis called "The Great Tradition" running from Burke and Cobbett through Romantic artists such as Coleridge and Wordsworth through to John Stuart Mill and Matthew Arnold and ultimately T. S. Eliot and Leavis himself. This "great tradition" affirmed an idea of culture which rejected numbers and measurement and kept the machine at bay. It sought to emphasize the superiority of human to the machine, and stressed that a machine was merely an expression of human ingenuity and was always inferior to the complex philosophical understanding of the human soul.

In this sense, the idea of the humanities described by critics such as Ronald Crane was itself a response to a Grand Challenge—it was an intellectual program designed to affirm the primacy of human thought and achievement and to resist the global challenges posed by the rise of technology and industry and their associated reductionism and quantification. In this context, it becomes evident why the arrival of computers caused such anxiety among humanities scholars in the 1950s and 1960s. The use of computers was a major breach in the anti-technological bastion of the humanities. If the superiority of human achievement and culture was to be maintained, it was vital to emphasize that the computer could only ever be a tool and that it was simply a means of enhancing the higher order theoretical and cultural reflections of the human scholar.

The humanities have been engaging Grand Challenges for a long time. Because modern study of the humanities was rooted in reactions to the growth of a commercial and technological society, the profound shifts to society and culture generated by digital and networking technologies challenge the way we conceive of the humanities. We may lament the rise of post-truth and wring our hands at the popularity of conspiracy theories fed by social media, but these troubling developments ask fundamental questions about how we engage with meaning and knowledge, issues at the heart of the humanities. Humanities computing and then digital humanities spent a long time persuading humanities scholars that computers have an important role in humanities scholarship. Now that humanities scholars have become addicted to computers and digital resources, the role of the digital humanities has switched to, on the one hand, promoting a more critical approach to digital methods and resources and, on the other, encouraging a more imaginative and creative engagement with digital methods.

What might humanities Grand Challenges look like? In the first instance, they will not be confined to the humanities alone. Grand Challenges are inherently complex with a large number of cross-connections across disciplines and organizational structures. Humanities will be only one of a number of fields tackling the challenge. Nevertheless, there are major challenges where the contribution of the humanities will be vital, and where the digital humanities can make a fundamental contribution.

At the forefront of these challenges is the health of the Internet. We have created a world in which our digital environment is as important for our well-being as clean air or pure water. We can see the dangers posed by our digital culture in every direction—whether it is the mental health of children damaged by social media algorithms, the threat to public health posed by anti-vaccination propaganda, or the political damage and social polarization caused by the manipulation of social media by governments and political parties. The enthusiasm with which many digital humanities practitioners (including myself) embraced social media in the late noughties now seems embarrassingly naive. Yet our early engagement with these media gave us a good understanding of the dangers and risks (Ross, Terras, Warwick, and Welsh 2011). I remember attending a seminar in 2011 in which the use by Putin supporters of Twitter bots to spread propaganda and discredit Putin's opponents in Russian elections was traced in chilling detail (Kelly, Barash et al. 2012). It seemed impossible to imagine that this could within five years be scaled up enough to have an impact on an American Presidential Election, but that is what happened.

These well-publicized problems are connected with the rise of "fake news" and "post-truth." It is sometimes suggested that these problems could be countered by large-scale fact checking and automated flagging and deletion of social media. There are regular calls from politicians and governments demanding that social media companies either delete or flag posts that are false, misleading, or wrong. These demands are naive, not only because of the sheer technical scale of the problem of monitoring billions of social media postings but also because of the difficulty of establishing what is right or wrong. Truth is rarely a simple matter of "yes" or "no," a binary field that can be flagged, and saying whether particular posts in social media are false or accurate can be a difficult task. In the UK, the expert Joint Committee on Vaccines and Immunisation was unable to establish that the health benefits of twelve–fifteen year olds receiving a Covid-19 vaccine outweighed the other risks, such as heart inflammation (JCVI 2021). The decision to offer vaccination to twelve–fifteen year olds was, in the UK, a political decision reflecting views on the wider best interests of children (Iacobucci 2021). Would a Facebook posting expressing doubts about vaccinating twelve–fifteen-year-old children be a piece of fake news? It is difficult to say, and doubt might be felt as to whether a Facebook advisory notice could cover all the issues. Identifying and eradicating fake news is a complex task, a true Global Challenge.

Among the most important initiatives launched in response to Internet security and privacy controversies is the Mozilla Foundation's monitoring of the health of the Internet.[9] Mozilla issued a prototype Internet health report in November 2017 and followed with the first full open-source report a year later.[10] The report assesses the health of the Internet with reference to such topics as privacy and security; openness; digital inclusion; web literacy and decentralization. At a technical level, these are all aims which the digital humanities should actively support. It is vital for example that digital humanities resists the increasing tendency for heritage institutions to rely on commercial partnerships in creating digital resources (Prescott 2014). Digital humanities practitioners should also urge universities to resist the corporate embrace of Microsoft, Google, and Amazon in provision of computing services in higher education.

But, if we are to have a truly healthy Internet, we need to go beyond these more narrowly technical measures to consider how we promote a healthy intellectual and cultural environment in our new digital world. We are clearly failing to do so in innumerable different ways. There is an interesting contrast here with the early days of radio. The Wild West of early radio broadcasting in the United States with its "chaos of the ether" and the "blasting and blanketing of rival programmes" echoed many of the problems of the Internet today (Briggs 1985, 17–20). In the UK, there was widespread concern that Britain should not follow the American model. Arthur Burrows, afterwards the first Director of Programmes for the BBC, warned in 1924 of "the dangers which might result in a diversely populated country of a small area like our own if the go-as-you please methods of the United States were copied" (Briggs 1985, 20), while an editorial in the *Manchester Guardian* in 1922 insisted that "broadcasting is of all industries the one most clearly marked out for monopoly. It is a choice between monopoly and confusion" (Briggs 1985, 33). The upshot was the creation of the BBC, financed by a levy on receiving equipment, and intended to ensure that the excesses of American commercial radio were avoided. Whatever misgivings we might feel about the BBC, it has made an important contribution to promoting cohesion in British society over the past century. It is tempting to imagine that the polarization caused by social media would be reduced if it was in the hands of a public corporation. However, the same solutions that worked with radio will not work with the Internet, not only because of the international character of the Internet but also because the imposition of a Reithian high culture is no longer a suitable or acceptable approach.

It is sometimes suggested that recent work in the humanities has fostered a relativistic outlook which is liable to promote post-truth (Calcutt 2016). This is misleading because the most celebrated post-truth conspiracies are driven by very positivistic reasoning. QAnon conspiracy theorists pile irrelevant fact on misleading detail to reach their bizarre conclusions. Birthers engage in detailed diplomatic examination of every aspect of Barak Obama's birth certificate, right down to the layers setting in the pdf, in attempting to "prove" it is a forgery (Prescott Forthcoming). "Post-truth" is perhaps rather more an overdose of an empirical positivistic view of truth—a belief in the ability of the individual to find hidden truths by extrapolating from supposed "facts," resulting in bizarrely misconceived conclusions, a tendency which has been very evident during the pandemic.

The digital humanities have deep links with librarians, archivists, and museum curators, and all these disciplines combine a profound respect for the importance of reliable evidence together with an awareness of the complexity of that evidence and the need for expert understanding in interpreting it. They are aware of the way truth shifts and unfolds as we explore who made a document, why, and for what purpose. They understand that a legal document might involve false names like Jane Roe, a pseudonym used to protect the privacy of Norma McCorvey, one of the parties in the celebrated Roe v. Wade abortion ruling in the United States (Talbot 2021). The use of the names Jane Roe and John Doe to conceal the identity of fictitious litigants is a practice that stretches back to the English Chancery of the fifteenth century. But the fact that Jane Roe is fictitious does not mean that the law case was a myth or that it is not important in the social history of America. Archival documents are full of fictions and common form which have a legal function but are otherwise completely invented. The type of land transaction known as a recovery by double voucher involved starting a sequence of fictitious lawsuits which were duly recorded in the property deed. The skill in interpreting such a document is in recognizing the elements that relate to an authentic land transfer (Prescott 2019).

Humanities scholars at their best understand the complexities of truth and the way in which its different layers of meaning shed light on various aspects of culture and society. How can that understanding of the complex nature of truth inform our digital culture and how can we (as Reith and others attempted to do with the BBC in the 1920s) develop an Internet which promotes popular awareness of cultural complexity, generates a more cohesive society, and leaves us better informed and educated? Here is a Grand Challenge indeed and one which the humanities can truly claim as its own. It stands in direct line of succession from the attempts of Coleridge, Ruskin, and Arnold to promote the life of the mind in the face of the Mammon of the Industrial Revolution.

This Grand Challenge will become more pressing as AI becomes more pervasive. Once again, it is hardly necessary to recapitulate examples of AI systems displaying racial and gender bias (Prescott In Press). There has been a stream of press stories. Word embeddings trained on Google news data complete the sentence "Man is to computer programmer as woman is to X" with the word "homemaker" (Bolukbasi et al. 2016). Another study used association tests automatically to categorize words as having pleasant or unpleasant associations and found that a set of African American names had more unpleasantness associations than a European American set. The same machine-learning program associated female names with words like "parent" and "wedding" whereas male names had stronger associations with such words as "professional" and "salary" (Caliskan, Bryson, and Narayanan 2017). Studies of commercial facial recognition systems deployed by police forces, law courts, and even educational institutions show that they do not accurately identify people of color (Buolamwini and Gebru 2018; Buolamwini 2019; Raji and Buolamwini 2019; Page 2020).

The deployment of AI systems by commercial firms and government agencies means that these problems are likely to become more acute in future. Many of the problems are caused by the way training sets of data for AI systems are developed. Humanities scholars are expert in assessing biases in large datasets such as censuses, library catalogs, or corpora of newspapers. They are exceptionally well placed to help develop training sets for AI systems which will avoid the biases which are at present all too evident. AI will shortly present the humanities with some of the grandest challenges of all.

Many other key technological challenges are also ones where the humanities can make fundamental contributions. The understanding and acquisition of language in a real-world environment is a fundamental issue for future robotics services. The ethics of our future relationship with robots and AI poses a myriad of complex issues. Do robots have rights? How do we need to adjust our understanding of rights in a robotic world? Addressing such Grand Challenges is not a distraction from the fundamental concerns of the humanities, but rather puts the humanities right back where they belong, at center stage in creating a more cohesive, educated, and equitable society.

NOTES

1. sdgs.un.org.
2. UKRI: Global Challenges Research Fund, www.ukri.org/our-work/collaborating-internationally/global-challenges-research-fund/.
3. European Union: Horizon 2020, ec.europa.eu/programmes/horizon2020/en/h2020-section/societal-challenges.

4. UK Advanced Research and Invention Agency, https://www.gov.uk/government/publications/advanced-research-and-invention-agency-aria-statement-of-policy-intent/advanced-research-and-invention-agency-aria-policy-statement.
5. UKRI: Global Challenges Research Fund, www.ukri.org/our-work/collaborating-internationally/global-challenges-research-fund/themes-and-challenges/.
6. SOLACE project, solace-research.com.
7. Global Research Network on Parliaments and People, grnpp.org.
8. The Quipu Project, interactive.quipu-project.com.
9. Mozilla Foundation Internet Health Report Overview, http://internethealthreport.org.
10. Mozilla Foundation 2018 Internet Health Report, https://internethealthreport.org/2018/.

REFERENCES

Birn, Anne-Emanuelle. 2005. "Gates's Grandest Challenge: Transcending Technology as Public Health Ideology." *The Lancet* 366 (9484) (August 6–12): 514–9. https://doi.org/10.1016/S0140-6736(05)66479-3.

Bolukbasi, Tolga, Kai-Wei Chang, James Zou, Venkatesh Saligrama, and Adam Kalai. 2016. "Man is to Computer Programmer as Woman is to Homemaker? Debiasing Word Embeddings." *29th Conference on Neural Information Processing Systems (NIPS).*

Bostic, Heidi. 2016. "The Humanities Must Engage Global Grand Challenges." *Chronicle of Higher Education*, March 30. http://chronicle.com/article/The-Humanities-Must-Engage/235902.

Briggs, Asa, 1985. *The BBC: The First Fifty Years*. Oxford: Oxford University Press.

Buolamwini, Joy. 2019. "Artificial Intelligence Has a Problem with Gender and Racial Bias. Here's How to Solve It." *Time Magazine*, February 7. https://time.com/5520558/artificial-intelligence-racial-gender-bias/.

Buolamwini, Joy and Timnit Gebru. 2018. "Gender Shades: Intersectional Accuracy Disparities in Commercial Gender Classification." Proceedings of the 1st Conference on Fairness, Accountability and Transparency. *PMLR* 81: 77–91. https://proceedings.mlr.press/v81/buolamwini18a.html.

Burdick, Anne, Johanna Drucker, Peter Lunenfeld, Todd Presner, and Jeffrey Schnapp. 2012. *Digital_ Humanities*. Cambridge, MA: MIT Press.

Calcutt, Andrew. 2016. "The Surprising Origins of 'Post Truth' – and How it was Spawned by the Liberal Left." *The Conversation*, November 18. https://theconversation.com/the-surprising-origins-of-post-truth-and-how-it-was-spawned-by-the-liberal-left-68929.

Caliskan, Aylin, Joanna J. Bryson, and Arvind Narayanan. 2017. "Semantics Derived Automatically from Language Corpora Contain Human-like Biases." *Science* 356: 183–6. https://doi.org/10.1126/science.aal4230.

Cech, Erin. 2012. "Great Problems of Grand Challenges: Problematizing Engineering's Understandings of Its Role in Society." *International Journal of Engineering, Social Justice, and Peace* 1 (2) (Fall): 85–94. https://doi.org/10.24908/ijesjp.v1i2.4304.

Coleridge, Samuel Taylor. 1815. "Letter to William Wordsworth, 30 May 1815." In Ernest Hartley Coleridge (ed.), *Letters of Samuel Taylor Coleridge, Vol II, eBook #44554, The Project Gutenberg*, gutenberg.org., 649.

Crane, Ronald S. 1967. *The Idea of the Humanities, and Other Essays, Critical and Historical*. Chicago, IL: University of Chicago Press.

Ferraro, Fabrizio, Dror Etzino, and Joel Gehman. 2015. "Tackling Grand Challenges Pragmatically: Robust Action Revisited." *Organization Studies* 36 (3): 363–90. https://doi.org/10.1177/0170840614563742.

Finnemann, Niels Ole. 2016. "Big Data and Grand Challenges. Digital Humanities in between the Humanities and the Anthropocene." Presentation at *Digital Humanities – Directions and Horizons*, Center for Computing and Communication, University of Copenhagen, July 8. https://curis.ku.dk/ws/files/163161892/Big_Data_and_Grand_Challenges_Presentation.pdf.

George, Gerard, Jennifer Howard-Grenville, Aparna Joshi, and Laszlo Tihanyi. 2016. "Understanding and Tackling Societal Grand Challenges through Management Research." *Academy of Management Journal* 59 (6): 1880–95. https://doi.org/10.5465/amj.2016.4007.

Goldstone, Andrew. 2017. "Teaching Quantitative Methods: What Makes It Hard (in Literary Studies)." 3 January. https://andrewgoldstone.com/blog/ddh2018preprint/.

Goldstone, Andrew. 2019. "Teaching quantitative methods: What makes it hard (in literary studies)". In: Gold, Matthew K. and Klein, Lauren F. (eds.). *Debates in the Digital Humanities, 2019*. Minneapolis, MN: University of Minnesota Press.

Grand Challenges Canada/Grand Défis Canada. 2011. *The Grand Challenges Approach*. Toronto: McLaughlin-Rotman Centre for Global Health.

Gümüsay, Ali Aslan, and Patrick Haack. 2020. "Tackling COVID 19 as a Grand Challenge." Digital Society blog, June 8. https://doi.org/10.5281/zenodo.3884986.

Howard-Grenville, Jennifer. 2020. "Grand Challenges, COVID 19 and the Future of Organizational Scholarship." *Journal of Management Studies* 58 (1): 254–8. https://doi.org/10.1111/joms.12647.

Iacobucci, Gareth. 2021. "Covid-19: Children Aged 12–15 Should Be Offered Vaccine, Say UK's Chief Medical Officers." *British Medical Journal* 374 (2248) (September 13).

JCVI. 2021. *JCVI statement on COVID-19 vaccination of children aged 12 to 15 years: 3 September 2021*. https://www.gov.uk/government/publications/jcvi-statement-september-2021-covid-19-vaccination-of-children-aged-12-to-15-years/jcvi-statement-on-covid-19-vaccination-of-children-aged-12-to-15-years-3-september-2021.

Kelly, John, Vladimir Barash, Karina Alexanyan, Bruce Etling, Robert Faris, Urs Gasser, and John G. Palfrey. 2012. *Mapping Russian Twitter*. Cambridge, MA: Berkman Center Research Publications (3). https://papers.ssrn.com/sol3/papers.cfm?abstract_id=2028158.

McCarty, Willard. 2014. "Getting There from Here: Remembering the Future of Digital Humanities: Roberto Busa Award lecture 2013." *Literary and Linguistic Computing* 29 (3) (September): 283–306. https://doi.org/10.1093/llc/fqu022.

Markman, Alan. 1965. "Litterae ex Machina. Man and Machine in Literary Criticism." *Journal of Higher Education* 36 (2) (February): 69–79.

Page, Rosalyn. 2020. "Spotlight on Facial Recognition after IBM, Amazon and Microsoft Bans." CMO. June 16. https://www.cmo.com.au/article/680575/spotlight-facial-recognition-after-ibm-amazon-microsoft-bans/.

Patten, Chris. 2021. "UK Funding Cuts Are a Slap in the Face for Science." *Financial Times*, April 7.

Prescott, Andrew. 2012. "An Electric Current of the Imagination: What the Digital Humanities Are and What They Might Become." *Journal of Digital Humanities* 1 (2). journalofdigitalhumanities.org/1-2/an-electric-current-of-the-imagination-by-andrew-prescott/.

Prescott, Andrew. 2014. "Dennis the Paywall Menace Stalks the Archives." http://digitalriffs.blogspot.com/2014/02/dennis-paywall-menace-stalks-archives.html.

Prescott, Andrew. 2015. "Beyond the Digital Humanities Centre: The Administrative Landscapes of the Digital Humanities." In *A New Companion to Digital Humanities*, edited by Susan Schreibman, Ray Siemens, and John Unsworth, 461–75. Oxford: Wiley-Blackwell.

Prescott, Andrew. 2019. "Tall Tales from the Archive." In *Medieval Historical Writing: Britain and Ireland, 500–1500*, edited by Jennifer Jahner, Emily Steiner, and Elizabeth M. Tyler, 356–69. Cambridge: Cambridge University Press.

Prescott, Andrew. In Press. "Bias in Big Data, Machine Learning and AI: What Lessons for the Digital Humanities?" *Digital Humanities Quarterly*.

Prescott, Andrew. Forthcoming. "Post-Truth, Digital Cultures and Archives." In David Thomas, Michael Moss, and Susan Stuart, *Post-Truth and Archives* (under review).

Raji, I. D. and J. Buolamwini. 2019. "Actionable Auditing: Investigating the Impact of Publicly Naming Biased Performance Results of Commercial AI Products." *AIES '19: Proceedings of the 2019 AAAI/ACM Conference on AI, Ethics, and Society*: 429–35. https://doi.org/10.1145/3306618.3314244.

Ross, Claire, Melissa Terras, Claire Warwick, and Andrew Welsh. 2011. "Enabled Backchannel: Conference Twitter Use by Digital Humanists." *Journal of Documentation* 67 (2): 214–37.
Scheinfeldt, Tom. 2013. "Where's the Beef? Does Digital Humanities Have to Answer Questions?" *Debates in the Digital Humanities*. University of Minnesota Press.
Simpson, David. 2009. *Wordsworth, Commodification, and Social Concern: The Poetics of Modernity*. Cambridge: Cambridge University Press.
Underwood, Ted and Sellers, Jordan. 2016. "The *longue durée* of literary prestige." *Modern Language Quarterly* 77 (3): 321–344.
Talbot, Margaret. 2021. "How the Real Jane Roe Shaped the Abortion Wars." *New Yorker*, September 13. https://www.newyorker.com/magazine/2021/09/20/how-the-real-jane-roe-shaped-the-abortion-wars.
Thompson, Andrew. 2021. "Cuts to International Research are a Wrong Turn for a Country Going Global." *Times Higher Education Supplement*, April 15. www.timeshighereducation.com/opinion/cuts-international-research-are-wrong-turn-country-going-global.
Williams, Raymond. 1961. *Culture and Society 1780–1950*. Harmondsworth: Penguin.
Williams, Raymond. 1976. *Keywords: A Vocabulary of Culture and Society*. London: Croom Helm.

CHAPTER THIRTY-SEVEN

Digital Humanities Futures, Open Social Scholarship, and Engaged Publics

ALYSSA ARBUCKLE, RAY SIEMENS
(UNIVERSITY OF VICTORIA), AND THE INKE PARTNERSHIP

Are academics alone responsible for the evolution of the digital humanities, and its future? Will the future of digital humanities be shaped by pieces in collections such as this, typically written for other academics? We think not, or at least, not entirely. Rather, we begin with the premise that, while the exact future of the digital humanities is ultimately unknowable, it will be shaped by a number of current and emerging forces—academic, individual, institutional, social, societal, and infrastructural among them. More than an academic thought experiment, the impact and influence of these broader forces draw on the interrelation of theory, praxis, and extra-academic involvement, and necessitate the involvement of all those who have a stock in that future. In this context, we are increasingly invested in the concept of *open social scholarship*, and how the digital humanities embraces, and may one day even fully embody, such a concept. Originating in partnered consultations among a group representing these broader perspectives, the term *open social scholarship* refers to academic practice that enables the creation, dissemination, and engagement of open research by specialists and non-specialists in accessible and significant ways.[1] Our contribution to the present volume suggests that open social scholarship supports many possible futures for the digital humanities, especially as its foundation incorporates a shift from notions of audience *for* academic work to publics engaged *by* and *in* that work.

CONTEXT: FROM SPEAKING TO AUDIENCES TO ENGAGING WITH PUBLICS

We understand the digital humanities as an evolving field. In what follows we highlight the influx of digital humanities engagement with open access, social media, public humanities, and other activities that deviate from earlier, more conventional forms of scholarly communication. In doing so, we

---

Our thinking in this chapter is influenced to some degree by our situation in pragmatic structures of the North American academic context, although we recognize that there are similarities across other geographic areas too. It is also informed at heart by our work with the community-based Implementing New Knowledge Environments (INKE) Partnership (inke.ca). Some elements of the thinking behind this work are drawn from Siemens' Alliance of Digital Humanities Organizations Zampolli Prize Lecture at the University of Lausanne on July 2, 2014 (Siemens 2016).

align with a conception of the digital humanities as a point of intersection between the concerns of the humanities, and all they might encompass, and computational methods. We perceive digital humanities as an accumulation of 1) computationally modeling humanistic data, 2) processes that provide tools to interact with that data (within humanistic frameworks), and 3) communication of resultant work on that data with those processes. All three of these areas have seen fast-paced development, adoption, and change over recent years. One of the most significant areas of change has been in and around how digital humanities practitioners encounter and communicate the work of the field.

What academics have typically called the *audience* for their work has changed alongside evolving points of academic interest in the humanities and is still changing in ways important to our community of practice in the digital humanities (Siemens 2016). Typically, audience has been implied rather than clearly articulated in the academic work of a field. Like a ghost, audience is both there and not there; persistent as an imagined receiver of address, but ambiguous in shape, size, and detail. In many ways, the primary audience of the digital humanities—like most fields—is understood to be other practitioners of the field. Perhaps a digital humanities scholar has a specific subset or group of people in mind as they undertake their digital scholarship (project- or publication-based), but often these audience members are still academics—as much as they may have differing opinions on the history, scope, and priorities of the field. There is also a presumption that this audience would somehow benefit from the work in question as a contribution to the field.

The notion of *publics* can be separated from that of *audience*, and in doing so presents a stronger, more useful way to talk about our connection with those served by, implicated in, and involved with our work. An elusive concept in some ways, *publics* may still be more useful to consider in the context of digital humanities futures. "Publics have become an essential fact of the social landscape, but it would tax our understanding to say exactly what they are" (2002, 413), Michael Warner considers in his piece "Publics and Counterpublics." He concludes that a public is constituted by virtue of its address, as well as by giving attention to the addressor. A public shows up and listens, passively or actively, and the active involvement of *publics* stems from the shift from a one-to-many discourse (one academic to an audience of many, for instance) to a collaborative relationship where multiple actors with a shared interest form a public through mutual investment and interaction. In this way, there is the potential for *publics* to be more involved than *audiences*.

The real, perceived, and potential relationships that exist between the academy and the broader publics it serves have seen increased attention and articulation. The value of the humanities and of humanities-based approaches to public engagement, for instance, is made evident in the emergence of the public humanities as a field unto itself. Groups like the Visionary Futures Collective take this notion as central to their work. Part of their mission statement reads: "This group believes that the study of human history and cultural expression is essential to a more just and meaningful society" (Visionary Futures Collective n.d.). Moreover, the group states: "We believe that higher education should be for the public good, and that the work of the humanities should be conducted for and with our communities" (n.d.). Kathleen Fitzpatrick emphasizes these ideals in *Generous Thinking: A Radical Approach to Saving the University* (2019), where she explicitly advocates for a values-based approach that surfaces and concretizes ideals of care, empathy, community, and receptivity. Ultimately, Fitzpatrick argues for a more humane academia both inside and out; an academia that understands and values its own community as well as considers and engages with broader communities. In solidifying and conceptualizing these relations, Fitzpatrick underscores that it is problematic to envision the public as a homogenous monolith and, moreover, as a singular group necessarily apart from academic institutions. As a concept, *publics* may resist simple definition; it is brought into being by processes of engagement, in discourse or otherwise, toward shared ends.

When moving from perceptions of delineated scholarly audience to deliberation with engaged publics, the potential *openness* of academic practice becomes an essential consideration. In practicing openness, scholars can carefully and explicitly consider their relation to and participation in publics, in discursive and collaborative communities. Such communities are not constructed by status (academic or otherwise); rather, they can be considered as collectives that come together with shared interests across the personal/professional continuum—perhaps even as communities of practice with academic, academic-aligned, and non-academic members. These publics embody positive, inclusive, and mutually beneficial relations between academic institutions and so-called *broader society*. Openness appears as a defining characteristic of the successful engagement of publics and, in this vein, public humanities scholars have much to offer the digital humanities. For instance, a key takeaway from public humanities scholars Wendy F. Hsu and Sheila Brennan is that digital humanists should not assume that simply putting research online constitutes public engagement. Brennan writes: "it is important to recognize that projects and research may be available online, but that status does not inherently make the work digital public humanities or public digital humanities" (2016, 384). Hsu argues that professional encounters with non-academic communities need to be understood as collaborative ventures or opportunities to communally share and create knowledge: "using the digital to learn from the public is a listening practice, one that yields more efficacious and engaged public humanities work" (2016, 281), she writes. According to both Brennan and Hsu, digital humanities researchers would be wise to consider publics at the inception of a research project rather than after the fact, or worse, as objectified subjects of examination or mere data points.

ACTION IN CONTEXT: DIGITAL HUMANITIES, ENGAGING ITS PUBLICS

Openness in academic practice is not necessarily a new consideration for the digital humanities community. Indeed, much digital humanities work already engages with publics and with increasing recent focus on critical social issues. According to Alison Booth and Miriam Posner, digital humanities is "most worthwhile if it also *promotes public engagement and humanistic knowledge and understanding*" (2020, 10; emphasis in original). Such a value statement indicates support for more open and more social digital work, rooted in humanities approaches to knowledge creation.

Many digital humanities projects are—at the very least—openly accessible for viewing on the Internet, and others consider *openness* in the context of public engagement and public service more explicitly as a clear part of their mandate. For instance, the *Torn Apart/Separados* project "aggregates and cross-references publicly available data to visualize the geography of Donald Trump's 'zero tolerance' immigration policy in 2018 and immigration incarceration in the USA in general" (n.d.) in order to raise awareness about immigration challenges and persecution.[2] The *American Prison Writing Archive,* directed by Doran Larson at Hamilton College, is a public database of prison writing that archives and presents submissions from currently and formerly incarcerated people, as well as those who work in the prison system (e.g., correctional officers, staff, administrators, and volunteers).[3] In doing so, the project humanizes incarcerated people and offers them a platform to share their voices and experiences. *The Digital Oral Histories for Reconciliation* project, led by Kristina Llewellyn, employs virtual reality in the service of reconciliation. Currently, the project is partnered with the Nova Scotia Home for Colored Children Restorative Inquiry to shed light on a residential school targeted at Black Nova Scotian children in the twentieth century (Philpott 2020).[4] Gretchen Arnold's *Nuisance Laws and Battered Women* project draws together research and interviews of domestic violence victims evicted from their homes under nuisance laws, with an aim of informing lawmakers who might consider changing legal practices.[5] Michelle Swartz and Constance Crompton lead the *Lesbian and Gay Liberation in Canada* project, which "is an interactive digital resource for the study of lesbian, gay, bisexual, and transgender (LGBT) history in Canada from 1964 to 1981" (n.d.).[6] *Lesbian and Gay Liberation in Canada* draws attention to the long history of queer activism, advocacy, and expression in Canada that is often overlooked. *The Atikamekw Knowledge, Culture and Language in Wikimedia Project* is a community-based project to develop a version of Wikipedia in the Atikamekw language, and in doing so, to increase the presence of Indigenous knowledge, culture, and language in the Wikimedia ecosystem.[7] These are a handful of examples, among many, of digital humanities projects that center and create publics around specific, evolving community concerns.[8]

Notably, the public-facing projects referenced above align with recent socially focused, critical calls for transition in our thinking about digital humanities in and of itself, and how we understand its assumptions, structures, and positions in varying publics. In 2012, Alan Liu encapsulated community discussions by asking "Where Is Cultural Criticism in the Digital Humanities?" at the same time as Tara McPherson posited "Why Are the Digital Humanities So White?" More recently, Roopika Risam has considered the idea of *locus*, arguing that digital humanities must shift its power centers from Global North locales like Canada, the United States, and the United Kingdom in order to facilitate actual diversity and inclusion (2016). Risam calls for local or regional concerns to be centered in digital humanities work, which will lead to "its global dimensions [being] outlined through an assemblage of the local" (2016, 359). David Gaertner has focused on the intersection between Indigenous literature and digital humanities, or rather the lack thereof.[9] On a similar note, Posner urges that calls to diversify the field and consider more seriously issues of race and gender are crucial, but not enough. "It is not only about shifting the focus of projects so that they feature marginalized communities more prominently," she writes, "it is about ripping apart and rebuilding the machinery of the archive and database so that it does not reproduce the logic that got us here in the first place" (2016, 35). Safiya Umoja Noble amplifies and extends these positions by foregrounding the social context of the digital humanities and suggests "if ever there were a place for digital humanists to engage and critique, it is at the intersection of neocolonial investments in information, communication, and technology infrastructures: investments that rest precariously on colonial history, past and present" (2018, 28). As Noble writes, the very positionality of the digital humanities—sitting, as it does, at the crossroads of humanistic inquiry and technology—comes with responsibility. These scholars, among many others, reflect a close-eye, critical consideration of the foundational concepts of the digital humanities, including its tools and technological contexts.[10]

The turn to more critical considerations reflects significant forces in the ongoing evolution of the digital humanities, including the shift from audience-thinking to publics-thinking. Rather than objectifying participants into delineated roles of either speaker or mass, imagined addressee, publics-thinking recognizes all as potentially participating, embodied subjects in a shared social plane. Such a transition requires a contextual understanding of and care for contemporary social, political, and economic conditions, including how such conditions affect different communities in different ways.

All of this considered, does our own thinking about publics and public engagement align, as fully as is ideal, with the products of our scholarship? Are digital humanities practitioners making the best use of open and accessible methods to engage broader, more diverse publics? Despite exemplary forays into public engagement, and at times transformative exploration and advancement of alternative practices to date, the digital humanities community can still be quite beholden to (and still perpetuates) a traditional model of audience-focused humanities academic publishing—replete with earlier assumptions about means, methods, and utility. This model of academic publishing relies on existing scholarly communication infrastructure: journals and monographs, largely, and the intellectual constructs they represent. Many digital humanists still follow the practice of producing a work of scholarship, publishing it in a standard, toll-access peer-reviewed journal, and passively waiting for this scholarship to reach an audience of other academics who have access to such specialized publication products. There is irony in upholding traditional publication norms that do not accurately reflect our own field's actual, evolving concerns and practices. Booth and Posner elaborate:

> While [digital humanities] may yield a peer-reviewed article, essay collection or journal, or single-author monograph, these are often based on datasets, programming, documentation of method and results, visualizations, and user interfaces, all of which are hard to encompass in one reading for review, not to mention that they remain living artifacts created by many hands, seldom finished or preserved on one publication date.
>
> (2020, 10)[11]

Beyond the disconnect between digital humanities *work* and digital humanities *publication*, it is necessary to draw attention to how conventional publishing practices delimit knowledge creation and sharing, especially within the context of public engagement. "Enabling access to scholarly work," Fitzpatrick argues, "does not just serve the goal of undoing its commercialization or removing it from a market-driven, competition-based economy, but rather is a first step in facilitating public engagement with the knowledge that universities produce" (2019, 148). With toll-access publishing there is little consideration for the publics who do *not* accept that it is reasonable for publicly funded research to be inaccessible. Moreover, there is little thought of publics whose own information consumption and knowledge creation activities occur in spaces very different than a proprietary and for-cost academic journal platform. As Fitzpatrick asserts: "If we hope to engage the public with our work, we need to ensure that it is open in the broadest possible sense: open to response, to participation, to more new thought, more new writing, and more kinds of cultural creation by more kinds of public scholars" (2019, 138).

Upholding a narrow scholarly communication focus limits the possibilities of the digital humanities, now and in future. Limitations do not stem only or even primarily from publication modalities, though. We believe that the digital humanities can more fully transition from a one-to-many mode of knowledge sharing to interaction done in-relation and in-community. Indeed, many digital humanities practitioners are already well embedded in publics that extend beyond and are not defined by the Ivory Tower, as noted above and evident in para-academic, academic-aligned, or non-academic partnerships.

As digital humanities practitioners look to the futures of our work, we suggest that a fuller embrace of open scholarship could ensure that digital humanists lead and further develop a collaborative and publicly responsive trajectory. In doing so, digital humanities could evolve more fully as a shared academic practice that enables the creation, dissemination, and engagement of open research by specialists and non-specialists in accessible and significant ways. This mode of engagement encapsulates the pursuit of more open, and more social, scholarly activities through knowledge mobilization, community training, public engagement, and policy recommendations in order to understand and address digital scholarly communication challenges. And it involves more publics in its pursuits—both those directly involved in and impacted by these areas.

In keeping with the many current and future paths for the digital humanities to take, open social scholarship manifests in many forms as well. Underlining open social scholarship is a commitment to sharing research in open access formats, the value of which Martin Paul Eve writes on earlier in this collection—even if there are challenges and even controversies around implementation to date. But open social scholarship is larger than open access publishing; it is an action-oriented approach to broadening the purview of the university, and to identifying, facilitating, and engaging with publics. In what follows we will focus on a subset of open social scholarship concerns our partnership has articulated as areas of interventionist activity—the commons, policy, training, and praxis—and reference how the INKE Partnership is engaging with these areas of development in digital and in-person contexts.

INTERVENTION: OPEN SOCIAL SCHOLARSHIP AND THE DIGITAL HUMANITIES

Responding to this context, our pragmatic and active intervention emanates from a long-standing, partnered community in and around the INKE Partnership.[12] This community involves digital humanities-focused academics from a number of areas and career paths. Community members share both a vested interest in the future of academic knowledge and its conveyance and a common belief that this future is aligned with engaging broad publics. Together, our partnership situates open social scholarship as a positive, conceptual intervention squarely within this larger context.

Action Area: Open Scholarship Commons

Academic conferences and events bring thousands of researchers together to share findings, refine ideas, and build collaborations. But these crucial events usually occur only once a year, and location, timing, expense, travel, and environmental concerns mean that such gatherings do not include all who could benefit from involvement. More and more scholarship is moving online, but researchers, partners, students, and members of the engaged public have varying levels of access to digital materials and the conversations around them—never mind comfort level and skill in navigating such materials.[13] The confluence of social, sharing technologies with evolving academic practice presents an opportunity to enrich digital humanities research, engagement, and impact for all those who participate in and are served by this valuable work.

The online research commons, for instance, is a chief avenue for facilitating more findable and usable academic research. An online research commons is a virtual space for a delineated community to connect, share, and collaborate. US-based new media, scholarly communication, and copyright scholars have researched and argued in favor of commons-based models for years and have reflected on the relative disadvantages of the corporate control of culture versus a decentralized system of commons-based knowledge production (Benkler 2006; Boyle 2008). Currently, commercial American sites like academia.edu and ResearchGate are the most popular platforms for sharing research—purportedly because of their social media style interfaces. But several scholars take issue with these for-profit models (Adema and Hall 2015; Duffy and Pooley 2017; Tennant 2017; Pooley 2018; Fitzpatrick 2020). Information scholars such as Julia Bullard also argue that the conscientious design of such systems is both critical and often overlooked (2019). She probes, in regard to platform design and construction: "what are the acceptable trade-offs regarding the intensity of labour in designing and maintaining a system consistent with open values and Canadian scholarship?" (2019). Regardless, Christine Borgman advocates for the commons as a viable open scholarly communication system in *Scholarship in the Digital Age: Information, Infrastructure, and the Internet* (2007), and John Willinsky even suggests the commons is an ideal model for scholarly communication when we consider research as a public good, to be shared widely for everyone's benefit in *The Access Principle: The Case for Open Access to Research and Scholarship* (2006). According to Peter Suber, an open access research commons avoids the *tragedy of the commons* (i.e., people taking more than their share, and not contributing back) because online research products are non-rivalrous: they do not diminish with access or use (2011). Of note, within the context of these discussions, the INKE Partnership's *Connection* cluster is currently developing an online research commons called the *Canadian Humanities and Social Sciences (HSS) Commons* (Winter et al. 2020): an in-development, national-scale, bilingual (French and English) network for HSS researchers in Canada to share, access, re-purpose, and develop scholarly projects, publications, educational resources, data, and tools.[14]

Action Area: Open Scholarship Policy

Many open access and open data policies are forward-looking but have proven challenging to implement in ways that meet the needs of all users and stakeholders. In part, this is due to the fast-paced nature and rapid evolution of open scholarship worldwide, and the mass of information, research, policies, and news media generated on the topic. The growing prominence of open scholarship developments and accessibility of the Internet presents an opportunity to streamline information processing and decision-making. An economy of scale approach where the scholarly community works together toward a shared understanding of open scholarship policy and best practices for finding, organizing, and presenting relevant information could harness such an opportunity. In response to these possibilities, the INKE Partnership's *Policy* cluster is collaborating on the *Open Scholarship Policy Observatory*, which collects research, tracks findings and national and international policy changes, and facilitates understanding of open social scholarship across Canada and internationally.[15]

Action Area: Open Scholarship Training

There are few training opportunities related to open scholarship or social knowledge creation, although millions of people engage with socially generated information daily. Training is required for academic specialists to learn how to share their research more broadly, as well as for engaged publics to increase their digital literacy and discover how to access and work with open scholarship. Emerging scholars also need training in how to use technology to engage with publics in meaningful ways. Dedicated, high-level open scholarship training could ensure that all who engage in socially created knowledge do so in productive and beneficial ways. The INKE Partnership engages these issues through the *Training* cluster and its *Open Social Scholarship Training Program.*

Action Area: Open Scholarship Praxis

Digital humanities projects have significant potential to facilitate closer collaboration between humanities and social sciences researchers and broader publics. Current digital humanities projects and initiatives can be found across the exclusive–inclusive spectrum; some are individually focused and others are much more social. This range from closed to open scholarship begs the question: which models for collaborative, open scholarly practices can effectively meet the interests and needs of engaged publics, and why? The INKE Partnership's *Community* cluster researches and develops public digital scholarship prototypes and initiatives in order to explore open publishing, scholarly communication, and citizen scholarship.

Action Areas in Sum

Taken together, these four areas—the commons, policy, training, and praxis—can increase positive impact by welcoming and fostering publics around humanities and social sciences work. Digital humanities research, prototyping, and publishing could be geared toward open, social approaches in order to produce knowledge output more effectively and with wider benefit. For instance, collaborating with communities to build interactive archival or storytelling experiences can facilitate and support publics over shared areas of interest. Modeling new ways to process, structure, and share digitized material brings digital humanities strengths to the access to and re-use of cultural material. Broadening the impact of academic interventions for multiple publics can diversify discursive communities in productive ways. This collection of action-oriented approaches, among others, suggest a future for the digital humanities that aligns with the values and promises of open social scholarship.

CONCLUSION: BROAD, REFLECTIVE SHIFTS TOWARD OPEN SOCIAL SCHOLARSHIP

By creating knowledge openly and socially, digital humanities researchers can address broader societal issues in relevant and timely ways and do so in relation with various publics. In recent times, there have been significant calls for more publicly oriented work in the humanities, and by extension, in the digital humanities. Internationally, many governments now require universities to justify their value and worthiness of public support, and education funding policy reflects this. In response, the potential for academic/public collaboration is substantial, but there is little understanding across academia of how exactly to implement such engagement with both efficacy and success. In fact, many hiring, tenure, and promotion guidelines still discourage the embrace of open scholarship practices (Alperin et al. 2019). Regardless, such a public commitment will lead to more citizen/scholar collaborations and an academic world that responds more directly to the publics it serves. Open social scholarship can open the door for specialists and non-specialists to interact with cultural materials and undertake digital humanities-based public engagement creatively.

As an evolving field, the digital humanities has already demonstrated its flexibility, reflexivity, and openness to expansion and growth. We consider the general commitment to open access, engagement with social media, and embrace of public humanities as evidence of an open, social trajectory for the field—especially as notions of audience change. Many digital humanities practitioners are committed, through their work, to openness, public engagement, and critical social issues. We believe such commitment allies and aligns with the theoretical undertones and pragmatic outcomes of open social scholarship. Where narrow research, development, and publishing practices necessarily delimit the scope of digital humanities work, engaging with networked, open knowledge creation provides more opportunities for collaboration, exploration, and growth.

Infrastructural, cultural, and institutional practices are interwoven with each other as well as enmeshed in the past, present, and future of an academic field. The challenges they bear are complex. Academic scholars will not be able to tackle and resolve these challenges on their own; this requires a holistic, strategic approach to revisioning the larger academic ecosystem. Open social scholarship activities may generate a more diverse, networked environment for creating and engaging with scholarship, diminishing perceived gaps between publics and the institutionalized research community, and increasing social engagement and broad access to scholarly outputs. Overall, this approach brings together communities of academics, experts, stakeholders, and publics around critical research, information, and policies. Situated in this context, we advocate that part of the approach to wide-scale change and evolution is for the digital humanities to continue to move from audience-thinking to publics-thinking; that is, for the digital humanities to embody open, social futures.

NOTES

1. This definition was first developed and articulated by the INKE Partnership and is cited on the "About INKE" page of the current website (n.d.). It is also referenced by Daniel Powell, Aaron Mauro, and Alyssa Arbuckle (2017).
2. See http://xpmethod.columbia.edu/torn-apart/; *Torn Apart/Separados* started as a collaboration between xpMethod, Borderlands Archives Cartography, Linda Rodriguez, and Merisa Martinez, with Moacir P. de Sá Pereira as lead developer.
3. See DHI, American Prison Writing at Hamilton College, http://apw.dhinitiative.org/.
4. See the DOHR website, http://www.dohr.ca/.
5. See http://nuisancelaws.org/.
6. See https://lglc.ca/.
7. See https://ca.wikimedia.org/wiki/Atikamekw_knowledge,_culture_and_language_in_Wikimedia_projects/.
8. Many of these projects have been recognized by Canadian Social Knowledge Institute (C-SKI) Open Scholarship Awards. For more on C-SKI, see https://c-ski.ca/.
9. Gaertner writes: "Indigenous [literature] scholars resist DH because the concerns Indigenous communities have about the expropriation of data have not been taken seriously. Those concerns will not be taken seriously until decolonial critique is actively installed at the foundations of DH theory and methodology and settler scholars need to start taking up some of this labour" (2017).
10. Additional notable sources here include Wendy Hui Kyong Chun's *Discriminating Data: Correlation, Neighborhoods, and the New Politics of Recognition* (2021), Catherine D'Ignazio and Lauren F. Klein's

Data Feminism (2020), and Noble's *Algorithms of Oppression: How Search Engines Reinforce Racism* (2018), all of which take up this work comprehensively.

11. Moreover, Booth and Posner suggest, such practices are not endemic to the digital humanities only, nor are they new to the broader meta-discipline of the humanities: "Creating and sharing such datasets may seem unprecedented in the humanities, though bibliographers, folklorists, musicologists, and others have collected and taxonomized in similar modes in the past" (2020, 20).
12. See Arbuckle et al. (Forthcoming) for a fuller engagement with the current and future directions of the INKE Partnership.
13. At the time of writing (2021), the Covid-19 pandemic's impact on how communities connect has shone an even brighter light on this situation.
14. See the prototype at hsscommons.ca, which is currently being developed in partnership with the Canadian Social Knowledge Institute, CANARIE, the Compute Canada Federation, the Electronic Textual Cultures Lab, the Federation for the Humanities and Social Sciences, the Modern Language Association's Humanities Commons, and others.
15. See Open Scholarship Policy Observatory, https://ospolicyobservatory.uvic.ca.

REFERENCES

Adema, Janneke and Gary Hall. 2015. *Really, We're Helping To Build This… Business: The Academia.edu Files*. http://liquidbooks.pbworks.com/w/page/106236504/The%20Academia_edu%20Files.

Alperin, Juan Pablo, Carol Muñoz Nieves, Lesley A. Schimanski, Gustavo E. Fischman, Meredith T. Niles, and Erin C. McKiernan. 2019. "How Significant are the Public Dimensions of Faculty Work in Review, Promotion, and Tenure Documents?" *eLife*.

Arbuckle, Alyssa, and Ray Siemens, with Jon Bath, Constance Crompton, Laura Estill, Tanja Niemann, Jon Saklofske, Lynne Siemens, and the INKE Partnership. Forthcoming. "An Open Social Scholarship Path for the Humanities." *Journal of Electronic Publishing*.

Benkler, Yochai. 2006. *The Wealth of Networks: How Social Production Transforms Markets and Freedom*. New Haven, CT, and London: Yale University Press.

Booth, Alison, and Miriam Posner. 2020. "Introduction: The Materials at Hand." *PMLA* 135 (1): 9–22.

Borgman, Christine. 2007. *Scholarship in the Digital Age: Information, Infrastructure, and the Internet*. Cambridge, MA: MIT Press.

Boyle, James. 2008. *The Public Domain: Enclosing the Commons of the Mind*. New Haven, CT, and London: Yale University Press.

Brennan, Sheila A. 2016. "Public, First." In *Debates in the Digital Humanities 2016*, edited by Matthew K. Gold and Lauren F. Klein, 384–9. Minneapolis, MN: University of Minnesota Press.

Bullard, Julia. 2019. "Knowledge Organization for Open Scholarship." *Pop! Public. Open. Participatory* 1(1).

Chun, Wendy Hui Kyong. 2021. *Discriminating Data: Correlation, Neighborhoods, and the New Politics of Recognition*. Cambridge, MA: MIT Press.

D'Ignazio, Catherine and Lauren F. Klein. 2020. *Data Feminism*. Cambridge, MA: MIT Press.

Duffy, Brooke Erin and Jefferson D. Pooley. 2017. "Facebook for Academics: The Convergence of Self-Branding and Social Media Logic on Academia.edu." *Social Media and Society*. https://doi.org/10.1177/2056305117696523.

Fitzpatrick, Kathleen. 2019. *Generous Thinking: A Radical Approach to Saving the University*. Baltimore, MD: Johns Hopkins University Press.

Fitzpatrick, Kathleen. 2020. "Not All Networks: Toward Open, Sustainable Research Communities." In *Reassembling Scholarly Communications*, edited by Martin Paul Eve and Jonathan Gray, 351–9. Cambridge, MA: MIT Press.

Gaertner, David. 2017. "Why We Need to Talk about Indigenous Literature and the Digital Humanities." *Novel Alliances: Allied Perspectives on Literature, Art, and New Media*, January 26. https://novelalliances.com/2017/01/26/indigenous-literature-and-the-digital-humanities/.

Hsu, Wendy. 2016. "Lessons on Public Humanities from the Civic Sphere." In *Debates in the Digital Humanities 2016*, edited by Matthew K. Gold and Lauren F. Klein, 280–6. Minneapolis, MN: University of Minnesota Press.

Implementing New Knowledge Environments (INKE) Partnership. n.d. "About INKE." https://inke.ca/about-inke/.

Lesbian and Gay Liberation in Canada. n.d. "About." https://lglc.ca/about.

Liu, Alan. 2012. "Where Is Cultural Criticism in the Digital Humanities?" In *Debates in the Digital Humanities*, edited by Matthew K. Gold, 490–509. Minneapolis, MN: University of Minnesota Press.

McPherson, Tara. 2012. "Why Are the Digital Humanities So White?" In *Debates in the Digital Humanities*, edited by Matthew K. Gold, 139–60. Minneapolis, MN: University of Minnesota Press.

Noble, Safiya Umoja. 2018. *Algorithms of Oppression: How Search Engines Reinforce Racism*. New York, NY: New York University Press.

Philpott, Wendy. 2020. "Collaboration Brings Stories of Institutional Racism and Abuse to Light." *Waterloo News*, February 12. https://uwaterloo.ca/news/collaboration-brings-stories-institutional-racism-and-abuse.

Pooley, Jefferson D. 2018. "Metrics Mania: The Case against Academia.edu." *The Chronicle of Higher Education*, January 7. https://www.chronicle.com/article/metrics-mania/.

Posner, Miriam. 2016. "What's Next: The Radical, Unrealized Potential of Digital Humanities." In *Debates in the Digital Humanities 2016*, edited by Matthew K. Gold and Lauren F. Klein, 32–41. Minneapolis, MN: University of Minnesota Press.

Powell, Daniel J., Aaron Mauro, and Alyssa Arbuckle. 2017. "Introduction." In *Social Knowledge Creation in the Humanities*, vol. 1, edited by Alyssa Arbuckle, Aaron Mauro, and Daniel J. Powell, 1–28. Arizona: Iter Academic Press and Arizona Center for Medieval and Renaissance Studies.

Risam, Roopika. 2016. "Navigating the Global Digital Humanities: Insights from Black Feminism." In *Debates in the Digital Humanities 2016*, edited by Matthew K. Gold and Lauren F. Klein, 359–67. Minneapolis, MN: University of Minnesota Press.

Siemens, Ray. 2016. "Communities of Practice, the Methodological Commons, and Digital Self-Determination in the Humanities." (Text of the Zampolli Prize Lecture, University of Lausanne, July 2.) *Digital Studies/Le Champ Numérique*. Reprinted in Constance Crompton, Richard Lane, and Ray Siemens, eds., *Doing Digital Humanities*. London: Routledge.

Suber, Peter. 2011. "Creating an Intellectual Commons through Open Access." In *Understanding Knowledge as a Commons*, edited by Charlotte Hess and Elinor Ostrom, 171–208. Cambridge, MA: MIT Press.

Tennant, Jonathan. 2017. "Who Isn't Profiting off the Backs of Researchers?" *Discover Magazine*, February 1. https://www.discovermagazine.com/technology/who-isnt-profiting-off-the-backs-of-researchers.

Torn Apart/Separados. n.d. "Textures." http://xpmethod.columbia.edu/torn-apart/volume/2/textures.html.

Visionary Futures Collective. n.d. "Visionary Futures Collective." https://visionary-futures-collective.github.io/.

Warner, Michael. 2002. "Publics and Counterpublics (Abbreviated Version)." *Quarterly Journal of Speech* (88) 4: 413–25.

Willinsky, John. 2006. *The Access Principle: The Case for Open Access to Research and Scholarship*. Cambridge, MA: MIT Press.

Winter, Caroline, Tyler Fontenot, Luis Meneses, Alyssa Arbuckle, Ray Siemens, and the ETCL and INKE Research Groups. 2020. "Foundations for the Canadian Humanities and Social Sciences Commons: Exploring the Possibilities of Digital Research Communities." *Pop! Public. Open. Participatory* 2. doi: 10.48404/pop.2020.05.

CHAPTER THIRTY-EIGHT

Digital Humanities and Cultural Economy

TULLY BARNETT (FLINDERS UNIVERSITY)

Digital humanities and culture go hand in hand. The stuff of digital humanities, the material and immaterial objects of study to which digital or computing tools and methods are applied, comes from the production of culture, so digital humanities is literally the analysis, by means of the use of digital methodologies, of the products or outputs of a cultural economy. But culture and cultural economy are not the same thing. Though they have complicated entanglements, cultural economy and its products can't be entirely conflated. Also, the increasing presence of quantitative data in the arts and culture sector, such as funding and ticketing stats, audience data, staff diversity data, board information, numerical indicators of quality or value, only serves to heighten tensions around the use of digital methodologies in the study of the cultural economy. Big data has opened new avenues for evidence-based decision-making in many domains and revealed elements of the cultural sector operating invisibly or less visibly, giving new insights that can shape real change in the world. Data, big and small, offers much to the cultural economy. This can be seen in the work of cultural organizations, individual artists, funding bodies, and others, around issues such as the representation of diverse humans in the arts labor market and its leadership (Brook et al. 2020), the make-up of the boardrooms of arts organizations (O'Brien et al. 2022), trends in artists' incomes and the pay gap between genders (Throsby et al. 2020), and the balance between quantity and quality in the arts (Gilmore et al. 2018). But there are also limits to the usefulness of data in the cultural sector and the political contexts in which data is collected, processed, and prosecuted needs to be acknowledged, interrogated, understood, and incorporated (Beer 2016; Phiddian et al. 2017). Too frequently data is used to constrain the understanding of value to economic impact. In looking at the relationship between digital humanities and cultural economy, this chapter focuses on questions around data in the cultural sector and its availability and implications for research. The chapter considers how the cultural economy might benefit in adopting digital humanities tools and methods and in collaborating with digital humanities researchers, but also cautions against some of the more problematic approaches to the data of the cultural economy which, without a critical approach, can harm the cultural sector. In acknowledging that for the cultural sector, in all its international expressions, the requirement for evidence bases for investment in arts and culture is only growing, the chapter asks what are the limits of data for the cultural economy? How do digital humanities tools and critical digital humanities approaches work together to tell us something new and useful about the mechanisms of culture?

Ultimately, this chapter argues that the cultural sector has adopted elements of digital humanities tools and thinking in useful ways—even if not explicitly identifying as digital humanities—and could benefit from the use of others. But those researchers working in, for, and with the cultural economy using or contemplating using digital humanities methods also need to ensure that they are learning from the *critical* digital humanities—the approach to digital humanities that incorporates a questioning of underlying assumptions about the digital tools, the infrastructures upon which they are based, and the very air that we breathe in the digital humanities. That is, there is an ethics of duty of care in the adoption of data science and digital humanities methodologies in the cultural economy that digital humanities researchers need to consider. This is because in studying the cultural economy researchers can lose sight of the fact that real lives and livelihoods are at stake and the production of arts and culture for the most part is conducted in vulnerable and underfunded environments. A component of the duty of care for researchers working in the cultural economy is to remind decision makers about the limits of data and that data is not neutral or uncomplicated, that big data is not big knowledge (Holden 2004, 21).

In exploring these issues, the chapter begins with the unavoidable defining of terms in cultural economy: a definitional problem the field shares with the digital humanities. For the cultural economy even more than for digital humanities, this work and infrastructure have pressing implications for a sector that is precarious and under-resourced. The chapter then looks at some of the scholarship around data in the critical digital humanities and the cultural economy, and concludes by discussing three case studies of work that sits in the crossover space between these two fields.

DEFINING CULTURAL ECONOMY

The cultural economy is a concept that is hard to pin down and yet my argument requires a solid definition. Thus, this chapter must spend what may seem like an inordinate amount of time defining terms. Readers with backgrounds in digital humanities may recognize both the need for definitional precision and the agony the navel-gazing produces (Terras et al. 2016). It is important to lay out these definitional territories or else we risk assuming that we are all speaking the same language.

Cultural economy is part of a suite of contested concepts attempting to assess, understand, measure, and describe the relationship between cultural and creative activities and the broader socio-political economy in which they are situated. Often the term "cultural economy" is used as an alternative to "creative industries." The term acknowledges that there are economic components to cultural and creative activity—as O'Connor points out, the cultural sector is organized into jobs, institutions, organizations, or clusters with income streams and artists and arts organizations buy things and sell things; and pay tax (2022)—but avoids identifying the sector explicitly as an industry in a way that elides the nuance and distinctiveness of the cultural sector. For some, the term industry is useful in identifying the cultural sector's economic contributions in ways similar to mining and primary industries, for example (Potts and Cunningham 2008, 239). In this view, when the cultural sector presents itself as a contributor to GDP alongside these other sectors then and only then it is taken seriously (Potts and Cunningham 2008). Others, such as Justin O'Connor, argue that the term industry is not appropriate for the cultural sector because the pursuit of profit is not its primary purpose (O'Connor 2016, 8; see also Galloway and Dunlop 2007). For O'Connor, the cultural sector is more like a service sector such as health or education and should be funded for the public contributions it provides to society (O'Connor 2022).

The terms "creative" and "cultural" are also disputed in relation to these economic factors, with quantitative research in creative industries tending to err on the side of including everything in the definition to the point of meaninglessness. This approach has been critiqued for including economic activity that wouldn't normally be included as work in arts and culture (accountants in arts organizations for example) and damaging the credibility of the argument. Galloway and Dunlop point to the "terminological clutter surrounding the term culture" (2007, 19) as another factor in the misunderstanding of arts and culture. "Culture" has a very wide set of definitions and practical uses. We hear a lot about the culture of organizations—meaning internal behavior patterns: a culture of bullying or a culture of respect, for example. Our Human Resources departments use "culture analytics" to reveal information about the culture inside an organization (where culture means norms, behaviors, satisfaction). For the purposes of this discussion, "arts and culture" is understood to mean the work of the sector including all art forms and the organizations and memory institutions that surround them.

In research terms, the study of cultural economy often refers specifically to the policies, practices, management, and business of producing cultural work: the political economy of culture. It's a focus on what lies at the infrastructural heart of the doing of cultural work—either in terms of the production of cultural artifacts or in terms of the policies that generate the conditions under which those cultural products can be generated. This might include local, state, federal, regional, international, formal, and informal cultural policies. It might include the study of the creative city and placemaking (Markusen 2013), of craft and maker communities (Luckman 2015), of the labor conditions under which creative work is produced distributed, supported, and platformized (Taylor and O'Brien 2017).

For the purposes of this chapter, cultural economy is understood to be the system and interconnected factors that underlie the production of cultural goods, services, and value, and the social, political, economic, and cultural context in which they and their consequences or flow-on effects are produced (O'Connor 2022). Cultural economy is the study of the political economy of arts and culture. It is the lived reality of those who work in the arts and cultural sector.

Cultural economy is also very different from "cultural economics"—the application of standard economics approaches to the material and immaterial artifacts of arts and culture. Cultural economists Towse and Hernández, for example, suggest that "cultural economics now offers expertise in the analysis of markets for a wide range of creative products" (2020, 1). For them:

> cultural economics uses economic theory and empirical testing in order to explain these many aspects of the creative economy. Microeconomics, welfare economics, public choice economics and macroeconomics may be used, and economists choose one or other of these bodies of theory as appropriate to analyse the topics they seek to explain.
>
> (2)

The field of cultural economics admits that the value of arts and culture can't be totally expressed in economic terms (Throsby 2001), and yet in studying that which is economic it uses traditional (neoliberal) methodologies to the exclusion of all other forms of value. That is, even though cultural economics explicitly acknowledges that not all value created by arts and culture is economic value it deliberately occludes those other forms of value (Klamer 2017; Meyrick et al. 2018). And in ignoring those other components as beyond scope, it can never fully understand the components it sees as in scope. Despite the apparent similarity between the terms "cultural economics" and "cultural economy," these approaches couldn't be more different.

Another field that could be confused with cultural economy, particularly when looking with a digital humanities lens, is "cultural analytics," which is concerned with the analysis of cultural data (Manovich 2020) and has nothing to do with the human resources focus on organizational behaviors mentioned above. Lev Manovich's 2020 book *Cultural Analytics* specifically identifies itself in relation to the analysis of contemporary culture. In his earlier explorations of the term, however, Manovich included historical as well as contemporary datasets applying a broad definition of "cultural" that included not only the Arnoldian definition of culture as "the best that has ever been said or thought" but also the cultural as everyday life (2020, 60–1). For Manovich cultural analytics draws on "methods from computer science, data visualization, and media art to explore and analyse different kinds of contemporary media and user interactions with them" (7). From a theoretical standpoint cultural analytics asks "how the use of such methods and large datasets of cultural media challenges our existing modern ideas about culture and methods to study it" (7). So, the work is practical and theoretical: what can we learn about arts and culture from using computing and visualization methods applied to large sets of cultural objects like digital photographs and what can we learn about how we study culture by using different methods?

Manovich sets the task at a grand scale, arguing that "we need to reinvent what it means to study culture" (18). Through the book's structure he identifies the three main elements of the task as: "studying culture at scale," "representing culture as data," and "exploring cultural data." While ostensibly this is a useful way to bring humanities disciplines into conversation with new methods and approaches in digital humanities and data science at both the methodological and theoretical levels, in practice, once again the work slips towards a narrower frame. Are we representing culture as data when we are using digital components only? Manovich subtitles his conclusion "Can We Think Without Categories?," further indicating an interrogation of the problems of data science for arts, culture, humanities, and cultural economy that is not borne out by the application of these ideas. Part of the problem here is another slippage of terminology and definition where digitally accessible culture comes to stand in for culture because of its convenient ubiquity. Approaching cultural processes and artifacts as data can lead us to ask the kinds of questions about culture that people who professionally write about it, curate it, and manage it do not normally ask today.

The way we define cultural economy, creative industries, and cultural analytics has implications for the kinds of research that can be conducted and the findings from those research approaches. As we'll see below, while the cultural economy does have a historical lens, its focus is on the pressing lived realities of a difficult political economy of culture. Conversely, largely because much cultural material is under copyright, Digital Humanities researchers tend to study available datasets that are historical rather than contemporary. This creates biases in the data; those datasets are the products of racist, sexist, homophobic, and transphobic gatekeeping. When we look to a future of artificial intelligence expressions trained on cultural material that has been funneled through this gatekeeping, the picture is grim (Horsley 2020). We have to ask ourselves whose voices are included in and excluded from the corpora we use to train AI and what effects does this have on the AI we produce?

SCHOLARSHIP AT THE CROSSROADS

There is surprisingly little explicit research on the application of digital tools to the contemporary arts and culture practices that form the cultural economy, and especially very little in the field that identifies as digital humanities. The index to Blackwell's 2004 *A Companion to Digital Humanities* (edited by Susan Schreibman, Ray Siemens, and John Unsworth) contains only one mention of culture, and that is for cultural heritage. This doesn't mean the book isn't concerned with culture—it explodes off every page—just that the word doesn't appear in the index, and therefore in the structuring of the work. By the book's 2016 reboot edition, this one reference had become a nuanced unpacking of cultural heritage in the index (thanks in large part to the presence of a strong chapter on digital cultural heritage from Sarah Kenderdine) but there remains little sense of the contemporary liveness of the production of arts and culture and the ways digital humanities might work with that. This is in large part driven by the copyright problem that haunts much digital humanities research: the stuff to hand to which scholars can apply digital humanities tools is largely the published and out-of-copyright works that have been digitized in some form or other. These orders of gatekeeping (publication, copyright, digitization, indexing) compound upon each other and shape the stories we can tell about arts and culture by using digital tools, methods, and approaches. The available material is triply limited. There's much research looking at the way what books are and are not published, what films are or are not made, what paintings are or are not exhibited, sold and canonized is filtered by gender, race, sexuality, class, ability, amongst many other markers, rather than as a reflection of artistic merit (Davis 2017). Work produced against these odds is also more likely to be excluded from analysis by the limitations of data preservation, management, and access. Yet the alternatives to researching published, out-of-copyright, and digitized cultural products have their own complications, as will be discussed below.

The 2020 *Routledge Companion to Digital Humanities and Art History* is a recent example of a limited research worldview where the stuff of research is restricted to historical datasets, although it begins by defining art history in a way that includes some version of contemporary arts practices and the political, social, and infrastructural realities in which they are embedded. In her introduction to the volume Kathryn Brown refers to Manovich's work on cultural analytics to argue that

> research techniques drawn from the digital humanities and from art history are related to each other and can be deployed in mutually illuminating ways. They both entail acts of seeing.
>
> (2020, 1)

For Brown, "human vision and computer vision are not so far apart" and yet the chapters tend towards the longer history of the visual arts again because of the limited availability of works to which digital humanities approaches can be legally or conveniently applied. These limits result in a distorted emphasis on out-of-copyright works on the one hand and on born-digital works freely shared on the Internet on the other.

Scholars have called for solutions to overcome some of these problems. For example, Deb Verhoeven's 2014 chapter "Doing the Sheep Good: Facilitating Engagement in Digital Humanities and Creative Arts Research" advocates for

> an expanded and inclusive view of digital humanities and creative arts research; in which computation and communication, method and media in combination enable us to explore the larger question of how we can work with technologies to co-produce, represent, analyse, convey and exchange knowledge.
>
> (2014, 219)

But here too there is a leap from humanities to creative arts research that conflates in ways that don't permit a clear view of the place of the fine arts, the creative arts, and the creative industries, i.e., the cultural economy, in a digital humanities landscape. Together for convenience isn't enough when the arts are marginalized within universities, within funding mechanisms, and within the public sphere. The work of Verhoeven and her team is discussed below as one of the few teams explicitly working at the intersection of digital humanities and cultural economy, tackling some of the limitations discussed above and exemplifying a mindset of research grounded in values and critical awareness.

Others in digital humanities refer explicitly to the arts and cultural sector, of course. Andrew Prescott argues that in the "dialogue between humanities scholar, scientist, and curator" that he sees sitting at the heart of the digital humanities "the creative artist also has a vital role" (2012) but this promise then slips straight to a discussion of "digital art" which he sees as "increasingly breaking down many familiar disciplinary barriers" (2012). That is, the framing of the research or argument includes contemporary creative arts and the work of the cultural economy but in practice it slips back to what datasets are conveniently available for study, i.e., digital art. This issue also arises in Manovich's work where the shallow fount of available material for research limits the broader horizon of the field. The question about digital humanities and cultural economy is about so much more than just the digital in the cultural sector, though the overlap between the two can be murky. This clearly requires more thinking and work from the digital humanities community.

The speed with which we are learning about big data, its affordances and challenges, its risks and biases, means we are updating our knowledge on the fly, even as we are collecting, storing, and using it. Yanni Loukissas opens his 2019 book *All Data Are Local: Thinking Critically in a Data-Driven Society* by talking about his experience working as a consultant on an IT Masterplan for New York's Metropolitan Museum of Art while he was a postgraduate student, which inspired him to consider "how visitors might use data to navigate the Met's vast holdings of American art" (2019, xii). He became aware at that early stage that "data is always curated" (xv). At the heart of Loukissas's work is the insight that "data are cultural artifacts, created by people and their dutiful machines, at a time, in a place, and with the instruments at hand for audiences that are conditioned to receive them" (1–2). In *Sorting Things Out,* Bowker and Star remind us that classifications, which underpin all data collection processes, "are powerful technologies. Embedded in working infrastructures they become relatively invisible without losing any of that power" (2000, 319). This is borne out in Safiya Umoja Noble's important work *Algorithms of Oppression* which demonstrates in detail the bias built into online search results and the ways they not only replicate the racism and sexism of the societies from which they emerge but reinforce those biases (2018). The international Indigenous Data Sovereignty movement is another example of critically analyzing the social and political effects of data. The movement acknowledges data as "a cultural and economic asset" (Walter et al. 2020, 10) and argues for a principles and governance approach to data about Indigenous peoples that acknowledges the gravity of effects around data, the historical misuse of data in the dehumanizing colonial project, and the rights of Indigenous people to maintain governance of their and their communities' data. These are some of the foundational works towards building a critical digital humanities that offer important lessons for a cultural economy seeking to use data effectively.

In their work on a critical approach to big data, boyd and Crawford highlight the creation of "a new kind of digital divide: the Big Data rich and the Big Data poor" (2012, 674). It is worth reflecting on how this might play out in specific and singular ways in the cultural sector—a sector always challenged by "efficiency dividends" (van Barneveld and Chiu 2018), political budget cuts, the whims of philanthropy, and cultural policy shifts, problems worsened by the 2008 Global Financial Crisis and now Covid-19 (Banks and O'Connor 2020). A sector with so few resources is at risk of vulnerability to the big four accounting firms, many of whose staff are genuinely interested in helping the arts and culture sector demonstrate their value and yet are determining the horizon of cultural value by spruiking methodologies of economic impact as the solution to the governmental and philanthropic devaluing of arts.

This then renders clearer at least a range of problems that scholars in digital humanities may face in developing trust with the cultural sector and in adhering to the ethical principle of all research: first do no harm (Cañas 2020). The use of datasets and digital humanities methodologies in the cultural economy encounters some significant and pressing problems which researchers need to be mindful of and attendant to when practicing digital humanities in cultural economy space.

CASE STUDY I: MAPPING THE CREATIVE INDUSTRIES

Every creative industries policy or strategy document begins with some kind of mapping exercise that purports to collect and visualize the particularities of the sector in order to understand what it looks like and how it operates—to read it from a distance. Since the 1980s when New Public Management and its version of citizens as consumers (O'Brien 2013, 117) gained traction in the UK, Australia, and the US amongst other places (Belfiore 2004; Crossick and Kaszynska 2014; Gilmore et al. 2017), a rolling set of instruments of account and trends in concepts and methodologies to underpin those instruments has been imposed upon the arts and culture sector. These include indicators, frameworks, 'targets and tracking' exercises, and most recently dashboards (Phiddian et al. 2017). As data science becomes more nuanced and wide-ranging, these approaches grow and change. The rise of creative industries rhetoric alongside new public management, that sees everything in terms of markets and is now inflected by data science techniques, has led to a perfect storm for a cultural sector systematically deprived of both actual funding and the government and corporate perception of its public value. By extending the legacy of this economically orientated rhetoric, a data and indicators approach further abstracts the value of arts and culture from its presumed place in a dubiously fabricated market.

While it's not hard to see the appeal of a data-driven approach to understanding and funding arts and culture, it is far from easy to express the limitations of such an approach in terms that policymakers can readily understand and act on. An early example of rhetoric and practical creative mapping exercises can be found in the 1998 UK Department for Culture, Media and Sport Creative Industries Mapping Document. The approach used in the data collection was purely economic and the researchers segmented the information by subsector, separating out Advertising, Antiques, Architecture, Crafts, Design, Fashion, Film, Leisure, Software, Music, Performing Arts, Publishing, and TV and Radio. This was an influential early exercise and as a result this approach was adopted and replicated internationally, including by Taiwan, New Zealand, Singapore, and Australia (Higgs and Cunningham 2008) despite, according to Hye-Kyung Lee, its arbitrary framework,

> ambiguous definitions and boundaries of "creative industries" but also its neoliberal implications, especially the market-centred reductive view of culture and romanticisation of cultural workers whose economic life tends to be notably perilous.
>
> (2017, 1078–9)

The link between digital humanities and cultural economy is a connection between theory and practice, between commentary and the lived experience of culture. You can't filter out anomalous data in real life—because it's real people and jobs and activity. Part of the responsibility of digital humanities researchers is to remind politicians and policymakers that the data is messy and trying to simplify the mapping further marginalizes the marginalized. We are getting better at this in many respects but arts and culture—the cultural economy—still bears the brunt of poorly thought through data exercises.

CASE STUDY II: THE KINOMATICS PROJECT AND ITS TEAM

Deb Verhoeven and her team of interdisciplinary collaborators invented the term Kinomatics for their large-scale project that "collects, explores, analyses and represents data about the creative industries" (Zemaityte et al. 2017, 10). The international team uses large datasets to examine the particularities of various sectors that make up the creative industries but with a strong focus, at least thus far, on spatial and temporal film industry data. This is a project that is designed to take advantage of the affordances of digital humanities methodologies and big data analysis to better see elements of the creative industries that might not be surfaced immediately, but it is also built on an acknowledgment of the limitations of the methodologies and approach. They use a network analysis approach that considers the institutional, social, and commercial work of the international cinema sector by using large datasets of cinema showing times (records in the millions) and then contextualizing that data with financial, spatial, and demographic data. But it is how the team approaches their data that is the most fascinating and unique aspect of the project. In an article in which the team analyzes their own practices and networks, they write:

> We respond to calls for scholars to reject the "counting culture" of enumerative self-auditing and instead consider care-full academic work: "What if we counted differently?"
>
> (Verhoeven et al. 2020)

The team is starting from a critical position, adopting a values-led and feminist research ethics approach to think about how metrics of quality can be better understood and applied in the academy. It remains to be seen to what extent this careful approach to counting is extended to the object of study, the creative industries, where similar problems exist and counting must indeed be done differently. Kinomatics may be an example of the critical digital humanities in action. The objects of study have pulses; they are beings not merely data points. The Kinomatics project respects this at each phase of the work.

CASE STUDY III: NETWORK ANALYSIS AND THE BOARDS OF ENGLISH ARTS AND CULTURE ORGANIZATIONS

In their paper "Who runs the arts in England? A social network analysis of arts boards" (2022), Dave O'Brien, Griffith Rees and Mark Taylor use data accessed through Companies House, the United Kingdom's registrar of companies situated with the Department for Business, Energy and Industrial Strategy, to map networks and relationships between organizations, identifying dominances by particular organizations and individuals within those networks. Their project uses an API that enables users to access relevant data (in this case data that falls under the Companies Act 2006) that is up to date and in a format that can be used for analysis. Like Verhoeven, O'Brien, Rees and Taylor are focused on a social justice agenda. In *Culture is Bad for You* (2020), Brook, O'Brien, and Taylor explore numerous forms of discriminations in the culture sector and its labor market using quantitative and qualitative methods to demonstrate the disturbing extent of social inequalities, including class, race, and gender inequalities, in the production and consumption of arts and culture. They point out the ways in which arts and culture is not a meritocracy and that the precarity of labor in the sector is not evenly distributed. O'Brien, Rees and Taylor build on this work looking at the underlying infrastructure of arts and culture through its governance structures—boards—looking for diversities and concentrations at that level that may explain the experiences of workers in the cultural sector (2022). Using the API, the authors constructed a "bimodal network" of 735 national portfolio organizations, i.e., those funded by Arts Council England, and from there look for "networks where companies are connected to companies, and members of the board of directors are connected to other members of boards of directors." Importantly, the research uses a series of "hops" which means they can look for more distant connections and reveal relationships normally hidden from view. Their analysis shows that "there are more than five times as many financial service organisations in the one-hop dataset as one would expect from a random sample of companies from Companies House in general" (17). They find "a clear dominance of business, financial and accounting services expertise" (2022). Here digital humanities tools, methods, and ways of thinking are put to use to identify elements affecting the governance of the cultural economy that aren't immediately visible or evidenced.

CONCLUSION

These are profoundly difficult times for the cultural sector with the Covid-19 pandemic stopping a lot of cultural production and performance in its tracks, on top of cultural policy and funding instability and political instability more broadly in many places around the world. This era may also reveal itself to be a difficult time for the digital humanities with universities around the world, upon which much of the digital humanities activities are dependent, in various forms of economic and ethical emergencies. How this plays out for a discipline like digital humanities remains to be seen. In many places, though, a drop in student numbers, particularly international students, has been an excuse to cut programs and staff in the international tertiary sector. This will certainly affect the disciplines on which digital humanities depends.

Cultural economy in study and practice is under threat too by constant cycles of policy programming that require different kinds of data and methodology in pursuit of shifting policy outcomes—excellence, accessibility, to name just two. Incessant defunding, efficiency dividends, and restructures are visited upon an already overtaxed and underpaid, largely precarious, sector.

Thus, there is an urgency and a valency in this field that is not felt to the same degrees in other disciplines in the humanities, creative arts, or allied social science disciplines. For example, questions such as who wrote the plays attributed to Shakespeare or what does it mean that Rembrandt erased the figure of a young child from his painting are not, I argue, fraught with the same contemporary precarities in the way that, for example, political interference in contemporary arts funding or the inclusion of diverse representatives on cultural organizations' governing boards might be. As researchers we are called upon to tread carefully with our uses of cultural datasets and be ever mindful of the ways our research data can be used against the production of next century's important works of art. This is not to suggest that the significant opportunities for advancement of knowledge about cultural economy through digital humanities methodologies and projects is not significant but that a cautious approach is required because of the very real consequences for people, projects, and organizations in the arts and culture sector.

Digital humanities methodologies offer many potential and meaningful insights for cultural economy researchers in the academy and organizations and individuals in the sector, and the many who cross between the two domains. But digital humanities' focus on historical content, a result of the copyright problem that limits what is legally and ethically available as datasets, means that digital humanities researchers are often not aware of or attendant to the political urgency that is life in the contemporary cultural economy (either as researchers or as workers, though many share the same precarious working conditions). At the same time, while we are increasingly aware of the data bias in terms of race, sexuality, gender, and class to list just a few, it is also crucial that more work is done on how data perpetuates biases against certain sectors. This is where DH and cultural economy researchers and workers in the cultural sectors can collaborate to ensure that digital humanities approaches are grounded in a critical approach, using all of the knowledge and vibrancy in the critical digital humanities to address key research questions that make a difference in the world.

REFERENCES

Banks, Mark and Justin O'Connor. 2020. "'A Plague upon Your Howling': Art and Culture in the Viral Emergency." *Cultural Trends* 30 (1): 1–16.

van Barneveld, Kristin and Osmond Chiu. 2018. "A Portrait of Failure: Ongoing Funding Cuts to Australia's Cultural Institutions." *Australian Journal of Public Administration* 77 (1): 3–18.

Beer, David. 2016. *Metric Power*. London: Palgrave Macmillan.

Belfiore, Eleonora. 2004. "Auditing Culture: The Subsidised Cultural Sector in the New Public Management." *International Journal of Cultural Policy* 10 (2): 183–202.

Bowker, Geoffrey C. and Susan Leigh Star. 2000. *Sorting Things Out: Classification and Its Consequences*. Cambridge, MA: MIT Press.

boyd, danah and Kate Crawford. 2012. "Critical Questions for Big Data: Provocations for a Cultural, Technological, and Scholarly Phenomenon." *Information, Communication & Society* 15 (5): 662–79.

Brook, Orian, Dave O'Brien, and Mark Taylor 2020. *Culture is Bad for You*. Manchester: Manchester University Press.

Brown, Kathryn. 2020. *The Routledge Companion to Digital Humanities and Art History*. New York, NY, and London: Routledge.

Cañas, T. 2020. "Ethics and Self-Determination." In *The Relationship is the Project*, edited by Jade Lillie, Kate Larson, Cara Kirkwood, and Jax Brown, 41–50. Melbourne, Vic: Blank Inc Books.

Crossick, Geoffrey and Patrycja Kaszynska. 2014. "Under Construction: Towards a Framework for Cultural Value." *Cultural Trends* 23 (2): 120–31.

Davis, Mark. 2017 "Who Are the New Gatekeepers?" *Publishing Means Business: Australian Perspectives*, edited by Aaron Mannion, Millicent Weber, and Katherine Day, 125–46. Clayton, Vic: Monash University Publishing.

Galloway, Susan and Stewart Dunlop. 2007. "A Critique of Definitions of the Cultural and Creative Industries in Public Policy." *International Journal of Cultural Policy* 13 (1): 17–31.

Gilmore, Abigail, Konstantinos Arvanitis, and Alexandra Albert. 2018. "Never Mind the Quality, Feel the Width": Big Data for Quality and Performance Evaluation in the Arts and Cultural Sector and the Case of "Culture Metrics." *Big Data in the Arts and Humanities: Theory and Practice*, 27–39. Boca Raton, FL: CRC Press.

Gilmore, Abigail, Hilary Glow, and Katya Johanson. 2017. "Accounting for Quality: Arts Evaluation, Public Value and the Case of 'Culture Counts.'" *Cultural Trends* 26 (4): 282–94.

Higgs, Peter and Stuart Cunningham. 2008. "Creative Industries Mapping: Where Have We Come from and Where Are We Going?" *Creative Industries Journal* 1 (1): 7–30.

Holden, John. 2004. *Capturing Cultural Value*. London: DEMOS. http://www.demos.co.uk/files/CapturingCulturalValue.pdf.

Horsley, Nicola. 2020. "Intellectual Autonomy after Artificial Intelligence: The Future of Memory Institutions and Historical Research." In *Big Data—A New Medium?*, edited by Natasha Lushetich, 130–44. New York, NY, and London: Routledge.

Klamer, Arjo. 2017. *Doing the Right Thing: A Value Based Economy*. London: Ubiquity Press.

Lee, Hye-Kyung. 2017. "The Political Economy of 'Creative Industries'." *Media, Culture & Society* 39 (7): 1078–88.

Loukissas, Yanni Alexander. 2019. *All Data Are Local: Thinking Critically in a Data-Driven Society*. Cambridge, MA: MIT Press.

Luckman, Susan. 2015. *Craft and the Creative Economy*. Basingstoke and New York, NY: Palgrave Macmillan.

Manovich, Lev. 2020. *Cultural Analytics*. Cambridge, MA: MIT Press.

Markusen, Ann. 2013. "Fuzzy Concepts, Proxy Data: Why Indicators Would Not Track Creative Placemaking Success." *International Journal of Urban Sciences* 17 (3): 291–303.

Meyrick, Julian, Robert Phiddian, and Tully Barnett. 2018. *What Matters? Talking Value in Australian Culture*. Clayton, Vic: Monash University Publishing.

Noble, Safiya Umoja. 2018. *Algorithms of Oppression: How Search Engines Reinforce Racism*. New York, NY: NYU Press.

O'Brien, Dave. 2013. *Cultural Policy: Management, Value and Modernity in the Creative Industries*. New York, NY, and London: Routledge.

O'Brien, Dave, Griffith Rees, and Mark Taylor. 2022. "Who Runs the Arts in England? A Social Network Analysis of Arts Boards." *Poetics*. https://doi.org/10.1016/j.poetic.2022.101646.

O'Connor, Justin. 2016. "After the Creative Industries: Why We Need a Cultural Economy." *Platform Papers* 47: 1–60.

O'Connor, Justin. 2022. *Reset: Art, Culture and the Foundational Economy*. https://resetartsandculture.com/wp-content/uploads/2022/02/CP3-Working-Paper-Art-Culture-and-the-Foundational-Economy-2022.pdf.

Phiddian, Robert, Julian Meyrick, Tully Barnett, and Richard Maltby. 2017. "Counting Culture to Death: An Australian Perspective on Culture Counts and Quality Metrics." *Cultural Trends* 26 (2): 174–80.

Potts, Jason and Stuart Cunningham. 2008. "Four Models of the Creative Industries." *International Journal of Cultural Policy* 14 (3): 233–47.

Prescott, Andrew. 2012. "An Electric Current of the Imagination: What the Digital Humanities Are and What They Might Become." *Journal of Digital Humanities* 1 (2).

Schreibman, Susan, Ray Siemens, and John Unsworth, eds. 2004. *A Companion to Digital Humanities*. Blackwell Companions to Literature and Culture 26. Malden, MA: Blackwell Publishing.

Taylor, Mark. Forthcoming. "Cultural Governance within and across Cities and Regions: Evidence from the English Publicly-Funded Arts Sector."

Taylor, Mark and Dave O'Brien. 2017. "'Culture Is a Meritocracy': Why Creative Workers' Attitudes May Reinforce Social Inequality." *Sociological Research Online* 22 (4): 27–47.

Terras, Melissa, Julianne Nyhan, and Edward Vanhoutte, eds. 2016. *Defining Digital Humanities: A Reader*. New York, NY, and London: Routledge.

Throsby, David. 2001. *Economics and Culture*. Cambridge: Cambridge University Press.

Throsby, David, Katya Petetskaya, and Sunny Y. Shin. 2020. "The Gender Pay Gap among Australian Artists: Some Preliminary Findings." Sydney: Australia Council for the Arts.

Towse, Ruth and Trilce Navarrete Hernández. 2020. *Handbook of Cultural Economics*. Cheltenham and Northampton, MA, USA: Edward Elgar Publishing.

Verhoeven, Deb. 2014. "Doing the Sheep Good." In *Advancing Digital Humanities: Research, Methods, Theories*, edited by Katherine Bode and Paul Longley Arthur, 206–20. Basingstoke: Palgrave Macmillan.

Verhoeven, Deb, Paul S. Moore, Amanda Coles, Bronwyn Coate, Vejune Zemaityte, Katarzyna Musial, Elizabeth Prommer, Michelle Mantsio, Sarah Taylor, and Ben Eltham. 2020. "Disciplinary Itineraries and Digital Methods: Examining the Kinomatics Collaboration Networks." *NECSUS European Journal of Media Studies* 9 (2): 273–98.

Walter, Maggie, Tahu Kukutai, Stephanie Russo Carroll, and Desi Rodriguez-Lonebear. 2021. *Indigenous Data Sovereignty and Policy*. Abingdon: Routledge.

Zemaityte, Vejune, Bronwyn Coate, and Deb Verhoeven. "Coming (eventually) to a cinema near you." *Inside Film: If* 177 (2017): 10–12.

CHAPTER THIRTY-NINE

Bringing a Design Mindset to Digital Humanities

MARY GALVIN (MAYNOOTH UNIVERSITY)

Back in 2004, Matthew Kirschenbaum predicted that the coming decade would bring "major challenges" for the digital humanities, particularly in "designing for interfaces (and designing interfaces themselves) outside of the 13- to 21-inch comfort zone of the desktop box" (540). Now, some two decades later, the challenge facing the digital humanities, or, as it is often abbreviated, "DH," is designing with and for society.

A discipline that has successfully embraced this societal focus has been design. Design combines theory and practice across a variety of sectors to deal with a "class of social system problems which are ill-formulated, where the information is confusing, where there are many clients and decision makers with conflicting values" (Churchman 1967, 141). The future of DH should follow design's lead.

Design is not new to DH. Many of DH's most prominent tools, interfaces, programs, and methodologies have seen contributions from designers, scholars, and practitioners working across communications design, interaction design, user experience design, and human–computer interaction (Drucker 2014; Schofield, Whitelaw, and Kirk 2017). Design has always been ubiquitous within DH, and so it is time that the latter discipline took a moment to reflect on its relationship to the former. Up until now, DH has regarded the discipline of design as a bit-part of the creative process, as something that one simply does in an effort to plan some aesthetic aspect or function of an artifact. But DH should no longer regard design as a way to just create some end product, but as a way to think; something epistemic rather than a matter of craft.

Right now, design in DH is project-specific, and so it needs to be re-oriented so that it becomes a fundamental part of all those disciplines which sit under the "big ten." To borrow a phrase from Kennedy, Carroll, and Francoeur (2013, 10), design should offer a "mindset not skill set." Mindsets are shaped by social contexts (Crum et al. 2017, 2) and they imply an emotional, behavioral, and cognitive response to something. A *disciplinary* mindset becomes less about the tools and methods used and more about the values underpinning that discipline, which I argue should now be in response to the current social setting. When taking a mindset approach, the tools become, to borrow terminology from service design, backstage operations, with the frontstage occupied by the knowledge generated by DH.

What is being advocated here in no way diminishes the ways of knowing that have come to define—and redefine—DH. The argument here is that DH scholarship needs to adopt a mindset which will help it—and all that is offered by leveraging computers in the service of the humanities—to better understand our changing society in a moral, ethical, and responsive manner. As Freidman and Hendry state, "all technologies, to some degree reflect, and reciprocally affect, human values" (2019, 1). A designer's mindset (DM) brings an awareness and responsibility to the digital tools we create, the tools and techniques of DH. The debate should not be "what is DH?," but rather "what are the *values* of DH?," and how do we ensure those values achieve something *good*?

Cross (1982) refers to "designerly" ways of knowing. DH knows its subject matter best, and a DM does not offer a list of prescribed projects, courses, and papers the DH community should follow. But a DM can offer a shared way of knowing, a cognitive framework that approaches all aspects of DH, from undergraduate teaching to project management, which ensures inclusivity and diversity, and more importantly, a focus on researching the *right* problems.

"A brilliant solution to the wrong problem can be worse than no solution at all: solve the correct problem" (Norman 2013, 218). If DH were to consider design less as a way to create and more as a way to think and research, the focus of the discipline would shift from tools and methods to solving complex problems involving multiple stakeholders, in human-centered ways. It is DH as design thinking[1] as opposed to simply design in DH.

Thanks to the rise of design thinking within academia, various disciplines now advocate for human-centered, solution-focused approaches to problem-solving, which is a positive thing. Design thinking has brought a heuristic intent to disciplines without the need for expertise or experience in the social sciences. However, across industry boardrooms, executive chatter has also embraced design thinking as a way of strategically innovating products and services. It is perhaps this exchange between academic and corporate worlds that has caused tension, with universities now adopting a recycled neoliberalism of an academic discipline, recycled, as it has been claimed and reclaimed by industry and academia since its naissance (again, much like what has happened with DH).

As far as DM's application is concerned, the Double Diamond is its most synonymous framework (see Figure 39.1). It is often referenced and updated to allow designers and non-designers (the latter largely influenced by the publication of *Change by Design* by Brown and Katz 2011) to use design thinking in their practices and processes. It is one of many design thinking models or frameworks; with others including Stanford Design School Model, IDEO Model, and the IBM Model (Hoffman 2016).

The Double Diamond was originally presented by the Design Council in 2004 as a visual description of the cognitive process (that is, DM) a designer goes through when approaching a problem. It outlines the divergent and convergent thinking phases of the designer across two diamonds: the problem space and the solution space. The problem space is where a problem must be discovered and defined, while the solution space is where solutions are developed and delivered.

Norman argues that designers do not try to search for a solution until they have determined the real problem, "and even then, instead of solving that problem, they stop to consider a wide range of potential solutions." It is only then that "they converge upon their proposal" (2013, 219). How does this translate to DH? It offers DH a framework to follow that militates against a tool-led approach that is solution rather than problem focused (possibly the wrong problem). Deciding whether a problem is "real" or "right" is all about taking time to understand the problem space and those within it: *who* are the stakeholders and *what* do we need to help them? Often times, research, especially research that includes a digital component, can begin with the digital piece (method, tool, solution). This leads to a lack of understanding of the problem.

Imagine a client approaches a designer and says "I want you to design a new communications software for my company because staff don't like it." The right problem to explore would be to

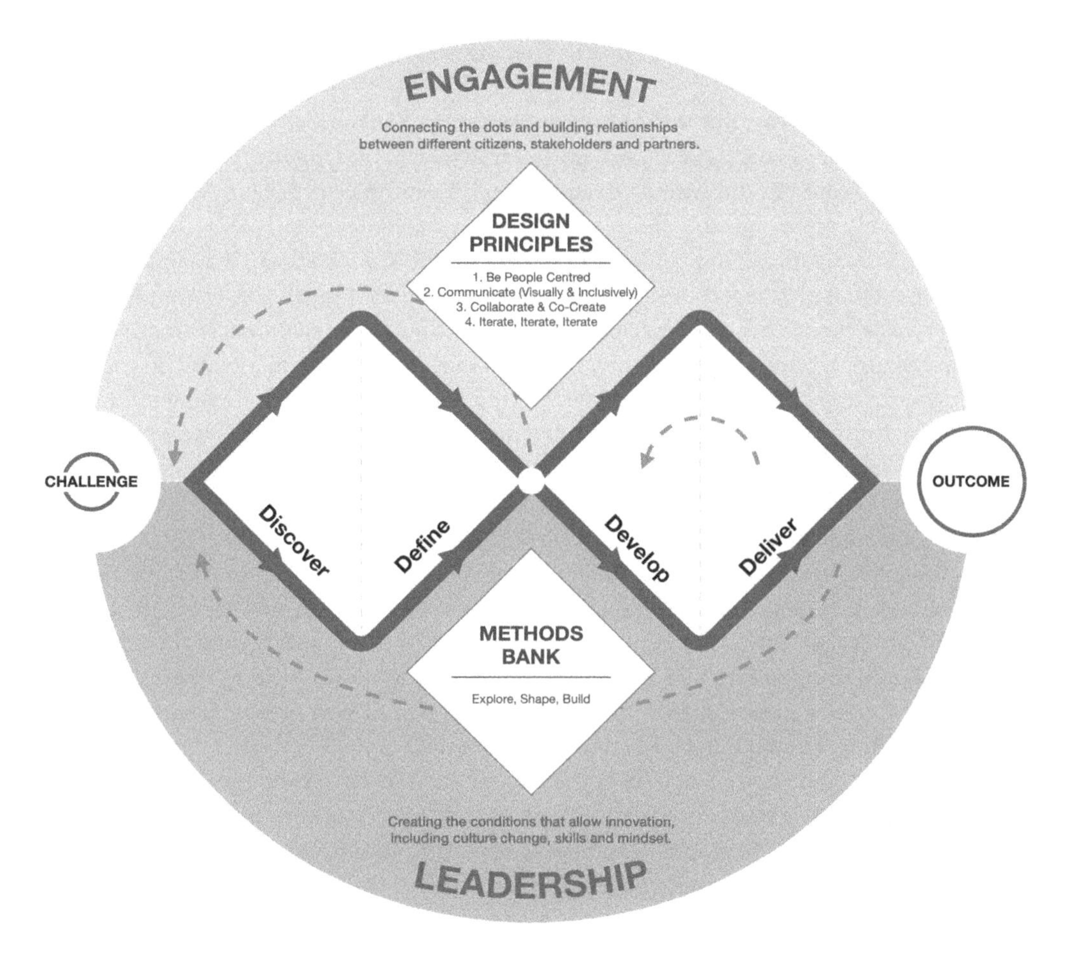

FIGURE 39.1 The Double Diamond, Design Council (2019). Credit: Design Council.

understand *why* the employees do not want to use the existing software. Maybe it is simply unfit-for-purpose, or maybe it has nothing at all to do with the software and instead has everything to do with a potentially problematic organizational culture, such as a lack of trust. Addressing such a lack would be the right problem to address. A technology-led approach would only lead to more wasted resources, with the same unhappy staff still unwilling to use whatever comes next.

Adopting design frameworks that support problem-led, as opposed to solution-led, approaches, can help DH scholars ensure that their research is tackling those problems most pertinent to our cultures and societies, rather than research questions that simply align with the digital tools and methods that drive DH.

In many respects, this essay is challenging all DH scholars to engage with broader societal issues, or as Rittel and Webber (1973) term, "wicked problems." Wicked problems are complex problems that are unique, have no clear problem definition, are multicausal and interconnected, have multiple stakeholders with conflicting agendas that straddle organizational and disciplinary boundaries, are connected, and their solutions are not right or wrong but better or worse, and inevitably they never completely solve the wicked problem at hand (Coyne 2005). Essentially, wicked problems are societal and systemic and design has been embraced as the discipline to turn these socially complex situations into preferred ones (Simon 1988). The future of DH should be one in which solving social problems is the priority.

Many DH scholars have addressed communities that have been institutionally and socially marginalized, those whose voices, faces, and outputs have been missing from our academic processes and priorities. Such work has done much to shed light on examples of racism, sexism, homophobia, transphobia, and ableism within the discipline. But attempts to decolonize or to diversify cannot be achieved by retrospectively adding content to existing work. For example, if a digital tool has been used to analyze a dataset that was not inclusive, updating that dataset will not undo the fact that the initial study was run without consideration for those that were excluded. Instead, the challenge for DH is to imagine a future where society's main issues are present by default so that *all output* will account for such failings. DH can begin achieving such a future by following a DM approach, which can help ensure the right stakeholders are included in the problem (research aim) space as well as offering a space to make. A DM approach allows us to prototype and test digital tools before they are published or disseminated, ensuring that these tools are addressing the right problem and, hence, holding the values of DH at their core.

In 2012, Johanna Drucker wrote that the challenge facing DH was "… to shift humanistic study from attention to the *effects* of technology … to a humanistically informed theory of the *making* of technology …" and called for the next phase of DH to "synthesize method and theory into ways of *doing as thinking*." Design offers DH a way to implement Drucker's vision. The Double Diamond Framework incorporates iteration and prototyping in its solution space, just as it militates against the researcher rushing in with a tool/solution-led approach.

While design does not offer a single definition that "adequately covers the diversity of ideas and methods gathered together under the label" (Buchanan 1992, 5), it does provide a starting point, a way for research to be responsive, for it to borrow tools and theories from other disciplines so that it can be relevant to societal needs, *real* societal problems.

NOTE

1 This idea of "design thinking" has become a point of contention in recent times, "having created a reductionist perspective of design which has simultaneously become a buzzword for innovation" (Baker and Moukhliss 2020, 307). Those working in Design argue that the intrigue and hype around design thinking have been appropriated without any real critical understanding. Some digital humanists, having seen "DH" become a buzzword itself, may empathize. Stop Design Thinking (@StopDThinking), a Twitter account which applauds tweets that compare design thinking to snake oil, recently responded to a new design thinking initiative with the following provocation: "Is this a) rearranging the chairs on the deck of the Titanic or b) rolling shit in glitter. Discuss." (June 23, 2021). The intention here is not to engage unnecessarily with the debate around design thinking, but it is important to be aware of the unease which exists.

REFERENCES

Baker, Fredrick W. and Sarah Moukhliss. 2020. "Concretising Design Thinking: A Content Analysis of Systematic and Extended Literature Reviews on Design Thinking and Human-Centred Design." *Review of Education* 8 (1): 305–33. https://doi.org/10.1002/rev3.3186.

Brown, Tim and Barry Katz. 2011. "Change by Design." *Journal of Product Innovation Management* 28 (3): 381–3. https://doi.org/10.1111/j.1540-5885.2011.00806.x.

Buchanan, Richard. 1992. "Wicked Problems in Design Thinking." *Design Issues* 8 (2): 5. https://doi.org/10.2307/1511637.

Churchman, C. West. 1967. "Guest Editorial: Wicked Problems." *Management Science* 14 (4): B141–42. http://www.opanalytics.ca/MORS/pdf/Churchman-WickedProblems.pdf.

Coyne, Richard. 2005. "Wicked Problems Revisited." *Design Studies* 26 (1): 5–17. https://doi.org/10.1016/j.destud.2004.06.005.

Cross, Nigel. 1982. "Designerly Ways of Knowing." *Design Studies*, 3 (4): 221–27. https://www.sciencedirect.com/science/article/abs/pii/0142694X82900400?via%3Dihub.

Crum, Alia J., Kari A Leibowitz, and Abraham Verghese. 2017. "Making Mindset Matter." *BMJ*, February: j674. https://doi.org/10.1136/bmj.j674.

Design Council. 2019. "What is the Framework for Innovation? Design Council's Evolved Double Diamond." https://www.designcouncil.org.uk/our-work/skills-learning/tools-frameworks/framework-for-innovation-design-councils-evolved-double-diamond/.

Drucker, Johanna. 2012. "Humanistic Theory and Digital Scholarship." In *Debates in the Digital Humanities*, edited by Matthew K. Gold. https://dhdebates.gc.cuny.edu/read/40de72d8-f153-43fa-836b-a41d241e949c/section/0b495250-97af-4046-91ff-98b6ea9f83c0#ch06.

Drucker, Johanna. 2014. "Knowledge Design." *Design and Culture* 6 (1): 65–83. https://doi.org/10.2752/175470814X13823675225117.

Friedman, Batya, and David G. Hendry. 2019. *Value Sensitive Design: Shaping Technology with Moral Imagination*. Cambridge, MA: MIT Press.

Hoffman, Libby. 2016. "10 Models for Design Thinking." *Medium* (blog). October 18. https://libhof.medium.com/10-models-for-design-thinking-f6943e4ee068.

Kennedy, Fiona, Brigid Carroll, and Joline Francoeur. 2013. "Mindset Not Skill Set: Evaluating in New Paradigms of Leadership Development." Edited by Gareth Edwards and Sharon Turnbull. *Advances in Developing Human Resources* 15 (1): 10–26. https://doi.org/10.1177/1523422312466835.

Kirschenbaum, Matthew G. 2004. "'So the Colors Cover the Wires': Interface, Aesthetics, and Usability." In *A Companion to Digital Humanities*, edited by Susan Schreibman, Ray Siemens, and John Unsworth, 523–42. Malden, MA: Blackwell Publishing. https://doi.org/10.1002/9780470999875.ch34.

Norman, Donald A. 2013. *The Design of Everyday Things, Revised and Expanded Edition*. Cambridge, MA: MIT Press.

Rittel, Horst W. J. and Melvin M. Webber. 1973. "Dilemmas in a General Theory of Planning." *Policy Sciences* 4 (2): 155–69. https://doi.org/10.1007/BF01405730.

Schofield, Tom, Mitchell Whitelaw, and David Kirk. 2017. "Research through Design and Digital Humanities in Practice: What, How and Who in an Archive Research Project." *Digital Scholarship in the Humanities* 32 (supplement 1): i103–20. https://doi.org/10.1093/llc/fqx005.

Simon, Herbert A. 1988. "The Science of Design: Creating the Artificial." *Design Issues* 4 (1/2): 67–82. doi: 10.2307/1511391.

Stop Design Thinking (@StopDThinking). 2021. "Is This a) Rearranging the Chairs on the Deck of the Titanic or b) Rolling a Shit in Glitter. Discuss." Twitter post, June 23. https://twitter.com/StopDThinking/status/1407511116813533188.

CHAPTER FORTY

Reclaiming the Future with Old Media

LORI EMERSON (UNIVERSITY OF COLORADO BOULDER)

In the face of entrepreneurs' and tech companies' attempts to over determine the shape of what's to come, how can we participate in (re)claiming communal ownership of the future? In what follows I think through a series of possible answers to this question by first unpacking why and how the past keeps getting eclipsed by an ever-receding future we seem to have little to no control over. I then propose six interrelated values we might take from old media: slow, small, open, cooperative, care, and failure. All six values are intentionally opposed to: ungrounded speculation; early adoption in the name of disruption, innovation, and progress; and convenient quick-fixes. Rather than recapitulate these same logics and claim my argument is wholly new or groundbreaking, and contrary to those who have been named as participating in the "dark side of DH" with practices that are "rooted in technocratic rationality or neoliberal economic calculus," my intention is to gather together tactics that many DH community members have already embraced and reframe them in relation to recovering past media traditions for the sake of a reimagined future (Chun et al. 2016).

I imagine this piece as being in quiet conversation with certain lines of thought on old or so-called dead media in media archaeology—for example, Friedrich Kittler's attempts to get around our inability to understand regimes embedded in contemporary technologies by way of excavations of regimes in old media; Wolfgang Ernst's attempts to escape from the grip of humanism by attending to the unique, material functioning of machines; and Siegfried Zielinski's notion of "variantology" or interruptions, "fractures and turning points" in the otherwise tidy fiction of the history of technology and his attempts to catalog as many instances of it as possible (Kittler 1990, 1999; Ernst 2013; Zielinski 2006). However, while these ideas have been deeply generative for media studies generally and also for those in the humanities who have needed a thoroughgoing justification for the importance of hands-on labs and centers dedicated to old media of all kinds, as of this writing it is not clear whether the particular German context of many of these works can be translated into a North American context where, in the last decade, the depth and scope of systemic/historical poverty, racism, sexism, transphobia, ableism, profound environmental degradation, and so on, have only become more clear. Jussi Parikka is right to assert that there is tremendous potential in media archaeology as an "innovative 21st-century arts and humanities discipline that investigates non-human temporalities and ... wants to address those material and cultural contexts and forces that are beyond our control"—but how do we tap that potential so that it more obviously attends to the socio-political realities of our time (Parikka 2012)?

While rarely intersecting with media archaeology and even more rarely working to excavate media archaeology's untapped potential, DH has produced an increasing number of works that are intensely engaged with socio-political realities and that, intentionally or not, respond to what Alan Liu named the primary deficit in the field: the absence of cultural criticism, or, a lack of reflection on "the relation of the whole digital juggernaut to the new world order" (Liu 2012). Just a few of the more recent examples of the latter include the DH project "Land Grab Universities" by Robert Lee et al. (2020), Roopika Risam and Kelly Baker Joseph's *The Digital Black Atlantic* (2021), and Elizabeth Losh and Jacqueline Wernimont's *Bodies of Information: Feminist Debates in Digital Humanities* (2019). Still, how do we bridge media archaeology's investment in old media and rethinking the history of technology with the wide range of digital humanities projects that may not interrogate "the digital" or what we mean when we say "old" or "new" but that certainly harness the capabilities of the digital as a way to bring to light local and global injustices?

Rather than provide a definitive answer to the foregoing question, in this chapter I explore how an approach such as the one coined by Lizzie O'Shea, "future histories," could bring these two fields closer together. For O'Shea and me, future histories are overlooked histories which also, crucially, bear the potential to lead us into alternative futures. They are also arguments in themselves "for what the future could look like, based on what kinds of traditions are worth valuing and which moments are worth remembering" (O'Shea 2021, 7). Future histories, then, are fundamentally political—they are the groundwork from which we can enact positive change today and tomorrow.

***

Twenty-first-century life in the West is now practically synonymous with self-appointed experts making seemingly neutral assertions on a daily basis about what's to come—assertions that are actually self-interested bids to own the future, or, more accurately, bids to own *our* future. Because they are necessary to maintaining the illusion that capitalism is and always will be right and good for all, these assertions are usually short on facts, long on speculation, and oblivious to the actual material conditions of an increasingly degraded everyday life. Take, for example, Elon Musk's claim that "If you get up in the morning and think the future is going to be better, it is a bright day. Otherwise, it's not" (Economy 2017). While it's laughable that positive thinking could make everyday life better now and in the future for those who are poor, sick, grieving, and so on, the perpetual generation of future-oriented rhetoric that's blind to material realities is still necessary to buoy up belief in the rightness of Musk's attempt to own the future of electric cars, space travel, solar energy, AI, public transportation, and whatever else pops into his mind on a day-to-day basis. Tech companies generally take a slightly different approach in their attempts to own the future, instead frequently declaring "the future is now!" and warning that unless we become "early adopters" of gadget *x* and embrace the same "tactics for disruptive thinking" as the tech companies relentlessly rolling out new devices, we will be "left behind" or "left out" of the coming future (Gutsche 2020). Regardless of the source, the result of these future-oriented rhetorical maneuverings is the same: the erosion of a democratic future.

Whether you're an investor or a tech company, a bundle of interrelated strategies for generating long-term profit underlies this rhetoric about the future. These strategies include an embrace of planned obsolescence—by intentionally designing and producing products with a built-in short life span that forces consumers to buy more products more quickly—often by preventing us from being able to open up and repair their products, also known as black boxing. In addition to contributing to environmental damage from a never-ending cycle of consumption and disposal, as well as cultivating a sense of passivity in the face of our inability to repair or understand our devices, planned obsolescence via black boxing has also helped create tech company monopolies and labor inequities at all levels and at a global scale (as the consumer/worker has to work harder to spend more, more often, and thus have a lower quality of life which also makes it impossible to have the luxury of time to think otherwise) (Grout and Park 2005).

While planned obsolescence and black boxing depend on each other, they also depend on an erasure of history, just in case we dare to question whether objects from the (near) past are really as quaint, cumbersome, inefficient, and even primitive as we're told they are. More, planned obsolescence and black boxing also result in our alienation from the materiality of everyday life and its objects. I know, for example, that the MacBook Pro laptop I'm using to write this chapter is a material object but, beyond that general sense, I am utterly locked out of its inner workings and mechanisms and, should I spill the coffee I'm sipping on it, there is almost certainly no recourse other than to replace the entire machine. Unlike the roughly eighty-year-old manual typewriter that sits next to my desk and to which I have a deep attachment because of the years I've spent understanding its sounds, behaviors, quirks, and charms, I have no such attachments to this laptop. After all, I've been told countless times that I should only keep it for three to five years. But, if it weren't for this supposedly "quirky" and eccentric habit I have of keeping old media around long after they should have been consigned to the trash heap, how would I know to even question the design of my laptop? Without access to the past, how would I ever be able to imagine an alternative present and therefore a future where our things are built with, for example, longevity, care, maintenance, and sustainability in mind?

I have written elsewhere about Apple's attempts to eradicate programmable computers with relentless releases of ever new versions of hermetically sealed iPads and iPhones that have short life spans and extremely limited options for repair (Emerson 2020a, 350). But Apple is far from the only corporation that attempts to erase histories of early personal computing which would allow us to know what computing was, could have been, and still could be. For example, in a blatant disregard for the decades when computers were some combination of kit or pre-assembled machine that was still open (both in terms of hardware and software) and extensible, Samsung released a video advertising their Galaxy Book in April 2021. Reminiscent of Apple's infamous ad for the Macintosh in 1984 which opens in a concrete environ meant to conjure up thoughts of a Soviet era Cold War bunker, Samsung's ad opens in another futuristic version of a concrete structure that, we're told, is the Museum of Laptops. The docent dressed in business attire stands in front of a sign that reads "The Laptop is History" as he encourages us to come with him on "a journey back to the time before we asked the question 'Why can't laptops be more like phones?'" (Samsung 2021). I can't be the only one who has never wished my laptop were more like my phone. Either way, if we had any awareness of the past we'd understand what programmable computers (ought to be able to) do, why they are important, and why a mobile phone can never be a substitute.

In 1967, just a few years after Marshall McLuhan first coined the term "information age" to describe the radical cultural shift underway because of changing technological media, he declared that "The past went that-a-way. When faced with a totally new situation, we tend always to attach ourselves to the objects, to the flavor of the most recent past. We look at the present through a rear view mirror. We march backwards into the future" (McLuhan and Fiore 1967). This assertion

FIGURE 40.1 Corona Standard manual typewriter from the late 1930s that sits beside author's desk. Credit: Lori Emerson.

that the present, and therefore the future, is saturated with the past—whether we are aware of it or not—is certainly accurate in principle. However, regardless of what is or ought to be the case, McLuhan could not anticipate the contours of twenty-first-century late capitalism and how it seems to depend on a methodical, persistent, and nearly instantaneous eradication of the past so that we are rarely aware of how the past informs nearly everything around us. In other words, now it's less that we "look at the present through a rear view mirror" and more that we barely have enough time to even register the present before it's replaced with another present—and another and another.

However, every once in a while we are awakened from this externally imposed collective amnesia and we stumble over something as seemingly magical and miraculous as a computing device from the 1970s or 1980s that still functions. Take, for example, the Altair 8800b computer from 1976 (Figure 40.2), (usually) still functioning at the Media Archaeology Lab at the University of Colorado Boulder—a hands-on lab that is a home for thousands of still-functioning media objects from the late nineteenth century to the present. Unlike the three- to six-year average life span of a contemporary

FIGURE 40.2 An Altair 8800b computer from 1976, housed in the Media Archaeology Lab. Credit: Libi Striegl.

computer and the even shorter average life span of a smartphone, and even contrary to our own experience of inevitably being unable to download system updates and each piece of software gradually becoming unusable, the forty-five-year-old Altair should not still work. But the moment we power it on, see the red LED lights flash, and hear the cooling fan begin its arduous work, the grip of the ideology of planned obsolescence loosens just a bit as we experience what should not be the case. And then, while we ponder the limitations on this 8-bit computer's ability to compute anything with any degree of speed, power, and even reliability we also cannot help but marvel at the fact that the spare circuit boards sitting next to the Altair suggest that it can and even has been opened up, altered, and repaired numerous times; we also marvel at the fact that this box with switches and no keyboard, mouse, or screen presents an alternative trajectory for computing that never came to pass but that could still happen. In other words, machines like the Altair 8800b show us an entirely different worldview that is not dominated by planned obsolescence or black boxing and one that does not revolve around speed and efficiency as the only viable criteria by which to measure the machine as an indication of "progress." Instead, the worldview embedded in the still-functioning Altair is one that values slowness, care, maintenance, environmental sustainability, openness, transparency, glitch/failure and therefore it's also one that—contrary to futurists like Elon Musk and tech companies like Apple and Samsung—values the material conditions of our everyday lives.

Thus, in the spirit of the Altair along with all the other machines of its vintage and even earlier, below is a list of six values old media offer us to reimagine a different way of life today and tomorrow. Although framed in terms of future histories, this piece was first inspired by Lisa Spiro's powerful 2012 essay "'This is Why We Fight': Defining the Values of the Digital Humanities" in which she lists values DH should embody or aspire to, including openness, collaboration, collegiality and connectedness, diversity, and experimentation. Throughout the writing process, I have also had in mind the document that Catherine D'Ignazio and Lauren Klein published as part of *Data Feminism*, "Our Values and Our Metrics for Holding Ourselves Accountable," as well as other value statements D'Ignazio and Klein pointed me to, such as those by the University of Maryland's African American History, Culture, and Digital Humanities Initiative and the University of Delaware's Colored Conventions Project (D'Ignazio and Klein 2020).

***

SLOW

Rob Nixon has documented the profoundly destructive power of what he calls "slow violence"—a "violence that occurs gradually and out of sight, a violence of delayed [environmental] destruction that is dispersed across time and space" (Nixon 2013). But, if we think about slowness in relation to contemporary digital technology, is it possible that embracing the slow and the inefficient could help intervene in the ideology of planned obsolescence and black boxing I discuss above? That if we intentionally embrace slow media we also reduce the slow violence taking place out of sight? If we can recognize the unique capabilities and affordances of, say, the 1976 Altair 8800b without comparing it to the processing power of, say, a 2021 MacBook Pro computer, we might be able to see what "wait times can teach us about the forces in life that have shaped our assumptions about time, efficiency, and productivity" (Farman 2018). What else might be possible outside of the relentless push toward productivity, consumption, and waste? How might slow and inefficient media decelerate the frenetic movement of electronics from our homes and offices to e-waste sites on the other side of the globe?

SMALL

The value of approaching the collection and interpretation of data in terms of the small and the local has been compellingly documented by Christine Borgman (2015), Yanni Alexander Loukissas (2019), and many others in both digital humanities and information science. We might also extend the value of small to individual media if, similar to the value of slow, we think in terms of adjusting our expectations of our machines so that they have more modest capabilities. However, the value of small starts to appear more compelling in the context of the networked systems connecting media. Given, for example, the state of our contemporary internet which is driven by the pursuit of profit by way of tracking, surveillance, and relentless expansion to every corner of the planet, what would be possible if we reverted to a culture of small networks populated by mostly local participants like the Bulletin Board Systems of the 1980s and early 1990s? These early, small networks seemed to create the possibility (granted, not always achieved) for meaningful community that extended into the offline as much as the online world (Emerson 2020b). Also, in part because of their slowness *and* their smallness, these early networks opened up opportunities for extended modes of engagement and discourse that are no longer acceptable in our current era dominated by an expectation of immediate, fast-paced communication.

OPEN

One reason it's so difficult to conceive of an internet other than the one we currently have is because, despite the fact that the internet is built on open-source software, the vast majority of us will never understand how the internet works at the levels of software, hardware, and infrastructure. Parts of the internet are nearly too complex for any one person to understand while other parts (such as the submarine cables it runs on) are practically inaccessible. Also, even though the internet is in theory open to anyone to use, censorship is alive and well in countries such as China, Syria, Iran, Egypt, and Sri Lanka; and further, many people who live in rural areas around the world often have to access the internet on their mobile phones, which severely limits the range of things one can effectively do online (from applying for jobs to taking classes online) (Raman et al. 2020). In short, the openness of the internet is more assertion than fact. By contrast, think again of a Bulletin Board System (BBS) from thirty years ago that could have been owned and run by an individual who originally purchased BBS software that came with extensive documentation about how it works and how to customize nearly every aspect of the network. Openness, then, is not only about transparency about how things work; it is also about accessibility and a more straightforward ability to understand how and for whom things work. The latter is especially important to attend to if we want to avoid perpetuating implicit power structures through the mere appearance of openness (Schneider 2021).

COOPERATIVE

Imagine if, like some early networks, our contemporary social media platforms were owned and run cooperatively rather than by corporations like Facebook or Twitter? Not only would we have the ability to determine the shape, scope, and functionality of our networks according to criteria that do not necessarily relate to potential profitability but so too would we be able to adopt cooperative governance structures for members, employees, management, and overseeing boards that have accountability to users/owners baked into them. While never explicitly named a cooperative, the Berkeley-based network Community Memory was collectively created by five people (Lee Felsenstein, Efrem Lipkin, Ken Colstad, Jude Milhon, and Mark Szpakowski) in 1972 by connecting a handful of scavenged teletype terminals—all installed in coop supermarkets, record stores, and libraries—to a donated Scientific Data Systems SDS-940 timesharing computer (Felsenstein n.d.). From 1972 until its demise in 1975, the collective provided a computerized version of an analog bulletin board whose messages could be read for free while posting cost twenty-five cents or more.

Admittedly, Community Memory may not provide a relevant model for financial sustainability in the twenty-first century but it has certainly paved the way for a growing number of decentralized, privacy-friendly networks such as Social.coop. More, according to the University of Wisconsin's Center for Cooperatives, coops are often based on the values of "self-help, self-responsibility, democracy, equality, equity, and solidarity" and since cooperative members also "believe in the ethical values of honesty, openness, social responsibility, and caring for others" (n.d.), the structure bears tremendous potential for creating equitable practices that are more attuned than conventional corporations to the complex intersections of different socioeconomic statuses.

FIGURE 40.3 Community Memory Terminal from 1975, housed in the Computer History Museum. Credit: Kathryn Greenhill.

CARE

The short life span of Community Memory raises another important issue: the need for care and repair of media of all kinds to ensure sustainability, in terms of the media themselves and also in terms of long-term environmental impact. Insofar as care and repair are both facets of maintenance, they are also, as Lee Vinsel and Andrew L. Russell put it, "the opposite of innovation" and therefore the opposite of whatever Elon Musk is perpetually dreaming up for the future. Care, repair, and maintenance are the bedrock of the "overlooked, uncompensated work" that "preserve[s] and sustain[s] the inheritance of our collective pasts" (Vinsel and Russell 2020). Related to the section above on cooperatives, currently, because most of us do not have a stake in any of the infrastructure that networks run on, we have very little to no control over networks' longevity. However, if we return once again to the example of the Altair 8800b, machines that are built to be opened up, repaired, and even extended have the potential for long life spans if we care for them and (know how to) repair them. As feminism has long established, care not only benefits human beings but it also—or it ought to also—extends to an "appreciation of context, interdependence, and vulnerability—of fragile, earthly things and their interrelation" (Nowviskie 2019). In other words, caring for and maintaining the Altair 8800b well beyond what any computer manufacturer would say is its recommended life span is not simply about a single computer or its fetishization; it is about seeing the deep and broad impacts, on both a local and planetary scale, of our relations with the living and nonliving things of the world.

FAILURE

If old media liberate us to embrace the slow, small, open, cooperative, and an ethics of care, then they also insist we embrace failure and glitch. When the Altair 8800b faulters and glitches, as it does on an almost weekly basis, we are challenged not to consign it to the trash or the recycling heap but to be open to the rewards of its failure. Since the 1990s, glitch artists have been showing us how to see aesthetic value in unprovoked and sometimes even intentional failure—but so too have queer theorists such as Jack Halberstam who reminds us that "under certain circumstances failing, losing, forgetting, unmaking, undoing, unbecoming, not knowing may in fact offer more creative … more surprising ways of being in the world" (Halberstam 2011). As I move toward ending this chapter, I would like to underscore that I, alongside many of the writers I have cited in this piece, am not merely writing nostalgically about machines from a bygone era—machines that sometimes, miraculously, work. I am talking more fundamentally about how old media provide one of many ways to imagine, as Halberstam declares queer studies also offers, "not some fantasy of an elsewhere, but existing alternatives to hegemonic systems" (Halberstam 2011). And it is often in that exquisite moment of failure that we catch a glimpse of future histories showing us alternatives to power or perhaps a way toward its undoing.

REFERENCES

Borgman, Christine. 2015. *Big Data, Little Data, No Data: Scholarship in the Networked World*. Cambridge, MA: MIT Press.

Chun, Wendy Hui Kyong, Richard Grusin, Patrick Jagoda, and Rita Raley. 2016. "The Dark Side of the Digital Humanities." In *Debates in the Digital Humanities*, edited by Matthew K. Gold and Lauren F. Klein, 493–509. Minneapolis, MN: University of Minnesota Press.

D'Ignazio, Catherine and Lauren Klein. 2020. "Our Values and Our Metrics for Holding Ourselves Accountable." Data Feminism.

Economy, Peter. 2017. "These 11 Elon Musk Quotes Will Inspire Your Success and Happiness." Inc. 24. https://www.inc.com/peter-economy/11-elon-musk-quotes-that-will-push-you-to-achieve-impossible.html#:~:text=%22When%20something%20is%20important%20enough,and%20you%20want%20to%20live.

Emerson, Lori. 2020a. "Interfaced." *Further Reading*, edited by Matthew Rubery and Leah Price. New York, NY and London: Oxford University Press.

Emerson, Lori. 2020b. "'Did We Dream Enough?' THE THING BBS as Social Sculpture." Phantom Threads: Restoring The Thing BBS. Rhizome. https://rhizome.org/editorial/2020/dec/16/did-we-dream-enough-the-thing-bbs/.

Ernst, Wolfgang. 2013. *Digital Memory and the Archive*, edited by Jussi Parikka. Minneapolis, MN: University of Minnesota Press.

Farman, Jason. 2018. *Delayed Response: The Art of Writing from the Ancient to the Instant World*. New Haven, CT: Yale University Press.

Felsenstein, Lee. (n.d). "Resource One/Community Memory – 1972–1973." http://www.leefelsenstein.com/?page_id=44.

Grout, Paul A. and In-Uck Park. 2005. "Competitive Planned Obsolescence." *The RAND Journal of Economics* 36 (3): 596–612.

Gutsche, Jeremy. 2020. *Create the Future + The Innovation Handbook: Tactics for Disruptive Thinking*. New York, NY: Fast Company.

Halberstam, Jack. 2011. *The Queer Art of Failure*. Durham, NC: Duke University Press.

Kittler, Friedrich. 1990. *Discourse Networks 1800/1900*. Translated by Michael Metteer. Stanford, CA: Stanford University Press.

Kittler, Friedrich. 1999. *Gramophone, Film, Typewriter*.Translated by Geoffrey Winthrop-Young and Michael Wutz. Stanford, CA: Stanford University Press.

Lee, Robert, Tristan Ahtone, Margaret Pearce, Kalen Goodluck, Geoff McGhee, Cody Leff, Katherine Lanpher, and Taryn Salinas. 2020. "Land Grab Universities." https://www.landgrabu.org/.

Liu, Alan. 2012. "Where is Cultural Criticism in the Digital Humanities?" *Debates in the Digital Humanities*, edited by Matthew K. Gold, 490–510. Minneapolis, MN: University of Minnesota Press.

Losh, Elizabeth and Jacqueline Wernimont. 2019. *Bodies of Information: Feminist Debates in Digital Humanities*. Minneapolis, MN: University of Minnesota Press.

Loukissas, Yanni Alexander. 2019. *All Data Are Local: Thinking Critically in a Data-Driven Society*. Cambridge, MA: MIT Press.

McLuhan, Marshall and Quentin Fiore. 1967. *The Medium is the Massage: An Inventory of Effects*. San Francisco, CA: HardWired.

Nixon, Rob. 2013. *Slow Violence and the Environmentalism of the Poor*. Boston, MA: Harvard University Press.

Nowviskie, Bethany. 2019. "Capacity through Care." In *Debates in the Digital Humanities 2019*, edited by Matthew K. Gold and Lauren F. Klein, 424–26. Minneapolis, MN: University of Minnesota Press.

O'Shea, Lizzie. 2021. *Future Histories: What Ada Lovelace, Tom Paine, and the Paris Commune Can Teach us About … Digital Technology*. New York, NY: Verso Books.

Parikka, Jussi. 2012. *What is Media Archaeology?* Malden, MA and Cambridge: Polity Press.

Raman, Ram Sundara, Prerana Shenoy, Katharina Kohls, and Roya Ensafi. 2020. "Censored Planet: An Internet-wide, Longitudinal Censorship Observatory." CCS '20: Proceedings of the 2020 ACM SIGSAC Conference on Computer and Communications Security, 49–66.

Risam, Roopika and Kelly Baker Josephs, eds. 2021. *The Digital Black Atlantic*. Minneapolis, MN: University of Minnesota Press.

Samsung. 2021. "The New Galaxy Book." Online advertisement. https://twitter.com/SamsungMobile/status/1387467870859374594?s=20.

Schneider, Nathan. 2021. "The Tyranny of Openness: What Happened to Peer Production?" *Feminist Media Studies*. 17 February.

Spiro, Lisa. 2012. "'This is Why We Fight': Defining the Values of the Digital Humanities." *Debates in the Digital Humanities*, edited by Matthew K. Gold, 16–35. Minneapolis, MN: University of Minnesota Press.

University of Wisconsin Center for Cooperatives. (n.d.). "Cooperative Principles." https://uwcc.wisc.edu/about-co-ops/cooperative-principles/.

Vinsel, Lee and Andrew L. Russell. 2020. *The Innovation Delusion: How Our Obsession with the New Has Disrupted the Work that Matters Most*. New York, NY: Currency.

Zielinski, Siegfried. 2006. *Deep Time of the Media: Toward an Archaeology of Hearing and Seeing by Technical Means*, translated by Gloria Custance. Cambridge, MA: MIT Press.

CHAPTER FORTY-ONE

The (Literary) Text and Its Futures

ANNE KARHIO (INLAND NORWAY UNIVERSITY OF APPLIED SCIENCES)

The rapidly shifting technological environment of the early twenty-first century prompts frequent speculation on either the death, or the fantastic future, of various media. Tools and platforms must either perish, or evolve and adapt in a system that is in a constant state of flux. The entire field of digital humanities is built on such shifting ground: it owes its existence to the interchange between text and technology, and the reader's and the writer's equally evolving relationship with the text as an alphabetic system of symbolic and material signification. Digital humanities emerged from scholarly curiosity concerning the adaptability of texts to radically new circumstances, and their multiple reconfigurations in computerized environments. Just as digital media have embraced visual, audiovisual, haptic, and immersive methods of communication, the role of the text, or the literary text, is continuously tested in relation to other modes of interaction and transmission. A good example of our perpetual preoccupation with its fate is *The Future Text Initiative*, started by Frode Hegland and a group of prestigious advisors, who in 2020 published the extensive volume titled *The Future of Text* with contributions from more than 170 scholars, artists, and authors. (A second volume on the topic was published during the editorial process of this present collection.) The aim of the project was to present "a collection of dreams for how we want text to evolve as well as how we understand our current textual infrastructures, how we view the history of writing, and much more" (*The Future of Text* n.d.). The book would also serve as a record of the "future past" of text: "as a record for how we saw the medium of text and how it relates to our world, our problems and each other in the early twenty first century" (n.p.). When the present is too close for a clear perspective, such a sprawling futuristic approach is a means of envisioning an imaginary retrospective view. *The Future of Text* follows its texts into archive shelves, databases, hard drives, digital platforms, human and non-human neural systems, paper, print, screen, code, and beyond. The closer we move, the more texts morph, mutate, and multiply, and the stranger our view of something formerly familiar becomes.

Scholars considering digital literary textuality have frequently addressed their topic through a framework or narrative that considers digital writing as something that occurred after print, or at least writing, or the centuries of handwritten and then mechanically printed texts that constitute the twenty-first-century popular idea of "text" in the first place—even in cases where screen-based writing has become the norm but often mimics or repeats the characteristics and functionality of writing on paper (or some other similar flat, non-digital interface). Few of us in academia would subscribe to a simple narrative of development from oral to print to digital, but this narrative nevertheless tends to haunt, if not quite determine, the discussion on digital literary textuality. There are occasional, and often significant, itineraries into orality or oral texts, like Walter Ong's much-quoted work on "secondary orality," where oral communication and transmission is tied to "the existence of devices that depend for their existence and functioning on writing and print" (Ong 2002 [1982], 10–11). Yet the concept of "text" itself is almost impossible to detach from the idea of written, alphabetic inscription, and emerged concurrently with it. This despite poststructuralist theory's recognition of various cultural phenomena as not-strictly-alphabetic texts, including "oral textuality" and oral literatures (Ong 2002 [1982], 10). The past as well as the future of the text is thus intimately tied to its visual materiality, and the potential of its renegotiation and adaptability to non-print, but also non-ocular or multimodal environments. Speech, text, image, and sound form new assemblages as texts interact with new technologies and are translated between humans and machines, code and natural language, or touch, vision, and sound.

Literary texts are particularly appealing for the speculative methods that *The Future of Text* also calls for: they build on language's capacity to convey multiple meanings, its denotative and connotative dimensions, or an unruly digestion of new associations—or the discarding of old ones—across historical and geographical distances. Literary language and literary texts enlighten, hide, disguise, deceive, and expose. As they travel across media, texts also reveal new features of themselves or, at times, reach a dead end outside their native medium. The novel as a visually inscribed document that requires sustained attention is more at home on the page than it is on a small screen—though it may quite comfortably find a new dwelling as recorded sound (as the recent increase in popularity of audiobooks suggests). Texts carved in stone call for more condensed expression but enjoy greater longevity, whereas the texts of instant messages may be here now, gone next minute—at least as far as their intended readers are concerned. Alternatively, direct messages on smartphones may end up having uncomfortably long afterlives in contexts that involve unintended audiences or authorities. The very real and not so distant futures of individual texts can be of more than marginal interest to those who wish their readership to remain limited. Texts are capricious, especially in the digital domain: their preservation or ephemerality is difficult to control, or may end up being controlled by parties whose interests do not coincide with those of their original creators. In a wider context, this applies to the fate of digitized as well as born-digital texts in a networked world, where data infrastructures and storage are managed by corporations and state bodies. What is the future of texts, in a very concrete sense, as they increasingly have no existence outside the digital domain?

Leonardo Flores writes how in digital environments the computer interface for texts enacts "a whole complex matrix of intentions and desires—the writer's, reader's and everyone else's who contributed to the workings of a computer" (2013, 88). In recent years, scholars and digital literary practitioners have increasingly highlighted the sometimes problematic economic and political dimensions of digital infrastructures and platforms. At the same time, they have revised their views on the material networks of authorship, dissemination, and reception of pre-digital (or post-digital) texts. Text is also data, and in the data economy the text of the code or the programming language, which mostly remains invisible to the user/reader of words on a computer screen, often does the bidding of desires that have little to do with the desire of a literary author, or a literary reader. In other words, inasmuch as "computers have intentions and desires encoded within them, and their external behaviors may be interpreted as expressions of those intentions" as Flores (2013, 88) notes, increasing attention is being paid to the agents, institutions, and socioeconomic groups who inscript their interests in the very make-up of computer and network design. Yet, importantly, there is a future-oriented dimension in such a focus on "intentions and desires." The question is not merely one of the intentions and desires behind the existing features of digital environments and texts, but also of the intentions and desires fueling ongoing transformations of textuality in the present, and suggesting multiple possible futures. *Whose* desires inform the future, or futures, of text? Many authors of born-digital or electronic literature have harnessed their own creative medium to reveal the future-building aspects of its design and function, or to employ it for new, unintended purposes.

A shift from oral to written storytelling and narrative memory also reflects a recalibrated attitude towards preservation and storage. The securing of shared tradition now takes place through a medium that relies less on collective memory and transmission, and more on a perception of seemingly reliable and accurate technological storage, detached from embodied communal practice and performance. But as the temporality of oral narration has been measured against the more stable spatial arrangement of visual signs and symbols on a page, the emergence of digital media has, for some, marked a return of sorts to the more temporal experience of continuous presence (as Ong, too, recognized in the context of secondary orality). It is perhaps more accurate to consider textuality in digital environments and interfaces as a question of reassembling the possibilities of temporal and flexible transmission in orality, and the (relative) fixity of the page, or its precedents (as listed by Flores): "stone, clay, papyrus, vellum, wood [...] and other materials that lend physical stability to the inscriptions they carry" (2013, 88). In addition to the new questions of preservation and forgetting prompted by digital texts and messages, the specific characteristics of digital texts may in the end have less to do with any of their immediately perceptible qualities than with their reliance on an in-built dynamic relationship between multiple layers of textuality and operationality, and consequently their distinctive "behaviour," as Flores describes it. Sound-based, non-verbal, or non-alphabetic interface experiences, too, rely on text-based code and programming languages whose readers may be machines or human programmers, and this layer of "text" in a digital image or narrative (which can be visual or auditory, as well as textual) remains detached from the visible narrative. But rather than make text and textuality marginal or redundant, the layered structures of online environments and their reliance on complex relations between power, desire, and interaction highlight the continuing importance of accounting for the numerous textual strata of networked, digital exchanges. When the text goes underground, we need new means of enticing it to view.

If poststructuralist theory trained a generation of scholars to read various aspects of social and cultural life and production as texts, regardless of whether or not they strictly consisted of written, alphabetic expression, more recently new materialist thought has recalibrated scholarly approaches to acknowledge how all text has some kind of material basis, regardless of medium. Literary scholars, too, are invited to consider how this materiality is also always social, cultural, and political, as well as aesthetically relevant. In the digital domain, the materiality of texts is entangled with the materiality of digital and network infrastructures at a global scale. The materiality and modality of any medium, whether oral/aural, print/visual or digital/electronic, is inseparable from social and historical desires and power relations that foreground certain forms of communication while marginalizing others. Each medium has a material but also a socio-political foundation, as do choices related to aesthetic and literary design and experimentation, even though some texts may manifest a larger degree of self-awareness of such dimensions than others. These questions underpin, for example, the long poem or sequence *Arbor Vitae* by the Irish neo-modernist poet Randolph Healey. It was inspired by the author's experience of discovering that his infant daughter was deaf, and subsequently becoming aware of the social and historical biases that allow different degrees of agency to citizens with differing means and abilities of communication: the marginalization of the deaf in twentieth-century Ireland, for example, was underpinned by an emphasis on "oralism," or oral and aural communication (Healy 1997), and gestural sign systems that allowed deaf communities to participate in cultural and social life were pushed aside. Thus a dominance of orality can be as partial and problematic as the "dominance of textuality" that preoccupies Ong, and how we envision the future of the text says much about our sometimes unacknowledged ties to communities, networks, and institutions that inform the production and preservation of texts. For Healy, such commitment to particular methods of signification has been characteristic of human societies since their emergence; the time frame of *Arbor Vitae* extends from the Upper Paleolithic cave paintings at Vallon-Pont-d'Arc, France, to digital code in the computer age. Different systems of communication and material transmissions may co-exist, but do not necessarily connect, and this break-up in transmission and translation highlights the social and politically drawn entanglements of textual as well as non-textual media. The history of the text, alongside the history of other media, reveals how the idea of "*the* text" is itself specific to cultural and historical situations.

What the considerable and growing array of different digital tools and platforms, and modes of transmission in digital environments facilitates, however, is a more multifaceted approach to how texts and textuality function, or behave, in relation to the needs and desires of different readers and users. Much of the discussion on digital texts and textuality has continued to focus on texts, letters, and typography on screen rather than in print interfaces, though more recently this perspective, too, has widened. Smart technologies, ubiquitous computing, and immersive, haptic environments are pushing digital communication to material interfaces that no longer require engagement with screens or visible text. Tactile, embodied, auditory, and mixed-reality experiences are suggesting multiple unforeseen, possible futures for literary texts, too. For example, Chloe Anna Milligan's discussion on forms of literary fiction based on "*touchable* texts," and "the digital text as an act of *embodied* cognition" (Milligan 2017, 3, 4; emphasis in the original) pushes the idea of text from visuality towards embodied interaction. Other sensory and perceptual approaches to textuality alongside, or beyond, visual perception will call for new "aesthetic uses of [novel media] interfaces for textual evolution" (6). Print culture evolved as a response to the potential of the paper interface of the printed page, and the entire social-material system of production and exchange within which books could be produced, printed, circulated, carried around by their owners, and read in various locations. The reader of a novel or a poem in a printed volume does not necessarily pay attention to all of the factors that have contributed to its physical existence and form. But printers and publishers, who must consider financial viability of their products, often demand that works be restricted to certain lengths or word counts due to lower production or transportation costs. In digital environments, authors of electronic literary works and texts for screens consider the forms and structures suited for this visual or haptic interface, yet they must do so within the confines of devices and platforms that were mainly designed for other kinds of uses, financially viable and profitable for their makers. Typewriters allowed poets to experiment with typography in new ways, though their design was motivated by accuracy, uniformity, and efficiency of writing and record-keeping. Authors drawn to the possibilities of the early telegraph had to consider the financial as well as spatial constraints of the limited length, as each letter and word added to the cost of a message, the price ultimately determined by the owner of the telegraph network and infrastructure.

The shapes and forms that texts assume now and in the future will continue to be a result of such coming together of economic, personal, creative, and other related factors. The future of texts is tied to the future of our "intentions and desires," and also, increasingly, to our needs and limited resources. Texts can serve (or disobey) several masters.

Crucially, if not the death of the text, we are currently witnessing a profound transformation of the culture of writing, exchanging, and reading texts. And as the sensory scope and modal range of communication widens and evolves, this also raises the question of how far we can extend the meaning of "text" for it to remain purposeful or useful. The word "text" in its current sense emerged alongside a culture of writing, and cultural practices related to textures and weaving, or different patterns of webs.[1] It has not gone unnoticed that these metaphors connect the text as written alphabetic expression with the material metaphor for textuality as a woven tissue or net in the digital World Wide Web, where verbal inscription and multidimensional textures of visual and aural transmission intertwine. But text in the digital age has also become multidimensional in the sense of its constant translation between the layers of the networked, digital interface. Texts emerge, display, and transform in response to the relationships between various strata of code and language, their existence in natural languages written and read by computers as well as humans, and, increasingly, artificial or hybrid cognitive systems. Inasmuch as the concept of "text" emerged alongside inscribed, signifying practices of writing and reading in a specific socioeconomic context also informed by practices of stitching, weaving, and designing, this raises the question of the role it will have as these systems and practices become something else.

Crucially, Flores highlights that "texts in digital media are informational patterns which are subject to manipulation and reconfiguration in computers" (2013, 89). They are also subject to manipulation and reconfiguration within digital networks and networked operations of data, which are a key determining factor in how we experience digital screen interfaces, how those interfaces are designed, and how our engagement with texts relates to wider processes of the digital economy and its infrastructures. The visible or otherwise perceptible interface is underpinned by what Andersen and Pold have described as the "metainterface": it is key to a system where "the computer's interface [...] becomes [both] omnipresent and invisible," and is "embedded in everyday objects and characterized by hidden exchanges of information between objects" (Andersen and Pold 2018, 5). Literary texts, too, are now circulated within this system, but can also emerge as responses to its processes, like Cayley and Howe's *How It Is in Common Tongues*. The work, where a script trawls the web for strings of text that match, yet do not directly quote from Samuel Beckett's *How It Is*, highlights for Andersen and Pold that the "text" of a literary work is never its own. Instead, the economic model of multinational technology giants and search engines like Google demonstrates how "language [is] a network of common text," and how these corporations can nevertheless monetize it by "managing the paradigmatic network of language" (2018, 64, 66). Cayley and Howe's project, and Andersen and Pold's study thus adopt a wider perspective on digital environments and networks to emphasize, as Flores also observes, that when it comes to texts in digital media "we cannot trust that what we see is what we get," and we cannot quite be certain as to "what conditions might change over time or through interaction" (Flores 2013, 90). In other words, this crucial tension between the visible and the invisible dimensions of digital texts is for many authors of digital literary works an aesthetic or formal concern, but also an ethical and political one that concerns the wider structures of global and corporate capital, social agency, and technological design.

Pip Thornton, in her scholarship and practice-based work, has similarly drawn attention to the role of linguistic data in online platforms and network environments, and highlighted how "[the] language that flows through the platforms and portals of the Web is increasingly mediated and manipulated by large technology companies that dominate the internet" (Thornton 2019, n.p.). This affects how texts emerging or discovered online, through searches, autocomplete functions, or suggested content are composed or prioritized in displayed results, and how the way in which "digitised language is structured is also dependant [sic] on the monetary value of words in the online advertising industry" (n.p.). Consequently, "[much] of the text that exists online is structured and restricted by digital processing systems, and/or created or optimised not for human readers, but for the algorithms that scrape text for the purposes of targeted advertising" (n.p.). Cayley in his much-quoted essay "The Code is not the Text (Unless It Is the Text)" emphasized the crucial distinction of the code of the (digital) literary text and the literary text itself, as "code and language require distinct strategies of reading" and do not serve similar purposes and functions: "there are divisions and distinctions between what the code is and does, and what the language of the interface text is and does" (Cayley 2002). Yet, as Andersen, Pold, and Thornton, as well as Cayley and Howe and others demonstrate, the future, or fate of literary as well as everyday language online is intrinsically connected to the workings of coded, programmed, and algorithmic substructures of text, which also contribute to the material existence of literary assemblages. Inasmuch as the digital text always relies on code, this relation is key to its life and its place in society and culture, even if it does not constitute its text per se in a semantic sense.

A growing number of discussions have tackled these at times contradictory aspects of the circulation of texts and images in the digital society and culture. But what they also highlight is that the design of interfaces and platforms, and the life of texts within them, is a result of identifiable desires and intentions that Flores, too, considers key to digital textuality in the context of digital poetry. Texts are written, processed and read in specific ways not because certain designs and environments are inevitable, or a result of some kind of "natural law" of technological progress as gradual improvement towards higher degrees of sophistication. In the above discussion I have briefly considered phenomena and examples from the past and the present to suggest that literary (and other) texts and textuality have copious possible futures, just as they have numerous co-existing presents. We may have to be willing to let go of pre-held conceptions as to what "the text" means, or even how it exists in the evolving networks of communication and transmission. So far there is no sign of texts—literary, journalistic, legal, documentary, personal, public, or simply pleasurably distracting—disappearing or being under threat of extinction. Instead, the manner in which texts relate to authorship and authority, the ways in which they are composed as records and documents, or the settings where they inform or entice readers in digital networks (human or non-human, individuals or communities) requires a willingness to let go, to take a leap of faith, and to recalibrate our received notions of textuality in reading and literacy.

NOTE

1. According to the *OED*, the "French *texte*, also Old Northern French *tixte*, *tiste* (12th cent. in Godefroy)" and "Latin *textus* [...] style, tissue of a literary work (Quintilian), lit. that which is woven, web, texture."

REFERENCES

Andersen, Christian Ulrik and Søren Bro Pold. 2018. *The Metainterface: The Art of Platforms, Cities, and Clouds*. Cambridge, MA: MIT Press.

Cayley, John. 2002. "The Code is not the Text (unless It Is the Text)." *Electronic Book Review*, September 10. https://electronicbookreview.com/essay/the-code-is-not-the-text-unless-it-is-the-text/.

Flores, Leonardo. 2013. "Digital Textuality and Its Behaviours." *Journal of Comparative Literature and Aesthetics* xxxvi (1–2): 85–102.

Future of Text, The. n.d. https://futuretextpublishing.com/.

Healy, Randolph. 1997. *Arbor Vitae*. Bray, Ireland: Wild Honey Press. http://www.wildhoneypress.com/BOOKS/AVCOM/AVMENU.html.

Hegland, Frode Alexander, ed. 2020. *The Future of Text*. https://doi.org/10.48197/fot2020a.

Milligan, Chloe Anna. 2017. "The Page Is a Touchscreen: Haptic Narratives and 'Novel' Media." *Paradoxa* (29): 1–17.

Ong, Walter J. 2002 [1982]. *Orality and Literacy: The Technologizing of the Word*. New York, NY, and London: Routledge.

Thornton, Pip. 2019. "Words as Data: The Vulnerability of Language in an Age of Digital Capitalism." *CREST Security Review Magazine*, March 29. https://crestresearch.ac.uk/comment/thornton-words-as-data/.

CHAPTER FORTY-TWO

AI, Ethics, and Digital Humanities

DAVID M. BERRY (UNIVERSITY OF SUSSEX)

With the rise of artificial intelligence (AI) and machine learning (ML) it is inevitable that they become both a focus of method within the digital humanities but also a question of practice and ethics. Across a range of disciplines, the impact of the automation of methods of analysis that AI promises raises specific challenges in each field in terms of implementation and methodological reflection. However, even as they raise questions of biases and errors, there is also a more fundamental obscurity problem that is raised by the particular explanatory deficit in relation to the difficulty of understanding what it is that an AI is doing when it classifies and analyzes the underlying research data. How then can we be sure that the algorithms act ethically and legally? We should note that there is a growing literature on digital technology, software, and related areas that are pertinent to the questions raised in this chapter. For example, studies of algorithms such as software studies/cultural analytics (Manovich 2001, 2008; Dix et al. 2003; Fuller 2003; Galloway 2006; Chun 2008, 2017; Hayles 2012; Bratton 2016; Montfort 2016; Berry 2021) and critical code studies (Marino 2006, 2020; Wardrip-Fruin 2009). Other relevant research includes work on new media studies and machine learning (Manovich 2010; Alpaydin 2016; Pasquale 2016, 2020; Domingos 2017), as well as more recent work on ethics and algorithmic bias (Chun 2011; Floridi 2013; Gray 2014; Floridi and Taddeo 2016; Han 2017; Noble 2018; Benjamin 2019; Amoore 2020). There is also an important related area analyzing data and surveillance (Browne 2015; D'Ignazio and Klein 2019; Zuboff 2019).

In this chapter, I walk through some of the issues that pertain to the field of digital humanities, particularly in light of AI's likely impact on ways of doing digital humanities, automation more generally, and ethics of artificial intelligence. In section one, I look to debates about how the automation of humanities approaches to their research objects change the nature of humanities research and the role of the humanities scholar in relation to digital humanities. In section two, I concentrate on how digital humanities currently uses artificial intelligence, mostly machine learning, but also some of the potentials for expansion of its methods more widely in data analysis and forms of distant reading. Finally, in section three I look at the potential problems raised by artificial intelligence and machine learning for the practices of digital humanists, how we might think about the ethics issues that are raised and bring forward key elements of the debates currently taking place.

AUTOMATION OF THE HUMANITIES

As far back as 1981, Steve Jobs, then CEO of Apple, famously called computers "Bicycles for the Mind," implying that they augmented the cognitive capacities of the user, making them faster, sharper, and more knowledgeable. He argued that when humans "created the bicycle, [they] created a tool that amplified an inherent ability … The Apple personal computer is a 21st century bicycle if you will, because it's a tool that can amplify a certain part of our inherent intelligence … [it] can distribute intelligence to where it's needed" (Jobs 1981, 8–9). Jobs was influenced by earlier work by Doug Engelbart and Vannevar Bush who were both concerned with creating "man-computer symbiosis" and "augmenting human intellect" using technology (Ekbia and Nardi 2017, 26, 7).[1] This has also been the major way in which digital humanities, particularly in its earlier incarnations as humanities computing, conceptualized creating digital technology to augment the capacities of the humanities researcher. This generally took the form of thinking about methods of performing text analysis using a number of different forms of calculation, particularly of large corpora or datasets. Augmentation was thought to extend the human capacity to do things, and in contrast automation replaced the human with an algorithm, both of which raise different ethical considerations.[2]

Much of the earlier work that discussed the question of automation and augmentation tended to assume rather basic forms of automation that could assist the humanist in order to bring order to large amounts of data.[3] This was through algorithms that could help sort, list, or classify in simple ways, such as through the application of statistical methods. For example, McCarty (2003, 1225) described the "methodological commons" the digital humanities shared, which he listed as database design, numerical analysis, imaging, text analysis, and music retrieval and analysis. He offered a high-level division of digital humanities labor in terms of three branches working respectively on algorithmic (e.g., "mechanical" data analysis), metalinguistic (e.g., encoding texts in TEI and related markup languages), and representational (e.g., visual and presentational techniques) aspects of digital humanities work.

To simplify, today we tend to group digital humanities work into the divisions of digital tools and digital archives. Digital tools are software methods for working with textual data, such as the infamous word cloud which provides a visualization based on the frequency of words in a text so that words that have greater frequency are visually represented in a larger size.[4] Digital tools augment the humanist to be able to select and manipulate datasets so that they can be grouped and re-sorted to look for patterns. In contrast, the creation of digital archives tends to be image-based or transcriptions of physical textual archives, often with diplomatic versions for each of reading and searching. In the case of digital archives, automation is embedded in databases that collect corpora together and allow various searches and filters to be applied. Some archives have attempted to use OCR (optical character recognition) to change the text-based images into searchable text, but generally the quality of automated OCR output has remained relatively unsatisfactory for unsupervised use and actually generated a requirement for a lot of human labor to correct the mistakes that it inevitably created. This drove a need for what has come to be called "crowd-sourcing" through the use of Internet technologies to mediate a relationship between a researcher and a purported public (Terras 2016, 420–39).[5] This has also raised the interesting idea of slow digitization that foregrounds the difficulty of producing digital archives (Prescott and Hughes 2018). Both of these generate questions over how humanists should undertake their practices and the ethical considerations that might guide their action. In other words, what digital humanists ought to do and how they should think about their practices reflexively as a discipline.

Today, many of these ways of working with text and other forms of humanities data have become standardized into software packages and websites. This allows them to be used by scholars who may not necessarily understand the underlying complexity of the algorithms, and who may be unaware of the ethical decisions that have been embedded in the code. This is similar to the way that computer scientists working on algorithmic systems bracket off complexity by studiously ignoring how the functions they depend on are implemented—that is, humanists "black box" the computational aspects of doing digital humanities. Programmers who create these systems are taught to construct and respect "walls of abstraction"—functional modules that can be invoked in standard and consistent ways, hiding complexities within. Within the digital humanities there is a growing move toward these black-boxed systems as their convenience has increased, such as with the Voyant Tools system which provides a number of different text analysis functions within one system.[6] Even though Voyant Tools is an open-source project and the code is available through GitHub, few digital humanists are likely to inquire deeply into the code itself or check the source code. This leaves open the possibility that digital humanists may not fully be aware what the code that operates on their data may be doing. That is to raise the issue of software mediation which becomes a potentially new and interesting research question for the digital humanist (Berry 2011b, 2012). It also gestures to some of the underlying ethical issues when using automated systems that may have norms embedded into the code which humanists may be unaware of.

This process of automation can also increase the tendency towards formalization made possible by "algorithmization" (for an example see Allison et al. 2011, 6). Digital humanists then run the risk of creating systems where few people know explicitly or understand how they work, or how they make decisions. This creates the danger that we rely on the automation and output of these systems and thereby are asked to trust the computer. This also raises the question of the creation and maintenance of digital humanities' intersubjective, historical, and philosophical concepts as manifested in and supported by these technologies (Liu 2012; Berry 2014). So, for example, assumptions can be made about gender or race, or about which books are more important than others, such as a literature canon (Benjamin 2019; Bordalejo and Risam 2019). Indeed, computational systems tend towards the generation of quantifiable hierarchies of entities, particularly when classifying, sorting, and filtering, and even the notion of such a hierarchy is, of course, an imposed framework of understanding (see Johnson 2018).[7] Similarly, because of the underlying programming logics of a particular software tool, certain design assumptions, such as objects, networks, collections, or even the presumed suitability for particular properties of entities to be valued over others, may have effects on the way in which those digital objects are subsequently manipulated by the software (Berry 2011a; Emerson 2014; Marino 2020; Chun 2021).

DIGITAL HUMANITIES AND ARTIFICIAL INTELLIGENCE

We might, therefore, think of the automation of the humanities as moving through a number of phases. The first involved bespoke software, often programmed by a humanist-coder, to solve a particular problem. The second was the generalization of these early software tools into packages which provide a "one-stop shop" and which involve little or no programming requirement, such as Voyant Tools. In the latest phase, we see the beginning of much more sophisticated machine-learning tools, again by humanist-coders who are required to customize the machine-learning software they use for a specific digital humanities project, for example the MALLET (MAchine Learning for LanguagE Toolkit) software.[8] We can probably assume therefore that we are on the cusp of a new set of packages that will further democratize access to machine learning for non-programming humanists over the next few years, such as Orange, which offers a visual programming experience for using machine learning.[9] Each of these distinct phases raises different kinds of ethical conundrums for the humanist, from questions of use, sharing, openness, respect for others, privacy, and inclusion to the ethics of crowdsourcing, managing, and working with controversial or situated knowledges, and even with the issue of absences and decolonization of digital humanities more broadly.

Digital humanists have already begun to experiment with algorithms that draw on artificial intelligence and machine learning such as the widespread use of topic models. These are algorithms that use machine-learning techniques to sort and classify textual data into "buckets" that can represent themes or concepts drawn from the underlying data (Blei 2012). Machine learning relies on a process whereby the programmer does not necessarily instruct the system by writing an algorithm, rather the system is shown the input and the output and taught to "extract automatically the algorithm for this task" (Alpaydin 2016, 17). So, for example, a humanist might show an image to a machine-learning system to teach it to generate a text description in order to standardize the descriptive text associated in a database of images. This adds another layer of complexity to the use of digital tools because when one utilizes a machine-learning system, there remains the strong possibility that the source code, which is at least available with systems like Voyant Tools, is actually not legible in quite the same way. Indeed, due to the way in which machine-learning systems work, they tend to encode their classifying processes within a table of numbers which represent relationships between entities. These tables of numbers can be very large and daunting for a human researcher to understand. One can see how these automating processes might cover over ethical issues by transferring them into the hashtables of the machine-learning system. Machine learning also has the power to automate larger parts of the humanities, replacing research work that was previously thought could only be the preserve of the human researcher such as identifying key concepts. So, while the humanist might have access to the source code, this becomes merely a preliminary layer upstream from the later classification learning process which is encoded through either supervised or unsupervised learning processes. This downstream classificatory structure then becomes much more difficult for the researcher to cognitively map or check. In fact, as the number of parameters used in the classificatory processing increase, so does the complexity of the tables or networks of numbers. For example, in 2020, the OpenAI Generative Pre-trained Transformer 3 (GPT-3) is described as having a capacity of over 175 billion machine-learning parameters and the texts it classifies produce tables that are extremely difficult to understand or, indeed, explain at the level of the models generated by the software. In relation to this, the idea of explainability has emerged as the idea that artificial intelligence systems should be able to generate a sufficient explanation of how an automated decision was made, representing or explaining, in some sense, its technical processing. With concerns over biases in algorithms there is an idea that self-explanation by a machine-learning system would engender trust in these systems.[10] Explanations are assumed to tell us how things work, thereby giving us the power to change our environment in order to meet our own ends and thereby raise interesting ethical questions (see Berry 2021).

As the capability of these systems to automatically classify even very large datasets expands, we see a growing use of computational systems to abstract, simplify, and visualize complex "Big Data." However, these often produce simplistic probabilistic or statistical models used to organize complex humanities data. Additionally, many of these systems allow researchers to analyze data without a hypothesis so they can "play" with the data until a preferred result is found. Humanists can throw their data into large-scale computing clusters and let machine-learning algorithms find patterns that might have evaded human capacities—they may also encourage a carelessness with the data and what it may mean to specific communities (Johnson 2018, 61). This ability to analyze large datasets can be combined with the desire to deepen or widen the corpus from which data is being extracted, meaning that often the entire set of texts on which the dataset rests is never read in total, or even partially, by a researcher. This is the "computational turn" towards what has been called "distant reading" and assumes the automation of the reading process to ascertain themes and concepts from within a set of texts (Berry 2011a; see Underwood 2019). In using these tools, it has been argued that a researcher is required to trust the output of the algorithm given to analyzing the data, and they may have little formal understanding of the meaning of the results calculated or their significance (see Da 2019; Eccles 2020 has an excellent summary of this debate).

The use of these digital tools is increasing across the humanities, but at different speeds depending on the discipline. Nonetheless, the rise in convenience and the ease of access to both digitized and born-digital content in online databases strongly creates affordances towards using these techniques (see Posner 2016, 2018). The application of searches, such as Boolean search, to the downloading of datasets into pre-packaged CSV files or Zip files and using artificial intelligence, has the potential to skew research because technology tends to foreground a selection which might hide absences in a dataset or assert a presumed neutrality (Johnson 2018, 68). It is certainly one of the reasons for the calls for a critical digital humanities that foregrounds reflexivity and social justice in research practice.[11] This approach argues that digital humanists should become more critical of working in a digital environment, and the ways in which automation and the digital can subtly affect research practices (Berry and Fagerjord 2017; Johnson 2018, 68–9; Dobson 2019).[12]

ETHICS OF ARTIFICIAL INTELLIGENCE AND DIGITAL HUMANITIES

In this final section I want to take these emergent issues around digital humanities research in terms of the automation and augmentation of humanities work to look at the ethical issues raised.[13] The literature concerned with ethics related to the digital is a huge and growing area and, of course, I can only provide a broad outline of some of the key issues for consideration. So, in this chapter I will be focusing on the specific issues raised by thinking about ethics and ethical practices in relation to the digital humanities (see Ess 2009; Berry 2012, 12; Floridi 2013; Bostrom and Yudkowsky 2014 for some of the wider issues; Mittelstadt et al. 2016). The question is: What ought to be the principles or normative frameworks for guiding decisions in digital humanities projects and research units?[14]

Within the digital humanities there are differing ways of thinking about the ethics which might attach to digital projects or indeed to practices of digital humanists and have only recently begun to be foregrounded within the field (see for example Kennedy 2019; Padilla 2019). There have been some important debates about the ethos of the digital humanities, such as detailed by Cecire (2011) who has criticized the supposed "'virtues' of digital humanities – which include collaboration, humility, and openness" and its tendency to self-narrate through its putative moral virtues rather than specific methodological or epistemological commitments (see also Spiro 2012). In this section I want to think about how ethics is seen in practice in relation to digital humanities through the lens of artificial intelligence. There has been a considerable amount of work on ethics in the related areas of digital media and Internet studies, together with areas like sociology, science and technology studies and in computer science itself, but much less in digital humanities (see Ess 2004: CPSR 2008, 134; Floridi 2010; AoIR 2021).[15]

When thinking about ethics, broadly speaking it is common to talk of three approaches to ethics, namely deontological, consequentialist, or virtue ethics (see van den Hoven 2010). Deontological approaches tend to bring forward the idea that choices cannot be made based on their result, rather the decision has to be in conformity with a moral norm or duties. This means that deontological approaches often are constructed in terms of maxims or principles that should be followed, in "compliance with the highest ethical principle: the Categorical Imperative," that is not to use people as means to ends (van den Hoven 2010, 69). This creates the sense of a shared set of rules, perhaps professional guidelines, that also allow others to complain or hold to account those who break these principles, for example the AoIR ethics guidelines for research (AoIR 2021). In contrast, consequentialist approaches hold that ethics should be assessed in relation to the consequences that actions bring about, one "ought to choose that course of action which maximizes utility compared to the alternatives" (van den Hoven 2010, 69). The more likely the actions that maximize that a given end is realized, that is that the moral definition of a good is achieved, the more right the decision (Anscombe 1958, 9). Virtue ethics, on the other hand, emphasizes the "virtues" or "moral character" of the actor. So, virtues and vices tend to be the key categories used to assess ethical action, and virtue itself is a matter of degree (Annas 2011).[16] So the "virtuous person also possesses a general intellectual capacity, practical wisdom" and this enables them to "identify the morally relevant features in every situation and determine the right course of action" (MacIntyre 2007; van den Hoven 2010, 68).[17]

Much of the existing literature on ethics and artificial intelligence tends to prioritize either the deontological or consequentialist approaches to thinking about the questions that the digital raises.[18] For example, in the case of data bias one of the key ways in which ethics tend to be thought about in relation to the digital is to locate it within the context of computational inequities or computational power. Injustice, for example, might be strongly linked to problems with the data, which some have argued can be addressed by more data, ethical data, or democratizing data sources (but see Presner 2015; Eubanks 2019, 299; D'Ignazio and Klein 2019, 60–1). This might raise deontological issues of how we prevent artificial intelligent systems using people as sources of data, or manipulating them, or alternatively to focus on the perceived consequences of particular machine-learning systems and the possible harms they might cause.

Within the digital humanities as a field, thinking about the idea of a set of ethical principles that might serve as a professional guide to action has been discussed but as yet no specific digital humanities guidance has been drawn up (but see Berne 2013; Rehbein 2018). Considering the breadth and depth of digital humanities work that has been undertaken over the past twenty years, it is notable that ethics has been under-theorized within the digital humanities itself to a remarkable extent (see Raley 2014 as an alternative way of formulating this).[19] When one wishes to undertake a digital humanities research project, a set of guidelines is missing, a serious lack when one might want to think through ethical concerns.[20] Instead, most ways of thinking about ethics in digital humanities tend towards a form of virtue ethics in terms of the kinds of character that digital humanists are thought to have, such as being "'nice': collegial, egalitarian, non-hierarchical, and able to work well with others" (Koh 2014, 95). While virtue ethics offers some means for thinking through the questions raised by challenging problems in undertaking digital humanities work, virtue ethics tends to articulate these questions in terms of aretaic concepts (that is, in terms of vices/virtues) or axiological concepts (that is, in terms of better/worse or good/bad) rather than deontological notions (such as the right/wrong action, duty, or obligation).[21] This emphasis on the "good will" of the individual researcher places responsibility on the good character of the individual to do the virtuous thing but is less clear on the professional expression of that requirement.[22] This lack of formalized standards of good practice is notable considering the close relationships digital humanities have with digital archives, for example (see eHumanities 2010), which tend to raise questions about decolonization, gender, race, class, and sexuality, not just in terms of representation but also in terms of the assumed categories inherited from physical archive practices or in using data scraped from the Internet (see Liu 2012; Johnson 2018; Berry 2019).[23] Considering that computer ethics itself has been thinking about questions of how computer technology should be used since at least the 1960s, and as Moor (1985) argues, a central task of such an ethics should be to formulate policies to guide our actions, the lack of these guidelines is puzzling.[24] Drawing from Rehbein (2018) we might outline three areas from which such a set of ethical guidelines for the digital humanities might be drawn,

1. **Moral issues in specific fields of research and in close relation to the objects of study** (e.g., personal data, attribution, publishing historical or web data that might have wider public consequences for individuals, neutrality, quantification).
2. **Moral aspects of digital humanities as a profession**, both in terms of members of the digital humanities community and externally for society or individuals outside this community. Examples include "do not fabricate, do not falsify, do not plagiarize, honour the work of others, give credit to all who supported you, name your co-authors, do not publish the same thing twice, and various other guidelines that many scholarly communities have given themselves" (Rehbein 2018, 23).
3. **The responsibility of an individual scholar as well as of the scholarly community at large within digital humanities to the larger society**, for example creating risk for future generations through the use of technologies, their environmental impact, or commitment to equality, diversity, and inclusion.

Within the context of artificial intelligence, a field which has itself only recently been subjected to some ethical critique (but see Agre 1997), its growing use by digital humanities raises the importance of developing a set of ethics, particularly for thinking about the use of these technologies within the field. From the ethics of creating digital tools and digital archives, the use of facial recognition software, automated processes of cataloging or text generation, to the use of crowdsourcing using the "artificial artificial intelligence" or "ghost work" of human task work systems, there is clearly an urgent need for digital humanities to be thinking seriously about ethics, both professionally and at the level of the individual scholar (see Bianco 2012, 97; Reid 2012, 363; Spiro 2012, 18;

Smithies 2017, 208–9; Gray and Suri 2019, 7).[25] The time is indeed ripe for the professional bodies in digital humanities to begin to develop such a set of ethics to build a strong foundation for future research, and to encourage a wider discussion of ethics within the digital humanities community.

ACKNOWLEDGMENTS

Thanks to Natalia Cecire, Charles Ess, David Golumbia, Lauren F. Klein, Alan Liu, Willard McCarty, Miriam Posner, and Andrew Prescott for their helpful suggestions and comments on this chapter.

NOTES

1. Engelbart himself was deeply influenced by Vannevar Bush's article *As We May Think*, published in *The Atlantic* and his notion of the "memex," an enlarged intimate mechanical supplement to a person's memory (Bush 1945), and J. C. R. Licklider's (1960) work on *Man-Computer Symbiosis*. As Licklider wrote using a very revealing metaphor:

   > the fig tree is pollinated only by the insect *Blastaphaga grossorum*. The larva of the insect lives in the ovary of the fig tree, and there it gets its food. The tree and the insect are thus heavily interdependent: the tree cannot reproduce without the insect; the insect cannot eat without the tree; together, they constitute not only a viable but a productive and thriving partnership. This cooperative "living together in intimate association, or even close union, of two dissimilar organisms" is called symbiosis. "Man-computer symbiosis" is a subclass of "man-machine systems" … The hope is that, in not too many years, human brains and computing machines will be coupled together very tightly, and that the resulting partnership will think as no human brain has ever thought and process data in a way not approached by the information-handling machines we know today.

   Licklider also makes the important distinction between automation and a form of augmentation, although he doesn't use the term, when he argues

   > "Mechanical extension" has given way to replacement of men, to automation, and the men who remain are there more to help than to be helped … [Whereas] to think in interaction with a computer in the same way that you think with a colleague whose competence supplements your own will require much tighter coupling between man and machine than is suggested by the example than is possible today. (Licklider 1960)

2. As Ceruzzi observes,

   > the term "automation" was coined at the Ford Motor Company in 1947 and popularized by John Diebold in a 1952 book of that title. Diebold defined the word as the application of "feedback" mechanisms to business and industrial practice, with the computer as the principal tool. He spoke of the 1950s as a time when "the push-button age is already obsolete; the buttons now push themselves." (Ceruzzi 2000, 32)

3. It is notable that the automation through artificial intelligence of formerly critical and cognitive practices by the humanities scholar raises particular ethical issues in terms of understanding and responsibility.
4. We should also note that different digital tools may have differing kinds of opacity associated with them. Their development and usage may require them to be critically analyzed depending on how this opacity is manifested.
5. Crowdsourcing is based on principles of outsourcing, raising related issues of labor inequality and exploitation.

6. See Voyant Tools, https://voyant-tools.org.
7. See Chapter 12, Critical Digital Humanities in this volume.
8. http://mallet.cs.umass.edu/index.php.
9. https://orangedatamining.com.
10. Within the field of AI there is now a growing awareness of this problem of opaque systems and a sub-discipline of "explainable AI" (XAI) has emerged and begun to address these very complex issues—although mainly through a technical approach.
11. I argue that it is important that "ethics" and "social justice" should not be seen as antithetical, and both can contribute towards an idea of a future digital humanities that is inclusive, diverse, ethical, and equitable (see Chapter 12, Critical Digital Humanities in this volume, but also see Metcalf et al. 2019, 471).
12. For example, the work of the #transformDH collective, which formed in 2011, and "offered a recent challenge to [digital humanities]. Members of the collective argued that the consolidation of digital humanities as a field presented university-affiliated laborers with an opportunity to embark on an explicitly antiracist, anti-ableist, radical and inclusive academic project" (Johnson 2018, 68).
13. Zuboff (1988) has argued that we might need to distinguish between *automation*, which relieves humans of physical effort, and *informating*, which integrates people into a higher level of intellectual involvement and productivity. Today, we are more likely to use the term *augmentation* than informating, but the idea of it being a better way of automating that "realizes human potential of the workforce as well as the technical potentialities of the computer" remains key.
14. Indeed, I argue that ethics is inseparable from reflexivity (Berry 2004, see Keane 2015, 23).
15. There is also an interesting attempt to reconceptualize the arts and humanities as "deep humanities" where one of the key contributions would be an attempt to engage with issues of ethics related to technology, sustainability and augmentation. This includes Ethics, understood as "ethical ways of conceiving and connecting with the Other all its planetary diversity; integrating ethics into STEM/STEM education, business, politics, planning, and policy;" Bias, "social, cultural, and technological, including algorithmic bias; common sense and critical thinking in ML/AI" and Design, "ethical, affective, inclusive, and sustainable but not human centered; instead of placing human at the center/top, Deep Humanities proposes to place the human in proportion to and in right relations with the nonhuman (animal, plant, machine)" amongst other topics (Krishnaswamy 2019). Indeed, a "deep digital humanities" might consider proposals from deep humanities and other critical attempts to think about technology and the humanities to develop its own disciplinary matters of concern.
16. Although related, due to space considerations the area of "information ethics" is not able to be included, but see Doyle (2010) for a critique of information ethics. Additionally, the area of "contractarianism" and ethics which stems from the work of the social contract philosophers, either through the work of Hobbes or Kant, and argues that ethics draws its force from mutual agreement or an idea of contract, is beyond the scope of the chapter (Rawls 1971; Gauthier 1986).
17. Due to limitations on space, it is impossible to do justice to MacIntyre's work on virtue ethics, but see Anscombe (1958) and MacIntyre (2007). There is a good summary of the position in the Stanford *Encyclopedia of Philosophy*, https://plato.stanford.edu/entries/ethics-virtue/.
18. See also the ethics of care, an important way of reconceptualizing ethics in relation to feminist thought. Held (2006) states that an ethics of care "builds relations of care and concern and mutual responsiveness to need on both the personal and wider social levels. Within social relations in which we care enough about one another to form a social entity, we may agree on various ways to deal with one another." See also D'Ignazio and Klein (2019, 257, fn39) for a discussion of this in relation to what they call "Data Feminism."
19. There is also an interesting question about the relationship between open access and the ethics which unfortunately I don't have space to discuss in this article. For example, are the technical guidelines and requirements of open access rooted in ethical practices within the digital humanities? Or is there an alignment of open access and computational capital which needs to be carefully considered in relation to the neoliberal university (see Golumbia 2016)?
20. For example, see Lennon (2014) for a thoughtful discussion about digital humanities and the ethics of funding.

21. Aretaic concepts are concerned with *arête* (excellence or virtue), *phronesis* (practical or moral wisdom), and *eudaimonia* (usually translated as happiness or flourishing) (see Annas 2011, 3). Axiological concepts are concerned with value, particularly intrinsic and extrinsic value, based on comparative analysis of, for example, which activity is better or worse than another (see Scheler 1973). In contrast, deontological concepts are concerned with moral rules or professional codes of practice, particularly in terms of duty and obligations and particularly associated with the work of Immanuel Kant and his conception of the "good will" which acts out of respect for the moral law (see Kant 1997).
22. Virtue ethics tends to derive from Christian notions of a "good will" and the reliance on virtue ethics within digital humanities is interesting in relation to its early historical links to biblical and theological scholarship (particularly philology and textual criticism). Early digital humanists, such as Father Roberto Busa, an Italian Jesuit who studied Thomas Aquinas's texts, were extremely influential on the early use of computers in humanities work and this connection perhaps raises suggestive questions about an under-theorized Christian ethical foundation to the discipline of digital humanities (see Fiormonte 2017 for an overview of this history, but also Rockwell and Sinclair 2020). It is notable that Aquinas has also been a very influential thinker on the development of some versions of virtue ethics (see MacIntyre 2007, x).
23. See also Chapter 12, *Critical Digital Humanities*, in this volume.
24. Although there are examples of good practice, such as https://www.colorado.edu/lab/dsrl/statement-ethics and http://mcpress.media-commons.org/offthetracks/part-one-models-for-collaboration-career-paths-acquiring-institutional-support-and-transformation-in-the-field/a-collaboration/collaborators'-bill-of-rights/.
25. Crowdsourcing has become increasingly mediated by platforms in the past decade, for example, Amazon's Mechanical Turk, https://www.mturk.com, Microworkers, https://www.microworkers.com, Clickworker, https://www.clickworker.com/clickworker/, Appen, https://appen.com/jobs/contributor/, and Upwork, https://www.upwork.com/i/how-it-works/client/.

REFERENCES

Agre, Philip E. 1997. "Toward a Critical Technical Practice: Lessons Learned in Trying to Reform AI." In *Social Science, Technical Systems, and Cooperative Work: Bridging the Great Divide*, edited by Geoffrey Bowker, Les Gasser, Susan Leigh Star, and Bill Turner, 131–58. Hillsdale, NJ: Erlbaum.

Allison, Sarah, Ryan Heuser, Matthew Jockers, Franco Moretti, and Michael Witmore. 2011. *Quantitative Formalism: An Experiment*. Literary Lab, Pamphlet #1, January 15.

Alpaydin, Ethem. 2016. *Machine Learning: The New AI*. Cambridge, MA: MIT Press.

Amoore, Louise. 2020. *Cloud Ethics: Algorithms and the Attributes of Ourselves and Others*. Durham, NC: Duke University Press

Annas, Julia. 2011. *Intelligent Virtue*. New York, NY: Oxford University Press.

Anscombe, G. E. M. 1958. "Modern Moral Philosophy." *Philosophy* 33 (124) (January): 1–19.

AoIR. 2021. Ethics. https://aoir.org/ethics/.

Benjamin, Ruha. 2019. *Race after Technology*. Cambridge: Polity.

Berne. 2013. Berne DHSS Declaration of Research Ethics in Digital Humanities. Written collaboratively at the Digital Humanities Summer School 2013. https://docs.google.com/document/d/1A4MJ05qS0WhNlL dlozFV3q3Sjc2kum5GQ4lhFoNKcYU/.

Berry, David M. 2004. "Internet Research: Privacy, Ethics and Alienation: An Open Source Approach." *Internet Research* 14 (4): 323–32.

Berry, David M. 2011a. "The Computational Turn: Thinking about the Digital Humanities." *Culture Machine* 12.

Berry, David M. 2011b. *The Philosophy of Software: Code and Mediation in the Digital Age*. London: Palgrave Macmillan.

Berry, David M. 2012. *Understanding Digital Humanities*. London: Palgrave Macmillan.

Berry, David M. and Anders Fagerjord. 2017. *Digital Humanities: Knowledge and Critique in a Digital Age*. Cambridge: Polity.

Berry, David M. 2014. *Critical Theory and the Digital*. New York, NY: Bloomsbury.
Berry, David M. 2019. "Critical Digital Humanities." *Conditiohumana.io*. https://conditiohumana.io/critical-digital-humanities/.
Berry, David M. 2021. "Explanatory Publics: Explainability and Democratic Thought." In *Fabricating Publics: The Dissemination of Culture in the Post-truth Era*, edited by Bill Balaskas and Carolina Rito. London: Open Humanities Press.
Bianco, Jamie "Skye." 2012. "This Digital Humanities Which Is Not One." In *Debates in the Digital Humanities*, edited by Matthew K. Gold, 96–112. Minneapolis, MN: University of Minnesota Press.
Blei, David M. 2012. "Topic Modelling and Digital Humanities." *Journal of Digital Humanities* 2 (1) (Winter).
Bordalejo, Barbara, and Roopika Risam, eds. 2019. *Intersectionality in Digital Humanities*. York: Arc University Press.
Bostrom, Nick and Eliezer Yudkowsky. 2014. "The Ethics of Artificial Intelligence." In *The Cambridge Handbook of Artificial Intelligence*, edited by Keith Frankish and William Ramsey, 316–34. Cambridge: Cambridge University Press.
Bratton, Benjamin. 2016. *The Stack: On Software and Sovereignty*. Cambridge, MA: MIT Press.
Browne, Simone. 2015. *Dark Matters: On the Surveillance of Blackness*. London: Duke University Press.
Bush, Vannevar. 1945. "As We May Think." *The Atlantic*, July.
Cecire, Natalia. 2011. "Introduction: Theory and the Virtues of Digital Humanities." *Journal of Digital Humanities* 1 (1) (Winter).
Ceruzzi, Paul E. 2000. *A History of Modern Computing*. Cambridge, MA: MIT Press.
Chun, Wendy H. K. 2008. "On 'Sourcery,' or Code as Fetish." *Configurations* 16: 299–324.
Chun, Wendy H. K. 2011. *Programmed Visions: Software and Memory*. Cambridge, MA: MIT Press.
Chun, Wendy H. K. 2017. *Updating to Remain the Same: Habitual New Media*. Cambridge, MA: MIT Press.
Chun, Wendy H. K. 2021. *Discriminating Data: Correlation, Neighborhoods, and the New Politics of Recognition*. Cambridge, MA: MIT Press.
CPSR. 2008. Computer Professionals for Social Responsibility. http://cpsr.org.
Da, Nan Z. 2019. "The Computational Case against Computational Literary Studies." *Critical Inquiry* 45 (3) (Spring).
D'Ignazio, Catherine and Lauren F. Klein. 2019. *Data Feminism*. Cambridge, MA: MIT Press.
Dix, Alan, Janet E. Finlay, Gregory D. Gregory Abowd, and Russell Beale. 2003. *Human Computer Interaction*. London: Prentice Hall.
Dobson, James E. 2019. *Critical Digital Humanities: The Search for a Methodology*. Urbana, IL: University of Illinois Press.
Domingos, Pedro. 2017. *The Master Algorithm: How the Quest for the Ultimate Learning Machine Will Remake Our World*. London: Penguin.
Doyle, Tony. 2010. "A Critique of Information Ethics." *Knowledge, Technology and Policy* 23: 163–75.
Eccles, Kathryn. 2020. "Digital Humanities." *The Year's Work in Critical and Cultural Theory* 28 (1): 86–101.
eHumanities. 2010. "Ethical Issues in Digital Data Archiving and Sharing." *eHumanities NL*. https://www.ehumanities.nl/ethical-issues-in-digital-data-archiving-and-sharing/.
Ekbia, Hamid R. and Bonnie Nardi. 2017. *Heteromation, and Other Stories of Computing and Capitalism*. Cambridge, MA: MIT Press.
Emerson, Lori 2014. *Reading Writing Interfaces: From the Digital to the Bookbound*. Minneapolis, MN: University of Minnesota Press.
Ess, Charles. 2004. "Revolution? What Revolution? Successes and Limits of Computing Technologies in Philosophy and Religion." In *A Companion to Digital Humanities*, edited by Susan Schreibman, Ray Siemens, and John Unsworth, 132–42. London: Wiley-Blackwell.
Ess, Charles. 2009. *Digital Media Ethics*. Cambridge: Polity.
Eubanks, Virginia. 2019. *Automating Inequality: How High-Tech Tools Profile, Police, and Punish the Poor*. New York, NY: St Martin's Press.

Fiormonte, Domenico 2017. *The Digital Humanities from Father Busa to Edward Snowden*. The Centre for the Internet and Society. https://cis-india.org/raw/the-digital-humanities-from-father-busa-to-edward-snowden.

Floridi, Luciano. 2010. *The Cambridge Handbook of Information and Computer Ethics*. Cambridge: Cambridge University Press.

Floridi, Luciano. 2013. *The Ethics of Information*. Oxford: Oxford University Press

Floridi, Luciano and M. Taddeo. 2016. "What is Data Ethics?" *Philosophical Transactions of the Royal Society A* 374: 20160360.

Fuller, Matthew. 2003. *Behind the Blip: Essays on the Culture of Software*. London: Autonomedia.

Galloway, Alexander. 2006. *Protocol: How Control Exists after Decentralization*. London: MIT Press.

Gauthier, David. 1986. *Morals by Agreement*. Oxford: Oxford University Press.

Golumbia, David. 2016. "Marxism and Open Access in the Humanities: Turning Academic Labor against Itself." *Workplace* 28: 74–114.

Gray, Kishonna L. 2014. *Race, Gender, and Deviance in Xbox Live: Theoretical Perspectives from the Virtual Margins*. New York, NY: Routledge.

Gray, Mary and Siddharth Suri. 2019. *Ghost Work: How Amazon, Google, and Uber Are Creating a New Global Underclass*. Boston, MA: Houghton Mifflin Harcourt Publishing Company.

Han, Byung-Chul. 2017. *Psychopolitics: Neoliberalism and New Technologies of Power*. New York, NY: Verso.

Hayles, N. Katherine. 2012. *How We Think: Digital Media and Contemporary Technogenesis*. Chicago, IL: University of Chicago Press.

Held, Virginia. 2006. *The Ethics of Care: Personal, Political, and Global*. Oxford: Oxford University Press.

van den Hoven, Jeroen. 2010. "The Use of Normative Theories in Computer Ethics." In *The Cambridge Handbook of Information and Computer Ethics*, edited by Luciano Floridi, 59–76. Cambridge: Cambridge University Press.

Jobs, Steve. 1981. "When We Invented the Personal Computer...." *Computers and People Magazine*, July–August.

Johnson, Jessica Marie. 2018. "Markup Bodies: Black [Life] Studies and Slavery [Death] Studies at the Digital Crossroads." *Social Text* 36 (4): 57–79.

Kant, Immanuel. 1997. *Groundwork of the Metaphysic of Morals*. Cambridge: Cambridge University Press.

Keane, Webb. 2015. *Ethical Life: Its Natural and Social Histories*. Princeton, NJ: Princeton University Press.

Kennedy, Mary Lee. 2019. "What Do Artificial Intelligence (AI) and Ethics of AI Mean in the Context of Research Libraries?" *Research Library Issues* (299): 3–13. https://doi.org/10.29242/rli.299.1.

Koh, Adeline. 2014. "Niceness, Building, and Opening the Genealogy of the Digital Humanities: Beyond the Social Contract of Humanities Computing." *differences* 25 (1): 93–106.

Krishnaswamy, Revathi. 2019. "What is Deep Humanities & Arts?" San José State University, https://web.archive.org/web/20210425063845/https://www.sjsu.edu/humanitiesandarts/Faculty_Development_Resources/What_Is_Deep_Humanities.html.

Lennon, Brian. 2014. "The Digital Humanities and National Security." *differences* 25: 132–55.

Licklider, J. C. R. 1960. "Man Computer Symbiosis." *IRE Transactions on Human Factors in Electronics*, March.

Liu, Alan. 2012. "The State of the Digital Humanities: A Report and a Critique." *Arts & Humanities in Higher Education* 11 (1–2): 8–41.

MacIntyre, Alasdair. 2007. *After Virtue: A Study In Moral Theory*. Notre Dame, IN: University of Notre Dame Press.

Mackenzie, Adrian. 2017. *Machine Learners: Archaeology of a Data Practice*, London: MIT Press.

Manovich, Lev. 2001. *The Language of New Media*. London: MIT Press.

Manovich, Lev. 2008. *Software Takes Command*. https://web.archive.org/web/20110411164341/http://lab.softwarestudies.com/2008/11/softbook.html.

Manovich, Lev. 2010. *Info-Aesthetics*. London: Bloomsbury.

Marino, Mark C. 2006. *Critical Code Studies*. https://web.archive.org/web/20130804160545/http://www.electronicbookreview.com/thread/electropoetics/codology.

Marino, Mark C. 2020. *Critical Code Studies*. Cambridge, MA: MIT Press.

McCarty, Willard. 2003. "Humanities Computing." In *Encyclopedia of Library and Information Science*, 2nd edn. New York, NY: Marcel Dekker.

Metcalf, Jacob, Emanuel Moss, and danah boyd. 2019. "Owning Ethics: Corporate Logics, Silicon Valley, and the Institutionalization of Ethics." *Social Research: An International Quarterly* 82 (2) (Summer): 449–76.

Mittelstadt, Brent Daniel, Patrick Allo, Mariarosaria Taddeo, Sandra Wachter, and Luciano Floridi. 2016. "The Ethics of Algorithms: Mapping the Debate." *Big Data & Society* (December 2016). https://doi.org/10.1177/2053951716679679.

Montfort, Nick. 2016. *Exploratory Programming for the Arts and Humanities*. Cambridge, MA: MIT Press.

Moor, James H. 1985. 'What is Computer Ethics?' *Metaphilosophy* 16 (4): 266–75.

Noble, Safiya Umoja. 2018. *Algorithms of Oppression*. New York, NY: New York University Press.

Padilla, Thomas. 2019. *Responsible Operations: Data Science, Machine Learning, and AI in Libraries*. Dublin, OH: OCLC Research. https://doi.org/10.25333/xk7z-9g97.

Pasquale, Frank. 2016. *The Black Box Society: The Secret Algorithms That Control Money and Information*. Cambridge, MA: Harvard University Press.

Pasquale, Frank. 2020. *New Laws of Robotics: Defending Human Expertise in the Age of AI*. Cambridge, MA: Harvard University Press.

Posner, Miriam. 2016. "What's Next: The Radical, Unrealized Potential of Digital Humanities." In *Debates in the Digital Humanities*, edited by Matthew K. Gold and Lauren Klein, 32–41. Minneapolis, MN: University of Minnesota.

Posner, Miriam. 2018. "Digital Humanities." In *The Craft of Criticism: Critical Media Studies in Practice*, edited by Mary Celeste Kearney and Michael Kackman, 331–46. New York, NY: Taylor & Francis.

Prescott, Andrew and Lorna Hughes. 2018. "Why Do We Digitize? The Case for Slow Digitization." *Archive Journal*. http://www.archivejournal.net/essays/why-do-we-digitize-the-case-for-slow-digitization/.

Presner, Todd. 2015. "The Ethics of the Algorithm: Close and Distant Listening to the Shoah Foundation Visual History Archive." In *History Unlimited: Probing the Ethics of Holocaust Culture*, edited by Claudio Fogu, Wulf Kansteiner, and Todd Presner, 175–202. Cambridge, MA: Harvard University Press.

Raley, Rita. 2014. "Digital Humanities for the Next Five Minutes." *differences* 25 (1): 26–45.

Rawls, John. 1971. *A Theory of Justice*. Cambridge, MA: Harvard University Press.

Rehbein, Malte. 2018. "It's our department: On Ethical Issues of Digital Humanities." *Core.ac.uk*. https://doi.org/10.25366/2018.41.

Reid, Alexander. 2012. "Graduate Education and the Ethics of the Digital Humanities." In *Debates in the Digital Humanities*, edited by Matthew K. Gold, 350–67. Minneapolis, MN: University of Minnesota Press.

Rockwell, Geoffrey and Stéfan Sinclair. 2020. "Tremendous Mechanical Labor: Father Busa's Algorithm." *Digital Humanities Quarterly*. http://www.digitalhumanities.org/dhq/vol/14/3/000456/000456.html.

Scheler, Max. 1973. *Formalism in Ethics and Non-Formal Ethics of Values: A New Attempt toward the Foundation of an Ethical Personalism*. Evanston, IL: Northwestern University Press.

Smithies, James. 2017. *The Digital Humanities and the Digital Modern*. London: Palgrave.

Spiro, Lisa. 2012. "'This Is Why We Fight': Defining the Values of the Digital Humanities." In *Debates in the Digital Humanities*, edited by Matthew K. Gold, 16–35. Minneapolis, MN: University of Minnesota Press.

Terras, Melissa. 2016. "Crowdsourcing in the Digital Humanities." In *A New Companion to Digital Humanities*, edited by Susan Schreibman, Ray Siemens, and John Unsworth, 420–39. London: Wiley-Blackwell.

Underwood, Ted. 2019. *Distant Horizons – Digital Evidence and Literary Change*. Chicago, IL: University of Chicago Press.

Wardrip-Fruin, Noah. 2009. *Expressive Processing: Digital Fictions, Computer Games, and Software Studies*. Cambridge, MA: MIT Press.

Zuboff, Shoshana. 1988. *In the Age of the Smart Machine*. New York, NY: Basic Books.

Zuboff, Shoshana. 2019. *The Age of Surveillance Capitalism: The Fight for a Human Future at the New Frontier of Power*. London: Profile Books.

CHAPTER FORTY-THREE

Digital Humanities in the Age of Extinction

GRAHAM ALLEN AND JENNIFER DEBIE (UNIVERSITY COLLEGE CORK)

Environmental Digital Humanities, or EcoDH as it is known, exists, but it does so patchily, uncertainly, in the shadows without clear definitions or borders. Go to a good many of the introductions of edited collections (the shop windows of the new discipline) that introduce readers to DH and you will either not find it present, or only briefly mentioned within another chapter as an emerging field.[1] On the other hand, for a subject to be treated to a PMLA cluster would suggest that it has come into the light of wider scholarly analysis, debate, and consideration.[2] The reasons for these discrepancies will be touched on later in this chapter, but for now we turn to Stephanie Posthumus and Stéfan Sinclair and their description of EcoDH as:

> ... a new and burgeoning area that still remains largely undefined, but that asserts the importance of the humanities in responding to the ecological crisis while leveraging new tools and technologies.
>
> (2014, 254)

They go on to admit that "encounters between the digital and ecocriticism remain few and far between" (255), before adding: "The digital humanities bring to ecocriticism the possibility of new interactions, new collaborations, while ecocriticism brings to digital humanities a different political lens for reading environments" (260). Sinclair and Veronica Poplawski seem to agree when they write: "Collaborative and interdisciplinary in nature, the digital humanities have much in common with the environmental humanities" (2018, 156). What both disciplines (Digital Humanities and Environmental or Eco Studies) share is that they exist on the avant-garde, the coalface (if you pardon the pun), where the humanities address contemporary culture and society. The obvious question to ask of the nascent discipline of EcoDH is how can it help the humanities and the post-humanities contribute to the global encounter with the Anthropocene?[3] Another, far deeper, more ontological question might be heard in that basic question: does the emergence of EcoDH offer the humanities a new chance of relevance in the era of the Anthropocene? That is, in an era in which what gets called "traditional" humanities finds it increasingly difficult to defend against the progressively stronger demands of instrumentalist accounts of research and teaching, is EcoDH putting the "traditional" humanities back in the fray? Or is it a transformation (or mutation, or reduction) of the "traditional" humanities into the "soft," or even sometimes "hard," sciences? Is this strange, yet to be fully realized phenomenon, EcoDH, a savior or a betrayal of the human(-ities)?

To pick out DH and Environmental Studies as particularly relevant to current global problems, whether separately or in their new conjunction as EcoDH, threatens to gloss over the manner in which they are, themselves, picked out by government-led funding bodies as the only relevant humanities subjects within a new instrumentalized academy. As Patrick Svensson puts it:

> The humanities are undergoing a set of changes which relate to research practices, funding structures, the role of creative expression, infrastructural basis, reward systems, interdisciplinary sentiment and the emergence of a deeply networked humanities both in relation to knowledge production processes and products. An important aspect of this ongoing transformation of the humanities is humanities scholars' increasing use and exploration of information technology as both a scholastic tool and a cultural object in need of analysis. Currently, there is a cumulative set of experiences, practices and models flourishing in what may be called digital humanities.
>
> (2009, n.p.)

What is most important in that shifting academic ground is the way Svensson sees IT as both a new tool of humanities studies and the object of Humanities research itself. If it is to not erase the socio-psycho-political transformations that digital technology is generating today, then there has to be a double, circular, self-reflexive aspect of all EcoDH work. Or, put another way, EcoDH is never simply the means of researching the world of "nature" (local losses, the sixth extinction event, global warming, deforestation, mass migration of humans and animals). It is always also the medium (already vastly varied in itself) in which most of these events are knowable, shared, and perhaps most importantly of all, created for intellectual consumption. We remind ourselves here of Derrida's admonition that:

> ... the technical structure of the *archiving* archive, also determines the structure of the *archivable* content even in its very coming into existence and in its relationship with the future. The archivization produces as much as it records the event. This is also our political experience of the so-called news media.
>
> (1996, 17)

Building on Derrida, Margaret Linley (2016) sees this point in terms of the ecological/environmental discourse through which critics often account for digital media:

> ... a central tenet of media ecology is that media are not mere tools but function rather both within and as environments, and thereby shape human perception and cognition, forms of discourse, and patterns of social behaviour. A medium's physical properties therefore define the nature of communication; any changes in modes of communication have an effect on individual perceptions, expressions, values, and beliefs, as well as those of the society as a whole.

It should be clear from the above that EcoDH offers not a stable methodology with which to do predictable work on the "real" world, but rather a multi-perspectival identity crisis that constantly shifts its ground in order not to forget about the ground-making, or ground-assuming, technology which in turn makes its work possible. Part of this stems from the nature of the researchers involved in EcoDH themselves. As previously stated, this is a new and nebulous field, meaning most participants in the conversation surrounding EcoDH come to it from other, related fields. For instance, the PMLA cluster mentioned early in this essay is affiliated with the MLA Commons group TC Ecocriticism and Environmental Humanities. This group consists of hundreds of members (450+ as of December 2021) with specialities ranging from Shakespearean scholars to marine biologists to artists. This interdisciplinary draw offers the opportunity for vast and fascinating collaboration on a global scale, but will also necessarily continue to contribute to the plural and thus undefined nature of EcoDH.

In her keynote address to the 2014 Digital Humanities conference, Bethany Nowviskie, one of the earliest proponents of EcoDH and herself a literary scholar who came to the forefront of DH as the field gained recognition in the 2010s, sums up the spirit of EcoDH when she asks:

> What is a digital humanities practice that grapples constantly with little extinctions and can look clear-eyed on a Big One? Is it socially conscious and activist in tone? Does it reflect the managerial and problem-solving character of our 21st-century institutions? Is it about preservation, conservation, and recovery—or about understanding ephemerality and embracing change? Does our work help us to appreciate, memorialize, and mourn the things we've lost? Does it alter, for us and for our audiences, our global frameworks and our sense of scale? Is it about teaching ourselves to live differently? Or, as a soldier of a desert war wrote in last autumn's New York Times, is our central task the task of learning how to die—not (as he put it) to die 'as individuals, but as a civilization' … in the Anthropocene?
>
> (2015, 15–16)

The ludicrous and yet quite serious answer to this litany of questions is, *yes*. As Nowviskie knows, the scholar of EcoDH is what Rosi Braidotti (2011) (after Deleuze and Guattari) calls *nomadic*, occupying many positions (often contradictory) at the same time. It is profoundly contradictory, for example, to bear witness to the Sixth Great Extinction Event and yet to spend one's career attempting to teach and inculcate Enlightenment values into our students. If our civilization is dying, for whom are we building our archives or even teaching our courses? If the dominant activity of Digital Humanities so far in its short life span has been archive building, the challenges facing an EcoDH archive are numerous and immensely challenging.

Jennifer K. Ladino, in her "What Is Missing?: An Affective Digital Environmental Humanities," argues:

> In this era of overwhelming species extinction and environmental degradation, a primary task for digital environmental humanities, or 'EcoDH,' scholars is to reimagine how to represent, cope with, and deploy loss in the service of a more just present and perhaps even a liveable future.
>
> (189)

The verbs ("represent, cope with, and deploy") are not necessarily compatible, although as Ladino shows they are all required. For Ladino, as for Nowviskie, a good many functions, incompatible when considered epistemologically, are brought into a potentially effective alliance when they are viewed and practiced on the level of affect. Ladino explains:

> As a scholar of affect, I'm concerned not only with how we imagine and represent life in the Anthropocene but also how we *feel* about it … An important project for the EcoDH, then, is to

> think more carefully about affect: the powerful, visceral, pre- or even non-cognitive feelings that arise and are transmitted in both virtual and actual environments.
>
> (189)

In her article, Ladino looks at Maya Lin's ongoing, multi-medium memorial of the sixth mass extinction, *What is Missing?*, how it "destabilizes the binary of hope and despair that so often structures the environmental discourse in the Anthropocene" and "points out other affects – notably solastalgia – that deserve more attention from EcoDH scholars" (192). Ladino goes on: "What interests me most about this memorial is its assumptions about how affect works to generate environmental activism" (195). "Solastalgia," she explains, "is Australian philosopher Glenn Albrecht's term for 'a form of homesickness one gets when one is still at home,' a nostalgia for a place as it's changing around you" (196). Reflecting on the power of such an eco-psychological effect, Ladino adds:

> Solastalgia is an appropriate way to mark environmental change and to empathize not with a changing planet – I don't think most humans can empathize with a planet – but with other people and perhaps with our own future selves via what we might call *anticipatory* solastalgia – a felt premonition of our eventual solastalgia as our own homes change under our feet.
>
> (197)

EcoDH presents us with an intense form of identity crisis as unsettling as any form of politico-cultural solastalgia. EcoDH's formation, which as we know from Derrida always involves violence, produces a hybrid discipline the intentions of which (the objects and objectives, the perspectives, and the orientation) are radically conflictual, aporetic even. And yet these intentions co-exist and support each other within the figure in question.

We have said that one of the strongest motives in DH's short history has undoubtedly been what Derrida calls "archivization": the impulse to gather, assemble, link, and display information normally associated with the "traditional" humanities, i.e., literature and literary texts, historical, archaeological, and cultural artifacts, and so on. The motives for building an archive may involve interpretation and analysis, but the strongest of all archivizing impulses are surely commemoration, *anamnesis* or calling to mind, along with the contradictory impulses of stabilization (fixing, establishing, monumentalizing) and a certain putting into play (activating intertextual relations, linking, contextualizing, and so on).

The motives or impulses behind ecological or environmental studies are obviously vast and wide and not possible to enumerate in an essay like this. It is our contention here, however, that when Eco Studies comes into a new identification with DH, there necessarily arises a question about its dark side, its inevitably and unremittingly (despite Ladino's affective aesthetic) pessimistic side, a side of EcoDH that has been present and visible since Nowviskie's keynote address in 2014 and her report on the Dark Mountain Project. To call this aspect of Eco Studies pessimistic does not do justice to its objects of study, or to the disillusion felt by many of its practitioners. Serious consideration of the ecological state of the planet, of where we are within the sixth extinction event (and the rapidity with which that event is spreading across the globe), of the unpreparedness and refusals built into the global capitalist politico-economic system with its insatiable dependence upon "growth," of the condition of a proliferating number of the world's poor, of the rising phenomenon of mass migration, and all the socio-cultural tensions that it brings, can only lead us to confront the indivisible tie between the human (and thus perhaps the new or post-humanist perspective) and death, or thanatos, or the Freudian death drive.

One might here think a little deeper into Derrida's account of what he calls archive fever, in particular the relation he draws between the archivization drive and the death drive; in other words the desire to assemble is not opposed to the desire to finish, complete, totalize, close-down, and thus forget:

> There would indeed be no archive desire without the radical finitude, without the possibility of a forgetfulness which does not limit itself to repression. Above all, and this is the most serious, beyond or within this simple limit called finiteness or finitude, there is no archive fever without the threat of the death drive, this aggression and destruction drive.
>
> (1996, 19)

Starkly, this brings to mind such projects as the Svalbard Global Seed Vault in Norway, a "Doomsday vault" of the world's food crops created to feed future-humans in the event of a global catastrophe which is, perversely, being threatened by the man-made global catastrophe of climate change. As warming temperatures threaten to flood the vault's contents with unprecedented snowmelt, we are reminded that even in this act of archiving and preserving the means of life for a future generation, the threat of archival erosion is omnipresent, and the actions of today may prove futile for the humans of tomorrow.

There is something of a strange, unfamiliar contradiction held in our figure of "EcoDH," this understanding of the archive and the death drive. It is a productive contradiction which throws both sides of its (contra)diction into relation, drawing together a desire to bear witness to the world (nature, life, ecology) and an equal desire to bear witness to death (the death drive, the abandon of death in the world).[4] A desire exists within the figure "EcoDh" to archive (what? the world, the local, locality, *bios*, *zoe*, those that live and those that are dying or dead). Death, thus, gets into the DH as well as the eco. Seen most clearly in any positive politics of eco-action (and how can there be an "EcoDH" without such a political drive?) there is a concomitant drive to forget. Especially, one might say, a desire (perhaps a necessity) to forget the dominance (the force, the power, the ubiquity of death) in the face of a willed optimism. This contradiction of intentions was no doubt already there within the name of humanism, as if we reduced digital humanism to initials so we could forget the embarrassments and the contradictions contained within the "traditional" name of humanism. An "EcoDH" worthy of the title must not always forget its last name, and in particular, as Braidotti suggests, the division forgotten within that name of the *bios* and the *zoe*. Braidotti compares and contrasts the "traditional" humanist notion of the *bios* (which is held in distinction from and ontologically superior to all other life forms on the planet) and the post-human notion of life figured by the classical trope of *zoe*. For her *zoe* reconnects what humanism sundered and always forgets; it establishes us squarely within the *post-anthropos* era of vitality and vitalism:

> Zoe-centered egalitarianism is, for me, the core of the post-anthropocentric turn: it is a materialist, secular, grounded and unsentimental response to the opportunistic trans-species commodification of Life that is the logic of advanced capitalism.
>
> (2011, 60)

And yet of course, and this we are saying should be seen as a typical EcoDH observation, in order to support life *qua zoe* here, Braidotti has to temporarily (intentionally) forget the fact of the general, unprecedentedly fast, perhaps irreversible extinction of that *zoe*-sphere she is promoting and even celebrating in her remarks. When we remember that force of death in the *zoe*-sphere, how do we read Braidotti's affirmation of life *qua zoe*?

EcoDH is a contradictory-loaded, contradiction-riddled, positive and negative, life-affirming and death-witnessing new perspective and discipline. It is at once an activism to save, an archive to preserve, and a memorial to remember the loss of a planet ever-changing. As Dipesh Chakrabarty says in his much cited "The Climate of History: Four Theses":

> it is no longer a question simply of man having an interactive relation with nature. This humans have always had, or at least that is how man has been imagined in a large part of what is generally called the Western tradition. Now it is being claimed that humans are a force of nature in the geological sense. A fundamental assumption of Western (and now universal) political thought has come undone in this crisis.
>
> (2009, 207)

This change is one first and foremost of the *zoe*-sphere. But it is also a change to the very heart of Western humanism and all the sciences including the social sciences which have arisen from it. Massive changes are occurring in our sense of what it means to be human in the world. We need, therefore, an EcoDH which does not seek to eradicate, or forget, or resolve the tensions and contradictions it sees in the world and within its own modes of making and seeing. What we need are disciplines (and here EcoDH is specially positioned) which can activate, reflect on, and illuminate the contradictions that structure our world and our discourses.

NOTES

1. See Grigar and O'Sullivan (2021); Berry and Fagerjord (2017); Schreibman, Siemens, and Unsworth (2016).
2. See Cohen and LeMenager (2016).
3. For a useful account of post-humanism, see Braidotti (2013).
4. We use that word "abandon" in the sense of "complete lack of inhibition or restraint."

REFERENCES

Berry, David M. and Anders Fagerjord. 2017. *Digital Humanities; Knowledge and Critique in the Digital Age*. Cambridge, MA: Polity Press.

Braidotti, Rosi. 2011. *Nomadic Subjects: Embodiment and Sexual Difference in Contemporary Feminist Theory*, 2nd edn. New York, NY: Columbia University Press.

Braidotti, Rosi. 2013. *The Posthuman*. Cambridge, MA: Polity Press.

Chakrabarty, Dipesh. 2009. "The Climate of History: Four Theses." *Critical Inquiry* 35: 197–222.

Cohen, Jeffrey Jerome and Stephanie LeMenager. 2016. "Assembling the Ecological Digital Humanities." *PMLA* 131 (2): 340–46.

Derrida, Jacques. 1996. *Archive Fever: A Freudian Impression*. Chicago, IL, and London: University of Chicago Press.

Grigar, Dene and James O'Sullivan, eds. 2021. *Electronic Literature as Digital Humanities: Contexts, Forms, and Practices*. London and New York, NY: Bloomsbury.
Ladino, K. Jennifer. n.d. "What is Missing? An Affective Digital Environmental Humanities." https://doi.org/10.5250/resilience.5.2.0189.
Linley, Margaret. 2016. "Ecological Entanglements of DH." In *Debates in the Digital Humanities 2016*, edited by Matthew K. Gold and Lauren F. Klein. Minnesota, NM: University of Minnesota Press. https://www.bookdepository.com/Debates-Digital-Humanities-2016-Matthew-K-Gold/9780816699544?ref=bd_ser_1_1.
Nowviskie, Bethany. 2015. "Digital Humanities in the Anthropocene." *Digital Scholarship in the Humanities* 30 (1) (December): 14–15. https://doi.org/10.1093/llc/fqv015.
Posthumus, Stephanie and Stéfan Sinclair. 2014. "Reading Environment(s): Digital Humanities Meets Ecocriticism." *Green Letters* 18 (3): 254–73. doi: 10.1080/14688417.2014.966737.
Schreibman, Susan, Ray Siemens, John Unsworth, eds. 2016. *A New Companion to Digital Humanities*. Oxford: John Wiley.
Sinclair, Stéfan, Veronica Poplawski, and Stephanie Posthumus. 2018. "Digital and Environmental Humanities: Strong Networks, Innovative Tools, Interactive Objects." *Resilience: A Journal of the Environmental Humanities* 5 (2) (Spring): 156–71. https://doi.org/10.5250/resilience.5.2.0156.
Svensson, Patrick. 2009. "Humanities Computing as Digital Humanities." In *Digital Humanities Quarterly* 3 (3). http://digitalhumanities.org/dhq/vol/3/3/000065/000065.html.

INDEX